Lecture Notes in Mathematics

Edited by A. Dold, B. Eckmann and F. Takens

Subseries: *Mathematica Gottingensis*

1408

T0215078

Wolfgang Lück

Transformation Groups and Algebraic K-Theory

Springer-Verlag

Berlin Heidelberg New York London Paris Tokyo Hong Kong

Author

Wolfgang Lück
Mathematisches Institut, Universität Göttingen
Bunsenstr. 3–5, 3400 Göttingen, Federal Republic of Germany

Mathematics Subject Classification (1980): 57SXX, 18F25, 57Q10, 57Q12, 18GXX, 20L15

ISBN 3-540-51846-0 Springer-Verlag Berlin Heidelberg New York
ISBN 0-387-51846-0 Springer-Verlag New York Berlin Heidelberg

Printing and binding: Druckhaus Beltz, Hemsbach/Bergstr.
2146/3140-543210 – Printed on acid-free paper

0. Introduction.

The main goal of algebraic topology is the translation of problems and phenomena from geometry to algebra. In favourable cases we obtain a computable algebraic invariant which decides a given geometric question. A classical example is the classification of compact connected closed orientable surfaces by the genus.

This book is devoted to the connection between transformation groups and algebraic K-theory. We shall construct invariants such as the equivariant finiteness obstruction, Whitehead torsion, and Reidemeister torsion taking values in algebraic K-groups. We define injections or isomorphisms to the algebraic K-groups from groups such as finiteness obstruction groups, Whitehead groups, representation rings, homotopy representation groups or units of the Burnside ring. These are used to answer questions of the following type:

When is a finitely dominated G-CW-complex G-homotopy equivalent to a finite G-CW-complex? Under which conditions is a G-homotopy equivalence between finite G-CW-complexes simple? Is a given equivariant h-cobordism trivial? When are two semilinear G-discs G-diffeomorphic? Under which conditions are the unit spheres of two orthogonal G-representations G-diffeomorphic? When are two oriented G-homotopy representations oriented G-homotopy equivalent? Is a given oriented G-homotopy representation oriented G-homotopy equivalent to the unit sphere of a complex G-representation?

These questions will be treated in detail. They are related to the general problem of classifying group actions on manifolds. This problem and in particular its connections to algebraic K-theory are the basic motiviation for this book. We concentrate on developing the algebra. The algebraic tools and techniques presented here have applications to G-manifolds besides the one to the questions above . They will not be worked out, as this would exceed the scope of this book, but are discussed in the comments.

Roughly speaking, most of the material of chapter I can be found in the literature whereas chapters II and III mainly contain unpublished work. The study of modules over a category was initiated by Bredon [1967], where an equivariant obstruction theory for extending G-maps was established, and by tom Dieck [1981], where the equivariant finiteness obstruction and the diagonal product formula were studied for finite groups and simply connected fixed point sets. The author wants to express his deep gratitude to Prof. Tammo tom Dieck for his encouragement and generous help.

The book is based on a course given by the author in the winter term 1986/87 and on the author's Habilitationsschrift, Göttingen 1989.

The author thanks Christiane Gieseking and Margret Rose Schneider for typing the manus cript.

We briefly summarize the main results and constructions.

0.1. Modules over a category.

Let Γ be a EI-category, i.e., Γ is a small category whose endomorphisms are isomorphisms. A $R\Gamma$-module M is a contravariant functor $\Gamma \longrightarrow$ R-MOD into the category of modules over the commutative ring R. The functor category MOD-$R\Gamma$ of $R\Gamma$-modules is abelian. We reduce the study of $R\Gamma$-modules, their K-theory and homological algebra to the study of $R[x]$-modules for $x \in \mathrm{Ob}\,\Gamma$ and $R[x]$ the group ring $R[\mathrm{Aut}(x)]$ by the Cofiltration Theorem 9.39. and Filtration Theorem 16.8. The Cofiltration resp. Filtration Theorem assigns to a projective $R\Gamma$-module P of finite tpye resp. $R\Gamma$-module M of finite length a natural cofiltration

$$P = P_n \longrightarrow P_{n-1} \longrightarrow \ldots \longrightarrow P_o = \{0\}$$

resp. natural filtration

$$\{0\} = M_o \longrightarrow M_1 \longrightarrow \ldots \longrightarrow M_n = M$$

such that the kernel of $P_i \longrightarrow P_{i-1}$ resp. cokernel of $M_i \longrightarrow M_{i+1}$ can be expressed in terms of $R[x]$-modules S_xP resp. Res_xM which themselves are naturally constructed from P resp. M. The Cofiltration Theorem implies the Splitting Theorem 10.34. for algebraic K-theory of $R\Gamma$-modules

$$K_n(R\Gamma) = \bigoplus_{\bar{x} \in \mathrm{Is}\,\Gamma} K_n(R[x])$$

where $\bar{x}$ runs over the set $\mathrm{Is}\,\Gamma$ of isomorphism classes of objects and $n \in \mathbb{Z}$. As a special case we obtain the well-known splitting of the equivariant Whitehead group of a G-space. For finite Γ and R a field of characteristic 0 the Filtratration Theorem gives a second Splitting Theorem 16.29. for algebraic K-theory of $R\Gamma$-modules. These two Splitting Theorems are related by a K-<u>theoretic Moebius inversion</u> 16.29. In geometry this corresponds to switching between the isovariant and equivariant setting, or between the two stratifications $\{X_H \mid H \subset G\}$ and $\{X^H \mid H \subset G\}$ of a G-space X . Besides the K-theoretic application we also obtain a computation of Ext-groups $EXT^n_{R\Gamma}(M,N)$ by a spectral sequence whose E_2-term is given by Ext-groups over the various group rings $R[x]$ (see 17.18. and 17.28.). We introduce and study <u>generalized Swan homomorphisms</u> in section 19.

The algebra of $R\Gamma$-modules for Γ the <u>discrete fundamental category</u> $\Pi/(G,X)$ (see 8.15.) of a G-space X is the main ingredient for constructing and computing certain algebraic invariants of G-spaces and the K-groups in which they take values.

0.2. <u>Invariants for G-spaces</u>.

Here is a list of the most important invariants we will construct for G-spaces and G-maps.

name	symbol	value group	defined for	page
Euler characteristic	$\chi^G(X)$	$U^G(X)$ resp. $U(G)$	finitely dominated G-space X	100, 278, 360
multiplicative Euler characteristic	$h\chi(X)$	$\prod_{(H)} \mathbb{Q}^*/\mathbb{Z}^*$	finitely dominated G-space X	368
	$m\chi(X)$	$\prod_{(H)} \mathbb{Q}^*/\mathbb{Z}^*$	special G-space X	368
	$h\chi(X)_{1/m}$	$\bar{C}(G)^* = \prod_{(H)} \mathbb{Z}/\|G\|^*$	finitely dominated G-space X	387
	$h\chi(f)_{1/m}$	$\bar{C}(G)^*$	G-map between finitely dominated G-spaces	387
finiteness obstruction	$o^G(X)$	$K_0(\mathbb{Z}\Pi/(G,X))$ resp. $K_0(\mathbb{Z}OrG)$	finitely dominated G-space X	278, 360
reduced finiteness obstruction	$\tilde{o}^G(X)$	$\tilde{K}_0(\mathbb{Z}\Pi/(G,X))$ resp. $K_0(\mathbb{Z}OrG)$	finitely dominated G-space X	278, 360
	$w^G(X)$	$Wa^G(X)$		52

name	symbol	value group	defined for	page
(equivariant) Whitehead torsion	$\tau^G(f)$	$Wh(\mathbb{Z}\Pi/(G,Y))$ resp. $Wh(\mathbb{Z}Or\,G)$	G-homotopy equivalence of finite G-CW-complexes resp. G-manifolds $f : X \longrightarrow Y$	284, 360
	$\tau^G_{geo}(f)$	$Wh^G_{geo}(Y)$		68
isovariant Whitehead torsion	$\tau^G_{Iso}(B,M,N)$	$Wh^G_{Iso}(M)$	Isovariant h-cobordism (B,M,N)	85
Reidemeister torsion	$\rho^G(X)$	$Wh(\mathbb{Q}Or\,G)$	Finite G-CW-complex with round structure X	362
	$\rho^G(M)$	$Wh(\mathbb{R}Or\,G)$	M a closed Riemannian G-manifold satisfying 18.43.	375
	$\rho^G_{PL}(M)$	$K_1(\mathbb{R}G)^{\mathbb{Z}/2}$	M a Riemannian G-manifold	376
reduced Reidemeister torsion	$\bar{\rho}^G(X)$	$\frac{K_1(\mathbb{Q}Or\,G)}{K_1(\mathbb{Z}_{(\lvert G\rvert)}Or\,G)}$	Finitely dominated G-space X with round structure	363
Poincaré torsion	$\rho^G_{PD}(M)$	$K_1(\mathbb{R}G)^{\mathbb{Z}/2}$	Riemannian G-manifold M	377

We compute the value groups in terms of algebraic K-groups of certain group rings and state sum, product, diagonal product, join and restriction formulas. The reduced finiteness obstruction is the obstruction for a finitely dominated G-space X to be G-homotopy equivalent to a finite G-CW-complex. The Whitehead torsion is the obstruction of a G-homotopy equivalence of finite G-CW-complexes to be simple. Both invariants are defined geometrically and algebraically and these two approaches are identified by isomorphisms $Wa^G(X) \longrightarrow \tilde{K}_o(\mathbb{Z}\Pi/(G,X))$ and $Wh^G_{geo}(X) \longrightarrow Wh(\mathbb{Z}\Pi/(G,X))$. Certain relations between these invariants are established. Roughly speaking, Whitehead torsion is the difference of Reidemeister torsion, the reduced Reidemeister torsion is a refinement of the finiteness obstruction.

0.3. Maps between geometric groups and K-groups.

We give a list of maps relating geometrically defined groups to algebraic K-groups. They connnect geometry with algebraic K-theory. We denote injections by $\rightarrowtail$ and isomorphisms by $\xrightarrow{\cong}$:

$\Phi : Wa^G(X) \oplus U^G(X) \xrightarrow{\cong} K_o(\mathbb{Z}\Pi/(G,X))$	283
$\tilde{\Phi} : Wa^G(X) \xrightarrow{\cong} \tilde{K}_o(\mathbb{Z}\Pi/(G,X))$	283
$\tilde{\Phi} : Wh^G_{geo}(X) \xrightarrow{\cong} Wh(\mathbb{Z}\Pi/(G,X))$	286
$\Phi : Wh^G_{Iso}(X) \xrightarrow{\cong} Wh^G_\rho(M)$	86
$\tilde{\Phi} : Wh^G(Y \times T^{n+1})^{\mathbb{N}^{n+1}} = K^G_{-n}(Y)_{geo} \xrightarrow{\cong} K^G_{-n}(\mathbb{Z}\Pi/(G,X))$	299
$\omega : A(G)^* \longrightarrow Wh(\mathbb{Z}Or\, G)$	131
$\overline{SW} : \overline{C}(G)^* \longrightarrow K_1(\mathbb{Q}Or\, G)/K_1(\mathbb{Z}_{(\|G\|)}Or\, G)$	385
$\overline{SW}_o : Inv(G) \rightarrowtail K_1(\mathbb{Q}Or\, G)/K_1(\mathbb{Z}_{(\|G\|)}Or\, G)$	386
$SW : \overline{C}(G)^* \longrightarrow K_o(\mathbb{Z}Or\, G)$	385
$\rho^G_{\mathbb{R}} : Rep_{\mathbb{R}}(G) \rightarrowtail Wh(\mathbb{Q}Or\, G)$	373
$\overline{\rho}^G : V^{ev}_{or}(G,Dim) \rightarrowtail K_1(\mathbb{Q}Or\, G)/K_1(\mathbb{Z}_{(\|G\|)}Or\, G)$	401
$\overline{\rho}^G : V^{ev}_{\kappa}(G,Dim) \rightarrowtail \kappa(G)$	404

0.4. Applications to geometry

We restate the Isovariant s-Cobordism Theorem 4.42. saying that isovariant h-cobordisms are classified by their isovariant Whitehead torsion. We relate the isovariant and equivariant setting by an homomorphism $\Phi : Wh^G_{Iso}(M) \longrightarrow Wh^G(M)$. Provided that the weak gap conditions 4.49. are satisfied, we show that Φ is injective and determine its image and thus get the Equivariant s-Cobordism Theorem 4.51. We give counter-examples to the Equivariant s-Cobordism Theorem 4.51. without the weak gap hypothesis in Example 4.56.

We prove for a finite group G of odd order that the transfer on K_o and Wh induced by the sphere bundle of a G-vector bundle vanishes under mild conditions (see 15.29.). These transfer maps appear e.g. in the comparison of isovariant and equivariant White-head groups and in the involution defined on them by reversing h-cobordisms.

We construct an homomorphism $\rho^G_{\mathbb{R}} : Rep_{\mathbb{R}}(G) \longrightarrow Wh(\mathbb{Q}Or\, G)$, $[V] \longrightarrow \rho^G(S(V \oplus V))$ and prove injectivity in 18.38. Hence spheres of real G-representations are classified up to G-diffeomorphism by Reidemeister torsion. This reproves de Rham's theorem that two

G-representations are RG-isomorphic if and only if their spheres are G-diffeomeorphic.

A G-homotopy representation X is a finite-dimensional G-CW-complex such that $X^H \simeq S^{n(H)}$ holds for $H \subset G$. Given two G-homotopy representations X and Y with $\dim X^H = \dim Y^H$ for all $H \subset G$, we want to determine the set $[X,Y]^G$ of G-homotopy classes of G-maps between them. If we have choosen a coherent orientation, then

$$DEG : [X,Y]^G \longrightarrow \prod_{(H)} \mathbb{Z}, \quad [f] \longrightarrow (\deg f^H)_{(H)}$$

is an injection. We give in Theorem 20.38. a set of congruences describing the image of DEG and hence $[X,Y]^G$ which can be computed from the difference of the reduced Reidemeister torsion $\bar{\rho}^G(Y) - \bar{\rho}^G(X)$ by generalized Swan homomorphisms. In particular we get that G-homotopy representations are classified up to oriented G-homotopy equivalence by an absolute invariant, the reduced Reidemeister torsion.

0.5. On the concept of the book.

We have tried to keep the book fairly self-contained. We give the definitions, results and proofs in full generality and illustrate them by examples. At the end of each section there is a comment where the material of the section is put into context with the work of other mathematicians, further applications are discussed and additional references are given. More information and results are contained in the exercises. We advise the reader to at least read through them.

This expansive way of writing means that the sections contain much more material and results in much larger generality than needed for the following sections of specific applications. Therefore we have tried to give the reader, who is only interested in a specific question, the possibility to pick out a single section and read it without knowing the others. Here is some advice for such a reader.

The chapters II and III are independent of chapter I. If one is interested in the algebra only, one may skip chapter I completely.

In chapter I one may begin with one of the sections 3, 4, or 5 directly as they are independent of one another and sections 1 and 2 are quite elementary.

Nearly all notions and results are stated for Lie groups G and proper G-actions without any assumptions about the connectivity of the fixed point sets. The notational and technical difficulties decrease considerably if G is a finite group and the fixed point sets are empty or simply connected. In this case a summary of the invariants defined for G-spaces in chapter II is given in section 18 including their basic properties. Moreover, section 8 is in this case of no importance, as everything takes place over the orbit category. In particular this restriction does no harm if one studies G-homotopy representations.

If one is interested in the finiteness obstruction and torsion only over the group ring resp. for the universal covering of a G-space without group action, one may directly begin with section 11 and 12 thinking of RΓ as RG, and similarly for the material about the Swan homomorphism for group rings and its lifting in section 19.

An experienced reader can start with section 18 without having looked at the previous sections since the necessary input from them is reviewed in the beginning of section 18. Although section 20. makes use of section 18 and 19, section 20 can be read without knowing section 18 and 19 because only the formal properties of Reidemeister torsion and Swan homomorphism but not their explicit constructions are needed.

Table of Contents

CHAPTER I

GEOMETRICALLY DEFINED INVARIANTS

Summary

In the first section we collect elementary facts about G-CW-complexes for G a topological group. We prove the Slice Theorem 1.37. and 1.38. and deal in Corollary 1.40 with path lifting along the projection $p : X \longrightarrow X/G$. In Theorem 1.23. we show for a G-CW-complex X that X is proper if and only if the isotropy group G_x is compact for all $x \in X$.

We prove the Equivariant Cellular Approximation Theorem 2.1. and the Equivariant Whitehead Theorem 2.4. in section 2. We give criterions for a G-space to have the G-homotopy type of a G-CW-complex in Corollary 2.8., Corollary 2.11. and Proposition 2.12. and examine G-push outs of G-maps and their connectivity in Lemma 2.13.

In section 3 we introduce the finiteness obstruction $w^G(Y) \in Wa^G(Y)$ of a finitely dominated G-space geometrically. We call Y finitely dominated if there is a finite G-CW-complex X and G-maps $r : X \longrightarrow Y$ and $i : Y \longrightarrow X$ satisfying $r \circ i \simeq_G \mathrm{id}$. Elements in the abelian group $Wa^G(Y)$ are represented by G-maps $f : X \longrightarrow Y$ with a finitely dominated G-space as source and Y as target. Addition is given on representatives by the disjoint union. The zero element is represented by $\emptyset \longrightarrow Y$ and $w^G(Y)$ by $\mathrm{id} : Y \longrightarrow Y$.

Theorem 3.2.

a) Let X be a finitely dominated G-space. Then X is G-homotopy equivalent to a finite G-CW-complex if and only if $w^G(X)$ vanishes.

b) The finiteness obstruction is a G-homotopy invariant.

c) The finiteness obstruction is additive on G-push outs. □

A typical situation, where the finiteness obstruction comes in, is the following. Suppose X is a finitely dominated G-space for which we want to construct a (compact smooth) G-manifold M with $M \simeq_G X$. As any such M is a finite G-CW-complex, the vanishing of $w^G(Y)$ is a necessary condition. Often constructions of G-spaces give finitely dominated G-spaces but not necessarily finite G-CW-complexes.

In section 4 we extend the geometric construction of Whitehead group and Whitehead torsion due to Cohen [1973] and Stöcker [1970] to the equivariant setting following Illman [1974]. A G-homotopy equivalence $f : X \longrightarrow Y$ between finite G-CW-complexes is called simple if it is G-homotopic to a composition of so called elementary expansions and collapses. It determines an element $\tau^G(f)$, its (equivariant) Whitehead torsion, in the Whitehead group $Wh^G(Y)$ by its mapping cylinder. Elements in $Wh^G(Y)$ are represented by pairs of finite G-CW-complexes (X,Y) such that the inclusion $Y \longrightarrow X$ is a G-homotopy equivalence. Addition is given by the G-push out along Y and the zero element by (Y,Y).

Theorem 4.8.

a) A G-homotopy equivalence $f : X \longrightarrow Y$ between finite G-CW-complexes is simple if and only if $\tau^G(f)$ vanishes.

b) $f \simeq_G g \Rightarrow \tau^G(f) = \tau^G(g)$

c) τ^G is additive on G-push outs.

d) $\tau^G(g \circ f) = \tau^G(g) + g_* \tau^G(f)$ □

Let $f : X \longrightarrow Y$ be a G-homeomorphism of finite G-CW-complexes. If G is trivial, f is simple by Chapman [1973]. This is not true for non-trivial G in general (see Example 4.25. and 4.26.).

If G is a compact Lie group and M a (compact, smooth) G-manifold, we define a preferred simple structure on M (cf. Illman [1978], [1983], Matsumoto-Shiota [1987]). Hence for any G-homotopy equivalence $f : M \longrightarrow N$ between G-manifolds its Whitehead torsion is defined. It vanishes if f is a G-diffeomorphism.

A cobordism (B,M,N) of G-manifolds is an isovariant h-cobordism resp. (equivariant) h-cobordism if the inclusions $M \longrightarrow B$ and $N \longrightarrow B$ are isovariant G-homotopy equivalences resp. G-homotopy equivalences.

We introduce the isovariant Whitehead group $Wh^G_{Iso}(M)$ and the isovariant Whitehead torsion $\tau^G_{Iso}(B,M,N)$ of an isovariant h-cobordism. We restate the Isovariant s-Cobordism Theorem 4.42. saying that $\tau^G_{Iso}(B,M,N) \in Wh^G_{Iso}(B,M,N)$ classifies isovariant h-cobordisms over M up to G-diffeomorphism relative M, if $\dim M_H/WH \geq 5$ (i.e. the dimension of any component of M_H/WH is not smaller than 5) for all $H \in \mathrm{Iso}\, M$ holds (see Browder-Quinn [1973]. Hauschild [1978], Rothenberg [1978]). We construct an homomorphism

4.43. $\Phi(M) : Wh^G_{Iso}(M) \longrightarrow Wh^G(M)$

satisfying $\Phi(M)(\tau^G_{Iso}(B,M,N)) = \tau^G(M \longrightarrow B)$ (see Proposition 4.44.). This leads to (cf. Araki-Kawakubo [1988])

Theorem 4.51. The Equivariant s-Cobordism Theorem. Let M be a G-manifold satisfying the weak gap condition 4.49. such that $\dim M_H/WH \geq 5$ holds for $H \in \mathrm{Iso}\, M$. Then

a) Any h-cobordism over M is an isovariant h-cobordism.

b) $\Phi(M)$ is split injective with a certain direct summand $Wh^G_\rho(M) \subset Wh^G(M)$ as image.

c) $Wh^G_\rho(M)$ classifies h-cobordisms over M up to G-diffeomorphism relative M.

Because of this result equivariant Whitehead torsion is important for the classification of G-manifolds. It is in general much easier to handle with the equivariant Whitehead torsion than with the isovariant one. If one wants to show that two G-manifolds M and N are G-diffeomorphic, the general strategy is to construct an h-cobordism (B,M,N) by equivariant surgery and then apply Theorem 4.51. As an illustration we mention the classification of semilinear discs M (i.e. G-manifolds M such that for $H \subset G$ the pair $(M^H, \partial M^H)$ is homotopy equivalent to $(D^k\ S^{k-1})$ for appropriate $k \geq 0$) due to Rothenberg [1978] in Theorem 4.55. Such M is classified by the $\mathbb{R}G$-isomorphism type of TM_x for $x \in M^G$ and the Whitehead torsion of $STM_x \longrightarrow M \setminus \mathrm{int}\, DTM_x$ up to G-diffeomorphism, provided that M satisfies the weak gap conditions 4.49. and $\dim M_H/WH \geq 6$ for $H \in \mathrm{Iso}\, M$. The assumption 4.49. in Theorem 4.51. is necessary (see Example 4.56.).

In section 5 we introduce the (equivariant) Euler characteristic $\chi^G(X) \in U^G(X)$ and show that it is a G-homotopy invariant and additive under G-push out in Theorem 5.4.

We define Euler ring $U(G) = U^G(\{point\})$ and Burnside ring $A(G)$ for G a compact Lie group.

All the invariants above satisfy sum formulas for G-push outs and are G-homotopy invariant. It turns out that they can be characterized just by these two formal properties as universal functorial additive invariants (see Theorem 6.7., 6.9., and 6.11.).

In section 7 we derive from this characterization a product formula 7.1., a restriction formula 7.25. and a diagonal product formula 7.26. for the finiteness obstruction, Whitehead torsion and Euler characteristic in a simple geometric manner. This requires a careful analysis for the problem how to assign to the restriction res X of a finite G-CW-complex X to a subgroup H of the compact Lie group G a H-simple structure (see 7.10.). This is a non-trivial question if G/H is not finite, see e.g. the case where X is a homogeneous G-space G/K. It is remarkable that the geometric description of the restriction formula for infinite G/H and of the diagonal product formula for infinite G are much easier than the algebraic one, we will develop in section 14.

The diagonal product formula is the main ingredient in constructing an homomorphism

7.39. $\quad \omega : A(G)^* \longrightarrow Wh^G(\{point\})$.

A unit in the Burnside ring represented by a G-self equivalence $f : SV \longrightarrow SV$ of the unit sphere of an orthogonal G-representation is sent by ω to $(1-\chi^G(SV))^{-1}\tau^G(f)$. No G is known where ω is non-trivial. This map appears in the study of G-homotopy representations in section 20 and in equivariant surgery (see Dovermann-Rothenberg [1988]).

If X is a finitely dominated G-space, $X \times S^1$ is G-homotopy equivalent to a finite G-CW-complex by the product formula. This can directly be seen from Mather's trick which is used to define a geometric Bass-Heller-Swan injection

7.34. $\quad \Phi : Wa^G(Y) \longrightarrow Wh^G(Y \times S^1)$.

It relates finiteness obstruction and Whitehead torsion (cf. Ferry [1981a],

Kwasik [1983]) and leads to a geometric definition of *equivariant negative* K-*groups* (Definition 7.36., Theorem 7.38.). These appear for example as obstruction groups for equivariant transversality in Madsen-Rothenberg [1985a].

In section 8 we deal with *lifting a* G-*action* to a $\tilde{G}$-action on the universal covering $\tilde{X}$ which extends the $\pi_1(X)$-action and covers the G-action, where $\tilde{G}$ is an extension of $\pi_1(X)$ and G (Theorem 8.1.). If G_o and $\tilde{G}_o$ denote the components of the identity, we study the square

8.8.
$$\begin{array}{ccc} \tilde{X} & \longrightarrow & \tilde{X}/\tilde{G}_o \\ \downarrow & & \downarrow \\ X & \longrightarrow & X/G_o \end{array}$$

In particular the space $\tilde{X}/\tilde{G}_o$ is important as it is simply connected and carries the action of the discrete group $\pi_o(\tilde{G}) = \tilde{G}/\tilde{G}_o$ (Lemma 8.9.). We will read off a lot of algebraic information from the $\pi_o(\tilde{G})$-space $\tilde{X}/\tilde{G}_o$. We derive from 8.8. an explicit description of the kernel and cokernel of $\pi_1(X) \longrightarrow \pi_1(X/G)$ in Proposition 8.10. (cf. Armstrong [1983]).

We organize the book-keeping of the components of the various fixed point sets including their fundamental groups, universal coverings, WH-actions and diagrams corresponding to 8.8. All these data are codified in the *discrete fundamental category* $\Pi/(G,X)$ (Definition 8.28.). The notion of the *cellular* $\mathbb{Z}\Pi/(G,X)$-*chain complex* (Definition 8.37.) is the main link between geometry and algebra. It is the composition of the *discrete universal covering functor* $\tilde{X}/ : \Pi/(G,X) \longrightarrow \{\text{CW-complexes}\}$ (Definition 8.30.) and the functor "cellular chain complex". The algebra of modules over a category is modelled upon it.

1. G-CW-complexes

We introduce the equivariant version of a CW-complex and collect its main properties. We will deal also with the case where G is not compact and treat proper actions. The Slice Theorem will be proved.

Convention 1.1. We always work in the category of compactly generated spaces (see Steenrod [1967] or Whitehead [1978], p. 17 - 21). We recall that a compactly generated space X is a Hausdorff space such that a subset $A \subset X$ is closed if and only if its intersection with any compact subset is closed. □

In this section G is a topological group (which is assumed to be compactly generated by convention 1.1). A topological group which is a Hausdorff space and locally compact is compactly generated. Examples are Lie groups (see Bredon [1972] I.1 for the definition). We always assume that a subgroup $H \subset G$ is closed.

Definition 1.2. Let (X,A) be a pair of G-spaces such that A/G is a Hausdorff space. A relative G-CW-complex structure on (X,A) consists of

a) a filtration $A = X_{-1} \subset X_0 \subset X_1 \subset X_2 \subset \ldots$ of $X = \bigcup_{n=-1}^{\infty} X_n$

and

b) a collection $\{e_i^n \mid i \in I_n\}$ of G-subspaces $e_i^n \subset X_n$ for each $n \geq 0$

with the properties:

i) X has the weak topology with respect to the filtration $\{X_n \mid n \geq -1\}$. i.e., $B \subset X$ is closed if and only if $B \cap X_n \subset X_n$ is closed for any $n \geq -1$,

ii) For each $n \geq 0$ there is a G-push-out

$$\begin{array}{ccc} \coprod_{i\in I_n} G/H_i \times S^{n-1} & \xrightarrow{\coprod_{i\in I_n} q_i^n} & X_{n-1} \\ \downarrow & & \downarrow \\ \coprod_{i\in I_n} G/H_i \times D^n & \xrightarrow{\coprod_{i\in I_n} Q_i^n} & X_n \end{array}$$

such that $e_i^n = Q_i^n(G/H_i \times \operatorname{int} D^n)$.

If A *is empty we call* X *a* G-CW-*complex*. □

The G-subspace X_n is the n-*skeleton* of (X,A). The G-subspaces e_i^n are called the *open cells*. The number n is its *dimension* and the conjugacy class of subgroups (H_i) its *type*. The map q_i^n is an *attaching map* and the pair $(Q_i^n, q_i^n) : G/H_i \times (D^n, S^{n-1}) \rightarrow (X_n, X_{n-1})$ a *characteristic map* for e_i^n. We call $\bar{e}_i^n := Q_i^n(G/H_i \times D^n)$ a *closed cell* and $\partial e_i^n = q_i^n(G/H_i \times S^{n-1})$ its *boundary*. We emphasize that the filtration and the open cells are part of the structure of a relative G-CW-complex but not the attaching or characteristic maps. An *isomorphism* $f : (X,A) \rightarrow (Y,B)$ *of relative* G-CW-*complexes* is a G-homeomorphism of pairs respecting the skeletal filtration and mapping open cells bijectively to open cells.

Here is a list of basic facts proved later.

1.3. The open and closed cells are already determined by the skeletal filtration. Namely, the open cells e_i^n are the G-components of $X_n \setminus X_{n-1}$, i.e. the lifts of the components of $(X_n \setminus X_{n-1})/G$. The closed cell $\bar{e}_i^n$ is the closure of e_i^n in X and $\partial e_i^n = \bar{e}_i^n \setminus e_i^n$. In particular e_i^n is open in X_n and $\bar{e}_i^n$ is closed in X. A subset $C \subset X$ is closed if and only if $C \cap A$ in A and $C \cap \bar{e}_i^n$ in $\bar{e}_i^n$ is closed. □

1.4. Let $H \subset G$ be normal and (X,A) a relative G-CW-complex such that A/H is a Hausdorff space. Then (X/H,A/H) has a canonical G/H-CW-structure if one of the following conditions holds:

a) H is compact.

b) G_x is compact for each $x \in X \setminus A$.

c) G/H is discrete.

Namely, the n-skeleton is X_n/H and the open n-cells are $\{e_i^n/H \mid i \in I_n\}$, if X_n is the n-skeleton and $\{e_i^n \mid i \in I_n\}$ the open n-cells of X. □

1.5. If (X,A) is a relative G-CW-complex, the inclusion $A \to X$ is a G-cofibration. □

A G-space X is *obtained from the G-space A by attaching n-dimensional cells* if there is a G-push-out

$$\begin{array}{ccc} \coprod_{i \in I} G/H_i \times S^{n-1} & \longrightarrow & A \\ \downarrow & & \downarrow \\ \coprod_{i \in I} G/H_i \times D^n & \longrightarrow & X \end{array}$$

1.6. Let A be a (compactly generated) G-space such that A/G is a Hausdorff space. Then A/G is also compactly generated (Steenrod [1967] 2.6). Let (X,A) be a G-pair with a G-filtration $\{X_n \mid n \geq -1\}$, $X_{-1} = A$, $X = \cup X_n$ such that X has the weak topology with respect to this filtration and X_n is obtained from X_{n-1} by attaching n-dimensional cells. Then X is compactly generated and (X,A) a relative G-CW-complex. □

Example 1.7. Let G be a finite group and X be a CW-complex. Suppose that G acts cell preserving on X, i.e. if e is an open cell of X, then ge is again an open cell and ge = e implies, that $l_g : e \to e \quad x \to gx$ is the identity, for all $g \in G$. Then X is a G-CW-complex (compare Bredon [1972] III.1). □

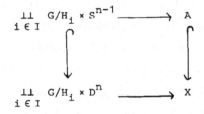

<u>Example 1.8.</u> Let G be a finite group and V be an orthogonal G-representation. The unit sphere SV has the following G-CW-complex structure. Choose a base $\{e_1,\ldots,e_m\}$. Let X be the convex hull of $\{\pm ge_i \mid g \in G,\ 1 \le i \le m\}$. Its boundary ∂X is G-homeomorphic to SV by radial projection so that it suffices to define a G-CW-complex structure on ∂X. Now there is a simplicial complex structure on ∂X such that G acts simplicially. Consider its first barycentric subdivision. Then for any simplex σ with $g\sigma = \sigma$ multiplication with g induces the identity on it (Bredon [1972] III 1.1). Now apply 1.7. □

1.9. Let G be a compact Lie group and M a G-manifold, i.e. M is a compact C^∞-manifold and G acts by a C^∞-map $G \times M \to M$. Then M has the structure of a G-CW-complex. Moreover, a G-triangulation can be constructed (see Illman [1983] p. 500). □

To prove the statements 1.3 to 1.6 we need some material about G-cofibrations and G-push-outs.

A G-map $i : A \to X$ is called a G-<u>cofibration</u> if $i(A)$ is closed in X and for any G-space Y, G-maps $f : X \to Y$ and $h : A \times I \to Y$ with $f \circ i = h_o$ there is a G-map $\bar{h} : X \times I \to Y$ satisfying $\bar{h} \circ (i \times id) = h$ and $\bar{h}_o = f$

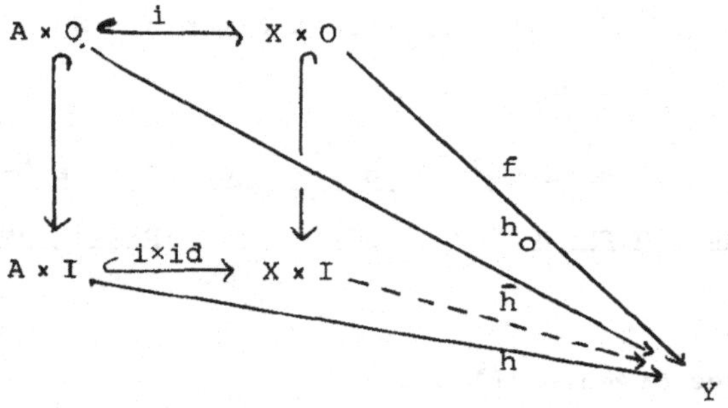

Let (X,A) be a G-pair. It is called a G-NDR (G-<u>neighbourhood deformation retract</u>) if there are G-maps $u : X \to I$ and $h : X \times I \to X$ satisfying:

i) $A = u^{-1}(0)$.

ii) $h_0 = id$.

iii) $h_t | A = id_A$ for $t \in I$.

iv) $h_1(x) \in A$ for $u(x) < 1$.

Lemma 1.10.

a) *If* $i : A \to X$ *is a* G-*cofibration then* i *is a closed embedding*, i.e. $i(A)$ *is closed in* X *and* $i : A \to i(A)$ *a* G-*homeomorphism*.

b) *Let* (X,A) *be a* G-*pair*. *Then* $i : A \to X$ *is a* G-*cofibration if and only if* (X,A) *is a* G-NDR.

Proof: analogous to the non-equivariant case (Strøm [1966], Steenrod [1967], 7.1.) □

A square

$$\begin{array}{ccc} A & \xrightarrow{f} & Y \\ {\scriptstyle j}\downarrow & & \downarrow{\scriptstyle J} \\ X & \xrightarrow{F} & Z \end{array}$$

of G-spaces is a G-*push-out* if for each pair of G-maps $f' : Y \to U$ and $j' : X \to U$ with $f'f = j'j$ there is a G-map $u : Z \to U$ uniquely determined by $uJ = f'$ and $uF = j'$.

Lemma 1.11. *Let* $j : A \to X$ *be a* G-*cofibration*. *Then a* G-*push-out* Z *is given by the adjunction space* $Y \cup_f X$ *which is compactly generated*. J *is a* G-*cofibration*.

Proof: For $G = 1$ see Steenrod [1967] 8.5. □

Let X be a G-space and $H \subset G$ a (closed) subgroup. For $x \in X$ let its *isotropy group* G_x be $\{g \in G \mid gx = x\}$. It is closed in G. Let the H-*fixed point set* X^H be $\{x \in X \mid G_x \supset H\}$, $X^{>H}$ be $\{x \in X \mid G_x \supset H, G_x \neq H\}$ and

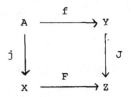

X_H be $\{x \in X \mid G_x = H\}$. Then X^H and $X^{>H}$ are closed in X and $X^H \setminus X^{>H} = X_H$. If H and K are subgroups we call H subconjugated to K if $gHg^{-1} \subset K$ holds for suitable g and write $(H) \subset (K)$. Define $X^{(H)} = \{x \in X \mid (G_x) \supset (H)\}$, $X^{>(H)} = \{x \in X \mid (G_x) \supset (H),\ (G_x) \neq (H)\}$ and $X_{(H)} = \{x \in X \mid (G_x) = (H)\}$. Then $X^{(H)}$, $X^{>(H)}$ and $X_{(H)}$ are G-subspaces and $X^{(H)} \setminus X^{>(H)} = X_{(H)}$.

For $H \subset G$ let NH be its *normalizer* $\{g \in G \mid gHg^{-1} = H\}$ and $WH = NH/H$ its *Weyl group*. The G-action on X induces WH-actions on X^H, $X^{>H}$ and X_H.

Lemma 1.12. Let $i : A \to X$ be a G-cofibration.

a) Its restriction to $H \subset G$ is a H-cofibration.

b) For $H \subset G$ the map $i^H : A^H \to X^H$ is a WH-cofibration.

c) The map $i/H : A/H \to X/H$ is a G/H-cofibration if A/H and X/H are Hausdorff spaces and $H \subset G$ normal.

Proof: Use Lemma 1.10. For G-NDR-s the statements are obvious. The assumption about A/H and X/H in c) guarantees that they are compactly generated (Steenrod [1967] 2.6). □

Lemma 1.13. Consider the G-push-out

$$\begin{array}{ccc} A & \xrightarrow{\;f\;} & Y \\ {\scriptstyle j}\downarrow & & \downarrow{\scriptstyle J} \\ X & \xrightarrow{\;F\;} & Z \end{array}$$

Assume that j is a G-cofibration.

a) The restriction to $K \subset G$ is a K-push-out.

b) Taking the H-fixed point set yields a WH-push-out.

c) Let $H \subset G$ be normal and A/H, X/H and Y/H be Hausdorff spaces. Then we get a G/H-push-out by dividing out the H-action.

Proof. We only verify c). We must show that $F/H + J/H : X/H + Y/H \to Z/H$ is an identification. This follows from the fact that for a H-map

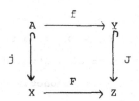

$u : A \to B$, which is an identification, also $u/H : A/H \to B/H$ is an identification.

$$\begin{array}{ccc} A & \xrightarrow{u} & B \\ {\scriptstyle pr_A}\downarrow & & \downarrow{\scriptstyle pr_B} \\ A/H & \xrightarrow[u/H]{} & B/H \end{array}$$

Namely, pr_A and $u/H \circ pr_A = pr_B \circ u$ are identifications. The map j/H is a G/H-cofibration by Lemma 1.12. c). □

Now we come to the proof of 1.3. to 1.6. Recall that X_n is the G-push-out

$$\begin{array}{ccc} \amalg G/H_i \times S^{n-1} & \xrightarrow{\amalg q_i} & X_{n-1} \\ \downarrow{\scriptstyle j} & & \downarrow \\ \amalg G/H_i \times D^n & \xrightarrow{\amalg Q_i} & X_n \end{array}$$

and j is a G-cofibration. Hence $X_{n-1} \to X_n$ is a G-cofibration. Now 1.5 follows from the equivariant version of Whitehead [1978] I. 6.3. In 1.6. the only problem is to show that X is compactly generated. One shows inductively using Lemma 1.11. that each X_n is compactly generated and applies Whitehead [1978] I. 6.3.

By Lemma 1.12. and 1.13. we have the G/H-push-out with j/H a cofibration

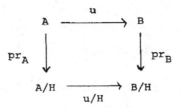

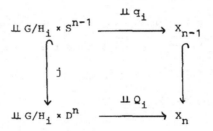

$$\begin{array}{ccc} \coprod (G/H_i)/H \times S^{n-1} & \xrightarrow{\coprod q_i/H} & X_{n-1}/H \\ \Big\downarrow{\scriptstyle j/H} & & \Big\downarrow \\ \coprod (G/H_i)/H \times D^n & \xrightarrow{\coprod Q_i/H} & X_n/H \end{array}$$

The conditions appearing in 1.4. guarantee that HH_i is closed in G (tom Dieck [1987], I.3.1.). Since H is normal, HH_i is a closed subgroup in G so that $(G/H_i)/H$ is G/H-homeomorphic to the compactly generated G/H-space $(G/H)/(HH_i/H)$. Now 1.4. follows using Whitehead [1978] I.6.4. to verify that X/H has the weak topology with respect to $\{X_n/H \mid n \geq -1\}$.

For 1.3. consider the push-out above for H = G

$$\begin{array}{ccc} \coprod S^n & \xrightarrow{\coprod q_i/G} & X_{n-1}/G \\ \Big\downarrow & & \Big\downarrow \\ \coprod D^n & \xrightarrow{\coprod Q_i/G} & X_n/G \end{array}$$

As above one shows that X_n/G is a Hausdorff space so that $Q_i/G(D^n) \subset X_n/G$ is compact and in particular closed. Then $\bar{e}_i^n = Q_i(G/H_i \times D^n) \subset X_n$ is closed in X_n. Now it is easy to prove 1.3.

<u>Example 1.14.</u> Let G be the multiplicative group of positive real numbers. Define an action

$$\rho : G \times \mathbb{R} \to \mathbb{R} \quad (g,r) \to gr$$

Then $\mathbb{R}/G$ is not a Hausdorff space. Let $q : G \times S^0 \to \mathbb{R}$ be the G-map induced from the inclusion $S^0 \subset \mathbb{R}$. Consider the G-push-out

$$\begin{array}{ccc} G \times S^{0} & \xrightarrow{\quad q \quad} & \mathbb{R} \\ {\scriptstyle j}\big\downarrow & & \big\downarrow{\scriptstyle J} \\ G \times D^{1} & \xrightarrow{\quad Q \quad} & X \end{array}$$

Then $Q(G \times D^1) \subset X$ is open but not closed in X since image q is open but not closed (compare 1.3.). □

Remark 1.15. In the definition of a G-CW-complex given f.e. in Illman [1974] 1.2. it is part of the definition of a G-CW-complex that $\bar{e}_i^n$ is closed in X. Hence our definition requires less. But we have proven by 1.3. that both definitions agree. However, we have gained that in 1.6. we can attach arbitrarily cells and have not to care whether $Q(G/H_i \times D^n)$ is always closed. This is very pleasant when we want to make a G-map highly connected by attaching cells. □

Now we need some basic facts about proper maps and proper actions.

A map $f : X \to Y$ is *proper* if f is closed and $f^{-1}(y)$ is compact for any $y \in Y$. A general reference for proper maps is Bourbaki [1961] I.10.. A map $f : X \to Y$ between compactly generated spaces is proper if and only if $f^{-1}(C)$ is compact for any compact $C \subset Y$ (use Bourbaki [1961] I.10.2. proposition 6).

Lemma 1.16.

a) Consider maps $f : X \to Y$ and $g : Y \to Z$.

i) If f and g are proper, then gf is proper.

ii) If gf is proper, f is proper.

iii) If gf is proper and f surjective, then g is proper.

b) If $f : X \to Y$ is proper and $B \subset Y$ then the induced map $f^{-1}(B) \to B$ is proper. If f is proper and $A \subset X$ is closed $f|A : A \to Y$ is proper.

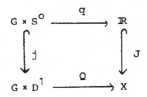

c) *If* $f : X \to Y$ *and* $f' : X' \to Y'$ *are proper then* $f \times f' : X \times X' \to Y \times Y'$ *is proper. If* $f : X \to Y$ *and* $g : X \to Z$ *are proper, then* $f \times g : X \to Y \times Z$ *is proper.*

d) *A projection* $pr : X \times Y \to Y$ *is proper if and only if* X *is compact.*

e) *Consider a push-out*

$$\begin{array}{ccc} A & \xrightarrow{f} & X \\ {\scriptstyle j}\downarrow & & \downarrow{\scriptstyle J} \\ Y & \xrightarrow{F} & Z \end{array}$$

If j *is a closed embedding and* f *is proper then* F *is proper.*

f) *Consider a pull-back*

$$\begin{array}{ccc} Z & \xrightarrow{F} & X \\ {\scriptstyle P}\downarrow & & \downarrow{\scriptstyle p} \\ Y & \xrightarrow{f} & A \end{array}$$

If f *is proper, then* F *is proper.*

g) *Let* $\{X_n \mid n \geq -1\}$ *resp.* $\{Y_n \mid n \geq -1\}$ *be a closed filtration for* X *resp.* Y *such that* X *resp.* Y *has the weak topology. Let* $f : X \to Y$ *be a map such that* $f(X_n \backslash X_{n-1}) \subset Y_n \backslash Y_{n-1}$ *holds for* $n \geq -1$. *Assume that each map* $f_n : X_n \to Y_n$ *is proper. Then* f *is proper.*

Proof: The verification of a), b), c) and d) given in Bourbaki [1961] I.10 is easily carried over to compactly generated spaces.

e) Given $C \subset Y$ the subset $F(C) \subset Z$ is closed if and only if $C \cup f^{-1}f(C \cap A) \subset Y$ is closed and $f(C \cap A) \subset X$ is closed.

f) Consider the commutative diagram

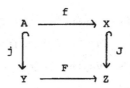

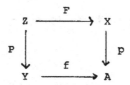

$$\begin{array}{ccc} X \times Y & \xrightarrow{\ id \times f\ } & X \times A \\ \uparrow{\scriptstyle F \times P} & & \uparrow{\scriptstyle id \times p} \\ Z & \xrightarrow{\ F\ } & X \end{array}$$

The map $id \times f$ is proper by assumption and c). From the explicit construction of a model for Z as the preimage of the diagonal under $f \times p : X \times Y \to A \times A$ we derive that $F \times P$ is a closed embedding and hence proper. Now apply a).

g) is left to the reader. □

Definition 1.17. A G-space X is *proper* if the map

$$\theta_X : G \times X \to X \times X \quad (g,x) \to (x,gx)$$

is proper. □

Lemma 1.18. If G is compact, any G-space is proper.

Proof: The projection $pr : G \times X \to X$ is proper (Lemma 1.16 d). Composing it with the G-homeomorphism $G \times X \to G \times X \quad (g,x) \to (g,gx)$, defines a proper map $p : G \times X \to X$ by Lemma 1.16 a. Then $\theta_X = pr \times p$ is proper by Lemma 1.16 c. □

Lemma 1.19. Let X be a proper G-space. Then X/G belongs also to the category of compactly generated spaces. We have for $x \in X$:

i) $G \to X \quad g \to gx$ *is proper.*

ii) G_x *is compact.*

iii) *The map* $G/G_x \to Gx \quad gG_x \to gx$ *is a G-homeomorphism.*

iv) *The orbit* Gx *is closed in* X.

Proof: Bourbaki [1961] III. 4.2. proposition 3 + 4. □

Lemma 1.19 shows that the G-space $\mathbb{R}$ of Example 1.14 is not proper.

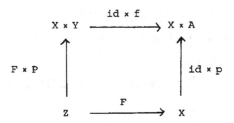

Lemma 1.20. Let X be a proper G-space and C a compact space with trivial G-action. Then any G-map $G/H \times C \to X$ is proper. In particular its image is closed in X.

Proof: The subgroup H is compact by Lemma 1.19. so that $G \to G/H$ is proper (Bourbaki [1961] III.4.1. cor. 2). Hence we can assume $H = 1$ by Lemma 1.16. a and c.

Let $D \subset X$ be the compact subset $f(\{e\} \times C)$ and $\bar{f} : C \to D$ be the induced map. By Lemma 1.16. the maps $id \times \bar{f} : G \times C \to G \times D$, $v : G \times D \to D \times X$ $(g,x) \to (x,gx)$ and $pr : D \times X \to X$ and hence their composition f are proper. □

Lemma 1.21. Consider the G-push-out such that Y and X_i are proper G-spaces, f_i is proper and j_i a G-cofibration,

$$\begin{array}{ccc} \coprod_{i \in I} A_i & \xrightarrow{\coprod_{i \in I} f_i} & Y \\ \Big\downarrow \coprod_{i \in I} j_i & & \Big\downarrow J \\ \coprod_{i \in I} X_i & \xrightarrow[F]{} & Z \end{array}$$

Then Z is a proper G-space.

Proof: If $C \subset Z$ is compact $C \cap F(X_i \setminus A_i) \neq \emptyset$ holds only for finitely many $i \in I$ as (X_i, A_i) is a G-NDR-pair. Hence we can assume that I is finite. Write $X = \coprod_{i \in I} X_i$. Consider the diagram

$$\begin{array}{ccc} G \times Z & \xrightarrow{\theta_Z} & Z \times Z \\ \Big\uparrow id \times (F \amalg J) & & \Big\uparrow F \times F \amalg J \times J \\ G \times (X \amalg Y) = G \times X \amalg G \times Y & \xrightarrow{\theta_X \amalg \theta_Y} & X \times X \amalg Y \times Y \end{array}$$

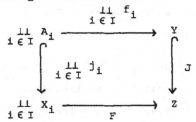

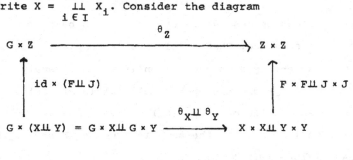

By Lemma 1.11. J is a G-cofibration and hence a closed embedding by Lemma 1.10. Then $J \times J$ is proper (Lemma 1.16. c). The map $F \times F$ is proper by Lemma 1.16. c and e. Since X and Y are proper G-spaces by assumption $\theta_Z \circ \mathrm{id} \times (F \amalg J) = (F \times F \amalg J \times J) \circ (\theta_X \amalg \theta_Y)$ is proper (Lemma 1.16. a). Then θ_Z is proper (Lemma 1.16. a). □

Lemma 1.22. *Let* $\{X_n \mid n \geq -1\}$ *be a closed filtration of* X *such that* X *has the weak topology. If each* X_n *is proper, then* X *is proper.*

Proof: Use Lemma 1.16. g and Steenrod [1967] 10.3. □

Theorem 1.23. *Let* (X,A) *be a relative* G-CW-*complex. Then* X *is a proper* G-*space if and only if* A *is proper and* G_x *compact for all* $x \in X \setminus A$. *In particular a* G-CW-*complex* X *is proper if and only if* G_x *is compact for each* $x \in X$.

Proof: One shows inductively that any X_n is proper. If H is compact, G/H is a proper G-space. By Lemma 1.16. the G-space $G/H \times D^n$ is proper. Now the induction step follows from Lemma 1.20. and 1.21. Finally apply Lemma 1.22. to show that X is proper. □

1.24. Let X be a free G-CW-complex. Then X is proper by Theorem 1.23. Since X is free, this is equivalent to $\mathrm{image}(\theta_X) \subset X \times X$ being closed and the map $\mathrm{image}(\theta_X) \to G \quad (x,gx) \to g$ being continuous (tom Dieck [1987] I.3.20). We will show in Theorem 1.37. that X is locally trivial so that $X \to X/G$ is a principal G-bundle (see Husemoller [1966], 4.2.2). By 1.4. X/G is a CW-complex. □

1.25. This process can be reserved. Let $p : E \to B$ be a principal G-bundle and $A \subset B$ a subspace. Then a relative CW-complex structure on (B,A) lifts to a G-CW-complex structure on $(E,p^{-1}(A))$. Namely, let E_n be $p^{-1}(B_n)$. If $\{e_i^n \mid i \in I_n\}$ are the open n-cells of (B,A), let $\{p^{-1}(e_i^n) \mid i \in I_n\}$ be the open n-cells of $(E,p^{-1}(A))$. Since B has the weak topology with respect to $\{B_n \mid n \geq -1\}$ the same is true for E and $\{E_n \mid n \geq -1\}$ (see Whitehead [1978], XIII.4.1).

Now B_n is the G-push-out

$$\begin{array}{ccc}
\coprod_{i \in I_n} S^{n-1} & \xrightarrow{\quad q \quad} & B_{n-1} \\
{\scriptstyle j}\downarrow & & \downarrow{\scriptstyle J} \\
\coprod_{i \in I_n} D^n & \xrightarrow{\quad Q \quad} & B_n
\end{array}$$

By the Lemma 1.26. below we obtain a G-push-out by the pull-back construction applied to $p_n : E_n \to B_n$

$$\begin{array}{ccc}
q^*E_n & \xrightarrow{\quad \bar{q} \quad} & E_{n-1} = J^*E_n \\
{\scriptstyle \bar{j}}\downarrow & & \downarrow{\scriptstyle \bar{J}} \\
Q^*E_n & \xrightarrow[\quad \bar{Q} \quad]{} & E_n
\end{array}$$

Since D^n is contractible, there is a G-homeomorphism of G-pairs $(Q^*E_n, q^*E_n) \to \coprod_{i \in I_n} G \times (D^n, S^{n-1})$ (Husemoller [1966], 5.10.3.). □

<u>Lemma 1.26. Consider the push-out with</u> j <u>a cofibration</u>

$$\begin{array}{ccc}
A & \xrightarrow{\quad f \quad} & Y \\
{\scriptstyle j}\downarrow & & \downarrow{\scriptstyle J} \\
X & \xrightarrow[\quad F \quad]{} & Z
\end{array}$$

<u>Let</u> $p : E \to Z$ <u>be a fibration. Then the pull-back construction yields a push-out with</u> $\bar{J}$ <u>a cofibration</u>.

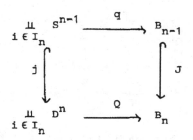

$$\begin{array}{ccc} f^*J^*E & \xrightarrow{\bar{f}} & J^*E \\ {\scriptstyle \bar{j}}\big\downarrow & & \big\downarrow{\scriptstyle \bar{J}} \\ F^*E & \xrightarrow[\bar{F}]{} & E \end{array}$$

<u>Proof:</u> The map $\bar{j}$ is a cofibration by Whitehead [1978] I.7.14. It remains to show that $\bar{J} \cup_{\bar{f}} \bar{F} : J^*E \cup_{f^*J^*E} F^*E \rightarrow E$ is an identification. By assumption $J \cup_f F : Y \cup_f X \rightarrow Z$ is an identification. By Steenrod [1967] 4.4. $id \times (J \cup_f F)$ and hence $(id \times J) \cup_{(id \times f)} (id \times F) : E \times B_1 \cup_{id \times f} E \times B_2 \rightarrow E \times B$ is an identification. Restricting it to $\{(e,p(e)) \mid e \in E\} \subset E \times B$ yields just $\bar{J} \cup_{\bar{f}} \bar{F}$. □

1.27. If (X,A) is a relative G-CW-complex and (Y,B) a relative G'-CW-complex then $(X,A) \times (Y,B)$ has a relative $G \times G'$-CW-complex structure. The k-skeleton $(X \times Y)_k$ is $\bigcup_{n+m=k} X_n \times Y_m$. Then $\{(X \times Y)_k \mid k \geq -1\}$ is a closed filtration of $X \times Y$ such that $X \times Y$ has the weak topology (Steenrod [1967] 10.3.).

If $\{e_i^n \mid i \in I_n\}$ are the open n-cells of X and $\{f_j^m \mid j \in J_m\}$ the open m-cells of Y then $\{e_i^n \times f_j^m \mid i \in I_n, j \in J_m, n+m=k\}$ are the open k-cells of $X \times Y$. The characteristic map for $e_i^n \times f_j^m$ is the product of the ones for e_i^n and f_j^m, if we identify $(G/H_i \times D^n) \times (G'/H_j' \times D^m)$ and $(G \times G')/(H_i \times H_i') \times D^{n+m}$. □

1.28. Let (X,A) be a relative G-CW-complex. A <u>relative</u> G-CW-<u>subcomplex</u> (Y,B) is a pair satisfying

i) Y is a G-subspace of X.

ii) B is a closed subspace of A.

iii) Y is the union of A and a collection of open cells whose boundaries also belong to Y.

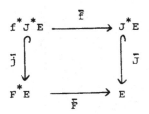

Then (Y,B) itself is a relative G-CW-complex with $Y_n = X_n \cap Y$ and Y is closed in X. If A and B are empty, (X,Y) is called a *pair of G-CW-complexes*. □

1.29. Consider the G-push-out

$$\begin{array}{ccc} A & \xrightarrow{f} & Y \\ \downarrow & & \downarrow \\ X & \xrightarrow{F} & Z \end{array}$$

Let (X,A) be a relative G-CW-complex. Then there is a relative G-CW-complex structure on (Z,Y) such that the relative homeomorphism (F,f) maps X_n to Z_n and open cells bijectively to open cells.

Assume that (X,A) is a pair of G-CW-complexes and f is *cellular* i.e. $f(A_n) \subset Y_n$ for all $n \geq 0$. Then we get the structure of a pair of G-CW-complexes on (Z,Y).

In particular X/A is a G-CW-complex if (X,A) is a pair of G-CW-complexes. Given a cellular G-map $f : X \to Y$ between G-CW-complexes its mapping cylinder and mapping cone get G-CW-structures by 1.27. and the construction above. □

1.30. Consider the space map(X,Y) of maps $X \to Y$ topologized as in Steenrod [1967] §5 . If X and Y are G-spaces, G acts on map(X,Y) by $f \to l_g \circ f \circ l_{g^{-1}}$ where l_g is multiplication with G. The G-fixed point set $\mathrm{map}(X,Y)^G \subset \mathrm{map}(X,Y)$ consists of all G-maps $f : X \to Y$. Consider the maps

$$\Phi : \mathrm{map}(G/H,X)^G \to X^H \quad \varphi \to \varphi(eH)$$

and

$$\psi : X^H \to \mathrm{map}(G/H,X)^G \quad x \to (\psi(x) : gH \to gx)$$

Use Steenrod [1967] 5.2. and 5.8. to show that they are continuous. Hence Φ and ψ are inverse homeomorphisms. □

We need some facts about homogenous spaces.

<u>Lemma 1.31.</u> <u>Let</u> H <u>and</u> K <u>be subgroups of</u> G.

a) <u>There is an equivariant map</u> $G/H \to G/K$ <u>if and only if</u> $(H) \subset (K)$ <u>holds</u>.

b) <u>If</u> $g \in G$ <u>and</u> $g^{-1}Hg \subset K$, <u>then we get a well defined</u> G-<u>map</u>

$$R_g : G/H \to G/K \quad g'H \to g'gK$$

c) <u>Every</u> G-<u>map</u> $G/H \to G/K$ <u>is of the form</u> R_g. <u>We have</u> $R_g = R_{g'}$ <u>if and only if</u> $g^{-1}g' \in K$ <u>holds</u>.

d) <u>Assume that</u> G <u>is compact or that</u> G <u>is a Lie group and</u> $H \subset G$ <u>compact</u>. <u>Then we have</u> $g^{-1}Hg \subset H \Rightarrow g^{-1}Hg = H$ <u>for any</u> $g \in G$ <u>and obtain a homeomorphism of topological groups</u>

$$WH \to \operatorname{map}(G/H,G/H)^G \quad gH \to R_{g^{-1}}$$

<u>Proof:</u> a), b) and c) are verified in tom Dieck [1987] I.1.14. The proof of d) for compact G can be found in Bredon [1972] 0.1.9. Let G be a Lie group and H compact. Suppose $g^{-1}Hg \subset H$ for $g \in G$. Then $g^{-1}Hg$ is a submanifold of H. This implies $g^{-1}Hg = H$ because for a connected submanifold M of a connected manifold N with dim M = dim M already M = N holds and H has finitely many components. Finally apply 1.30. □

<u>Example 1.32.</u> We want to illustrate by this example that the conditions in Lemma 1.31. d) are necessary.

Let $G \subset GL(2, \mathbb{R})$ be the Lie group of matrices over $\mathbb{R}$ of the shape

$$A = \begin{pmatrix} a & b \\ 0 & a^{-1} \end{pmatrix} \quad a,b \in \mathbb{R},\ a \neq 0$$

Denote by $H \subset G$ the subgroup of all matrices A with $a = 1$ and $b \in \mathbb{Z}$. One easily checks

$$\begin{pmatrix} a & b \\ 0 & a^{-1} \end{pmatrix} \begin{pmatrix} 1 & n \\ 0 & 1 \end{pmatrix} \begin{pmatrix} a & b \\ 0 & a^{-1} \end{pmatrix}^{-1} = \begin{pmatrix} 1 & a^2 n \\ 0 & 1 \end{pmatrix}$$

Hence $AHA^{-1} \subset H$ is equivalent to $a^2 \in \mathbb{Z}$ whereas $AHA^{-1} = H$ is equivalent to $a = \pm 1$. This gives a counterexample with G a Lie group.

If we substitute $\mathbb{R}$ by the p-adic rationals $\hat{\mathbb{Q}}_p$ and $\mathbb{Z}$ by the p-adic numbers $\hat{\mathbb{Z}}_p$ we obtain a counterexample where H is compact. □

The next result is one of the main properties of Lie groups.

<u>Theorem 1.33.</u> <u>Let</u> G <u>be</u> <u>a</u> <u>Lie</u> <u>group</u> <u>and</u> H <u>and</u> K <u>be</u> <u>subgroups</u>. <u>Suppose</u> <u>that</u> H <u>is</u> <u>compact</u>.

<u>Then</u> G/K^H <u>is</u> <u>the</u> <u>disjoint</u> <u>union</u> <u>of</u> <u>its</u> WH-<u>orbits</u> <u>or</u>, <u>equivalently</u>, $(G/K^H)/WH$ <u>is</u> <u>discrete</u>. <u>If</u> G <u>is</u> <u>compact</u>, G/K^H <u>is</u> <u>the</u> <u>disjoint</u> <u>union</u> <u>of</u> <u>finitely</u> <u>many</u> WH-<u>orbits</u>.

<u>Proof:</u> If G is compact this is shown in Bredon [1972] II.5.7. By inspecting the proof we see that it works also for G a Lie group and compact H if the result in Bredon [1972] II.5.6. is still true. But this is verified for G a Lie group and H compact in Montgomery-Zippin [1955], p. 216. □

1.34. Let H be a subgroup of G. If (X,A) is a relative H-CW-complex, then (ind X, ind A) $= G \times_H (X,A)$ has a canonical relative G-CW-complex structure. This follows from the identity $G \times_H H/K = G/K$. □

1.35. Let G be a Lie group and H a subgroup with dim H = dim G. Consider the restriction res G/K of the homogenous G-space G/K. Since G/H is discrete (res G/K)/H is discrete. Hence res G/K is a disjoint union of homogenous H-spaces and has therefore a canonical H-CW-structure. This carries over to G-CW-complexes. If (X,A) is a relative G-CW-complex then res (X,A) has a canonical relative H-CW-structure.

The assumption $\dim H = \dim G$ is essential. For example, a homogenous G-space G/K has exactly one G-CW-structure. If $\dim G \geq 1$ it is not obvious that G/K just as a space has a CW-complex structure and that there is even a canonical one. We will have to deal at several places with the problem that the restriction of a G-CW-complex to a subgroup H with $\dim H < \dim G$ has no obvious H-CW-structure. □

1.36. Let G be a Lie group and $H \subset G$ compact. For each $K \subset G$ we have a canonical WH-CW-structure on G/K^H by Theorem 1.33. Hence for any relative G-CW-complex (X,A) there is a canonical relative WH-CW-structure on (X^H, A^H). □

We make some remarks about slices. Let G be a topological group and X a G-space. A <u>slice</u> S at $x \in X$ is a G_x-subspace $S \subset X$ such that GS is an open neighbourhood for x and $\varphi : G \times_{G_x} S \to GS$ a G-homeomorphism. Then GS is called a <u>tube</u> around the orbit Gx.

<u>Theorem 1.37. Slice Theorem for G-CW-complexes.</u>

<u>Let G be a topological group and (X,A) a relative G-CW-complex. Assume that A is proper, there is a slice $a \in S \subset A$ in A for any $a \in A$ and G_x is compact for each $x \in X \setminus A$. Then there is a slice S at x in X for each</u> $x \in X$.

<u>Proof:</u> Let $x \in X$ be given. Choose $n \geq -1$ such that x lies in X_n but not in X_{n-1}. We construct inductively for $m = n, n+1, \ldots$ G_x-subsets $S_m \subset X_m$ such that $S_{m+1} \cap X_m = S_m$, GS_m is open in X_m and $\varphi_m : G \times_{G_x} S_m \to GS_m$, $g, y \to gy$ is a G-homeomorphism. Notice for the sequel that X is a proper G-space by Theorem 1.23.

The induction begin $m = n$ follows for $n = -1$ from the assumption about A. If $n \geq 0$ holds there is an open cell e^n containing x. Since e^n is open in X_n and G-homeomorphic to $G/G_x \times \operatorname{int} D$ it suffices to find a slice around any point $(gG_x, y) \in G/G_x \times \operatorname{int} D$. But for any open neighbourhood

U of y in int D the G_x-set $gG_x \times U$ is a slice.

We come to the induction step from $m-1$ to m for $m-1 \geq n$. Consider the G-push-out

$$\begin{array}{ccc} \coprod_{i \in I} G/H_i \times S^{m-1} & \xrightarrow{\coprod_{i \in I} q_i} & X_{m-1} \\ \downarrow & & \downarrow \\ \coprod_{i \in I} G/H_i \times D^m & \xrightarrow{\coprod_{i \in I} Q_i} & X_m \end{array}$$

Let $U_i \subset G/H_i \times S^{m-1}$ be $q_i^{-1}(S_{m-1})$. Define $V_i \subset G/H_i \times D^m$ as $\{(gH_i,tu) \mid (gH_i,u) \in U_i,\ 1/2 < t \leq 1\}$. Notice that U_i and V_i are G_x-subsets and V_i is G_x-homeomorphic to $U_i \times]1/2,1]$. Define S_m as the union $S_{m-1} \cup \bigcup_{i \in I} Q_i(V_i)$. We have by construction $S_m \cap X_{m-1} = S_{m-1}$.

By assumption $\varphi_{m-1} : G \times_{G_x} S_{m-1} \to GS_{m-1}$ is a G-homeomorphism. The following diagram commutes

$$\begin{array}{ccc} G \times_{G_x} U_i & \xrightarrow{\varphi_{U_i}} & GU_i \\ \downarrow{\scriptstyle id \times_{G_x} q_i | U_i} & & \downarrow{\scriptstyle q_i | GU_i} \\ G \times_{G_x} q_i(U_i) & \xrightarrow[\varphi_{m-1} | G \times_{G_x} q_i(U_i)]{} & q_i(GU_i) \end{array}$$

The map φ_{U_i} is bijective and continuous, $\varphi_{m-1} | G \times_{G_x} q_i(U_i)$ a G-homeomorphism and the vertical maps are proper by Lemmata 1.16. and 1.20. since

G_x is compact and therefore $Y \to Y/G_x$ is proper for any G_x-space Y (see tom Dieck [1987] I.3.6.). Then φ_{U_i} is proper by Lemma 1.16. a) and hence a G-homeomorphism. Since V_i is G-homeomorphic to $U_i \times]1/2,1]$ also φ_{V_i} is a G-homeomorphism. Because $GS_{m-1} \subset X_{m-1}$ is open, $GU_i \subset G/H_i \times S^{m-1}$, $GV_i \subset G/H_i \times D^m$ and $GS_m \subset X_m$ are open and we have a G-push-out

$$\begin{array}{ccc} \coprod_{i \in I} GU_i & \longrightarrow & GS_{m-1} \\ \downarrow & & \downarrow \\ \coprod_{i \in I} GV_i & \longrightarrow & GS_m \end{array}$$

Now the inverse maps of φ_{m-1}, φ_{U_i}, φ_{V_i} induce a G-map $GS_m \to G \times_{G_x} S_m$ such that both compositions with $\varphi_m : G \times_{G_x} GS_m \to GS_m$ are the identity. Hence φ_m is a G-homeomorphism onto the open subset $GS_m \subset X_m$. This finishes the induction step.

Now we give the final limit argument. Of course we define our slice $S \subset X$ as the G_x-subspace $\bigcup_{m \geq n} S_m$. Let $\Phi : G \times_{G_x} S \to GS$ be the obvious G-map. Since $GS \cap X_m$ is GS_m, $GS_m \subset X_m$ is open and X has the weak topology with respect to $\{X_m \mid m \geq n\}$, GS is open in X and GS has the weak topology with respect to $\{GS_m \mid m \geq n\}$. Hence the collection of G-maps $GS_m \xrightarrow{\Phi_m^{-1}} G \times_{G_x} S_m \to G \times_{G_x} S$ induce an inverse $GS \to G \times_{G_x} S$ of Φ. Hence Φ is the desired G-homeomorphism. □

A Hausdorff space X is <u>completely regular</u> if for any $x \in X$ and neighborhood U there is a continuous function $f : X \to [0,1]$ with $f(x) = 0$ and $f(X \setminus U) = 1$. A Hausdorff space X is completely regular if and only if it is homeomorphic to a subspace of a compact space (Tychonoff, see Schubert [1964] I.9.2., Satz 1).

<u>Theorem 1.38. Slice Theorem.</u>

<u>Let</u> G <u>be a Lie group and</u> X <u>a completely regular proper</u> G-<u>space</u>. <u>Then there is a slice at</u> x <u>for any</u> $x \in X$.

<u>Proof:</u> Palais [1961]. The definition of a proper G-space given there and the one we use agree by tom Dieck [1987], I.3.21. Also the definition of a slice in Palais [1961] and our are equivalent by Palais [1961], p. 306. A proof for compact G can also be found in Bredon [1972], II.5.4., Montgomery-Yang [1957], Mostow [1957]. □

The slice theorem has fundamental meaning for the theory of transformation groups (see for example tom Dieck [1987] I.5.). We are especially interested in path lifting.

<u>Proposition 1.39. Let</u> X <u>be a proper completely regular</u> G-<u>space and</u> G <u>be a Lie group</u>. <u>Then any path</u> $u : I \to X/G$ <u>can be lifted to a path</u> $v : I \to X$, i. e. $p \circ v = u$ <u>for the projection</u> $p : X \to X/G$.

<u>Proof:</u> For compact G this is proved in Montgomery-Yang [1957] or Bredon [1972] II.6.2. We show how we can reduce the problem to compact G. Because X is proper X/G is again compactly generated and especially a Hausdorff space (Lemma 1.19). Since image u is compact we can assume without loss of generality that image u lies in GS for a slice S. Then

GS/G is homeomorphic to S/G_x so that it suffices to lift the path along $S \to S/G_x$. But G_x is compact. □

Corollary 1.40. _Let G be a topological group and X a proper G-space such that G/G_x is path connected for some $x \in X$. Assume either that X is a G-CW-complex or that G is a Lie group and X completely regular. Then:_

a) $p_* : \pi_1(X,x) \to \pi_1(X/G,xG)$ _is surjective._

b) _If X is simply connected then also X/G._ □

Corollary 1.40. is sharpened in Proposition 8.10 and 8.12.

Proposition 1.41. _Let G be a topological group and (X,A) a relative G-CW-complex. Let $B \subset A$ be a G-subspace and U an open G-neighbourhood of B in A such that $i : B \to U$ is a strong G-deformation retraction (i.e. there is a G-map $r : U \to B$ with $r \circ i = id$ and $i \circ r \simeq_G id$ rel. B). Consider an open G-neighbourhood V of B in X with $B \subset \text{clos}\, U \subset V$._

Then there is an open neighbourhood W of B in X with $B \subset U \subset W \subset \text{clos}\, W \subset V$ such that $B \to W$ is a strong G-deformation retraction.

Proof: We leave the proof to the reader as the technique is the same as in the proof of Theorem 1.37. There we have described how to thicken a G-subset $S_{m-1} \subset X_{m-1}$ to a G-subset $S_m \subset X_m$ such that $S_{m-1} \to S_m$ is a strong G-deformation retraction. See also Lundell-Weingram [1969] II. 6.1. in the non-equivariant case. □

We close this section with some terminology. A relative G-CW-complex (X,A) is n-_dimensional_ if $X = X_n$ holds and _finite-dimensional_ if it is n-dimensional for some n. It is _finite_ if there are only finitely many open cells. We call it _skeletal-finite_ if each X_n is finite. In the literature this is often named of finite type but we prefer skeletal-finite to avoid confusion with the notion "of finite orbit type" introduced later. If G and A are compact, then (X,A) is finite, if and only if X is compact. One should notice that G is a zero-dimensional finite G-

CW-complex but just as a space G is not a zero-dimensional or finite CW-complex in general.

Comments 1.42. The notion of a CW-complex is due to J. H. C. Whitehead [1949] and can be found in nearly any text book on algebraic topology. Its extension to the equivariant case is carried out in tom Dieck [1987] II.1. + 2., Illman [1974] and Matumoto [1971]. We have compared their definitions with our in Remark 1.15.

Of course the main interesting case is the one of a compact Lie group where a lot of the proofs above are much simpler. For example, the set $Q_i^n(G/H \times D^n)$ is compact and hence closed (compare 1.3.). Therefore the reader might wonder why we also include the more general case. One reason is the following. Consider a G-CW-complex X with fundamental group $\pi = \pi_1 X$. Then there is an extension of Lie groups $1 \to \pi \to \widetilde{G} \to G \to 1$ and a $\widetilde{G}$-CW-structure on $\widetilde{X}$ such that the $\widetilde{G}$-action on $\widetilde{X}$ extends the π-action on $\widetilde{X}$ and covers the G-action. If π happens to be infinite $\widetilde{G}$ is not compact. However, if G acts properly on X, then $\widetilde{G}$ acts properly on $\widetilde{X}$. Therefore we are forced not only to study actions of compact Lie groups, but proper actions of Lie groups, since the passage to the universal covering is very important. We will see that most of the important properties of G-CW-complexes for compact Lie groups carry over to proper G-CW-complexes for Lie groups. Proper actions of non-compact groups are interesting for their own right and appear in the literature (see for example Bourbaki [1961] III.4., Connolly-Kozniewski [1986], [1988], Connolly-Prassidis [1987].

Exercises 1.43.

1. A G-CW-complex X is finite if and only if X/G is compact.

2. Let X be a G-CW-complex and Y a H-CW-complex. Define their join $X * Y$ by $CX \times Y \cup_{X \times Y} X \times CY$ if CX is the cone over X. Show that there is a canonical $G \times H$-CW-complex structure on it.

3. Equip SO(3)/SO(2) with at least two different WSO(2)-CW-complex-structures.

4. Let G be a finite p-group and X a finite G-CW-complex. Show for the Euler-characteristic χ

$\chi(X) \equiv \chi(X^G) \mod p$

If G is a torus prove $\chi(X) = \chi(X^G)$.

5. Let G be a n-dimensional compact Lie group. If X is G-CW-complex of dimension m, then the singular homology groups $H_i(X)$ vanish for $i > n + m$. Is the converse true?

6. Let G be a compact Lie group acting on the (compactly generated) space X. Then:

a) X/G is compactly generated.

b) $p : X \to X/G$ is proper.

c) $G \times X \to X \quad (g,x) \to g \cdot x$ is proper.

7. If H is a (closed) subgroup of G and X a proper G-space then X is a proper H-space.

8. Let $f : X \to Y$ be a surjective proper G-map. If X is a proper G-space then also Y.

9. Let H be a subgroup of G. Then G acts properly on G/H if and only if H is compact.

10. The action of $SL(2,\mathbb{C})$ on the Riemann sphere $S^2 = \mathbb{C} \cup \{\infty\}$

$$\begin{pmatrix} a & b \\ c & d \end{pmatrix}, z \to \frac{az+b}{cz+d}$$

is not proper.

11. Let G be a Lie group and X a proper completely regular G-space. Assume that all orbits have type G/H. Then the orbit map $X \to X/G$ is the projection of a fibre bundle with fibre G/H and structure group WH.

12. Let G be a Lie group and $f : X \to Y$ a bijective map between completely regular proper G-spaces. If $f/G : X/G \to Y/G$ is a homeomorphism then f is a G-homeomorphism.

13. Let G be a Lie group and X a completely regular G-space. If X is proper, $X^{(H)}$ is closed in X for all $H \subset G$.

14. Prove Corollary 1.40. in the case of a G-CW-complex. Hint: $X_1/G \to X/G$ is 1-connected.

2. Maps between G-CW-complexes.

We consider G-maps between G-CW-complexes. We state an equivariant Cellular Approximation Theorem and Whitehead Theorem and give criterions for a G-map to be a G-homotopy equivalence. We deal with G-CW-approximations and G-spaces of the G-homotopy type of a G-CW-complex like G-ENR-s.

A G-map $f : (X,A) \rightarrow (Y,B)$ between relative G-CW-complexes is called cellular if $f(X_n) \subset Y_n$ holds. Let G be a topological group.

Theorem 2.1. (Cellular Approximation Theorem)

a) *Let (X,A) and (Y,B) be relative G-CW-complexes and $f : (X,A) \rightarrow (Y,B)$ be a G-map. Suppose that f restricted to the relative G-CW-subcomplex (X',A') is cellular.*

Then f is G-homotopic relative X' to a cellular G-map.

b) *Let f and $g : (X,A) \rightarrow (Y,B)$ be cellular G-maps which are G-homotopic. Then there is a cellular G-homotopy between them.*

Proof: See Whitehead [1978] II.4.6. for the non-equivariant case. A proof for G-CW-complexes is given in tom Dieck [1987], II.2. □

The set of isotropy groups of a G-space X is Iso $X = \{G_x \mid x \in X\}$. It is closed under conjugation because $G_{gx} = gG_xg^{-1}$. Let Con G be the set $\{(H) \mid H \subset G\}$ of conjugacy classes of subgroups of G. Given $\mathfrak{F} \subset$ Con G we call X of orbit type $\mathfrak{F}$ if $\{(H) \mid H \in \text{Iso } X\} \subset \mathfrak{F}$ holds. We say X has finite orbit type if it is of orbit type $\mathfrak{F}$ for finite $\mathfrak{F}$.

Recall that a map $f : X \rightarrow Y$ is n-connected if $\pi_i(f) : \pi_i(X,x) \rightarrow \pi_i(Y,fx)$ is bijective for $i < n$ and surjective for $i = n$ for all $x \in X$. It is a weak homotopy equivalence if $\pi_i(f)$ is bijective for all $i \geq 0$ and $x \in X$.

In the following let H_* be an additive generalized homology theory satisfying the Eilenberg-Steenrod axioms except the dimension axiom and fulfills additionally the additivity axiom $(H_*(\coprod_I X_i) = \bigoplus_I H_*(X_i)$ for arbitrary I) (see Whitehead [1978] XII.6.). It is obvious what *homological n-connected* and *weak homology equivalence* means.

Consider $\mathfrak{F} \subset \mathrm{Con}\, G$ and a function $\nu : \mathfrak{F} \to \mathbb{Z}$. We call a G-map $f : X \to Y$ (homological) $(\mathfrak{F}, \nu)$-*connected* if f^H is (homological) $\nu(H)$-connected for each $(H) \in \mathfrak{F}$. If $\mathfrak{F} = \mathrm{Con}\, G$ and ν the constant function with value n, we say, that f is (G,n)-*connected*. If f^H is a weak homology resp. homotopy equivalence for each $(H) \in \mathfrak{F}$ we call f a *weak* $\mathfrak{F}$ -G-*homology* resp. -*homotopy equivalence*. In the case $\mathfrak{F} = \mathrm{Con}\, G$ we omitt $\mathfrak{F}$.

The definition of a G-CW-complex is settled in such a way that the following technique can be carried out (see 1.6.).

2.2. Consider a G-map $f : A \to Y$, a subset $\mathfrak{F} \subset \mathrm{Con}\, G$ and a function $\nu : \mathfrak{F} \to \mathbb{N}_0$. Assume that A/G is a Hausdorff space. Suppose that we are given for each $(H) \in \mathfrak{F}$ a set $S(H) = \{u(i,H) \mid i \in I(H)\}$ such that $u(i,H) \in \pi_{\nu(H)}(f^H, x)$ for some $x \in X^H$ holds. Represent each u(i,H) by a diagram

$$\begin{array}{ccc} S^{\nu(H)-1} & \xrightarrow{\hat{q}(i,H)} & A^H \\ \downarrow & & \downarrow f^H \\ D^{\nu(H)} & \xrightarrow{\hat{Q}(i,H)} & Y^H \end{array}$$

This is the same as a diagram of G-spaces

$$\begin{array}{ccc} G/H \times S^{\nu(H)-1} & \xrightarrow{q(i,H)} & A \\ \downarrow & & \downarrow f \\ G/H \times D^{\nu(H)} & \xrightarrow{Q(i,H)} & Y \end{array}$$

Let $F : X \to Y$ be given by the G-push-out where $\coprod$ runs over $\{i \mid i \in I(H), (H) \in \mathcal{F}\}$.

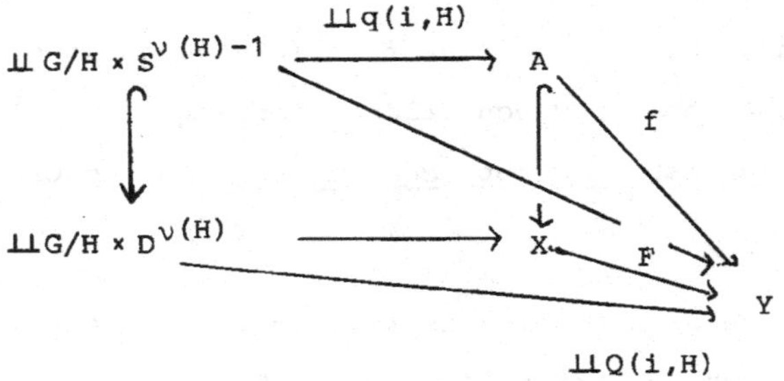

Then (X,A) is a relative G-CW-complex (1.6.). We say that F is *obtained from f by attaching cells* according to $\mathcal{F}$, ν, $\{S(H) \mid (H) \in \text{Con } G\}$. If A is already a G-CW-complex we can assume by the Cellular Approximation Theorem 2.1. that image $q(i,H)$ lies in $A_{\nu(H)-1}$. Then (X,A) is even a pair of G-CW-complexes.

Let $\nu^- : \mathcal{F} \to \mathbb{Z}$ be the function $\nu^-(H) = \nu(H)-1$. Suppose that f is $(\mathcal{F}, \nu^-)$-connected. Moreover assume for any $(H) \in \mathcal{F}$ and $x \in A^H$ that a set of generators of $\pi_{\nu(H)}(f^H, x)$ is contained in $S(H)$. Then $F : X \to Y$ is $(\mathcal{F}, \nu)$-connected (compare Whitehead [1978], p. 211 - 216).

If we do such a process or even an iteration of it, we say that *we make f highly connected by attaching cells*. □

If $f : Y \to Z$ is a G-map we denote by $[f]$ its G-homotopy class. Let $[X,Y]^G$ be the set of G-homotopy classes of G-maps $X \to Y$. If (X,A) is a relative G-CW-complex, let dim X^H be the integer n, if $X^H = X^H_n$ but $X^H_n \neq X^H_{n-1}$ holds. If no such n exists we write dim $X^H = \infty$.

Proposition 2.3. Let $\mathfrak{F} \subset \mathrm{Con}\, G$ *and a function* $\nu : \mathfrak{F} \to \mathbb{Z}$ *be given. Then the following statements for a G-map* $f : Y \to Z$ *are equivalent provided that* Y/G *is a Hausdorff space.*

i) f *is* $(\mathfrak{F}, \nu)$*-connected.*

ii) *Let* X *be a G-CW-complex of orbit type* $\mathfrak{F}$ *and* $f_* : [X,Y]^G \to [X,Z]^G$ *be the induced map. Then* f_* *is bijective if* dim $X^H < \nu(H)$ *for all* $(H) \in \mathfrak{F}$ *holds and surjective if* dim $X^H \leq \nu(H)$ *for all* $(H) \in \mathfrak{F}$ *is valid.*

iii) *There is a relative G-CW-complex* $(\bar{Y}, Y)$ *and an extension* $\bar{f} : \bar{Y} \to Z$ *of* f *such that each cell* $G/H \times D^n$ *satisfies* $(H) \in \mathfrak{F}$ *and* $n > \nu(H)$ *and* $\bar{f}$ *is a weak* $\mathfrak{F}$*-G-homotopy equivalence.*

Proof: i) $\Rightarrow$ iii). Make f highly connected by attaching cells.

iii) $\Rightarrow$ i). It suffices to check that (X^H, X^H_n) is n-connected for a G-CW-complex X. This is easily reduced to the case $X = X_{n+1}$. Then X^H is given as a push-out

$$\begin{array}{ccc} A & \longrightarrow & X^H_n \\ \downarrow & & \downarrow \\ B & \longrightarrow & X^H_{n+1} \end{array}$$

such that A is a strong deformation retraction of a neighbourhood in B and (B,A) is n-connected. Now the result follows from Blakers-Massey excision theorem (see tom Dieck-Kamps-Puppe [1970], p. 211.).

ii) $\Rightarrow$ i). Use the identification $[G/H \times S^n, X]^G = [S^n, X^H]$.

i) $\Rightarrow$ ii) see tom Dieck [1987], II.2.6. □

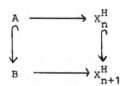

In particular any G-space Y has a G-CW-*approximation* (X,f) i.e. a G-CW-complex X together with a weak G-homotopy equivalence f. Namely, apply Proposition 2.3. to $\mathfrak{F}$ = Con G, $\nu = -1$ and $\emptyset \to Y$. A space X has a natural CW-approximation by the geometric realization of the associated semi-simplicial complex (see Lamotke [1968] p. 218, May [1967] 16.6., Milnor [1957]. This is carried over to the equivariant case in Matumoto [1984].

Theorem 2.4. The Equivariant Whitehead Theorem.

Let $f : X \to Y$ be a G-map between G-CW-complexes such that f^H is a weak homotopy equivalence for all $H \in \mathrm{Iso}\, X \cup \mathrm{Iso}\, Y$. Then f is a G-homotopy equivalence.

Proof: The map $f_* : [Y,X]^G \to [Y,Y]^G$ is bijective by Proposition 2.3. Hence there is $g : Y \to X$ with $f \circ g \simeq_G \mathrm{id}_Y$. Since also $g_* : [X,Y]^G \to [X,X]^G$ is bijective there is $h : X \to Y$ with $g \circ h \simeq_G \mathrm{id}_X$. Hence $g \circ f \simeq_G g \circ f \circ g \circ h \simeq_G g \circ h \simeq_G \mathrm{id}_X$ so that f has the G-homotopy inverse g. □

Assumption 2.5. Assume for the G-space X:

i) X has finite orbit type.

ii) $X^{>(H)} \to X^{(H)}$ is a G-cofibration for $H \in \mathrm{Iso}\, X$.

iii) $X_H \to X_H/WH$ is a numerable principal WH-bundle in the sense of Dold [1963], i.e. locally trivial over an open cover, which has a subordinate locally finite partition of unity, and X_H is a proper WH-space.

There is the following variant of Theorem 2.4.

Theorem 2.6. *Let G be a compact Lie group. Let $f : X \to Y$ be a G-map between G-spaces satisfying 2.5. Then f is a G-homotopy equivalence if and only if $f^H : X^H \to Y^H$ is a homotopy equivalence for any $H \in \mathrm{Iso}\, X \cup \mathrm{Iso}\, Y$.*

Proof: See tom Dieck [1979] 8.2.4. In the proof assumption 2.5. ii) is demanded for any $H \subset G$ and f^H is supposed to be a homotopy equivalence for all $H \subset G$. By inspecting the arguments one recognizes that our assumptions are sufficient. □

Remark 2.7. Theorem 2.6. seems also to be true for G an arbitrary Lie group if X and Y are proper completely regular G-spaces since the existence of slices is still true (Theorem 1.38.). □

Corollary 2.8. *Let G be a compact Lie group and Y be a G-space satisfying assumption 2.5. Then Y has the G-homotopy type of a G-CW-complex X with Iso X = Iso Y if and only if Y^H has the homotopy type of a CW-complex for any $H \in$ Iso Y.*

Proof: By Proposition 2.3. there is a G-CW-complex X of orbit type $\mathfrak{F} = \{(H) \mid H \in \text{Iso } Y\}$ and a weak $\mathfrak{F}$-G-homotopy equivalence $f : X \to Y$. Since X^H and Y^H are homotopic to CW-complexes for $H \in$ Iso Y the map f^H is a homotopy equivalence for any $H \in$ Iso Y. Then f is a G-homotopy equivalence by Theorem 2.6. □

Remark 2.9. Let G be a compact Lie group. A G-ENR (Euclidean Neighbourhood Retract) is a G-space which is G-homeomorphic to a G-retract of some open G-subset in a G-representation. A G-ENR satisfies assumption 2.5. (see tom Dieck [1979] 8.2.5.). A finite G-CW-complex and a compact smooth G-manifold are compact G-ENR-s. All these statements can be derived from the next result. □

Proposition 2.10. *Let G be a compact Lie group. Let X be a G-space which is separable metric and finite-dimensional. Then X is a G-ENR if and only if X is a locally compact G-space of finite orbit type and X^H is a (non-equivariant) ENR for any $H \in$ Iso X.*

Proof: Jaworowski [1976]. □

Corollary 2.11. *Let G be a compact Lie group. Then a G-ENR has the G-homotopy type of a G-CW-complex of finite orbit type.*

Proof: Each ENR has the homotopy type of a CW-complex (Milnor [1959]). Now apply Corollary 2.8. □

Proposition 2.12. *Let G be a topological group and Y a G-space. Consider a subset* $\mathcal{F} \subset \mathrm{Con}\, G$. *Then Y has the G-homotopy type of a G-CW-complex X of orbit type* $\mathcal{F}$ *if and only if Y is dominated by a G-CW-complex X of orbit type* $\mathcal{F}$ *(i.e., there are G-maps* $r : X \to Y$ *and* $i : Y \to X$ *with* $r \circ i \simeq_G id_Y$)

Proof: Choose such domination (X,r,i) of Y. By attaching cells $G/H \times D^n$ with $(H) \in \mathcal{F}$ we get an extension $\bar{r} : \bar{X} \to Y$ of $r : X \to Y$ such that $\bar{r}$ is a weak $\mathcal{F}$-G-homotopy equivalence and $(\bar{X},X)$ a pair of G-CW-complexes. Let $\bar{i} : Y \to \bar{X}$ be the composition of i with the inclusion. Then $\bar{r} \circ \bar{i} \simeq_G id_Y$. The G-map $\bar{i} \circ \bar{r} : \bar{X} \to \bar{X}$ is a weak $\mathcal{F}$-G-homotopy equivalence and $\bar{X}$ a G-CW-complex of orbit type $\mathcal{F}$. Then $\bar{i} \circ \bar{r}$ and hence $\bar{r}$ is a G-homotopy equivalence by the Equivariant Whitehead Theorem 2.4. □

Lemma 2.13. *Let G be a topological group. Consider the commutative diagram of G-spaces with G-cofibrations i and j.*

$$\begin{array}{ccccc} X_2 & \xleftarrow{\;i\;} & X_o & \longrightarrow & X_1 \\ \downarrow{\scriptstyle f_2} & & \downarrow{\scriptstyle f_o} & & \downarrow{\scriptstyle f_1} \\ Y_2 & \xleftarrow{\;j\;} & Y_o & \longrightarrow & Y_1 \end{array}$$

Let $f : X \to Y$ *be the G-map induced on the G-push-outs of the rows. Assume for* f_o, f_1 *and* f_2 *one of the following properties*

a) f_i *is* $(\mathcal{F},\nu)$*-connected.*

b) f_i *is homological* $(\mathcal{F},\nu)$*-connected.*

c) f_i *is a weak* $\mathcal{F}$*-G-homotopy equivalence.*

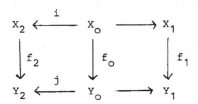

d) f_i is a weak $\mathcal{F}$-G-homology-equivalence.

e) f_i^H is a homotopy equivalence for any $H \subset G$ with $(H) \in \mathcal{F}$.

f) f_i is a G-homotopy equivalence.

Then f has the same property.

Proof:

a) We reduce a) to c). Suppose a) holds for each f_i. Then one can construct using Proposition 2.3. a new diagram with $\bar{i}$ a G-cofibration

$$\begin{array}{ccccc} \bar{X}_2 & \xleftarrow{\ \bar{i}\ } & \bar{X}_o & \longrightarrow & \bar{X}_1 \\ \downarrow \bar{f}_2 & & \downarrow \bar{f}_o & & \downarrow \bar{f}_1 \\ Y_2 & \xleftarrow{\ j\ } & Y_o & \longrightarrow & Y_1 \end{array}$$

such that $\bar{f}_i$ is a weak $\mathcal{F}$ -G-homotopy equivalence extending f_i and $(\bar{X}_i, X_i)$ is a relative G-CW-complex with cells $G/H \times D^n$ satisfying $(H) \in \mathcal{F}$ and $n > \nu(H)$ for $i = 0,1,2$, Then the same is true for the G-push-out $\bar{f} : \bar{Y} \rightarrow X$ and $f : Y \rightarrow X$ by c). Hence f is $(\mathcal{F}, \nu)$-connected by Proposition 2.3.

b) This follows from Lemma 1.12. and 1.13., the Map Excision Theorem for a homology theory (Whitehead [1978] XII.6.7.) and the long homology sequence of pairs.

c) Because of Lemma 1.12. + 1.13. we can assume $G = \{1\}$. Now apply Bousfield-Kan [1972] XII 3.1. and XII 4.2.

d) follows from b).

e) Brown [1968] p. 249 or tom Dieck [1971] Lemma 1.

f) The non-equivariant proof in e) carries over directly. □

Proposition 2.14. Let G be a compact Lie group and $f : X \rightarrow Y$ be a G-map between G-spaces satisfying assumption 2.5. Assume for any $H \in \mathrm{Iso}\, X \cup \mathrm{Iso}\, Y$ that one of the following statements holds

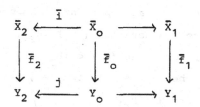

a) f^H *is* n-*connected*.

b) f^H *is homological* n-*connected*.

c) f^H *is a weak homotopy equivalence*.

d) f^H *is a weak homology equivalence*.

e) f^H *is a homotopy equivalence*.

Then this holds for any $H \subset G$. *In the case* e) *we know already from Theorem* 2.6. *that* f *is a* G-*homotopy equivalence*.

Proof: The proof is done using an important technique, *induction over the orbit bundles*. Choose a numeration $\{(H_1),(H_2),\ldots,(H_r)\}$ of $\{(H) \mid H \in \text{Iso } X \cup \text{Iso } Y\}$ such that $(H_i) \subset (H_j) \Rightarrow i \geq j$ holds. The induction runs over r. The begin r = 0 is trivial, the step from r - 1 to r is done as follows. Write $H = H_r$.

Let $\bar{X}$ be $\bigcup_{i=1}^{r-1} X^{(H_i)}$ and $\bar{Y}$ be $\bigcup_{i=1}^{r-1} Y^{(H_i)}$. For any $H \subset G$ the H-fixed point set X^H is $\bigcap_{g \in H} X^g$ where X^g is the preimage of the diagonal under $X \to X \times X \quad x \to (x,gx)$. Hence X^H is closed. Since G is compact $X^{(H)} = G \cdot X^H$ is closed because $G \times X \to X$ is a closed map (Bredon [1972] I.1.2.). Therefore $X^{(H)}, X^{>(H)}$ and $\bar{X}$ are closed G-subspaces of X. Moreover, $X = \bar{X} \cup X^{(H)}$ and $X^{>(H)} = \bar{X} \cap X^{(H)}$ so that we have the G-push-out

2.15.
$$\begin{array}{ccc} X^{>(H)} & \longrightarrow & \bar{X} \\ \downarrow & & \downarrow \\ X^{(H)} & \longrightarrow & X \end{array}$$

Let $\bar{f}$, $f^{>(H)}$ and $f^{(H)}$ be the G-maps induced by f so that f is their G-push-out. The G-spaces $X^{>(H)}$ and $\bar{X}$ have at least one orbit type less than X. By induction hypothesis $\bar{f}^K$ and $(f^{>(H)})^K$ have property a), b), c), resp. d) for any $K \subset G$. Because of Lemma 2.13. it suffices to prove for any $K \subset G$ that $(f^{(H)})^K$ has property a), b), c) resp. d) for each $K \subset G$.

Consider the square

2.16.
$$\begin{array}{ccc} G/H^K \times_{WH} X^{>H} & \longrightarrow & (X^{>(H)})^K \\ \downarrow & & \downarrow \\ G/H^K \times_{WH} X^H & \longrightarrow & (X^{(H)})^K \end{array}$$

where the horizontal maps send (gH,x) to gx. Since $(X^{>(H)})^K$ and $(X^{(H)})^K$ are closed in X and $G \times X \to X$ $(g,x) \to gx$ is closed (Bredon [1972] I.1.2.) the map $G/H^K \times_{WH} X^H \amalg (X^{>(H)})^K \to (X^{(H)})^K$ is closed and especially an identification. Now it is easy to check that 2.16. is a push-out so that $(f^{(H)})^K$ is the push-out of id $\times_{WH} f^H$, id $\times_{WH} f^{>H}$ and $(f^{>(H)})^K$. The map $(f^{>(H)})^K$ has property a), b), c) resp. d) by induction hypothesis and also f^H and $f^{>H}$ since $(X^{>(H)})^H$ is $X^{>H}$. If we can show that then id $\times_{WH} f^H$ and id $\times_{WH} f^{>H}$ satisfy a), b), c) resp. d) then an application of Lemma 2.13. finishes the proof of Proposition 2.14. But this claim follows from the lemma below applied to $X = G/H^K$ since G/H^K is a compact free smooth WH-manifold and hence a free WH-CW-complex by 1.9. □

Lemma 2.17. Let G be a topological group and X a G-CW-complex of orbit type $\mathcal{F} \subset$ *Con G. Let G operate on X from the right. Consider a G-map* $f : Y \to Z$ *between proper G-spaces satisfying one of the following statements for all* $(H) \in \mathcal{F}$.

a) $f/H : Y/H \to Z/H$ *is n-connected.*

b) f/H *is homological n-connected.*

c) f/H *is a weak homotopy equivalence.*

d) f/H *is a weak homology equivalence.*

Then id $\times_G f : X \times_G Y \to X \times_G Z$ *has the same property.*

Proof: We show inductively over n that Lemma 2.17. holds if $X = X_n$. We leave it to the reader to carry out the final limit argument using Milnor [1962] or Whitehead [1978] XIII 1.3. Notice that Y and Z are proper so that Y/H and Z/H are again compactly generated by Lemma 1.19.

We have the G-push-out

$$\begin{array}{ccc} \coprod_I G/H_i \times S^{n-1} & \longrightarrow & X_{n-1} \\ \downarrow j & & \downarrow \\ \coprod_I G/H_i \times D^n & \longrightarrow & X_n \end{array}$$

with j a G-cofibration. Crossing it with Y yields again a G-push-out with $j \times id$ a G-cofibration (Steenrod [1967] 4.4. and 7.3.). Since $(G/H_i \times S^{n-1}) \times_G Y$ is homeomorphic to $Y/H_i \times S^{n-1}$, the spaces $(G/H_i \times S^{n-1}) \times_G Y$ and $(G/H_i \times D^n) \times_G Y$ are compactly generated by Lemma 1.19. We can assume that $X_{n-1} \times_G Y$ is compactly generated by induction hypothesis. Hence we get by Lemma 1.12. and 1.13. a push-out with $j \times_G id$ a cofibration

$$\begin{array}{ccc} (\coprod_I G/H_i \times S^{n-1}) \times_G Y & \longrightarrow & X_{n-1} \times_G Y \\ \downarrow j \times_G id & & \downarrow \\ (\coprod_I G/H_i \times D^n) \times_G Y & \longrightarrow & X_n \times_G Y \end{array}$$

Because of Lemma 2.13. and the induction hypothesis it suffices to show that

$$id \times_G f: (\coprod_I G/H_i \times D^n) \times_G Y \longrightarrow (\coprod_I G/H_i \times D^n) \times_G Z$$

satisfies a), b), c) resp. d). But this follows from the assumption about f/H_i since $(G/H_i \times D^n) \times_G Y$ is naturally homotopy equivalent to Y/H_i. □

Corollary 2.18. Let G be a compact Lie group and Y be a G-space satisfying assumption 2.5. Then there is a G-CW-approximation (X,f) by a G-CW-complex of finite orbit type satisfying Iso X = Iso Y.

Proof: Because of Proposition 2.3. there is a G-CW-complex X of orbit type $\mathcal{F} = \{(H) \mid H \in \text{Iso } Y\}$ and a weak $\mathcal{F}$-G-homotopy equivalence $f : X \to Y$. Then f is a weak G-homotopy equivalence by Proposition 2.14. □

Comments 2.19. A variant of Theorem 2.6. is shown for G-ANR-s in James-Segal [1978]. Corollary 2.11. is proved for G-ANR-s in Murayama [1983] 13.3. More information about G-spaces of the homotopy type of a G-CW-complex can be found in Waner [1980a].

A lot of results of this section are of the type that a G-map f is a (weak) G-homotopy equivalence if f^H is a (weak) homotopy equivalence for any $H \subset G$ occuring as isotropy group. In fact, some work is done to ensure that it suffices to look at isotropy groups only. One reason is to obtain Corollary 2.18. as we need finite orbit type to apply the important technique of induction over the orbit bundle. Another application is equivariant surgery. Consider a G-map $f : M \to N$ between G-manifolds such that for simplicity M^H and N^H are connected for each $H \subset G$. By equivariant surgery we can achieve at best that f^H is a weak homotopy equivalence for each $H \in \text{Iso } M = \text{Iso } N$. Hence we need an extra result telling us that f is a G-homotopy equivalence.

In the case of a finite group G there is for any $K \subset G$ an $H \in \text{Iso } M = \text{Iso } N$ with $M^K = M^H$ and $N^K = N^H$ so that obviously f^K is a weak homotopy equivalence for all $K \subset G$. However, if G is a compact Lie group, one cannot argue in this fashion. A counter-example is $G = SO(3)$ and M the sphere in $\mathbb{R}^3 \oplus \mathbb{R}^3$ where G acts on each $\mathbb{R}^3$ in the obvious way. Then there is no $H \in \mathrm{Iso}\, M$ with $M = M^H$. □

Exercises 2.20.

1. Let X be the subspace of $\mathbb{R}^2$ given by the union of $\{(x, \sin \frac{\pi}{2x}),\ x \in [0,1]\} \cup \{0\} \times [-2,1] \cup [0,1] \times \{-2\} \cup \{1\} \times [-2,1]$. Show using Čech cohomology that the projection onto a point $p : X \to \{*\}$ is a weak homotopy equivalence but not a homotopy equivalence. Does X have the homotopy type of a CW-complex? (see Wallace [1970] p. 232).

2. Consider the G-push-out with j a G-cofibration:

$$\begin{array}{ccc} A & \xrightarrow{f} & Y \\ {\scriptstyle j}\downarrow & & \downarrow{\scriptstyle J} \\ X & \xrightarrow{F} & Z \end{array}$$

Assume that A,X,Y have the G-homotopy type of a G-CW-complex resp. finite G-CW-complex resp. finite dimensional G-CW-complex resp. skeletal-finite G-CW-complex. Then the same is true for Z.

Is the analogous statement for weak G-homotopy type true?

3. Let $\mathcal{F} \subset \mathrm{Con}\, G$ be given. Consider a G-CW-complex Y of orbit type $\mathcal{F}$ such that Y^H is contractible for all $(H) \in \mathcal{F}$.
   a) If X is any G-CW-complex of orbit type $\mathcal{F}$ then there is precisely one G-homotopy class of G-maps $X \to Y$.
   b) If Y' has the same properties as Y then Y and Y' are G-homotopy-equivalent.
   c) Assume for any $H,K \subset G$ that $(H),(K) \in \mathcal{F}$ implies $(H \cap K) \in \mathcal{F}$. Let A be $\coprod_{(H) \in \mathcal{F}} G/H$ and Y the countable infinite join $* A$. Then Y is a G-CW-complex of orbit type $\mathcal{F}$ such that Y^H is contractible for

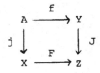

$(H) \in \mathcal{F}$.

(compare tom Dieck [1974], tom Dieck [1987] I.6., Elmendorf [1983])

4. Let G be a compact Lie group and M a compact smooth G-manifold. Show that M has the G-homotopy type of a finite G-CW-complex using the following hints (see also 1.9.).

Use induction over the orbit bundle. Let H be minimal in the usual ordering of the orbit types. Then $M_{(H)} = M^{(H)}$ is a closed G-submanifold in M. Use the smooth G/H-bundle $G/H \to M_{(H)} \to M_{(H)}/G$ (see Bredon [1972] VI. 2.5.) and the fact that the (non-equivariant) smooth compact manifold $M_{(H)}/G$ has the homotopy type of a finite CW-complex to show that $M_{(H)}$ is G-homotopic to a finite G-CW-complex. Let ν be the normal bundle of $M_{(H)}$ in M. By induction hypothesis the sphere bundle $S\nu$ and hence the disc bundle $D\nu$ have the G-homotopy type of a finite G-CW-complex. Identify $D\nu$ with a tubular neighbourhood of $M_{(H)}$ in M (Bredon [1972] VI. 2.2.). If M' is $M \setminus \text{int } D\nu$, we can view M as the G-push-out of $D\nu \leftarrow S\nu \to M'$. Now apply the induction hypothesis to M' and exercise 2 above.

5. The product of a G-cofibration with a H-cofibration is a $G \times H$-cofibration. The product of a G-push out with j a G-cofibration and a H-space is a $G \times H$-push-out with $j \times id$ a $G \times H$-cofibration.

6. Disprove or prove for a map $f : Y \to Z$.
   a) If $f_* : [X,Y] \to [X,Z]$ is bijective for any finite-dimensional CW-complex X then f is a weak homotopy equivalence.
   b) as in a) but only for finite X.

7. Let $X^+ \subset \mathbb{R}^2$ be the cone of $\{(\frac{1}{n},0) \mid n = 1,2,3,\dots\} \cup \{(0,0)\}$ over $(0,1)$ and $X^- = \{(a,b) \in \mathbb{R}^2 \mid (-a,-b) \in X\}$. Show
   a) The inclusion $j : (0,0) \to X^{\pm}$ is no cofibration.
   b) The projections $p^{\pm} : X^{\pm} \to \{(0,0)\}$ are homotopy equivalences.

c) The map $p^+ \vee p^- : X^+ \vee X^- \longrightarrow \{(0,0)\}$ induced between the push-outs of $X^+ \xleftarrow{j} \{(0,0)\} \xrightarrow{j} X^-$ and $\{(0,0)\} \xleftarrow{id} \{(0,0)\} \xrightarrow{id} \{(0,0)\}$ is not a homotopy equivalence (compare Lemma 2.13.).

8. Give an example of a $\mathbb{Z}/2$-space which does not satisfy assumption 2.5.

9. We say that two G-spaces X and Y have the same weak G-homotopy type if there is a G-CW-complex Z together with weak G-homotopy equivalences $Z \to X$ and $Z \to Y$. Show that this defines an equivalence relation. Is the following relation $\sim$ an equivalence relation?
$X \sim Y \Leftrightarrow$ there is a weak G-homotopy equivalence $X \to Y$.

10. Let G be a compact Lie group and X be a G-space satisfying assumption 2.5. Suppose for $H \in \text{Iso } X$ that X^H is simply connected and the singular homology groups $H_i(X^H)$ are finitely generated for $i \geq 0$. Then X has a skeletal-finite CW-approximation (Y,f) with Iso X = Iso Y. □

3. The geometric finiteness obstruction

We recall that a G-CW-complex is finite if it has only finitely many open cells e_i^n, $n \geq 0$, $i \in I_n$. A finite domination (X,r,i) of a G-space Y consists of a finite G-CW-complex X and G-maps $r : X \to Y$ and $i : Y \to X$ satisfying $r \circ i \simeq_G \mathrm{id}$. If Y has such a finite domination Y is finitely dominated. In this section we deal with the following question.

Problem 3.1. When is a finitely dominated G-space G-homotopy equivalent to a finite G-CW-complex? □

We have already shown in Proposition 2.12. that a finitely dominated G-space has the G-homotopy type of a G-CW-complex of finite orbit type.

The approach to problem 3.1., we describe in this section, is geometric. It is motivated by the geometric treatment of the equivariant Whitehead torsion we explain in section four. Later we also give algebraic approaches and show that both agree. The advantage of the geometric treatment is that it is completely elementary and all the formal properties of the equivariant finiteness obstruction can be derived quickly.

The goal of this section is to construct a functor from the category of G-spaces into the category of abelian groups

$$Wa^G : \{\text{G-spaces}\} \to \{\text{abel. gr.}\}$$

and a function w^G assigning an element $w^G(X) \in Wa^G(X)$ to any finitely dominated G-space such that the following holds:

Theorem 3.2.

a) Obstruction property.

Let X be a finitely dominated G-space. Then X is G-homotopy equivalent to a finite G-CW-complex if and only if $w^G(X)$ vanishes.

b) Homotopy invariance.

i) If $f : X \to Y$ is a G-homotopy equivalence between finitely dominated G-spaces, then $f_* : Wa^G(X) \to Wa^G(Y)$ sends $w^G(X)$ to $w^G(Y)$.

ii) $f \simeq_G f' \Rightarrow f_* = f'_*$.

c) Additivity.

If the following diagram is a G-push-out of finitely dominated G-spaces and k a G-cofibration

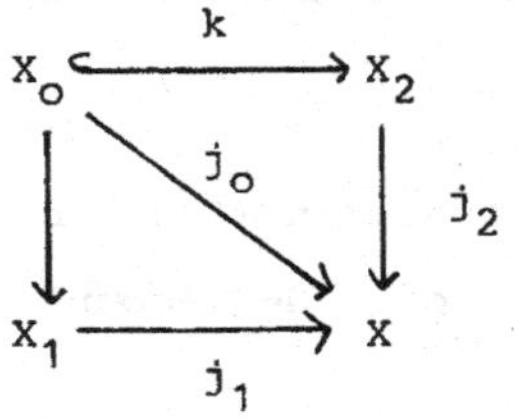

then:

$$w^G(X) = j_{1*}w^G(X_1) + j_{2*}w^G(X_2) - j_{o*}w^G(X_o). \quad \square$$

Given a G-space Y, consider the set of all G-maps $f : X \to Y$ with a finitely dominated G-space X as source. We define an equivalence relation $\sim$. Namely, $f_o : X_o \to Y$ and $f_4 : X_4 \to Y$ are equivalent if there is a commutative diagram

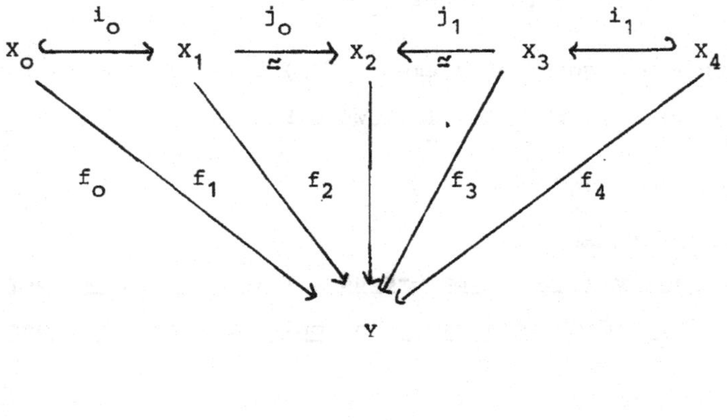

such that j_o and j_1 are G-homotopy equivalences, i_o and i_1 are inclusions and (X_1, X_o) and (X_3, X_4) are finite relative G-CW-complexes. In other words, X_1 and X_3 is obtained from X_o and X_4 by attaching finitely many cells. Obviously $\sim$ is symmetric and reflexive. The main difficulty in the proof of Theorem 3.2 is the verification of transitivity.

We introduce some notation. We symbolize a diagram

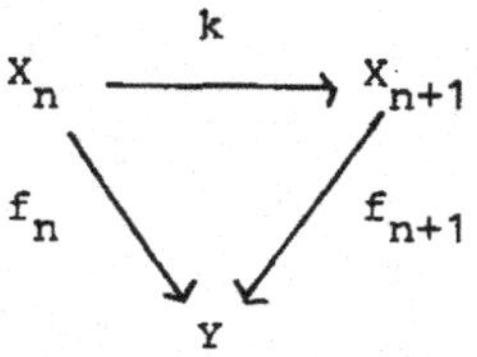

by $\subset$ resp. $\rightarrow$ if k is an inclusion and (X_{n+1}, X_n) a finite relative G-CW-complex resp. k is a G-homotopy equivalence. If k points in the other direction we write of course $\supset$ resp. $\leftarrow$. The diagram defining $\sim$ corresponds to the chain $\subset \rightarrow \leftarrow \supset$. Hence we have to show that $\subset \rightarrow \; \leftarrow \supset \subset \rightarrow \leftarrow \supset$ can be reduced to $\subset \rightarrow \leftarrow \supset$ without changing the ends. This can be done by a sequence of operations which again do not alter the maps at the ends:

1) $\supset \subset \Rightarrow \subset \supset$

Use the G-push-out to substitute

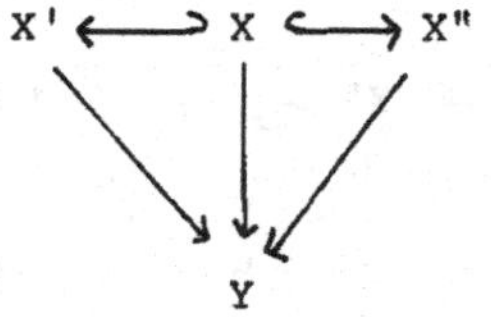

by

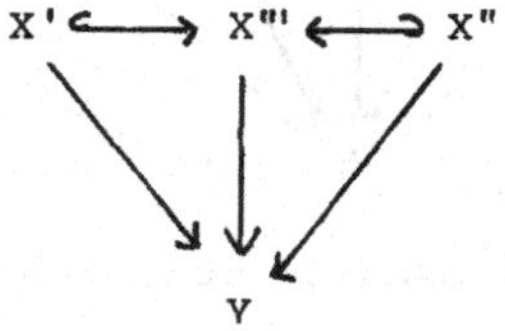

2) $\supset \to \Rightarrow \to \supset$

$\leftarrow \subset \Rightarrow \subset \leftarrow$

As in 1)

3) $\leftarrow \to \Rightarrow \to \leftarrow$

Glue the mapping cylinders together

4) $\subset \subset \Rightarrow \subset$

$\supset \supset \Rightarrow \supset$

$\to \to \Rightarrow \to$

$\leftarrow \leftarrow \Rightarrow \leftarrow$

obvious

5) $\to \subset \Rightarrow \leftarrow \subset \to \leftarrow$

$\to \subset$ stands for

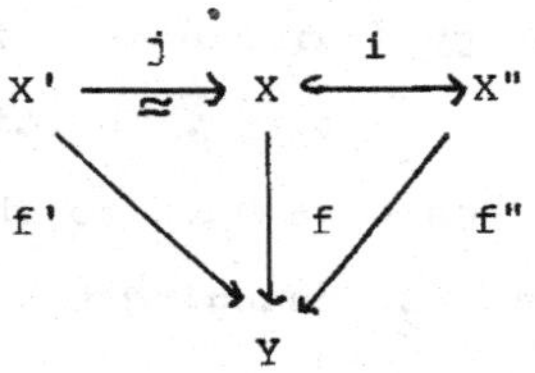

Let k be a G-homotopy inverse of j. Since i is a G-cofibration by 1.5. there is a G-homotopy $h : X'' \times I \to Y$ with $h_o | X = f' \circ k$ and $h_1 = f''$. Now consider

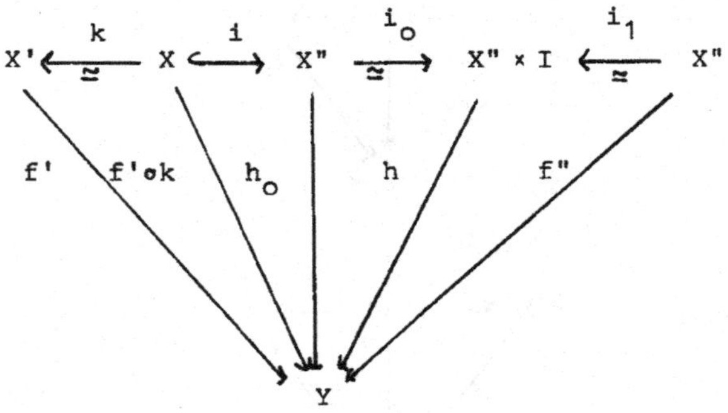

These operations make sense because of the following conclusions of 1.5,

1.29 and Lemma 2.13.

3.3. Consider the G-push-out with j the inclusion of a relative G-CW-complex

$$\begin{array}{ccc} A & \xrightarrow{f} & Y \\ {\scriptstyle j}\big\downarrow & & \big\downarrow{\scriptstyle J} \\ X & \xrightarrow[F]{} & Z \end{array}$$

If f is a G-homotopy equivalence then F is a G-homotopy equivalence. If (X,A) is finite, finite-dimensional, resp. skeletal finite relative G-CW-complex then also (Z,Y).

The following calculation finishes the proof that $\sim$ is an equivalence relation

$$\subset \to \leftarrow \supset \subset \to \leftarrow \supset$$

$$\subset \to \leftarrow \subset \supset \to \leftarrow \supset$$

$$\subset \to \subset \leftarrow \to \supset \leftarrow \supset$$

$$\subset \leftarrow \subset \to \leftarrow \leftarrow \to \to \leftarrow \supset \to \supset$$

$$\subset \subset \leftarrow \to \leftarrow \to \leftarrow \to \supset \supset$$

$$\subset \to \leftarrow \leftarrow \to \to \leftarrow \supset$$

$$\subset \to \leftarrow \to \leftarrow \supset$$

$$\subset \to \to \leftarrow \leftarrow \supset$$

$$\subset \to \leftarrow \supset$$

Now we define $Wa^G(Y)$ as the set of equivalence classes of such maps into Y. Disjoint union defines the structure of an abelian semi-group on it with zero element represented by $\emptyset \to Y$. Given $[f] \in Wa^G(Y)$ represented by $f : X \to Y$ we have to construct an inverse. Choose a domination (Z,r,i) of X. Let C_i and C_r be the mapping cylinders. There is a a G-map $F : C_i \to X$ with $F \mid X = id_X$ and $F \mid Z = r$. Then an inverse of $[f]$ is given by the class of the composition

$$C_i \cup_X C_i \xrightarrow{F \cup_X F} X \xrightarrow{f} Y$$

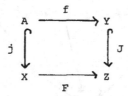

Namely, one easily constructs a commutative diagram (use Lemmata 4.11., 4.17., 4.18.)

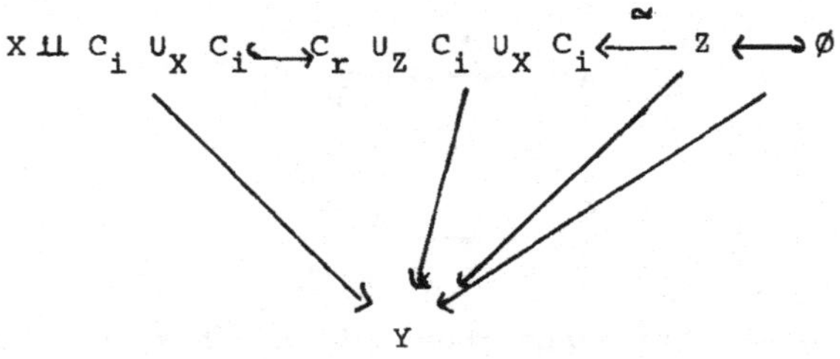

A map $g : Y \to Z$ induces an abelian group homomorphism $g_* : Wa^G(Y) \to Wa^G(Z)$ by composition. Hence we have defined a covariant functor

3.4. $Wa^G : \{G\text{-spaces}\} \to \{\text{abel. gr.}\}$

Define the <u>geometric</u> <u>finiteness</u> <u>obstruction</u> of a finitely dominated G-space

3.5. $w^G(X) \in Wa^G(X)$

by the class of $id : X \to X$.

<u>Proof of theorem 3.2.</u>

a) The if-statement is the non-trivial part. Suppose $w^G(X) = 0$ for the finitely dominated G-space X. Then there is a commutative diagram

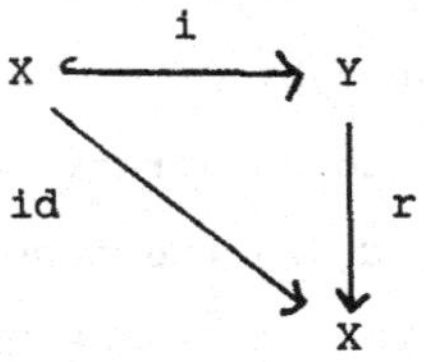

such that (Y,X) is a finite relative G-CW-complex and Y is G-homotopy equivalent to a finite G-CW-complex Z. The mapping cylinder C_i is a G-subspace of C_r such that (C_r, C_i) is a finite relative G-CW-complex. Choose a G-homotopy equivalence $g : C_i \to Z$. Consider the G-push out

$$\begin{array}{ccc} C_i & \hookrightarrow & C_r \\ g \downarrow \simeq & & \simeq \downarrow g' \\ Z & \hookrightarrow & Z' \end{array}$$

By 3.3. the G-map g' is a G-homotopy equivalence and (Z',Z) a finite relative G-CW-complex. Hence Z' is G-homotopy equivalent to X and to a finite G-CW-complex Z" by the Cellular Approximation Theorem 2.1.

b) is obvious

c) Given such a G-push-out, choose a finite domination (Z,r,i) of X_o. Let

$$C_i \cup_{X_o} C_i \xrightarrow{F \cup_{X_o} F} X_o \xrightarrow{j_o} X$$

be the inverse of $[j_o]$ constructed above. Hence we have to show in $Wa^G(X)$

$$[j_1] \amalg [j_2] \amalg [j_o \circ F \cup_{X_o} F] = [j_1 \cup_{j_o} j_2]$$

We leave it to the reader to construct a commutative diagram using Lemmata 4.11., 4.17. and 4.18.

$$\begin{array}{ccccc} X_1 \amalg C_i \cup_{X_o} C_i \amalg X_2 & \hookrightarrow & X_1 \cup_{X_o} C_r \cup_Z C_i \cup_{X_o} C_i \cup_Z C_r \cup_{X_o} X_2 & \xleftarrow{\simeq} & X_1 \cup_{X_o} X_2 \\ & {\scriptstyle j_1 \amalg j_o \circ F \cup_{X_o} F \amalg j_2} \searrow & \downarrow & \swarrow {\scriptstyle j_1 \cup_{j_o} j_2} & \\ & & X & & \end{array}$$

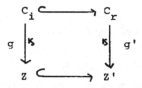

This finishes the proof of Theorem 3.2. □

The construction above, at least as an abelian semi-group, makes also sense in other situations. Namely, let D^G be the functor

3.6. D^G : {G-spaces} $\to$ {abel. semi-gr.}

we get by a construction analogous to the one for Wa^G if we consider G-maps $f : X \to Y$ for X a G-space of the G-homotopy type of a G-CW-complex and change the equivalence relation by

$X_o \subset X_1 \Leftrightarrow (X_1, X_o)$ is a relative G-CW-complex which is finite-dimensinal resp. skeletal-finite

$X_1 \xrightarrow{f} X_2 \Leftrightarrow$ f is a G-homotopy equivalence

Moreover, we define for a G-space X of the G-homotopy type of a G-CW-complex

3.7. $d^G(X) \in D^G(X)$

by the class of $id : X \to X$. The verification that $D^G(X)$ is a well-defined abelian semi-group is the same as for $Wa^G(X)$.

Proposition 3.8. *A G-CW-complex X is G-homotopy equivalent to a finite-dimensional resp. skeletal finite G-CW-complex if and only if $d^G(X) \in D^G(X)$ vanishes.* □

We will later show that a G-CW-complex of finite orbit type is finitely dominated if and only if it is finite-dimensional and skeletal finite. Hence we obtain in principal also an invariant to decide whether a G-space is finitely dominated. We will deal with this problem later.

The main difficulty, however, is that we cannot prove the existence of inverse elements in $D^G(X)$. In particular the proof for additivity also breaks down. The following example shows that we get a contradiction if $D^G(X)$ were an abelian group and fulfilled additivity.

<u>Example 3.9.</u> Let X be a finitely dominated G-CW-complex. Consider the two G-push-outs

$$\begin{array}{ccc} X \times \partial I & \hookrightarrow & X \times I \\ \downarrow & & \downarrow \\ X \times I & \longrightarrow & X \times S^1 \end{array} \qquad \begin{array}{ccc} \emptyset & \hookrightarrow & X \times 0 \\ \downarrow & & \downarrow \\ X \times 1 & \longrightarrow & X \times \partial I \end{array}$$

We get from additivity and homotopy invariance that $w^G(X \times S^1)$ vanishes in $Wa^G(X \times S^1)$. Hence $X \times S^1$ is G-homotopy equivalent to a finite G-CW-complex. We will deal with this phenomenon more extensively later.

If $D^G(X)$ were an abelian group and additivity were true, the same argument would prove for any G-CW-complex X that $d^G(X \times S^1)$ vanishes. Hence $X \times S^1$ would be up to homotopy finite-dimensional. Since $X \times \mathbb{R} \to X \times S^1$ is a covering and $\mathbb{R}$ contractible, X itself would be G-homotopy equivalent to a finite-dimensional G-CW-complex, a contradiction. □

If we restrict ourselves to G-CW-complexes X which are dominated by a finite-dimensional resp. skeletal-finite G-CW-complex we would get an abelian group and additivity as for Wa^G. However, we would obtain the zero-functor since we later prove that such X has already the G-homotopy type of a finite-dimensional resp. skeletal-finite G-CW-complex.

Now we indicate how the computation of $Wa^G(Y)$ is reduced to the case $G = 1$. We define an homomorphism

3.10. $\varphi(H) : Wa^1(EWH \times_{WH} Y^H) \to Wa^G(Y)$

as the composition

$$Wa^1(EWH \times_{WH} Y^H) \xrightarrow{(1)} Wa^{WH}(EWH \times Y^H) \xrightarrow{(2)} Wa^{WH}(Y^H) \xrightarrow{(3)} Wa^{NH}(Y^H) \xrightarrow{(4)}$$

$$Wa^G(G \times_{NH} Y^H) \xrightarrow{(5)} Wa^G(Y)$$

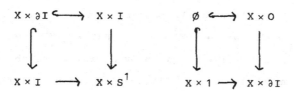

Here and elsewhere EG is the classifying space and $G \to EG \to BG$ the universal principal G-bundle of a topological group G (see Husemöller [1966], 5.10.5). It is determined by the property that EG is contractible. The map (1) is given by the pull-back construction applied to the WH-principal bundle $WH \to EWH \times Y^H \to EWH \times_{WH} Y^H$ (see 1.25.). It is an isomorphism, an inverse is given by dividing out the group action (1.24.). The homomorphism (2) and (5) are induced by the canonical maps $EWH \times Y^H \to Y^H$ and $G \times_{NH} Y^H \to Y$. Restriction with $NH \to WH$ and induction with $NH \to G$ defines (3) and (4). We later prove

Theorem 3.11.

a) *There is a natural isomorphism*

$$\bigoplus_{(H) \in \mathrm{Con}\, G} \varphi(H) : \bigoplus_{(H) \in \mathrm{Con}\, G} Wa^1(EWH \times_{WH} Y^H) \to Wa^G(Y)$$

b) *Let Z be a space such that $\pi_1(Z,z)$ is finitely presented for all $z \in Z$. Then there exists a natural isomorphism*

$$\bigoplus_{C \in \pi_0(Z)} \tilde{K}_0(\mathbb{Z}\pi_1(C)) \to Wa^1(Z) \quad \square$$

Hence the computation of $Wa^G(Y)$ is reduced to the computation of reduced projective class groups of integral group rings.

Remark 3.12. We have already mentioned that a compact G-CW-complex X is finite. However a compact finitely dominated G-space X is not necessarily G-homotopy equivalent to a finite G-CW-complex although it is G-homotopy equivalent to a finitely dominated G-CW-complex. Namely, any finitely dominated CW-complex is homotopy equivalent to a compact space (see Ferry [1981 b]). Moreover there are compact locally smooth topological G-manifolds with non-vanishing finiteness obstruction (see Dovermann-Rothenberg [1988], Quinn [1982]). □

Comments 3.13. The geometric approach to the finiteness obstruction can be found in Lück [1987 b]. We will later give an algebraic treatment and show that they agree. Other references for an algebraic approach to the equivariant finiteness obstruction are Andrzejewski [1986], Baglivo [1978], tom Dieck [1981], Iizuka [1984], Kwasik [1983] and Lück [1983]. They are based on Wall [1965] and Wall [1966] where the non-equivariant finiteness obstruction is introduced. Wall's articles seem to be motivated by Swan [1960 b].

In Swan [1960 b] the finiteness obstruction plays a role in the construction of a finite free G-CW-complex X for a finite group G such that X is homotopic to S^n. By homological algebra finitely dominated free G-CW-complexes X with $X \simeq S^n$ are established and the finiteness obstruction comes in to decide whether X can be choosen to be finite. This is essential if one wants to substitute X by a G-manifold and finally by the standard sphere with a free G-action. This leads to the space form problem (see for example Madsen-Thomas-Wall [1976]).

Further examples where finiteness obstructions naturally appear are the theory of ends of manifolds (Quinn [1979], [1982], Siebenmann [1965]), actions on discs (Oliver [1975], [1976], [1977], [1978]), equivariant surgery (Oliver-Petrie [1982]), the theory of homotopy representations (tom Dieck-Petrie [1982]), existence of equivariant handle decompositions for topological G-manifolds (Steinberger-West [1985])
The construction of the finiteness obstruction makes also sense in the controlled setting using the ideas in Chapman [1983], section 3. This is related to the construction in Chapman ([1983], section 1 by a controlled Bass-Heller-Swan homomorphism (cf. section 7).

Exercises 3.14.

1. Consider the G-push-out with j a G-cofibration

$$\begin{array}{ccc} A & \xrightarrow{f} & Y \\ {\scriptstyle j}\big\downarrow & & \big\downarrow{\scriptstyle J} \\ X & \xrightarrow[F]{} & Z \end{array}$$

If A,X and Y are finitely dominated then Z is finitely dominated.

2. Let G be a compact Lie group and X a G-space satisfying assumption 2.5. Suppose for any $H \in \text{Iso}\, X$ that there is a WH-homotopy equivalence of pairs $(X^H, X^{>H}) \to (Z,Y)$ into a finite relative WH-CW-complex. Show that X is a G-homotopy equivalent to a finite G-CW-complex.

3. Let G be a path-connected topological group and X a G-CW-complex such that X_1 consists of a single G-fixed point x. Consider the $\mathbb{Z}$-chain complex C whose differentials are boundary operators in exact sequences of triples

$$\ldots \to \pi_{n+1}(X_{n+1}, X_n, x) \to \pi_n(X_n, X_{n-1}, x) \to \ldots$$

Show that $H_*(C)$ is the singular homology of X/G.

4. Let X be a finite dimensional T^n-CW-complex of finite orbit type such that X^H is simply connected and $H_*(X^H;\mathbb{Z})$ finitely generated for $H \subset G$ Show that X is T^n-homotopy equivalent to a finite T^n-CW-complex.

Hint: $\widetilde{K}_0(\mathbb{Z}) = \{0\}$

5. Let $G \to E \to E/G$ be a principal G-bundle. Define inverse isomorphisms

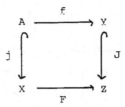

$$Wa^G(E) \overset{\cong}{\longleftrightarrow} Wa^1(E/G)$$

by dividing out the group action and the pull back construction.

6. Let G be a path-connected topological group and X be a free simply connected finitely dominated G-space. Suppose that $H_i(X/G)$ is finitely generated for $i \geq 0$ and zero for large i. Then X is G-homotopy equivalent to a finite G-CW-complex.

7. Let X be a finitely dominated CW-complex which is not homotopy equivalent to a finite CW-complex. Let G be a compact Lie group with $\dim G \geq 1$.

   Show that $G \times X$ is a finitely dominated G-CW-complex but not G-homotopy equivalent to a finite G-CW-complex, whereas its restriction to any finite subgroup H is H-homotopy equivalent to a finite H-CW-complex.

8. Let H be a (closed) subgroup of the compact Lie group G. Let X be a G-CW-complex and res X its restriction to H. If X is finite, skeletal finite, finite-dimensional resp. finitely dominated then res X has the H-homotopy type of a H-CW-complex with the same property.

9. Let X be a simply connected finitely dominated G-CW-complex and Y a finite free G-CW-complex. Then $X \times Y$ with the diagonal G-action is G-homotopy equivalent to a finite G-CW-complex.

10. Let X be a finitely dominated G-CW-complex. Show that X is G-homotopy equivalent to a finite-dimensional G-CW-complex (Hint: Consider $X \times S^1$)

4. The geometric Whitehead torsion

We introduce the equivariant version of a simple homotopy equivalence and define geometrically the equivariant Whitehead group and the obstruction for a G-homotopy equivalence to be simple, its equivariant Whitehead torsion. The main properties like homotopy invariance and additivity are verified. We also deal with simple structures (resp. simple G-homotopy type), especially on G-manifolds. We state the equivariant s-cobordism theorem and relate isovariant Whitehead torsion to equivariant Whitehead torsion. Let G be a topological group.

4.A. Geometric construction of Whitehead group and Whitehead torsion.

4.1. Consider the G-push out

$$\begin{array}{ccc} A & \xrightarrow{q} & X \\ {\scriptstyle j}\big\downarrow & & \big\downarrow{\scriptstyle J} \\ B & \xrightarrow{Q} & Y \end{array}$$

Suppose that (B,A) is a pair of G-CW-complexes, j is the inclusion and q is cellular. If Y is equipped with a G-CW-complex structure isomorphic to the one defined in 1.29 we call this G-push out a <u>cellular</u> G-<u>push-out.</u> □

<u>Example 4.2.</u> Let $f : X \to Y$ be a G-map. Its <u>mapping</u> <u>cylinder</u> Cyl(f) is defined as the G-push out

$$\begin{array}{ccc} X & \xrightarrow{f} & Y \\ {\scriptstyle i_o}\big\downarrow & & \big\downarrow{\scriptstyle j} \\ X \times I & \xrightarrow{\bar{f}} & Cyl(f) \end{array}$$

Let $i : X \to Cyl(f)$ be the canonical inclusion $\bar{f} \circ i_0$. The <u>mapping</u> <u>cone</u> of f is the G-push-out

$$\begin{array}{ccc} X & \longrightarrow & \{*\} \\ {\scriptstyle i}\big\downarrow & & \big\downarrow \\ Cyl(f) & \longrightarrow & Cone(f) \end{array}$$

The canonical inclusions $i : X \to Cyl(f)$, $j : Y \to Cyl(f)$ and $* \to Cone(f)$ are G-cofibrations by Lemma 1.11 and j is a G-homotopy equivalence by Lemma 2.13. The canonical retraction $r : Cyl(f) \to Y$ of j is given by

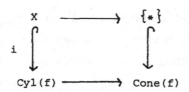

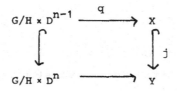

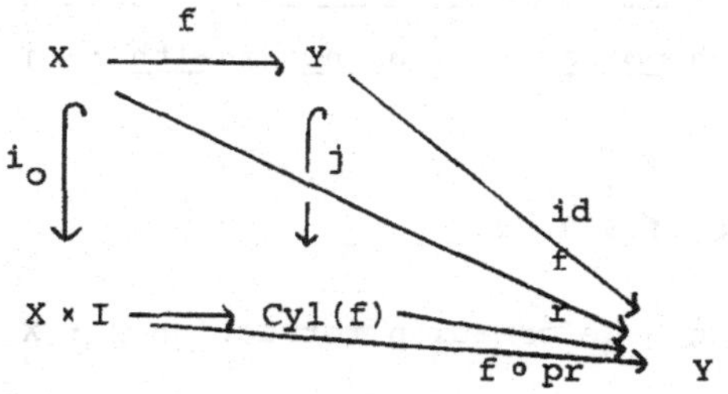

If X and Y are G-CW-complexes and f is cellular, equip Cyl(f) and Cone(f) with the G-CW-complex structure making the two G-push-outs above cellular. Here we use the G-CW-complex structure on $X \times I$ of 1.27. Notice that all the canonical inclusions and retractions above are cellular. □

Let $n \geq 1$ be given. We equip (D^n, D^{n-1}) with the following structure of a pair of CW-complexes. The zero-skeleton is a point and the n-2-skeleton is S^{n-2} obtained by attaching a (n-2)-cell trivially. We get D^{n-1} from S^{n-2} by attaching a (n-1)-cell and the (n-1)-skeleton S^{n-1} of D^n by attaching one more (n-1)-cell using the identity $S^{n-2} \to S^{n-2}$ in both cases. We end up with D^n by attaching a n-cell to S^{n-1} by the identity. Consider the G-push-out

$$\begin{array}{ccc} G/H \times D^{n-1} & \xrightarrow{\;q\;} & X \\ \big\downarrow & & \big\downarrow{\scriptstyle j} \\ G/H \times D^n & \longrightarrow & Y \end{array}$$

such that X is a G-CW-complex and $q(G/H \times S^{n-2}) \subset X_{n-2}$ and $q(G/H \times D^{n-1}) \subset X_{n-1}$ holds. Notice that q needs not to be cellular. Nevertheless the process described in 1.29. defines a G-CW-structure on Y such that (Y,X) is a pair of G-CW-complexes. Notice that Y is obtained from X by attaching a (n-1)-cell and n-cell in a specific way. If Y has a G-CW-complex structure isomorphic to the one above we call $j : X \to Y$ an elementary expansion.

Lemma 4.3. Let $j : A \to X$ be a G-cofibration. Then j is a G-homotopy equivalence if and only if there is a strong G-deformation retraction $r : X \to A$, i.e. a G-homotopy equivalence r with $r \circ j = id_A$ and $j \circ r \simeq_G id_X$ rel j(A).

Proof: Whitehead [1978] I.5.9. □

Hence there is a strong G-deformation retraction $r : X \to A$ if $j : A \to X$ is an elementary expansion. If $\bar{r}$ is a second one r and $\bar{r}$ are G-homotopic relative j(A) because of $r \simeq_G r \circ j \circ \bar{r} = \bar{r}$ rel j(A). We call any such strong G-deformation $r : X \to A$ an elementary collapse.

Let $f : X \to X'$ and $g : A' \to A$ be isomorphisms of G-CW-complexes and $j : A \to X$ and $r : X \to A$ G-maps. Then j is an elementary expansion if and only if $f \circ j \circ g'$ is, and analogously for r as an elementary collapse.

A G-map $(j,id) : (X,A) \to (Y,A)$ between pairs of G-CW-complexes is an expansion relative A if it is a finite composition of elementary expansions.

$$X = X_o \to X_1 \to X_2 \to \qquad \to X_r = Y$$

We symbolize this by $X \nearrow Y$ rel A.

A G-map $(r,id) : (Y,A) \to (X,A)$ between pairs of G-CW-complexes is a collapse relative A if it is a finite composition of elementary collapses

$$Y = Y_o \to Y_1 \to Y_2 \to \dots \to Y_r = X$$

We write $Y \searrow X$ rel A.

A finite composition of expansions and collapses relative A

$$X = X_o \nearrow X_1 \searrow X_2 \dots \searrow X_r = Y$$

is called a *formal deformation relative* A and symbolized by $X \nearrow\!\!\!\searrow Y$ rel A.

Definition 4.4. A G-*map* $(f,id) : (X,A) \to (Y,A)$ *between pairs of* G-CW-*complexes is a simple* G-*homotopy equivalence relative* A *if it is* G-*homotopic relative* A *to a formal* G-*deformation relative* A.

We speak of a simple G-*homotopy equivalence if* A *is empty*. □

We want to study in this section

Problem 4.5. When is a G-homotopy equivalence simple? □

We will construct a covariant functor

4.6. $Wh^G : \{G\text{-CW-compl.}\} \to \{\text{abel. gr.}\}$

and a function assigning to any G-homotopy equivalence $f : X \to Y$ between finite G-CW-complexes an element

4.7. $\tau^G(f) \in Wh^G(Y)$

called its *Whitehead torsion*, such that the following holds.

Theorem 4.8.

a) *Obstruction property.*
   A G-*homotopy equivalence* $f : X \to Y$ *between finite* G-CW-*complexes is simple if and only if* $\tau^G(f) \in Wh^G(Y)$ *vanishes.*

b) *Homotopy invariance.*
   i) *If* $f,g : X \to Y$ *are* G-*homotopy equivalences between finite* G-CW-

complexes then $f \simeq g \Rightarrow \tau^G(f) = \tau^G(g)$

ii) If $f, g : X \to Y$ are G-homotopic then $f_* = g_* : Wh^G(X) \to Wh^G(Y)$.

c) Additivity

Consider the following map between cellular G-push-outs of finite G-CW-complexes such i_1 and k_1 are inclusions of G-CW-complexes and f_o, f_1, f_2 and f G-homotopy equivalences

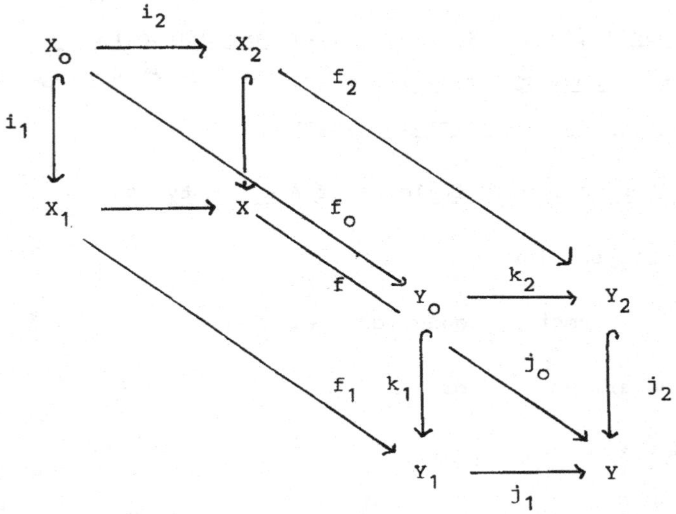

Then $\tau^G(f) = j_{1*}\tau^G(f_1) + j_{2*}\tau^G(f_2) - j_{o*}\tau^G(f_o)$.

d) Logarithmic Property

Let $f : X \to Y$ and $g : Y \to Z$ be G-homotopy equivalences between finite G-CW-complexes. Then $\tau^G(g \circ f) = \tau^G(g) + g_*\tau^G(f)$. □

Some preparations are needed.

Lemma 4.9. Consider the diagram of G-spaces

$$\begin{array}{ccccc} A & \xrightarrow{q} & X & \longrightarrow & U \\ j\downarrow & I & \downarrow & J \quad II & \downarrow \\ B & \xrightarrow[Q]{} & Y & \longrightarrow & V \end{array}$$

Let I be the left, II the right and III the outer square.

a) Let I be a G-push-out. Then II is a G-push-out if and only if III is a G-push-out.

b) Let I be a cellular G-push-out with respect to j, i.e. j is the inclusion of G-CW-complexes. Then II is a cellular G-push-out with respect to J if and only if III is a cellular G-pusn-out with respect to j.

d) Let j and q be inclusions of G-CW-complexes. Then I is a cellular G-push-out with respect to j if and only if I is a cellular G-push-out with respect to q.

Proof: left to the reader. □

Lemma 4.10. Let (X,A) and (Y,A) be pairs of G-CW-complexes with $X \curlywedge Y$ rel A and $f : A \to B$ be a cellular G-map. Then

$$B \cup_f X \curlywedge B \cup_f Y \text{ rel } B$$

Proof: It suffices to treat the case where Y is obtained from X by an elementary expansion. Now apply Lemma 4.9. several times to the diagram

$$\begin{array}{ccccc}
 & & A & \xrightarrow{f} & B \\
 & & \downarrow & & \downarrow \\
G/H \times D^{n-1} & \longrightarrow & X & \longrightarrow & B \cup_f X \\
\downarrow & & \downarrow & & \downarrow \\
G/H \times D^{n} & \longrightarrow & Y & \longrightarrow & B \cup_f Y
\end{array}$$

□

Lemma 4.11. Let $f : X \to Y$ be a cellular G-map between G-CW-complexes and A a G-CW-subcomplex of X

a) If (X,A) is relatively finite, then

$\mathrm{Cyl}(f|A) \nearrow \mathrm{Cyl}(f)$

b) If X is finite, then

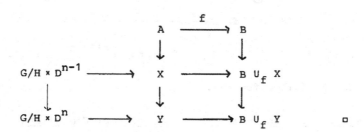

$Y \nearrow Cyl(f)$

c) *If* (X,A) *is relatively finite, then*

$A \times I \cup X \times \{k\} \nearrow X \times I$ for $k = 0,1$

d) *If* X *is finite then*

$X \times \{k\} \nearrow X \times I$ for $k = 0,1$

Proof: a) It suffices to treat the case where $X \setminus A$ contains only one cell

$$\begin{array}{ccc} G/H \times S^{n-1} & \longrightarrow & A \\ \downarrow & & \downarrow \\ G/H \times D^n & \longrightarrow & X \end{array}$$

Crossing it with I yields again a cellular G-push-out. Recall that the mapping cylinder is defined as a cellular G-push-out. By Lemma 4.9. we get a cellular G-push-out

$$\begin{array}{ccc} G/H \times (S^{n-1} \times I \cup_{S^{n-1}} D^n) & \longrightarrow & Cyl(f|A) \\ \downarrow & & \downarrow \\ G/H \times D^n \times I & \longrightarrow & Cyl(f) \end{array}$$

Notice that $(D^n \times I, S^{n-1} \times I \cup_{S^{n-1}} D^n)$ and (D^{n+1}, D^n) are isomorphic.

b), c) and d) follow from a). □

Lemma 4.12. Let f *and* $g : A \to B$ *be cellular* G-*maps and* (X,A) *a relative finite pair of* G-CW-*complexes. If* f *and* g *are* G-*homotopic, then:*

$$B \cup_f X \curlywedge B \cup_g X \text{ rel } B$$

Proof: Let $h : A \times I \to B$ be a cellular G-homotopy between f and g (Theorem 2.1.). We get from Lemma 4.10. and 4.11. c)

$$B \cup_f X = B \cup_h (X \cup_{A\times 0} A \times I) \nearrow B \cup_h (X \times I) \text{ rel } B$$

and

$$B \cup_g X = B \cup_h (X \cup_{A\times 1} A \times I) \nearrow B \cup_h (X \times I) \text{ rel } B \quad \square$$

Lemma 4.13. Let (X,A) and (Y,A) be relatively finite pairs of G-CW-complexes such that the inclusions are G-homotopy equivalences.

a) Then $(X \cup_A Y, A)$ is a relative finite pair of G-CW-complexes such that the inclusion is a G-homotopy equivalence.

b) $X \curlywedge X'$ rel A and $Y \curlywedge Y'$ rel A implies $X \cup_A Y \curlywedge X' \cup_A Y'$ rel A

Proof: a) Lemma 2.13.

b) follows from Lemma 4.10. □

Lemma 4.14. Let $A \subset B \subset X$ be a triple of G-CW-complexes such that the inclusions are G-homotopy equivalences and (X,A) relatively finite. Let $r : B \to A$ be any strong G-deformation retraction. Then

$$X \curlywedge B \cup_A (A \cup_r X) \text{ rel } A$$

Proof: Let $i : A \to B$ be the inclusion. We get from Lemma 4.12.

$$X = B \cup_{id} X = B \cup_{i \circ r} X \text{ rel } B$$

The claim follows from

$$B \cup_{i\circ r} X = B \cup_A (A \cup_r X) \text{ rel } A \quad \square$$

Now we can define $Wh^G(A)$ of a G-CW-complex A. Namely, we consider the equivalence relation on all relative finite pairs of G-CW-complexes (X,A) with $A \hookrightarrow X$ a G-homotopy equivalence given by $(X,A) \curlywedge (Y,A)$ rel A. Let $Wh^G(A)$ be the set of equivalence classes. Addition is given by $(X,A) + (Y,A) = (X \cup_A Y, A)$. It is well-defined by Lemma 4.13. The class of (A,A) is the zero-element. An inverse of the class of (X,A) is given as follows.

Choose a strong G-deformation retraction $r : X \to A$. Let $p : A \times I \to A$ be the projection. Our candidate is $(A \cup_r (A \cup_p \mathrm{Cyl}(r)), A)$. We have

$$A \times I \nearrow \mathrm{Cyl}(r) \qquad \text{(Lemma 4.11. a)}$$

$$A \cup_p \mathrm{Cyl}(r) \searrow A \text{ rel } A \qquad \text{(Lemma 4.10.)}$$

Consider the triple $(A, X, A \cup_p \mathrm{Cyl}(r))$. All inclusions are G-homotopy equivalences. We get from Lemma 4.14.

$$A \nearrow\!\!\searrow A \cup_p \mathrm{Cyl}(r) \nearrow\!\!\searrow X \cup_A (A \cup_r A \cup_p \mathrm{Cyl}(r)) \text{ rel } A$$

This finishes the construction of the abelian group $Wh^G(A)$. If $f : A \to B$ is a G-map, define $f_* : Wh^G(A) \to Wh^G(B)$ by $(X,A) \to (B \cup_f X, B)$. This is well-defined by Lemma 4.10. so that Wh^G becomes a covariant functor {G-CW-compl.} $\to$ abelian groups.

Definition 4.15. Let $f : X \to Y$ *be a* G-*map between finite* G-CW-*complexes. Define its Whitehead torsion*

$$\tau^G(f) \in Wh^G(Y)$$

by the class of $(Y \cup_g \mathrm{Cyl}(g), Y)$ *for any cellular* G-*map* g *which is* G-*homotopic to* f. *We call* $Wh^G(Y)$ *the Whitehead group of* Y. □

The proof that Definition 4.15. makes sense needs some preparation.

Lemma 4.16. Let (X,A) *be a relative finite pair of* G-CW-*complexes and* $f : X \to Y$ *be a cellular* G-*map. If* $A \nearrow X$ *holds, we get*

$$X \cup_A \mathrm{Cyl}(f|A) \nearrow \mathrm{Cyl}(f)$$

Proof: It suffices to treat the case of an elementary expansion

$$\begin{array}{ccc} G/H \times D^{n-1} & \longrightarrow & A \\ \downarrow & & \downarrow \\ G/H \times D^n & \longrightarrow & X \end{array}$$

We get from Lemma 4.9. a G-push-out

$$\begin{array}{ccc} G/H \times (D^n \times_{S^{n-1}} (S^{n-1} \times I)) & \longrightarrow & X \cup_A \mathrm{Cyl}(f|A) \\ \downarrow & & \downarrow \\ G/H \times D^n \times I & \longrightarrow & \mathrm{Cyl}(f) \end{array} \quad \square$$

Lemma 4.17.
Let $X_o \xrightarrow{f_o} X_1 \xrightarrow{f_1} X_2 \longrightarrow \ldots \xrightarrow{f_{n-1}} X_n$ *be cellular G-maps between finite G-CW-complexes. Denote by* $f : X_o \to X_n$ *their composition. Then*

$$\mathrm{Cyl}(f) \curlywedge \mathrm{Cyl}(f_o) \cup_{X_1} \mathrm{Cyl}(f_1) \cup_{X_2} \ldots \mathrm{Cyl}(f_{n-1}) \text{ rel } X_o \amalg X_n$$

Proof: We only treat the case $n = 1$. Let $p : \mathrm{Cyl}(f_o) \to X_1$ be the projection and $g : \mathrm{Cyl}(f_o) \to X_2$ be the composition $f_1 \circ p$. Notice that $g|X_1 = f_1 \circ f_o$. We get

$$X_1 \nearrow \mathrm{Cyl}(f_o) \qquad \text{(Lemma 4.11. b)}$$

$$\mathrm{Cyl}(g) \searrow \mathrm{Cyl}(f_o) \cup_{X_1} \mathrm{Cyl}(f_1) \qquad \text{(Lemma 4.16.)}$$

$$\mathrm{Cyl}(f_1 \circ f_o) \nearrow \mathrm{Cyl}(g) \qquad \text{(Lemma 4.11. a)} \quad \square$$

Lemma 4.18. *Let* f *and* $g : X \to Y$ *be G-homotopic cellular G-maps between finite G-CW-complexes. Then*

$$\mathrm{Cyl}(f) \curlywedge \mathrm{Cyl}(g) \text{ rel } X \amalg Y$$

Proof: Let $h : X \times I \to Y$ be a cellular G-homotopy between f and g. We have by Lemma 4.11. d)

$$X \times \{k\} \nearrow X \times I$$

Now Lemma 4.16. implies

$$\mathrm{Cyl}(f) \cup_{X \times \{0\}} X \times I \nearrow \mathrm{Cyl}(h) \searrow \mathrm{Cyl}(g) \cup_{X \times \{1\}} X \times I$$

Now apply Lemma 4.10. to the projections $X \times I \to X$. $\square$

Lemma 4.12. and Lemma 4.18. ensure that Definition 4.15. makes sense. Notice that $\tau^G(f)$ can be written as

4.19. $$\tau^G(f) = g_*(\mathrm{Cyl}(g), X)$$

for any cellular G-map g with $g \simeq_G f$.

Proof of Theorem 4.8.

b) Lemma 4.12. and Lemma 4.18.

a) Let $f : X \to Y$ be a simple G-homotopy equivalence. We show $\tau^G(f) = 0$ Without loss of generality we can assume that f is cellular. Simple means that f is G-homotopic to a formal deformation

$$X = X_o \nearrow X_1 \searrow X_2 \nearrow \dots X_n = Y$$

Let $g_i : X_i \to X_{i+1}$ be the corresponding G-map and $g : X \to Y$ the composition of the g_i - s. We prove

4.20. $$\mathrm{Cyl}(g_i) \searrow X_i$$

If g_i is an elementary expansion, 4.20. follows from Lemma 4.10. and 4.11. d)

$$X_i = X_i \times \{o\} \nearrow X_i \times I \nearrow X_i \times I \cup_{g_i} X_{i+1} = \mathrm{Cyl}(g_i)$$

If g_i is an elementary collapse Lemma 4.16. implies 4.20.

$$X_i = X_i \times \{o\} \nearrow X_{i+1} \times I \cup X_i = \mathrm{Cyl}(g_i | X_{i+1}) \cup X_i \nearrow \mathrm{Cyl}(g_i)$$

We conclude from Lemma 4.17. and Lemma 4.18.

$$\mathrm{Cyl}(f) \curlywedge \mathrm{Cyl}(g) \curlywedge \mathrm{Cyl}(g_o) \cup_{X_1} \dots\dots \mathrm{Cyl}(g_{n-1}) \text{ rel } X$$

An iterated application of 4.20. yields

$$\mathrm{Cyl}(g_o) \cup_{X_1} \mathrm{Cyl}(g_1) \cup_{X_2} \dots \mathrm{Cyl}(g_{n-1}) \curlywedge X \text{ rel } X$$

This proves $(\mathrm{Cyl}(g), X) = 0$ in $Wh^G(X)$ and by 4.19. $\tau^G(f) = 0$.

Now consider a G-homotopy equivalence $f : X \to Y$ between finite G-CW-complexes with $\tau^G(f) = 0$. We want to show that f is simple. We can assume f to be cellular. Since $f_*: Wh^G(X) \to Wh^G(Y)$ is an isomorphism $\tau^G(f) = 0$ implies

$$Cyl(f) \curlywedge X \text{ rel } X$$

Hence the inclusion $i : X \to Cyl(f)$ is a formal deformation. We know from Lemma 4.11. b) that the inclusion $Y \to Cyl(f)$ is a formal deformation so that the projection $p : Cyl(f) \to Y$ is G-homotopic to a formal deformation. Hence $f = p \circ i$ is G-homotopic to a formal deformation meaning that f is simple.

d) Consider the G-homotopy equivalences $f : X \to Y$ and $g : Y \to Z$ between finite G-CW-complexes. We claim $\tau^G(g \circ f) = \tau^G(g) + g_*\tau^G(f)$. Consider the triple $(Cyl(f) \cup_Y Cyl(g), Cyl(f), X)$. If $r : Cyl(f) \to X$ is a retraction, we get in $Wh^G(X)$ by Lemma 4.14.

$r_*(Cyl(f) \cup_Y Cyl(g), Cyl(f)) + (Cyl(f), X) = (Cyl(f) \cup_Y Cyl(g), X)$

If $j : Y \to Cyl(f)$ is the inclusion, we get from Lemma 4.17.

$r_*j_*(Cyl(g), Y) + (Cyl(f), X) = (Cyl(g \circ f), X)$

By Theorem 4.8. b) $r_*j_* = f_*^{-1}$, since $r \circ j \circ f \simeq_G id$. Applying g_*f_* to the equation above yields

$$\tau^G(g) + g_*\tau^G(f) = \tau^G(g \circ f)$$

c) Since $f_*: Wh^G(X) \to Wh^G(Y)$ is an isomorphism by Theorem 4.8. b) it suffices to prove in $Wh^G(X)$

$(Cyl(f), X) = (X \cup_{X_1} Cyl(f_1), X) + (X \cup_{X_2} Cyl(f_2), X) - (X \cup_{X_o} Cyl(f_o), X)$

By an iterated application of Lemma 4.9. we obtain a diagram such that all squares in it are cellular G-push-outs

4.21.

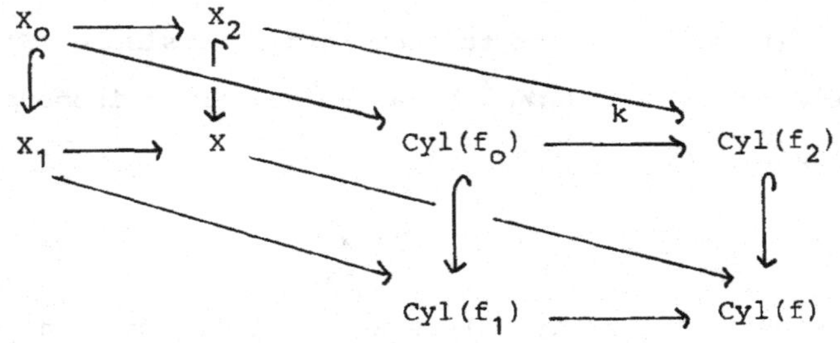

By Lemma 4.11. b) we have

$$\mathrm{Cyl}(f_2) \nearrow \mathrm{Cyl}(k : \mathrm{Cyl}(f_o) \to \mathrm{Cyl}(f_2))$$

and by Lemma 4.10.

$$\mathrm{Cyl}(f) \nearrow \mathrm{Cyl}(f) \cup_{\mathrm{Cyl}(f_2)} \mathrm{Cyl}(k)$$

Hence we obtain from Lemma 4.10. in $Wh^G(X)$

$$(X \cup_{X_2} \mathrm{Cyl}(f_2), X) = (X \cup_{X_2} \mathrm{Cyl}(k), X)$$

$$(\mathrm{Cyl}(f), X) = (\mathrm{Cyl}(f) \cup_{\mathrm{Cyl}(f_2)} \mathrm{Cyl}(k), X)$$

Write $Y = \mathrm{Cyl}(f) \cup_{\mathrm{Cyl}(f_2)} \mathrm{Cyl}(k)$, $Y_2 = X \cup_{X_2} \mathrm{Cyl}(k)$, $Y_1 = X \cup_{X_1} \mathrm{Cyl}(f_1)$ and $Y_o = X \cup_{X_o} \mathrm{Cyl}(f_o)$. Then we must show

4.22. $$(Y,X) = (Y_1,X) + (Y_2,X) - (Y_o,X)$$

We get from 4.21. and Lemma 4.9. that Y_o, Y_1, Y_2 are G-subcomplexes of Y with $Y = Y_1 \cup Y_2$ and $Y_o = Y_1 \cap Y_2$.

Let $r : Y_o \to X$ be a retraction. Then we get from Lemma 4.14. in $Wh^G(X)$

$$(Y,X) = r_*(Y,Y_o) + (Y_o,X)$$

$$(Y_1,X) = r_*(Y_1,Y_o) + (Y_o,X)$$

$$(Y_2,X) = r_*(Y_2,Y_o) + (Y_o,X)$$

Because of $Y = Y_1 \cup Y_2$ and $Y_o = Y_1 \cap Y_2$ we have in $Wh^G(Y)$

$$(Y,Y_o) = (Y_1,Y_o) + (Y_2,Y_o)$$

These equations implies

$$(Y,X) - (Y_o,X) = (Y_1,X) - (Y_o,X) + (Y_2,X) - (Y_o,X)$$

and hence 4.22.

This finishes the construction of the Whitehead group of a G-CW-complex and the Whitehead torsion of a G-homotopy equivalence between finite G-CW-complexes and the proof of Theorem 4.8. □

Let Q be the Hilbert cube $\prod_{i=o}^{\infty} I$.

Theorem 4.23. *Let $f : X \to Y$ be a map between finite CW-complexes. Then f is a simple homotopy equivalence if and only if $f \times id : X \times Q \to Y \times Q$ is homotopic to an homeomorphism.*

Proof: Chapman [1973]. □

Corollary 4.24. *Topological invariance of (non-equivariant) Whitehead torsion.*

Let $f : X \to Y$ be a homeomorphism between finite CW-complexes. Then f is a simple homotopy equivalence. □

The topological invariance of the Whitehead torsion does not hold in the equivariant case. We give an outline of two counter examples.

Example 4.25. The starting point is the question whether two topologically conjugated G-representation of a finite group G are already linearly isomorphic. This is true if G has odd order (Hsiang-Pardon [1982], Madsen-Rothenberg [1985a]. Counterexamples for G of even order are constructed in Cappell-Shaneson [1982], Cappell-Shaneson-Steinberger-Weinberger-West [1988]

Choose a G-homeomorphism $f : V \longrightarrow W$ between G-representations, which are not linearly isomorphic. Substituting V by $V_o = V \oplus V \oplus \mathbb{R} \oplus \mathbb{R}$ and W by $W_o = W \oplus W \oplus \mathbb{R} \oplus \mathbb{R}$ yields a G-homeomorphism $f_o : SV_o \to SW_o$ between the spheres of two G-representations such that V_o and W_o are not linearly isomorphic, WH acts orientation preserving on SV_o^H and SW_o^H and $\chi(SV_o^H) = \chi(SW_o^H) = 0$ for all $H \subset G$. Under these circumstances the Reidemeister torsion $\rho(SV_o)$ and $\rho(SW_o)$ is defined. We explain it later. Assume that $f_o : SV_o \to SW_o$ has trivial Whitehead torsion. Since $\rho(V_o) - \rho(W_o)$ is a function of $\tau^G(f_o)$, we have $\rho(V_o) = \rho(W_o)$. Later we reprove the result of de Rham that $\rho(V_o) = \rho(W_o)$ implies that V_o and W_o are linearly isomorphic, a contradiction. Hence f_o is a G-homeomorphism $SV_o \to SW_o$ but not simple. □

Example 4.26. The Hauptvermutung says that two homeomorphic simplicial complexes are already PL-homeomorphic (see Rourke-Sanderson [1972] I for the definition of a PL-space and PL-homeomorphism). A counterexample is given in Milnor [1961] (see also Stallings [1968]). In Milnor [1961] two finite G-CW-complexes X and Y together with a G-homeomorphism are constructed such that the Reidemeister torsion $\rho(X)$ and $\rho(Y)$ are defined and do not agree. Hence f is not a simple G-homotopy equivalence. □

We can extend Wh^G to a covariant functor

4.27. $\quad Wh^G : \{G\text{-spaces}\} \to \{\text{abel. gr.}\}$

Namely, define $Wh^G(Z) = \lim Wh^G(X)$ for a G-space Z, where the limit runs over all G-CW-approximations (X,f) of Z.

4B. Simple structures on G-spaces

Given a G-space Z, consider all pairs (X,f) consisting of a finite G-CW-complex X and a G-homotopy equivalence $f : X \to Z$. We call (X,f) and (Y,g) equivalent if $g_* \tau^G(g^{-1} \circ f) \in Wh^G(Z)$ vanishes. This is a well-defined equivalence relation by Theorem 4.8. A simple structure on a G-space Z is an equivalence class ξ of such pairs (X,f).

Let $f : (Z_o,\xi_o) \to (Z_1,\xi_1)$ be a G-homotopy equivalence between G-spaces with simple structures. Choose representatives (X_i,f_i) for ξ_i. Define

4.28. $\tau^G(f) \in Wh^G(Z)$

by $\tau^G(f) := f_{1*}\tau^G(f_1^{-1} \circ f \circ f_o)$. This is well-defined by Theorem 4.8.

Regard the G-push-out with j_1 a G-cofibration

4.29.
$$\begin{array}{ccc} Z_o & \xrightarrow{\ j_2\ } & Z_2 \\ {\scriptstyle j_1}\big\downarrow & & \big\downarrow \\ Z_1 & \longrightarrow & Z \end{array}$$

Assume that Z_i has a simple structure ξ_i for $i = 0,1,2$. We want to assign to Z a simple structure ξ.

Choose a commutative diagram

4.30.
$$\begin{array}{ccccc} X_1 & \xleftarrow{\ i_1\ } & X_o & \xrightarrow{\ i_2\ } & X_2 \\ {\scriptstyle f_1}\big\downarrow & & {\scriptstyle f_o}\big\downarrow & & {\scriptstyle f_2}\big\downarrow \\ Z_1 & \xleftarrow[\ j_1\ ]{} & Z_o & \xrightarrow[\ j_2\ ]{} & Z_2 \end{array}$$

satisfying

i) X_i is a finite G-CW-complex.
ii) i_1 is an inclusion of finite G-CW-complexes and i_2 is cellular.
iii) f_i is a G-homotopy equivalence.
iv) (X_i,f_i) represents ξ_i.

Let $f : X \to Z$ be the G-map induced on the G-push-outs. Then X is a finite G-CW-complex and f is a G-homotopy equivalence. Let ξ be the simple structure ξ on Z given by (X,f). If Z is equipped with ξ we call 4.29. a

G-push-out of G-spaces with simple structure.

The following construction guarantees that a diagram 4.30. exists. Lemma 4.32. ensures that ξ does only depend on ξ_o, ξ_1 and ξ_2 but not on the choices above.

4.31. Consider the G-push-out 4.29. Assume that we are given G-CW-approximations (X_i, f_i) for Z_i $(i = 0,1,2)$. Choose a cellular G-map $g_i : X_o \to X_i$ satisfying $f_i \circ g_i \simeq_G j_i \circ f_o$ for $i = 1,2$. Let $k_i : X_o \to \mathrm{Cyl}(g_i)$ be the canonical inclusion and $pr_i : \mathrm{Cyl}(g_i) \to X_i$ be the canonical projection. We have $f_i \circ pr_i \circ k_i \simeq_G j_i \circ f_o$. Since k_i is a G-cofibration we can change $f_i \circ pr_i$ G-homotopically into $f_i' : \mathrm{Cyl}(g_i) \to Z_i$ such that the following diagram commutes

$$\begin{array}{ccccc}
Z_1 & \xleftarrow{\;j_1\;} & Z_o & \xrightarrow{\;j_2\;} & Z_2 \\
\uparrow{\scriptstyle f_1'} & & \uparrow{\scriptstyle f_o'} & & \uparrow{\scriptstyle f_2'} \\
\mathrm{Cyl}(g_1) & \xleftarrow[\;k_1\;]{} & X_o & \xrightarrow[\;k_o\;]{} & \mathrm{Cyl}(g_2)
\end{array}$$

Then $(\mathrm{Cyl}(g_i), f_i')$ is a G-CW-approximation of Z_i. If Z_i has the G-homotopy type of a G-CW-complex then f_1', f_o' and f_2' are G-homotopy equivalences (Theorem 2.4.). If X_i is finite, skeletal-finite, finite-dimensional resp finitely dominated then also $\mathrm{Cyl}(g_i)$. If X_i is finite and f_i a G-homotopy equivalence, (X_i, f_i) and $(\mathrm{Cyl}(g_i), f_i')$ define the same simple structure on Z_i by Lemma 4.11. b). □

Consider the commutative diagram

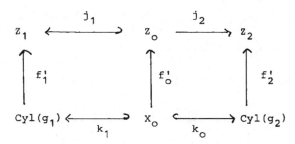

$$
\begin{array}{ccccc}
X_1 & \overset{i_1}{\hookleftarrow} & X_o & \xrightarrow{i_2} & X_2 \\
\downarrow f_1 & & \downarrow f_o & & \downarrow f_2 \\
Z_1 & \overset{j_1}{\hookleftarrow} & Z_o & \xrightarrow{j_2} & Z_2 \\
\uparrow g_1 & & \uparrow g_o & & \uparrow g_2 \\
Y_1 & \overset{k_1}{\hookleftarrow} & Y_o & \xrightarrow{k_2} & Y_o
\end{array}
$$

Assume

a) X_i and Y_i are finite G-CW-complexes for $i = 0,1,2$.

b) i_1 and k_1 are inclusions of finite G-CW-complexes and j_1 a G-cofibration.

c) f_i and g_i are G-homotopy equivalences for $i = 0,1,2$.

Let Z be the G-push-out

$$
\begin{array}{ccc}
Z_o & \overset{j_1}{\hookrightarrow} & Z_1 \\
{\scriptstyle j_2}\downarrow & \overset{l_o}{\searrow} & \downarrow{\scriptstyle l_1} \\
Z_2 & \xrightarrow[l_2]{} & Z
\end{array}
$$

Define X and Y similarly. Let $f : X \to Z$ and $g : Y \to Z$ be the induced maps.

__Lemma 4.32. Then__ X __and__ Y __are finite__ G-CW-__complexes and__ f __and__ g __are__ G-__homotopy equivalences__. __Moreover we have in__ $Wh^G(Z)$

$$g_*\tau^G(g^{-1}\circ f) = l_{1*}g_{1*}\tau^G(g_1^{-1}\circ f_1) + l_{2*}g_{2*}\tau^G(g_2^{-1}\circ f_2) - l_{o*}g_{o*}\tau^G(g_o^{-1}\circ f_1^{-1})$$

__Proof:__ We can assume that i_2, j_2 and k_2 are also G-cofibrations, otherwise substitute them by the inclusion of mapping cylinders. Now we can construct a diagram

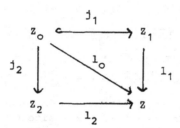

$$\begin{array}{ccccc} Z_1 & \overset{j_1}{\longleftrightarrow} & Z_o & \overset{j_2}{\hookrightarrow} & Z_2 \\ \downarrow h_1 & & \downarrow h_o & & \downarrow h_2 \\ Y_1 & \overset{k_1}{\longleftrightarrow} & Y_o & \overset{k_2}{\hookrightarrow} & Y_2 \end{array}$$

such that h_i is a G-homotopy inverse for g_i. Let $h : Z \rightarrow Y$ be the induced G-homotopy equivalence. We get from Theorem 4.8.

$$h_*^{-1}\tau^G(h \circ f) =$$
$$l_{1*}h_{1*}^{-1}\tau^G(h_1 \circ f_1) + l_{2*}h_{2*}^{-1}\tau^G(h_2 \circ f_2) -$$
$$l_{o*}h_{o*}^{-1}\tau^G(h_o \circ f_o) =$$
$$l_{1*}g_{1*}\tau^G(g_1^{-1} \circ f_1) + l_{2*}g_{2*}\tau^G(g_2^{-1} \circ f_2) -$$
$$l_{o*}g_{o*}\tau^G(g_o^{-1} \circ f_o)$$

$$g_*\tau^G(g^{-1} \circ h^{-1}) =$$
$$l_{1*}g_{1*}\tau^G(g_1^{-1} \circ h_1^{-1}) + l_{2*}g_{2*}\tau^G(g_2^{-1} \circ h_2^{-1}) -$$
$$l_{o*}g_{o*}\tau^G(g_o^{-1} \circ h_o^{-1}) = 0$$

$$g_*\tau^G(g^{-1} \circ f) = g_*\tau^G(g^{-1} \circ h^{-1} \circ h \circ f) =$$
$$g_*\tau^G(g^{-1} \circ h^{-1}) + g_*g_*^{-1}h_*^{-1}\tau^G(h \circ f) =$$
$$l_{1*}g_{1*}\tau^G(g_1^{-1} \circ f_1) + l_{2*}g_{2*}\tau^G(g_2^{-1} \circ f_2) - l_{o*}g_{o*}\tau^G(g_o^{-1} \circ f_o) \quad \square$$

Now one easily extends Theorem 4.8. for finite G-CW-complexes to spaces with simple structure.

Theorem 4.33.

a) Obstruction property

Let $f : (X,\xi) \rightarrow (Y,\eta)$ be a G-homotopy equivalence between G-spaces with simple structures. Then the following statements are equivalent.

i) $\tau^G(f) = 0$.

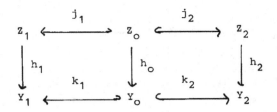

ii) There are representatives (A,u) and (B,v) of ξ and η such that $v^{-1} \circ f \circ u : A \to B$ is simple.

iii) For any representatives (A,u) and (B,v) of ξ and η the composition $v^{-1} \circ f \circ u : A \to B$ is simple.

b) Homotopy invariance

i) $f \simeq g \Rightarrow \tau^G(f) = \tau^G(g)$

ii) $f \simeq g \Rightarrow f_* = g_*$

c) Additivity

Consider the commutative diagram of G-space with simple structures

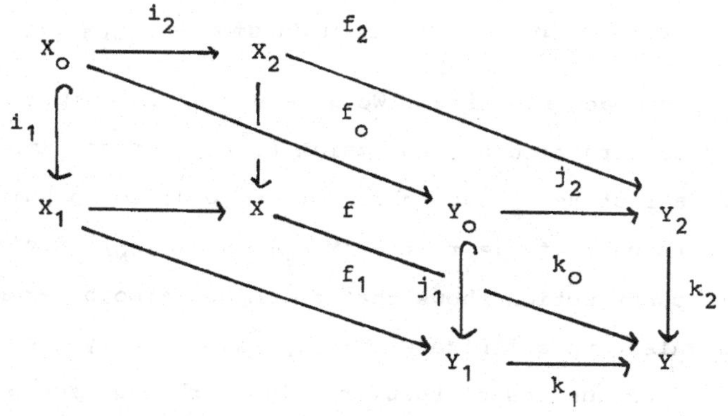

Assume that the sequences are G-push-outs of G-spaces with simple structures. Let i_1 and j_1 be G-cofibrations and f_o, f_1, f_2 and f be G-homotopy equivalences. Then:

$$\tau^G(f) = k_{1*}\tau^G(f_1) + k_{2*}\tau^G(f_2) - k_{o*}\tau^G(f_o)$$

d) Logarithmic property

Let $f : (X,\xi) \to (Y,\eta)$ and $g : (Y,\eta) \to (Z,\mu)$ be G-homotopy equivalences between G-spaces with simple structures. Then we have:

$$\tau^G(g \circ f) = \tau^G(g) + g_*\tau^G(f) \quad \square$$

4.34. If G is a finite group, the existence and uniqueness of a smooth equivariant triangulation of a compact smooth G-manifold M (possibly with non-empty boundary) is proved by Illman [1978]. Hence M has a prefered simple structure ξ_M. If $f : M \to N$ is a G-homotopy equivalence between smooth compact G-manifolds, we can define its Whitehead torsion.

4.35. $$\tau^G(f) \in Wh^G(N)$$

by $\tau^G(f : (M,\xi_M) \to (N,\xi_N))$. We have then for any G-diffeomorphism $f : M \to N$ that $\tau^G(f)$ vanishes. □

If G is a compact Lie group the existence of an equivariant triangulation of a compact smooth G-manifold is shown in Illman [1983]. The necessary uniqueness statement can be derived from Matumoto-Shiota [1987].

4.36. Let G be a compact Lie group. We give now an in comparison with 4.34 elementary construction how to assign to any compact smooth G-manifold M a simple structure ξ_M such that for any G-diffeomorphism $f : M \to N$ the Whitehead torsion $\tau^G(f) := \tau^G(f : (M,\xi_M) \to (N,\xi_N))$ vanishes. In particular this construction shows that a compact smooth G-manifold M is G-homotopy equivalent to a finite G-CW-complex. The only input will be the existence and uniqueness of tubular neighbourhoods (see Bredon [1972] VI 2) and the non-equivariant triangulation theorem, or in other words that we already know such a construction for G = 1.

We use induction over the orbit types $\{(H_1),(H_2),\ldots,(H_r)\} = \{(H) \in \mathrm{Con}\, G \mid H \in \mathrm{Iso}\, M\}$. Notice that M has finite orbit type (see tom Dieck [1987] I.5.11.). The induction runs over r. In the begin r = 1 write $H = H_1$. Then we have a G/H-fibre bundle with WH as structure group $p : M \to M/G$ (see Bredon [1972] II.5.8.).

Let $f : X \to Y$ be a G-homotopy equivalence between G-spaces such that $X \to X/G$ and $Y \to Y/G$ are G/H-fibre bundles with WH as structure group. Suppose that X/G and Y/G are finite CW-complexes and $f/G : X/G \to Y/G$.

is simple. Then the CW-structures on X/G and Y/G lift to G-CW-complex structures on X and Y (compare 1.25.) and $f : X \to Y$ is a simple G-homotopy equivalence. Hence any simple structure $\xi_{M/G}$ lifts uniquely to a simple structure ξ_M on M. Define ξ_M by this process where $\xi_{M/G}$ on the non-equivariant compact smooth manifold M/G comes from a triangulation. Let $f : M \to N$ be a G-diffeomorphism. Then $f/G : M/G \to N/G$ is a diffeomorphism so that $\tau^{\{1\}}(f/G : (M/G, \xi_{M/G}) \to (N/G, \xi_{N/G}))$ vanishes. Hence $\tau^G(f : (M,\xi_M) \to (N,\xi_N))$ is zero. This finishes the induction begin.

In the induction step from $r-1$ to r we write $H = H_1$. Then $M_{(H)} = M^{(H)} = G \cdot M^H$ is a closed G-submanifold of M (see Bredon [1972] VI 2.5.). Let ν be the normal bundle of $M_{(H)}$ in M. A <u>tubular neighbourhood</u> of $M_{(H)}$ in M is a G-embedding

$$\Phi : \nu \to M$$

such that Φ restricted to the zero section induces the identity $M_{(H)} \to M_{(H)}$ and the differential of Φ at the zero section induces the identity $\nu \to \nu$ (compare Bröcker-Jänich [1973], 12.10.). Choose an equivariant Riemannian metric on ν so that the sphere bundle $S\nu$ and disc bundle $D\nu$ are defined. Let $\bar{M}$ be $M \setminus \Phi(\mathrm{int}\, D\nu)$ Then M is the G-push-out

4.37.

$$\begin{array}{ccc} S\nu & \xrightarrow{\Phi|S\nu} & \bar{M} \\ \downarrow & & \downarrow \\ D\nu & \xrightarrow[\Phi|D\nu]{} & M \end{array}$$

Moreover, $D\nu$ is the G-push-out

$$\begin{array}{ccc} S\nu & \xrightarrow{p} & M_{(H)} \\ \downarrow & & \downarrow \\ S\nu \times I & \xrightarrow{p} & D\nu \end{array}$$

where p is the projection and $P(y,t) = t \cdot y$. By induction hypotheses there are simple structures $\xi_{S\nu}$ on $S\nu$ and $\xi_{\bar{M}}$ on $\bar{M}$. Equip $D\nu$ and M with simple structures $\xi_{D\nu}$ and ξ_M such that the two G-push-outs above become G-push-outs of spaces with simple structures. Notice that then $(M_{(H)}, \xi_{M_{(H)}}) \to (D\nu, \xi_{D\nu})$ has torsion zero. We must show that this is independent of the choice of the Riemannian metric and the tubular neighbourhood.

Let $S\nu'$, $D\nu'$, M_o' and $\Phi' : \nu \to M$ be induced by a second choice. Then there is a G-diffeotopy $\psi : M \times I \to M$ and a pair of G-diffeomorphisms $(F,f) : (D\nu, S\nu) \to (D\nu', S\nu')$ such that $\psi_o = \mathrm{id}$ and $\psi_1 \circ \Phi = \Phi' \circ F$ holds and $F(t \cdot y) = t \cdot F(y)$ is valid for $t \in [0,1]$ and $y \in D\nu$ (compare Bröcker-Jänich [1973] 12.13, Bredon [1972], p. 3.12.). Consider the commutative diagrams

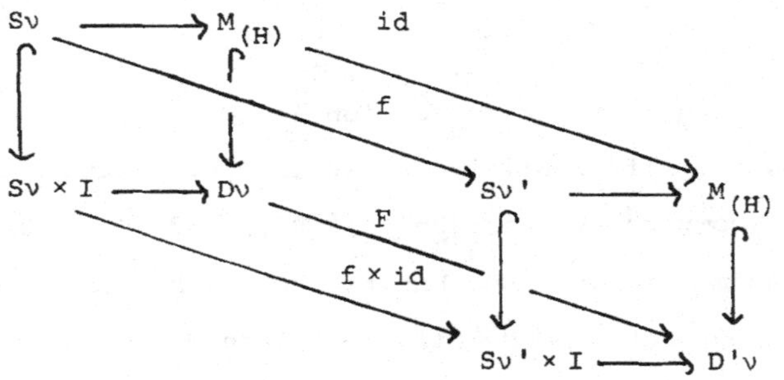

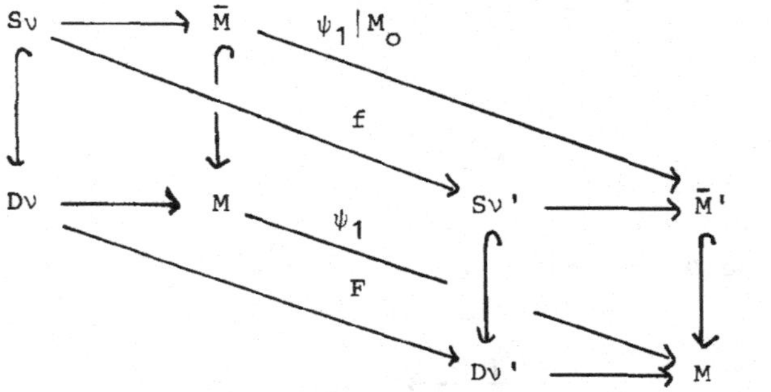

Now an iterated application of Theorem 4.33. shows that the induced

simple structure ξ_M and $\xi_{M'}$ agree since the induction hypotheses applies to $S\nu$, $S\nu'$, M, $\bar{M}'$ and $M_{(H)}$.

Let $f : M \to N$ be a G-diffeomorphism. We want to show $\tau^G(f) = 0$. Let ν_M and ν_N be the normal bundles of $M_{(H)}$ in M and $N_{(H)}$ in N and $\nu_f : \nu_M \to \nu_N$ the bundle map induced by f. Choose tubular neighbourhoods $\Phi_M : \nu_M \to M$ and $\Phi_N : \nu_N \to N$ and Riemannian metrics such that $\Phi_N \circ \nu_f = f \circ \Phi_M$ and ν_f is an isometry. Now the claim follows from Theorem 4.33. applied to the diagrams

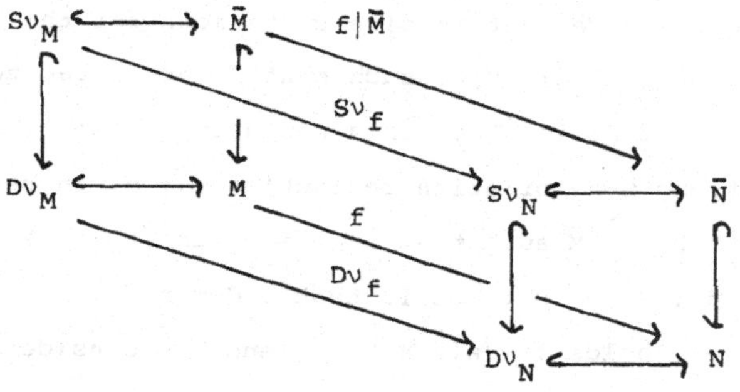

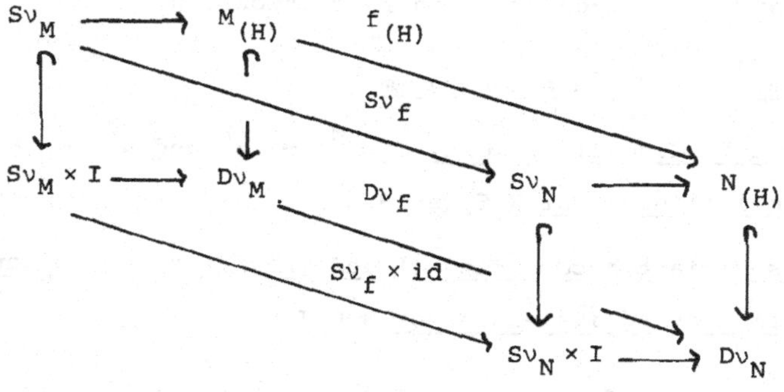

and the induction hypotheses applied to $f_{(H)}$, $f|\bar{M} : \bar{M} \to \bar{N}$ and $S\nu_f$.

We leave it to the reader to check that the choice of the numeration $\{(H_1),\ldots,(H_r)\}$ does not play a role. □

4.D. Isovariant and equivariant s-cobordism theorems

Next we come to the equivariant s-cobordism theorem which is an important tool for classification of G-manifolds. In the sequel G is a compact Lie group. Recall that a G-manifold M is a compact smooth manifold together with a smooth G-action. We will make frequently use of the simple structure we have assigned to a G-manifold M in 4.36.

4.38. A _cobordism_ (B,M,N) between the G-manifolds M and N is a G-manifold B together with embeddings $i_M : M \to \partial B$ and $i_N : N \to \partial B$ such that $\partial B = \text{image } i_M \cup \text{image } i_N$ and $i_M(\partial M) = \text{image } i_M \cap \text{image } i_N = i_N(\partial N)$ holds. In the sequel we often identify M with image i_M. We call (B,M,N) an h-_cobordism_ if the inclusions $M \to B$ and $N \to B$ are strong G-deformation retractions. We want to give criterions for the existence of a G-diffeomorphism $f : B \to M \times I$ such that f restricted to M is the inclusion $M \to M \times \{0\}$. Notice that this would imply that $i : M \to B$ is an isovariant strong G-deformation retraction (i. e. there is an isovariant G-map $r : B \to M$ such that $r \circ i = id$ and $i \circ r$ is isovariantly G-homotopic to $id : B \to B$). Recall that a G-map $f : X \to Y$ is _isovariant_ if $G_x = G_{fx}$ holds for all $x \in X$. Hence we consider _isovariant_ h-_cobordisms_ (B,M,N), i. e. a cobordism (B,M,N) such that $M \to B$ and $N \to B$ are isovariant G-deformation retractions. □

Lemma 4.39. Let (B,M,N) _be a cobordism._

a) _It is an_ h-_cobordism if and only if_ $M^H \to B^H$ _and_ $N^H \to B^H$ _are weak homotopy equivalences for_ $H \in \text{Iso } B$.

b) _It is an isovariant_ h-_cobordism if and only if_ $M_H \to B_H$ _and_ $N_H \to B_H$ _are weak homotopy equivalences for_ $H \in \text{Iso } B$.

Proof: a) This follows from Theorem 2.4. and Lemma 4.3.

b) Hauschild [1978] Satz V.3. □

Consider an isovariant h-cobordism (B,M,N). Define the _isovariant Whitehead group_

4.40 $$Wh^G_{Iso}(M) = \bigoplus_{\substack{(H)\in \mathrm{Con}\, G\\ H \in \mathrm{Iso}\, M}} Wh^{WH}(M_H)$$

Inductively over the number of orbit types of M, we define the <u>isovariant Whitehead torsion</u>

4.41 $$\tau^G_{Iso}(B,M,N) \in Wh^G_{Iso}(M) .$$

If B has only one orbit type (H) then $B^H = B_H$ is a (compact) free WH-manifold. Let $\tau^G_{Iso}(B,M,N) \in Wh^G_{Iso}(M) = Wh^{WH}(M_H)$ be the equivariant Whitehead torsion $\ell^{-1}_{H*}\tau^{WH}(\ell_H)$ for the inclusion $\ell_H : M_H \longrightarrow B_H$. In the induction step choose $H \in \mathrm{Iso}\, B$ such that $K \in \mathrm{Iso}\, B$, $(H) \subset (K)$ implies $(H) = (K)$. Then $B_{(H)} = B^{(H)}$ is a (compact) G-submanifold of B and analogously for $M_{(H)} = M^{(H)} \subset M$ and $N_{(H)} = N^{(H)} \subset N$ (see Bredon [1972] VI.2.5.). Let ν_B be the normal bundle of $B_{(H)}$ in B and define ν_M and ν_N analogously. Identify the disc bundles with tubular neighbourhoods. Define $\bar{B} = B\setminus \mathrm{int}\, D\nu_B$, $\bar{M} = M \setminus \mathrm{int}\, D\nu_M$ and $\bar{N} = (N\setminus \mathrm{int}\, D\nu_N) \cup S\nu_B$. Since $\bar{B}_K \to B_K$, $\bar{M}_K \to M_K$ and $\bar{N}_K \to N_K$ are homotopy equivalences for $K \in \mathrm{Iso}\, \bar{B}$, we obtain from Lemma 4.39. an isovariant h-cobordism $(\bar{B},\bar{M},\bar{N})$ with one orbit type less. By induction hypotheses $\tau^G_{Iso}(\bar{B},\bar{M},\bar{N}) \in Wh^G_{Iso}(\bar{M})$ is defined. Let $i_*: WH^G_{Iso}(\bar{M}) \longrightarrow Wh^G_{Iso}(M)$ be given by the inclusion $i : \bar{M} \to M$ and $j : Wh^{WH}(M_H) \longrightarrow Wh^G_{Iso}(M)$ the obvious split injection. Now define

$$\tau^G_{Iso}(B,M,N) = i_*(\tau^G_{Iso}(\bar{B},\bar{M},\bar{N})) + j\circ\ell^{-1}_{H*}(\tau^{WH}(\ell_H : M_H \longrightarrow B_H))$$

We leave it to the reader to show that this is independent of the various choices like (H), ν_M, ...

<u>Theorem 4.42.</u> (<u>The Isovariant</u> s-<u>Cobordism Theorem</u>). <u>Let</u> M <u>be a</u> G-<u>manifold such that the dimension of any component of</u> M_H/WH <u>for</u> $H \in \mathrm{Iso}\, M$ <u>is not smaller than</u> 5.

a) (B,M,N) <u>is an isovariant</u> h-<u>cobordism with</u> $\tau^G_{Iso}(B,M,N)=0$, <u>if and only if there is a</u> G-<u>diffeomorphism</u> $\phi : (B,M,N) \to (M\times I, M\times\{0\}, \partial M\times I \cup M\times\{1\})$

such that $\Phi|M$ *is the identity* $M \to M \times \{0\}$.

b) *Each element in* $Wh^G_{Iso}(M)$ *can be realized as* $\tau^G_{Iso}(B,M,N)$ *for an isovariant h-cobordism* (B,M,N).

c) $Wh^G_{Iso}(M)$ *classifies G-diffeomorphism classes relative* M *of isovariant h-cobordisms* (B,M,N).

Proof: This follows by induction over the orbit type from the non-equivariant case. See also Browder-Quinn [1973] 1.4., Hauschild [1978] V.4., Rothenberg [1978] 3.4. □

Now we want to relate isovariant Whitehead torsion and equivariant Whitehead torsion. We will define for an isovariant h-cobordism (B,M,N) an homomorphism

4.43. $$\Phi(M) : Wh^G_{Iso}(M) \to Wh^G(M)$$

and prove

Proposition 4.44. $\Phi(M) : Wh^G_{Iso}(M) \to Wh^G(M)$ *sends* $\tau^G_{Iso}(B,M,N)$ *to the equivariant Whitehead torsion* $\ell_*^{-1}\tau^G(\ell)$ *of the inclusion* $\ell : M \to B$

We define $\Phi(M)$ and prove Proposition 4.44. simultanously by induction over the orbit type. In the induction step choose $H \in \text{Iso}\, B$ and define $(\bar{B},\bar{M},\bar{N})$ as in the definition of $\tau^G_{Iso}(B,M,N)$. Let

4.45. $$k : Wh^{WH}(M_H) \to Wh^G(M)$$

be induced by restriction with $NH \to WH$, induction with $NH \to G$ and the obvious inclusion $G \times_{NH} M_H = M_{(H)} \to M$. If M has only one orbit type, let $\Phi(M)$ be k. The normal bundle of $M_{(H)}$ in M is denoted by ν_M. The pull-back construction with $S\nu_M$ defines a transfer homomorphism $trf' : Wh^G(M_{(H)}) \to Wh^G(S\nu_M)$ (see sec. 15). Notice that for any finite G-CW-complex X together with a G-map $f : X \to M_{(H)}$ the pull-back $f^*S\nu_M$ carries a canonical simple structure. Namely, this is true for $X = G/L \times D^n$ as then $f^*S\nu_M$ is a G-manifold, G-diffeomorphic to $G \times_L SV \times D^n$

for an appropriate L-representation V. Composing trf' with the homomorphisms given by restriction with NH $\to$ WH, induction with NH $\to$ G and the inclusion $S\nu_M \to \bar{M}$ or $S\nu_M \to M$ yields

4.46.
$$\overline{trf} : Wh^{WH}(M_H) \to Wh^G(\bar{M})$$
$$trf : Wh^{WH}(M_H) \to Wh^G(M)$$

Define $\Phi(M)$ by the following diagram if $i : \bar{M} \to M$ is the inclusion and $j : Wh^{WH}(M_H) \to Wh^G_{Iso}(M)$ the obvious split injection

4.47.
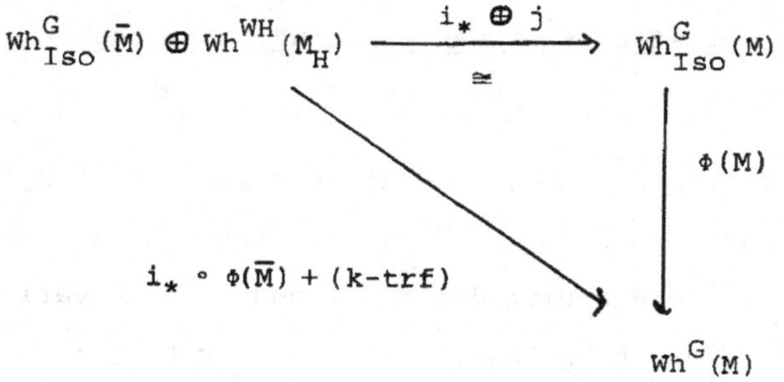

For the proof of Proposition 4.44. consider the commutative diagram with G-push-outs of G-spaces with simple structure as squares

4.48.
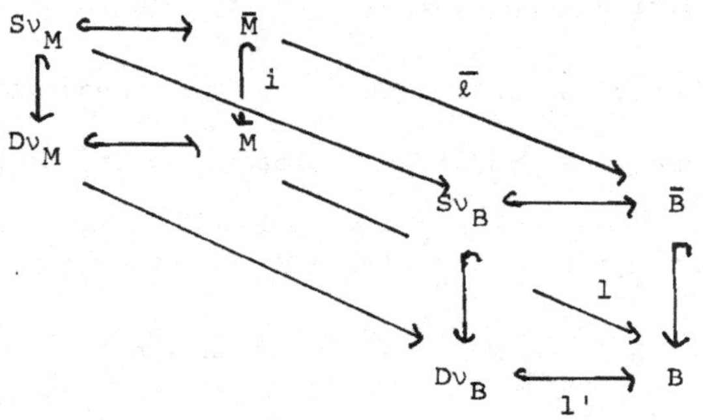

We get from Theorem 4.33.

$$\ell_*^{-1}\tau^G(\ell : M \to B) =$$

$$\ell_*^{-1} \circ \ell'_*\tau^G(D\nu_M \to D\nu_B) + i_* \circ \bar{\ell}_*^{-1}\tau^G(\bar{\ell} : \bar{M} \to \bar{B}) - trf \circ \ell_{H_*}^{-1}\tau^{WH}(\ell_H : M_H \to B_H) =$$

$$k \circ \ell_{H*}^{-1}\tau^{WH}(\ell_H : M_H \to B_H) + i_* \circ \Phi(\overline{M})(\tau^G_{Iso}(\overline{B},\overline{M},\overline{N})) - trf \circ \ell_{H*}^{-1}\tau^{WH}(\ell_H : M_H \to B_H)$$
$$= \Phi(M)(\tau^G_{Iso}(B,M,N)) .$$

This finishes the proof of Proposition 4.44. □

If N is a manifold, let dim N be $\max\{\dim C \mid C \in \pi_0(N)\}$

Definition 4.49. A G-manifold M satisfies the weak gap conditions if for $H,K \in \mathrm{Iso}\, M$, $H \subsetneq K$ and components $C \in \pi_0(M^H)$ and $D \in \pi_0(M^K)$ with $D_K \neq C_H \neq \emptyset$, $D \subset C$ the inequality $\dim G/NK^H + \dim D + 3 \le \dim C$ holds. □

Lemma 4.50. If M satisfies the weak gap conditions then $M_H \to \hat{M}^H$ is 2-connected for $H \in \mathrm{Iso}\, M$, if $\hat{M}^H$ is $\cup\{C \in \pi_0(M^H) \mid C_H \neq \emptyset\}$

Proof: Numerate $\{(K) \in \mathrm{Con}\, G \mid (H) \subsetneq (K),\ K \in \mathrm{Iso}\, M\} = \{(K_1),(K_2),\dots,(K_r)\}$ such that $(K_i) \subset (K_j)$ implies $i \ge j$. It suffices to show for any compact non-equivariant manifold N with $\dim N \le 2$ and non-equivariant map $f : N \to M^H$ with $f(\partial N) \in M_H$ that f is homotopic to $g : N \to M^H$ relative ∂N satisfying $g(N) \in M_H$. We show inductively for $n = 0,1,\dots,r$ that f is homotopic relative ∂N to $f_n : N \to M^H$ such that $f_n(N) \subset M^H \setminus (\bigcup_{i=1}^{n} M^{(K_i)} \cap M^H))$ holds. Then g can be choosen as f_r. The begin $n = 0$ is trivial, the induction step from $n-1$ to n done as follows.

By induction hypotheses the intersection of the compact sets $f_{n-1}(N)$ and $\bigcup_{i=1}^{n-1} M^{(K_i)} \cap M^H$ is empty. We can find a closed subset $A \subset M^H$ satisfying $\bigcup_{i=1}^{n-1} M^{(K_i)} \cap M^H \subset \mathrm{int}\, A$ and $A \cap f_{n-1}(N) = \emptyset$. Consider f_{n-1} as a map $N \to M^H \setminus A$. Now $M^H \setminus A \cap \bigcup_{i=1}^{n} M^{(K_i)} \cap M^H$ is contained in $M_{(K_n)} \cap M^H$

Notice that $M^H_{(K_n)}$ is diffeomorphic to $G/K_n^H \times_{WK_n} M_{K_n}$. If D is a path component of M_{K_n} we have $\dim G/K_n^H \times_{WK_n} (WK_n \cdot D) = \dim G/K_n^H - \dim WK_n + \dim D = \dim G/NK_n^H + \dim D$. If C is the component of M^H containing D, we have $\dim(G/K_n^H \times_{WK_n} (WK_n \cdot D)) + \dim N < \dim C$. By transversality we can change $f_{n-1} : N \to M^H \setminus A$ relative ∂N into $f_n : N \to M^H \setminus A$ satisfying $f_n(N) \cap M^H_{(K_n)} = \emptyset$. This finishes the induction step. □

<u>Theorem 4.51. The Equivariant s-Cobordism Theorem. Let</u> (B,M,N) <u>be an</u> h-<u>cobordism such that</u> M <u>satisfies the weak gap conditions</u> 4.49 <u>and</u> $\dim M_H/WH \geq 5$ <u>holds for</u> $H \in \operatorname{Iso} M$. <u>Then</u>

a) (B,M,N) <u>is an isovariant</u> h-<u>cobordism</u>.

b) $\Phi(M) : Wh^G_{Iso}(M) \to Wh^G(M)$ <u>is split injective with image</u> $Wh^G_\rho(M)$ (see 4.54.).

c) $\tau^G(M \to B)$ <u>vanishes if and only if</u> (B,M,N) <u>is</u> G-<u>diffeomorphic to</u> $(M \times I, M \times \{0\}, \partial M \times I \cup M \times \{1\})$ <u>relative</u> $M = M \times \{0\}$.

d) <u>Any element in</u> $Wh^G_\rho(M)$ <u>can be realized as</u> $\tau^G(M \to B)$.

e) $Wh^G_\rho(M)$ <u>classifies</u> G-<u>diffeomorphism classes relative</u> M <u>of</u> h-<u>cobordisms over</u> M. □

<u>Proof:</u> a) We use induction over the orbit type and choose $(H) \in \operatorname{Con} G$, $\nu_M, \ldots$ as above. If we take in the diagram 4.48. the K-fixed point sets for $K \in \operatorname{Iso} \bar{B}$ we get a commutative diagram with push-outs as squares

4.52.

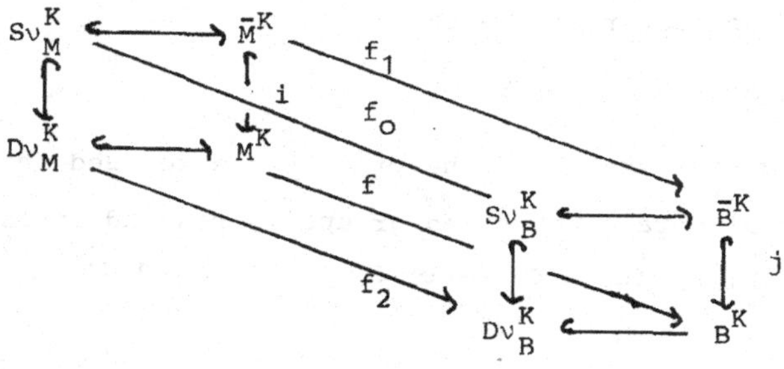

Since $M \to B$ is a G-homotopy equivalence, f_o, f_2 and f are homotopy equivalences because of $\nu_B \mid M_{(H)} = \nu_M$ and the covering homotopy theorem (see Bredon [1972] II.7.4.). By Lemma 4.50. $i : \bar{M}^K \to M^K$ and $j : \bar{B}^K \to B^K$ are 2-connected since $\bar{M}_K \to M_K$ and $\bar{B}_K \to B_K$ are homotopy equivalences. We want to show that f_1 is a weak homotopy equivalence. Suppose for simplicity that B^K is connected, the general case is done just componentwise. Let $\widetilde{B}^K$ be the universal covering. Pulling it back to each space appearing in 4.52. yields again a diagram with push-outs as squares and cofibrations as maps (Lemma 1.26.)

4.53.

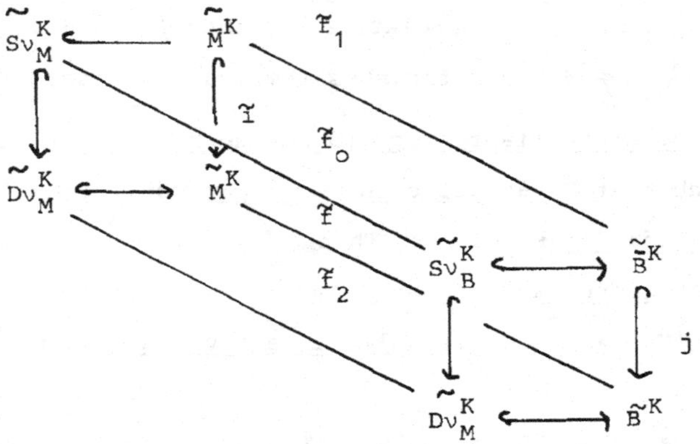

By a Mayer-Vietoris argument $\tilde{f}_1$ is a weak homology equivalence since $\tilde{f}_o$, $\tilde{f}_2$ and $\tilde{f}$ are homotopy equivalences. Since i, j, f_1 and f are 2-connected $\tilde{\bar{M}}^K$ is the universal covering of $\bar{M}^K$. By Whiteheads Theorem (see Whitehead [1978] IV.7.13. $\tilde{f}_1$ and hence f_1 is a weak homotopy equivalence. Since $\bar{M}^K \rightarrow \bar{B}^K$ is a weak homotopy equivalence for $K \in \text{Iso}\ \bar{B}$, we get from Lemma 4.39. a) that $(\bar{B}, \bar{M}, \bar{N})$ is an h-cobordism. By the induction hypotheses $(\bar{B}, \bar{M}, \bar{N})$ is even an isovariant h-cobordism. Because of Lemma 4.39. b) also (B, M, N) is an isovariant h-cobordism.

b) follows from c)

c) We can only give a sketch of the proof since we need the description of $Wh^G(X)$ by algebraic Whitehead groups introduced in section 14. There we establish for a G-space X an isomorphism

$$Wh^G(X) = \bigoplus_{(H)} \bigoplus_{C \in \pi_o(X^H)/WH} Wh(\pi_1(EWH(C) \times_{WH(C)} C)$$

where $WH(C)$ is the isotropy group of $C \in \pi_o(X^H)$ under the WH-action, $EWH(C)$ the universal $WH(C)$-principal bundle and $Wh(\mathbb{Z}\pi_1(EWH(C) \times_{WH(C)} C))$ the algebraic Whitehead group of the integral group ring of $\pi_1(EWH(C) \times_{WH(C)} C)$. Under this isomorphism the image $\Phi(M)$ is

4.54. $$Wh_\rho^G(M) = \bigoplus_{(H)} \bigoplus_{\substack{C \in \pi_o(M^H)/WH \\ C_H \neq \emptyset}} Wh(\pi_1(EWH(C) \times_{WH(C)} C))$$

One shows inductively over the orbit types that $\Phi(M)$ is injective with image $Wh_\rho^G(M)$. In the induction step we obtain a commutative diagram using the notation above

$$\begin{array}{ccc}
Wh_{Iso}^G(\bar{M}) \oplus Wh^{WH}(M_H) & \xrightarrow{\ i_* \oplus j\ } & Wh_{Iso}^G(M) \\
\Big\downarrow \begin{pmatrix} \Phi(\bar{M}) & \overline{trf} \\ 0 & id \end{pmatrix} & & \\
Wh_\rho^G(\bar{M}) \oplus Wh^{WH}(M_H) & & \Big\downarrow \Phi(M) \\
\Big\downarrow i_* \oplus k & & \\
Wh_\rho^G(M) & \xrightarrow{\ J\ } & Wh^G(M)
\end{array}$$

Here k and $\overline{trf}$ are given by the homomorphisms 4.45. and 4.46., j and J are the canonical split injections and i_* induced by the inclusion $i : \bar{B} \to B$. Since $i^K : \bar{B}^K \to B^K$ is 2-connected for $K \in Iso\ \bar{B}$ by Lemma 4.50., $i_* \oplus k$ is an isomorphism. This diagram and the induction hypothesis applied to $\Phi\ \phi(\bar{M})$ show that $\phi(M)$ is injective with image $Wh_\rho^G M$. □

As an illustration we give the smooth classification of semilinear discs. A <u>semi-linear</u> G-<u>disc</u> is a G-manifold M such that $(M^H, \partial M^H)$ is homotopy equivalent to (D^k, S^{k-1}) for appropriate $k \geq 0$ if $H \subset G$ is a (closed) subgroup of the compact Lie group G. Choose $x \in M^G$. Then the tangent space TM_x at x is a G-representation and there is an equivariant embedding

$\alpha : DTM_x \to M$ given by an exponential map. Let $\bar{M}$ be $M \setminus \text{int}\ \alpha(DTM_x)$. Suppose that M satisfies the weak gap conditions. Then $\bar{M}^K$ is simply connected for $K \subset G$. Using Lefschetz-Poincaré-duality, Lemma 4.39. and Theorem 4.42. we conclude that $(\bar{M}, \alpha(STM_x), \partial M))$ is an isovariant h-cobordism, whose G-diffeomorphism type rel$\alpha(STM_x)$ is independent of the choice of x and α. Here we use that M^G is simply connected. Moreover, the linear isomorphism class of TM_x is independent of x. Hence we can assign to the semi-linear disc M two algebraic invariants $([TM_x], pr_*\tau^G(\alpha(STM_x) \to \bar{M})) \in Rep_{\mathbb{R}}(G) \oplus Wh^G(*)$ where $pr_* : Wh^G(\bar{M}) \to Wh^G(*)$ is induced by the projection $pr : \bar{M} \to *$ and is an isomorphism since $\bar{M}^K$ is simply connected for $K \subset G$. Moreover, these invariants depend only on the G-diffeomorphism class. They even classify semi-linear discs as shown in Rothenberg [1978].

<u>Theorem 4.55. Let</u> M <u>and</u> N <u>be semi-linear discs satisfying the weak gap conditions and</u> $\dim M_H/WH \geq 6$, $\dim N_H/WH \geq 6$ for $H \in$ Iso M, Iso N. <u>Then</u> M <u>and</u> N <u>are</u> G-<u>diffeomorphic if and only if</u> $([TM_x], pr_*\tau^G(STM_x \to \bar{M}))$ <u>and</u> $([TN_y], pr_*\tau^G(STN_y \to \bar{N}))$ <u>in</u> $Rep_{\mathbb{R}}(G) \oplus Wh^G(*)$ <u>agree</u>. □

We have already mentioned that $Wh^G(*)$ is $\bigoplus_{(H)} Wh(\pi_o(WH))$ if $Wh(\pi_o(WH))$ is the algebraic Whitehead group of $\mathbb{Z}\pi_o(WH)$.

<u>Example 4.56.</u> We show by a counterexample that the weak gap conditions 4.49 are necessary for the Equivariant s-Cobordism Theorem 4.51. Notice that it suffices to construct a G-manifold M such that $\Phi(M) : Wh^G_{Iso}(M) \longrightarrow Wh^G_\rho(M)$ is not injective. Then we can find by the Isovariant s-Cobordism Theorem 4.42. an isovariant non-trivial h-cobordism (B,M,N) with $\tau^G(M \to B) = 0$.

Let G be $\mathbb{Z}/2$ and $p \geq 5$ be a prime number. Denote the 2-dimensional free G-representation by V and the m-dimensional trivial G-representation by $\mathbb{R}^m$ for some $m \geq 5$. Consider the embedding $f : S^1 \longrightarrow SV \times \mathbb{R}^m$, $z \longrightarrow z^p, (z, 0, \ldots, 0)$ where we identify $SV = S^1 \subset \mathbb{C}$ and $\mathbb{R}^m = \mathbb{C} \oplus \mathbb{R}^{m-2}$. The normal bundle of f is stably trivial and hence trivial for dimension reasons. Moreover $G \times S^1 \longrightarrow SV \times \mathbb{R}^m$, $(g,z) \longrightarrow g\cdot f(z)$ is a G-embedding since G acts freely on SV. Hence we can find a G-embedding $F : G\times S^1 \times D^m \longrightarrow SV \times \mathbb{R}^m$ whose restriction to $S^1 = \{1\}\times S^1 \times \{0\}$ is f. We identify $SV \times \mathbb{R}^m$ with $S(V \oplus \mathbb{R}^{m+1})_G$. Define a G-manifold M by the G-push out

$$\begin{array}{ccc} G\times S^1\times S^{m-1} & \longrightarrow & S(V\oplus\mathbb{R}^{m+1}) \setminus F(G\times S^1\times \operatorname{int} D^m) \\ \downarrow & & \downarrow \\ G\times D^2\times S^{m-1} & \longrightarrow & M \end{array}$$

Hence M is the result of equivariant surgery on $f : S^1 \longrightarrow S(V \oplus \mathbb{R}^{m+1})$. As $S(V \oplus \mathbb{R}^{m+1})$ and $S(V\oplus \mathbb{R}^{m+1})^G = S(\mathbb{R}^{m+1})$ are simply connected and $f : S^1 \longrightarrow S(V \oplus \mathbb{R}^{m+1})_G \simeq S^1$ has degree p, we get $\pi_1(M) = \pi_1(M^G) = \{1\}$ and $\pi_1(M_G) = \mathbb{Z}/p$. This implies $\pi_1(EG \times_G M) = G$ and $\pi_1(M_G/G) = \pi_1(EG\times_G M_G)$ is an extension of $\mathbb{Z}/p$ and G, actually, it is $\mathbb{Z}/p \times G$. If H is a finite group, the rank of Wh(H) is the difference of $\mathbb{R}$-conjugacy classes and $\mathbb{Q}$-conjugacy classes of elements in G (see Oliver [1988]). Recall that $h_1, h_2 \in H$ are $\mathbb{R}$-conjugated if the elements h_1 and h_2 or the elements h_1 and h_2^{-1} are conjugated, and that h_1 and h_2 are $\mathbb{Q}$-conjugated if and only if the cyclic subgroups $\langle h_1\rangle$ and $\langle h_2\rangle$ they generate are conjugated. Hence we get $Wh(\pi_1(M^G)) = Wh(\pi_1(EG\times_G M)) = \{0\}$ and $Wh(\pi_1(EG\times_G M_G)) \neq \{0\}$. This implies by the next result that $Wh^G_{Iso}(M) \neq \{0\}$ and $Wh^G_\rho(M) = \{0\}$ and we get the desired counterexample.

Let G be a compact Lie group and N be a G-manifold with $\operatorname{Iso} N = \{G,\{1\}\}$ such that N and N^G are connected. Let $i : EG\times_G N_G \longrightarrow EG\times_G N$ be the inclusion. We get from 4.40., 4.54., the definition of Φ and the results of section 14

4.57
$$\begin{array}{ccc} Wh^G_{Iso}(N) & \xrightarrow{\ \Phi(N)\ } & Wh^G_\rho(N) \\ \| & & \| \\ Wh(\pi_1(N^G)) \oplus Wh(\pi_1(EG\times_G N_G)) & \xrightarrow[\begin{pmatrix} id & 0 \\ trf & i_* \end{pmatrix}]{} & Wh(\pi_1(N^G)) \oplus Wh(\pi_1(EG\times_G N)) \end{array}$$

□

Finally we mention three technical results

Lemma 4.58. Consider the cellular G-push-outs

$$\begin{array}{ccc} \coprod_{j=1}^{r} G/H \times S^{n-1} & \xrightarrow{\quad q_i \quad} & A \\ \big\downarrow & & \big\downarrow \\ \coprod_{j=1}^{r} G/H \times D^{n} & \xrightarrow{\quad Q_i \quad} & B_i \end{array}$$

for $i = 0,1$. Suppose that q_0 and q_1 are G-homotopic. Then $B_0 \simeq B_1$ rel A.
Proof: Lemma 4.12. □

Proposition 4.59.

Let (X,Z) be a pair of G-CW-complexes which is relatively finite. Suppose that the inclusion is a G-homotopy equivalence. Let k be an intege with $k \geq \dim(X,Z) - 1$. Then there is a second pair (Y,Z) satisfying:

a) $(X,Z) \simeq (Y,Z)$ rel Z.

b) The relative G-CW-complex (Y,Z) has only cells in dimension k and k+

c) For an appropriate numeration $\{e_i^k \mid i = 1,2,\ldots,n\}$ and $\{e_i^{k+1} \mid i=1,\ldots$ of the open k- and (k+1)-cells of the relative G-CW-complex (Y,Z) we have $n = m$. Moreover, for each i there is a point $z_i \in Z$ and characteristic maps $(P_i,p_i) : G/H_i \times (D^k,S^{k-1}) \to Y_k$ and (Q_i,q_i) : $G/H_i \times (D^{k+1},S^k) \to Y_{k+1}$ such that $p_i(eH_i \times S^{k-1}) = \{z_i\}$ and $Q_i(eH_i \times S_+^k) = \{z_i\}$ where S_+^k is the upper hemi-sphere. □

Proof: Illman [1974]. □

Proposition 4.60. Let G be a compact Lie group and M_1, M_2 and N be G-manifolds with $\dim M_1 = \dim M_2 = 1 + \dim N$. Consider G-embeddings $i : N \to \partial M$ and $j : N \to \partial M_2$ and the G-push out

$$\begin{array}{ccc} N & \xrightarrow{\quad i \quad} & M_1 \\ {\scriptstyle j}\big\downarrow & \bar{j} & \big\downarrow{\scriptstyle \bar{i}} \\ M_2 & \xrightarrow{\qquad} & M \end{array}$$

Equip M with the structure of a G-manifold for which the maps $\bar{i}$ and $\bar{j}$ become G-embeddings (straighten the angle). Put on N, M_1, M_2 and M the simple G-structure constructed in 4.36. Then this G-push out is a G-push out of G-spaces with simple G-structures.

Proof: By induction over the orbit bundle the claim is reduced to the case of one orbit type which follows from the non-equivariant case. □

Comments 4.61. The geometric approach to the non-equivariant Whitehead torsion is given in Cohen [1973] and Stöcker [1970] and is generalized to the equivariant case in Illman [1974]. The notion of a simple G-homotopy equivalence is due to Whitehead [1939], [1941], [1949], [1952]. Later we will give an algebraic treatment.

More information about triangulations of G-manifolds can be found in Illman [1978], [1983] and Matumoto-Shiota [1987].

The non-equivariant s-Cobordism Theorem is proved in Barden [1963], Kirby-Siebenmann [1977] Essay II, Mazur [1963], Stallings [1968] for dim M ≥ 5. For dim M = 4 it fails in the smooth category (Donaldson [1987]) but is still true for "good" fundamental groups in the topological category (Freedman [1982], [1983]). Counterexamples for dim M = 3 are constructed in Cappell-Shaneson [1985].

The equivariant s-Cobordism Theorem 4.51. has analogoues in the PL- and TOP-category and is an important tool in equivariant surgery and the classification of G-manifolds. Let M and N be two G-manifolds, for which one wants to show that they are G-diffeomorphic.

The general strategy consists of three steps, firstly one constructs a cobordism (B,M,N) and a degree one normal map $(B,M,N) \to (M \times I, M \times 0, M \times 1)$, secondly changes it into a simple G-homotopy equivalence and finally applies the Equivariant s-Cobordism Theorem to the resulting h-cobordism (B',M,N). For more information see Araki [1986], Araki-Kabakubo [1988], Browder-Quinn [1975], Dovermann-Rothenberg [1988], Hauschild [1978], Lück-Madsen [1988a], [1988b], Rothenberg [1978], Steinberger [1988],

Steinberger-West [1985]. Controlled and bounded versions can be found in Anderson-Munkholm [1988], Pedersen [1986], Quinn [1979], [1982].

A different counterexample than 4.56. for the Equivariant s-Cobordism Theorem 4.51. without weak gap conditions can be obtaind from the construction of $\mathbb{Z}/p$-actions on S^n with non-trivial knots $S^{n-2} \longrightarrow S^n$ as fixed point sets proved by Giffen [1966] and Summers [1975] (cf. Kawakubo [1986]). □

<u>Exercises</u> 4. 62.

1. Suppose $X \wedge Y$. Then there is a finite G-CW-complex Z satisfying $X \nearrow Z \searrow Y$

2. Suppose that $A \subset X_o \subset X$ is a triple and $A \subset Y_o$ is a pair of finite G-CW-complexes and $f : X_o \rightarrow Y_o$ a cellular simple G-homotopy equivalence with $f|A = id$. Let Y be the G-push-out

$$\begin{array}{ccc} X_o & \xrightarrow{f} & Y_o \\ \downarrow & & \downarrow \\ X & \xrightarrow{F} & Y \end{array}$$

Then F is a simple G-homotopy equivalence and $Y \wedge X$ relative A.

3. Is $pr: DV \rightarrow \{*\}$ simple for a G-representation V?

4. Let $f : X \rightarrow Y$ be a G-homotopy equivalence between finite G-CW-complexes. Then $f \times id : X \times S^1 \rightarrow Y \times S^1$ is a simple G-homotopy equivalence.

5. Extend the definition of the Whitehead torsion to G-homotopy equivalences $(F,f) : (X,A) \rightarrow (Y,B)$ of pairs of G-CW-complexes which are relatively finite. How much carries over from Theorem 4.8.? If A and B are finite, does $\tau^G(F) = \tau^G(f) + \tau^G(F,f)$ hold?

6. Let $G \rightarrow E \rightarrow E/G$ be a principal G-bundle. Define by dividing out

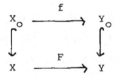

the group action and the pull-back-construction inverse isomorphisms

$$Wh^G(E) \xleftrightarrow{\cong} Wh(E/G)$$

7. Prove the topological invariance of the equivariant Whitehead torsion for free G-actions.

8. Suppose we have assigned to any compact Lie group G and any (smooth compact) G-manifold M a simple structure η_M such that the following holds

   i) $f : M \to N$ is a G-diffeomorphism. Then $\tau^G(f : (M,\eta_M) \to (N,\eta_N)) = 0$.

   ii) Let M_1 and M_2 be compact smooth G-manifolds and $i : M_o \to \partial M_1$ and $M_o \to \partial M_2$ G-embeddings of G-manifolds of the same dimension. The G-push-out $M = M_1 \cup_{M_o} M_2$ carries a canonical G-manifold structure. Then we have a G-push-out of G-spaces with simple structures

$$\begin{array}{ccc} (M_o,\eta_{M_o}) & \hookrightarrow & (M_1,\eta_{M_1}) \\ \downarrow & & \downarrow \\ (M_2,\eta_{M_2}) & \hookrightarrow & (M,\eta_M) \end{array}$$

   iii) If M has only one orbit type, η_M and ξ_M agree, where ξ_M is defined in 4.36.

   iv) The projection $pr : DV \to \{*\}$ has torsion zero with respect to η_{DV} for any G-representation V.

   Show that η_M agrees with ξ_M defined in 4.36.

9. Let G be a finite group. Show that the simple structures on a G-manifold M defined in 4.34. and 4.36. agree.

10. For which groups G semi-linear discs M satisfying the weak gap-conditions and $\dim M_H/MH \geq 6$ for $H \in \mathrm{Iso}\, M$ are classified by the dimensions $\dim M^H$, $H \subset G$?

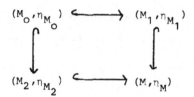

11) Let $p \geq 5$ be a prime number and G be $\mathbb{Z}/p$. Let V be a 2-dimensional free orthogonal G-representation and $\mathbb{R}^6$ be the trivial G-representation. Put $M = S(V \oplus \mathbb{R}^6)$. Show

a) $Wh^G_{Iso}(M) = \{0\}$

b) $Wh^G_\rho(M) = Wh^G(M) = Wh(\mathbb{Z}G) \neq \{0\}$

c) $\Phi : Wh^G_{Iso}(M) \longrightarrow Wh^G_\rho(M)$ is not surjective.

d) Any isovariant h-cobordism over M is trivial.

12) Let G be a compact Lie group with $\dim G > 0$. Consider a G-manifold M with $\mathrm{Iso}\, M = \{G,\{1\}\}$ possibly not satisfying the weak gap conditions 4.49. Suppose that M^G and M are connected. Show that $\Phi : Wh^G_{Iso}(M) \longrightarrow Wh^G(M)$ is injective (cf. Kawakubo [1988].

(Hint: If $3+\dim M^G \leq \dim M$ is not true, the inclusion $i : EG \times_G M_G \longrightarrow EG \times_G M$ has a section up to homotopy.)

13) Let G be a finite group and (B,M,N) an h-cobordism such that $\tau^G(M \to B)$ vanishes. Show the existence of a G-representation SV such that the h-cobordism $(B \times SV, M \times SV, N \times SV)$ is trivial (cf. Araki-Kawakubo [1988]).

5. The Euler characteristic

We introduce the notion of the equivariant Euler characteristic, study its main properties and define the Burnside and Euler ring of a compact Lie group.

Let G be a topological group. Given a G-space X let $\Pi_o(G,X)$ be the category having as objects G-maps $x : G/H \to X$. A morphism σ from $x : G/H \to X$ to $y : G/K \to X$ is a G-map $\sigma : G/H \to G/K$ with $y \circ \sigma \simeq_G x$. Let Is $\Pi_o(G,X)$ be the set of isomorphism classes $\bar{x}$ of objects x.

<u>Definition 5.1.</u> <u>We call</u> $\Pi_o(G,X)$ <u>the component category of</u> X. <u>Let</u> $U^G(X)$ <u>be the free abelian group generated by</u> Is $\Pi_o(G,X)$. □

We often think of an element η in $U^G(X)$ as a function η: Is $\Pi_o(G,X) \to \mathbb{Z}$ which takes only for finitely many elements values different from zero. By induction $U^G(X)$ becomes a covariant functor U^G : {G-spaces} → {abel. gr.}.

Given a G-space X and $x : G/H \to X$, let $X^H(x)$ be the path component of X^H containing x. By 1.30. we obtain a natural isomorphism sending the class of x to the one of $X^H(x)$

5.2.
$$\text{Is } \Pi_o(G,X) \to \coprod_{(H) \in \text{Con } G} \pi_o(X^H)/WH$$

In the sequel R is a ring such that the rank rk M of a finitely generated R-module is defined. We require $\text{rk } M_1 - \text{rk } M_o + \text{rk } M_2 = 0$ for any exact sequence of finitely generated R-modules $0 \to M_1 \to M_o \to M_2 \to 0$. Let H_* be a homology theory with values in the category of R-modules satisfying the Eilenberg-Steenrod axioms including the dimension axiom (see Whitehead [1978], XII.6.). We want to define for a G-pair (X,A) its equivariant Euler characteristic $\chi^G(X,A) \in U^G(X)$ provided that $A \to X$ is a G-cofibration and A and X are finitely dominated. Consider a G-CW pair (X,A).

If e_i^n is an open cell of X, let $u_i^n \in \operatorname{Is}\Pi_o(G,X)$ be given by $Q_i^n|\, G/H_i \times *$ for any characteristic map Q_i^n of e_i^n. Given $\bar{x}, \bar{y} \in \operatorname{Is}\Pi_o(G,Y)$ we write $\bar{x} \leq \bar{y}$ if there is a morphism $x \to y$. For $x\colon G/H \to X$ let $X^{(H)}(x) \subset X$ be the union of the open cells e_i^n satisfying $\bar{x} \leq u_i^n$. We have $u_i^n \leq y$ for any point $y\colon G/G_y \to X$ in ∂e_i^n. Hence $X^{(H)}(x)$ is a G-CW-subcomplex of X (see 1.28). If the homology $H_*(Y,B)$ is finitely generated, the ordinary Euler characteristic $\chi(Y,B)$ is defined by $\sum_n (-1)^n \operatorname{rk}_R H_n(Y,B)$. Let (X,A) be a pair of finitely dominated G-CW-complexes. Let $\chi^G(X,A) \in U^G(X)$ be the function sending $\bar{x}$ represented by $x\colon G/H \to X$ to $\chi((X^{(H)}(x),\ X^{(H)}(x) \cap (X^{>(H)} \cup A))/G)$. This definition can be extended to a G-NDR-pair (X,A) of finitely dominated G-spaces. Namely, we can choose a pair of finitely dominated G-CW-complexes (X',A') and a G-homotopy equivalence of pairs $f\colon (X',A') \to (X,A)$ by Proposition 2.12 and define $\chi^G(X,A) := f_*\chi^G(X',A')$.

<u>Definition</u> 5.3. <u>We call $\chi^G(X,A) \in U^G(X)$ the equivariant Euler characteristic.</u> □

We want to show that this is well-defined and the following results

<u>Theorem</u> 5.4.

a) <u>Homotopy invariance</u>

i) $f \simeq_G g\colon X \to Y \Rightarrow f_* = g_*\colon U^G(X) \to U^G(Y)$

ii) If $f\colon (X,A) \to (Y,B)$ is a G-<u>homotopy equivalence of</u> G-NDR <u>pairs of finitely dominated</u> G-<u>spaces then</u> $f_*\chi^G(X,A) = \chi^G(Y,B)$.

b) <u>Consider the following commutative diagram of finitely dominated</u> G-<u>spaces such that the squares are</u> G-<u>push outs and</u> i_1, j_1, l_o, l_1, l_2 <u>and</u> l <u>are</u> G-<u>cofibrations</u>

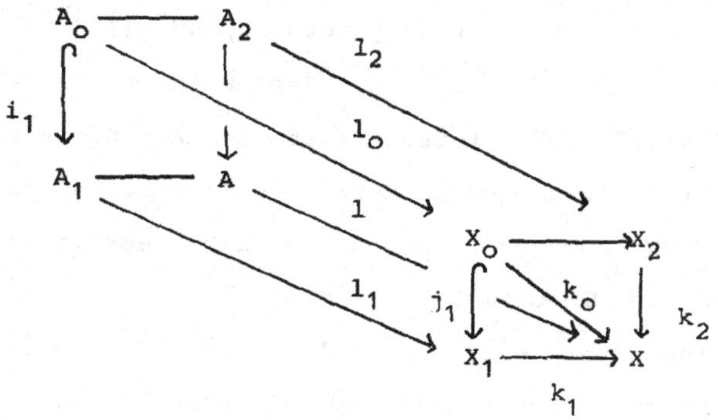

Then: $\chi^G(X,A) = k_{1*}\ \chi^G(X_1,A_1) + k_{2*}\ \chi^G(X_2,A_2) - k_{o*}\chi^G(X_o,A_o)$

c) Let (X,A) be a pair of finitely dominated G-spaces such that $i : A \to X$ is a G-cofibration. Then

$$\chi^G(X) = \chi^G(X,A) + i_*\ \chi^G(A)$$

□

Suppose that (X,A) is a pair of finitely dominated G-CW-complexes such that $X\setminus A$ is relatively finite.

Let

5.5. $\beta_n(X,A) \in U^G(X)$

send $\bar{x} \in \operatorname{Is} \Pi_o(G,X)$ to the number of open cells $e_i^n \subset X\setminus A$ with $\bar{x} = u_i^n$.

Lemma 5.6. $\chi^G(X,A) = \sum_{n\geq o} (-1)^n\ \beta_n(X,A)$ □

Consider a pair of finitely dominated G-CW-complexes (X,A) such that $X\setminus A$ is relatively finite. Then $(X^{(H)}(x),\ X^{(H)}(x) \cap (X^{>(H)}(x) \cup A))/G$ is a relatively finite pair of CW-complexes and only for finitely many $\bar{x} \in \operatorname{Is} \Pi_o(G,X)$ the complement is not emptv. Hence $H_*((X^{(H)}(x),\ X^{(H)}(x) \cap (X^{>(H)}(x) \cup A))/G)$ is zero for all except a finite number of elements

$\bar{x} \in \mathrm{Is}\, \Pi_o(G,X)$ and always finitely generated, and $\chi((X^{(H)}(x), X^{(H)}(x) \cap (A \cup X^{>(H)}))/G)$ is $\Sigma\, (-1)^n\, \beta_n(X,A)(\bar{x})$ (see Dold [1972] V.5.9.). Hence $\chi^G(X,A)$ is defined in this case and Lemma 5.6. is true. If (X,A) is any pair of finitely dominated G-CW-complexes it is dominated by a pair of finite G-CW- complexes. Now one easily checks that Definition 5.3. makes sense and Theorem 5.4a is true. It suffices to show b) in the case where the G-push outs are cellular G-push outs of finitely dominated G-CW-complexes. For simplicity we assume $A = \emptyset$. Fix $x : G/H \to \Pi_o(G,X)$. Let J_i for $i = 0, 1, 2$ be $\{\bar{y} \in \mathrm{Is}\, \Pi_o(G,X_i) \mid \overline{k_i \circ y} = \bar{x}\}$ and C_i be the direct sum of cellular chain complexes $\bigoplus_{J_i} C^c((X_i^{(H)}(y), X_i^{(H)}(y) \cap X^{>(H)})/G)$. We obtain a based exact sequence $0 \to C_o \to C_1 \oplus C_2 \to C^c((X^{(H)}(x), X^{(H)}(x) \cap X^{>(H)})/G) \to 0$. This implies for $\varepsilon_o = -1$, $\varepsilon_1 = \varepsilon_2 = +1$

$$\chi((X^{(H)}(X), X^{(H)}(X) \cap X^{>(H)})/G) =$$

$$\sum_{i=o}^{2} \varepsilon_i \cdot \sum_{J_i} \chi((X_i^{(H)}(Y), X_i^{(H)}(Y) \cap X_i^{>(H)})/G)$$

This shows Theorem 5.4 b and Theorem 5.4 c follows □

5.7. Sometimes we want to drop the condition that (X,A) has the G-homotopy type of a finitely dominated G-CW-complex. Then one needs the assumptions:

i) G is a Lie group.

ii) X is proper and completely regular and has finite orbit type.

iii) The inclusions $A^{>H} \to A^H$, $X^{>H} \to X^H$ and $A^H \cap X^{>H} \to X^H$ are WH-cofibrations and $i : A \to X$ is a G-cofibration.

iv) X^H/WH is the topological sum of finitely many path components or H_* is singular homology.

v) $H_n((X^H, X^{>H} \cap A^H)/WH)$ is finitely generated for $n \geq 0$ and zero for large n

Then $\pi_o(X^H)/WH \to \pi_o(X^H/WH)$ is a bijection for $H \subset G$ by Corollary 1.40.
Hence each $x : G/H \to X$ determines a path component $C \subset X^H/WH$.
We have $H_*((X^H, X^{>H} \cup A^H)/WH) = \oplus H_*((C, C \cap (X^{>H} \cup A^H))/WH)$ if $\oplus$ runs over $C \in \pi_o(X^H/WH)$ because of iv) (see Whitehead [1978] XII. 6.8., Schubert [1964]. IV. 1.7. Satz 4.). Now define $\chi^G(X,A) \in U^G(X)$ by $\chi^G(X,A)(\bar{x}) = \chi((C, C \cap (X^{>H} \cup A^H))/WH)$. Then the analogue of Theorem 5.4 is true because of the Map Excision Theorem (Whitehead [1978] XII 6.7). If X and A are furthermore finitely dominated G-spaces, this definition and Definition 5.3. agree. One can also use a cohomology theory H^* satisfying the Eilenberg-Steenrod axioms including the dimension axiom (see Whitehead [1978] XII.7) if in iv) the first condition is safisfied □

Let $\mathfrak{F}$ be a subset of $S(G) = \{H \subset G \mid H$ a (closed) subgroup$\}$ closed under conjugation ($H \in \mathfrak{F}$, $g \in G \Rightarrow g^{-1}Hg \in \mathfrak{F}$) and intersection ($H,K \in \mathfrak{F} \Rightarrow H \cap K \in \mathfrak{F}$) and containing G. Consider G-spaces X of the G-homotopy type of a finite G-CW-complex Y satisfying Iso $Y \subset \mathfrak{F}$. Define equivalence relations

5.8 $$X \sim Y \Leftrightarrow \chi(X^H) = \chi(Y^H) \text{ for all } H \subset G$$

5.9 $$X \sim Y \Leftrightarrow \chi(X^H/WH) = \chi(Y^H/WH) \text{ for all } H \subset G$$

where χ is the ordinary Euler characteristic $\chi(Z) = \Sigma(-1)^n \, rk_R \, H_n(Z)$. Let $A(G,\mathfrak{F})$ resp. $U(G,\mathfrak{F})$ be the set of equivalence classes. It becomes an associative commutative ring with unit G/G by disjoint union and cartesian product with diagonal G-action. Here we use that for a compact Lie group G and finite G-CW-complexes X and Y with Iso X and Iso $Y \subset \mathfrak{F}$ the G-space $X \times Y$ is G-homotopy equivalent to a finite G-CW-complex Z with Iso $Z \subset \mathcal{F}$. If X is a finite G-CW-complex with Iso $X \subset \mathcal{F}$, let [X] be its class in $A(G,\mathcal{F})$ or $U(G,\mathcal{F})$. If Y is any finite CW-complex with $\chi(Y) = -1$, the inverse of [X] under addition is $[X \times Y]$. If $\mathcal{F}$ is S(G) we write A(G) resp. U(G).

Definition 5.10. A(G) is called the Burnside ring and U(G) the Euler ring of the compact Lie group G. □

Let * be the point with trivial G-action. Let

5.11. $$\Phi : U^G(*) \to U(G)$$

be the homomorphism sending $\eta : \text{Con } G \to \mathbb{Z}$ to $\sum_{(H) \in \text{Con } G} \eta(H) \circ [G/H]$

Define

5.12. $$\psi : U(G) \to U^G(*)$$

by $\psi([X]) = pr_*(\chi^G(X))$ for $pr : X \to *$ the canonical projection.

Proposition 5.13. Φ and ψ are well-defined inverse isomorphisms.

Proof: Obviously Φ is well-defined. Surjectivity follows from the next formula in U(G) for a finite G-CW-complex X

5.14. $$[X] = \sum_{(H) \in \text{Con } G} \chi((X^H, X^{>H})/WH) \cdot [G/H]$$

One proves 5.14. inductively over the number of cells using Theorem 5.4. Next we prove injectivity. Let $\sum_{(H) \in \text{Con } G} \eta(H) \cdot [G/H]$ be zero in U(G). Assume that there is an $(K) \in \text{Con } G$ with $\eta(K) \neq 0$. Choose (K) with $\eta(K) \neq 0$ such that $(K) \subset (H)$, $(K) \neq (H)$ implies $\eta(H) = 0$. Then we get

$$\sum_{(H) \in \text{Con } G} \eta(H) \cdot \chi((G/H^K)/WK) = \eta(H)$$

but this sum must be zero, a contradiction. Hence Φ is a well-defined isomorphism. One easily checks $\Phi \circ \psi = \text{id}$. □

We mention the so called <u>character maps</u>

5.15.
$$ch : U(G) \to \prod_{\text{Con } G} \mathbb{Z} \quad [X] \to (\chi(X^H/WH))_{(H)}$$

$$ch : A(G) \to \prod_{\text{Con } G} \mathbb{Z} \quad [X] \to (\chi(X^H))_{(H)}$$

Both are homomorphisms of abelian groups, but only the second is a ring homomorphism in general.

<u>Comments 5.16.</u> The equivariant Euler characteristic will appear in the various product formulas for the finiteness obstruction and the Whitehead torsion.

The Burnside ring of a finite group is introduced in Dress [1969] as the Grothendieck construction applied to the semi-ring of finite G-sets. Its definition for a compact Lie group is given in tom Dieck [1975]. For a detailed treatment of the Burnside and Euler ring we refer to tom Dieck [1979] and [1987].

The Burnside ring A(G) is isomorphic to the stable cohomotopy of spheres in dimension zero ω_G^0 (see Segal [1971], tom Dieck [1979] 8.5.1.). Let X be a G-homotopy representation (e.g. the unit sphere in a G-representation) with dim $X^G \geq 1$. If Iso X is closed under intersection and X satisfies the weak gap conditions, $[X,X]^G$ is isomorphic to A(G, Iso X) by the equivariant Lefschetz index (see Lück [1986a]). More information about the (unstable) G-homotopy classes of G-maps between G-homotopy representations can be found in tom Dieck [1987] II. 4. and II. 10., Laitinen [1986], Rubinsztein [1973], Tornehave [1982].

The Burnside ring plays an important role in induction theory (see Dress [1973] and [1975], tom Dieck [1979] 6., tom Dieck [1987] IV. 8.9.). A lot of modules like stable equivariant homology groups are modules over the Burnside ring and it is useful to study their localizations at the prime ideal of the Burnside ring (see tom Dieck [1987]

IV. 4 + 10).

We will characterize the Euler ring by a universal property in the next section. Further references for equivariant Euler characteristics are Brown [1974], [1975], [1982]. For equivariant Lefschetz indices see Laitinen-Lück [1987]. □

Exercises 5.17.

1. Let SO(3) act on S^2 by evaluation at (0,0,1). Compute $U^{SO(3)}(S^2)$ and $\chi^{SO(3)}(S^2)$.

2. Let H be a (closed) subgroup of the compact Lie group G and X be a G-space. Let $\Phi(X) : U^G(X) \to U^H(\text{res } X)$ send the base element given by $x : G/K \to X$ to the image of $\chi^H(\text{res } G/K)$ under $U^H(\text{res } X) : U^H(\text{res } G/K) \to U^H(\text{res } X)$. Show that we obtain a well-defined natural transformation $U^G \to U^H \circ \text{res}$. It is uniquely determined by the property that $\Phi(X)(\chi^G(X)) = \chi^H(\text{res } X)$ holds for any finite G-CW-complex X

3. Compute the ring $A(A_5)$.

4. Show that we still get U(G) if we use instead of 5.9. the equivalence relation $X \sim Y \Leftrightarrow \chi(X^H/WH_o) = \chi(Y^H/WH_o)$ for all $H \subset G$. Here WH_o is the component of the unit.

5. The map $U(G) \to A(G)$ $[X] \to [X]$ is a well-defined epimorphism and a bijection if G is finite.

6. $\{[G/H] \mid (H) \in \text{Con } G, \text{ WH finite}\}$ is a $\mathbb{Z}$-base for A(G).

7. Let G be finite. The character map $ch : A(G) \to \prod_{\text{Con } G} \mathbb{Z}, [X] \to (\chi(X^H) \mid (H) \in \text{Con } G)$ is an injective ring homomorphism. Its cokernel is finite of order $\prod_{(H) \in \text{Con } G} |WH|$.

8. Let G be finite. Prove the existence of numbers n(H,K) with the following property. Let $m = (m(K) \mid (K) \in \text{Con } G)$ be an element in $\prod_{\text{Con } G} \mathbb{Z}$. It lies in the image of the character map if and only if

for any $(H) \in \mathrm{Con}\, G$ the congruence

$$\Sigma\, n(H,K) \cdot m(K) \equiv 0 \quad |WH|$$

holds, where the sum runs over $(K) \in \mathrm{Con}\, WH$ such that H is normal in K and K/H cyclic. Moreover, $n(H,H)$ is 1.

9. If X is a finite G-CW-complex, let $\eta^G(X)$ be the element $\Sigma(-1)^n$ $[H_n(X,\mathbb{Q})]$ in the rational representation ring $\mathrm{Rep}_{\mathbb{Q}}(\pi_0 G))$ Show that we obtain a well-defined homomorphism

$$A(G) \rightarrow \mathrm{Rep}_{\mathbb{Q}}(\pi_0(G)) \quad [X] \rightarrow \eta^G(X)$$

and that it is an isomorphism if and only if G is the product of a torus and a cyclic group.

10. Prove that we obtain an homomorphism from the additive group of the real representation ring into the multiplicative group of units in the Burnside ring

$$\mathrm{Rep}_{\mathbb{R}}(G) \rightarrow A(G)^*$$

by $[V] \rightarrow [G/G] - [SV]$.

11. Show that the kernel of $U(G) \rightarrow A(G)\ [X] \mapsto [X]$ is the nilradical of $U(G)$.

6. Universal functorial additive invariants

We characterize the invariants introduced above like finiteness obstruction, Whitehead torsion and Euler characteristic by a universal property involving only homotopy invariance and additivity.

We recall the notion of a <u>category with cofibrations and weak equivalences</u> defined in Waldhausen [1985] 1.2. A category $\mathcal{C}$ is <u>pointed</u> if it has a distinguished initial object and a distinguished terminal object $*$, i.e. $\mathrm{Mor}(\emptyset,X)$ and $\mathrm{Mor}(X,*)$ consist of precisely one element for any object X in $\mathcal{C}$ (In Waldhausen [1985] furthermore $\emptyset = *$ is assumed but we drop this condition). A category with cofibrations and weak equivalences is a small pointed category $\mathcal{C}$ together with subcategories co $\mathcal{C}$ and $w\mathcal{C}$ satisfying:

Cof 1: Isomorphisms in $\mathcal{C}$ are cofibrations (i.e. belong to co $\mathcal{C}$).

Cof 2: $\emptyset \to X$ is a cofibration for all objects X.

Cof 3: If $i : X_o \to X_1$ is a cofibration and $f : X_o \to X_2$ any morphism then the following push-out exists and j is a cofibration

$$\begin{array}{ccc} X_o & \xrightarrow{\ f\ } & X_2 \\ {\scriptstyle i}\big\downarrow & & \big\downarrow{\scriptstyle j} \\ X_1 & \xrightarrow{\ g\ } & X \end{array}$$

Weq 1: Isomorphisms are weak equivalences.

Weq 2: If the following diagram

$$\begin{array}{ccccc} X_1 & \overset{i_1}{\hookleftarrow} & X_o & \longrightarrow & X_2 \\ {\scriptstyle f_1}\big\downarrow & & {\scriptstyle f_o}\big\downarrow & & \big\downarrow{\scriptstyle f_2} \\ Y_1 & \overset{j_1}{\hookleftarrow} & Y_o & \longrightarrow & Y_2 \end{array}$$

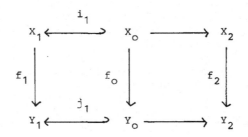

commutes, i_1 and j_1 are cofibrations and f_o, f_1 and f_2 are weak equivalences, then the map $f : X \to Y$ between the push-outs is a weak equivalence.

Examples are given later in this section.

The following notion is well-known. An <u>additive invariant</u> (A,a) for a category $\mathcal{C}$ with cofibrations and weak equivalences is an abelian group A together with a function assigning to an object X an element $a(X) \in A$ such that the following holds:

a) <u>Homotopy invariance</u>.

If $f : X \to Y$ is a weak equivalence then $a(X) = a(Y)$.

b) <u>Additivity</u>.

Let the commutative diagram

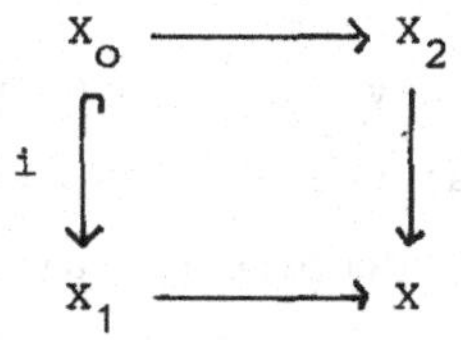

be a push-out with i a cofibration. Then $a(X) = a(X_1) + a(X_2) - a(X)$.

c) <u>Normalization</u>.

$a(\emptyset) = 0$

We call an additive invariant (U,u) <u>universal</u> if for any additive invariant (A,a) there is a map $\Phi: U \to A$ uniquely determined by the property that $\Phi(u(X)) = a(X)$ holds for any object X.

This can be generalized by substituting A by a functor. A <u>functorial additive invariant</u> (A,a) for $\mathcal{C}$ consists of a functor $A : \mathcal{C} \to \{\text{abel.gr.}\}$ and a function associating to an object X an element $a(X) \in A(X)$ satisfying.

a) Homotopy invariance.

If $f : X \to Y$ is a weak equivalence then $A(f)(a(X)) = a(Y)$.

b) Additivity.

Let the commutative diagram

$$\begin{array}{ccc} X_o & \longrightarrow & X_2 \\ {\scriptstyle i}\downarrow & \searrow^{k_o} & \downarrow{\scriptstyle k_2} \\ X_1 & \xrightarrow{k_1} & X \end{array}$$

be a push out and i be a cofibration, then:

$$a(X) = A(k_1)(a(X_1)) + A(k_2)(a(X_2)) - A(k_o)(a(X_o))$$

c) Normalization.

$a(\emptyset) = 0$

We call a functorial additive invariant (U,u) universal if for any functorial additive invariant (A,a) there is exactly one natural transformation $\Phi : U \to A$ satisfying $\Phi(X)(u(X)) = a(X)$ for all objects X.

Each additive invariant can be regarded as a functorial additive invariant. We can assign to a functorial additive invariant (A,a) an additive invariant $(\widehat{A},\widehat{a})$ by $\widehat{A} := A(*)$ and $\widehat{a}(X) := A(pr : X \to *)(a(X))$.

Theorem 6.1.

a) There is a universal functorial additive invariant (U,u) unique up to natural equivalence.

b) There is a universal additive invariant unique up to isomorphism. It is given by $(\widehat{U},\widehat{u})$.

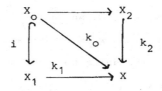

c) Given an object $Y \in \mathcal{C}$, let $\mathcal{C}(Y)$ be the category with cofibrations and weak equivalences having objects over Y, i.e. morphisms $f : X \to Y$ with target Y, as objects. Let $(U(Y),u_Y)$ be the universal additive invariant. Given a morphism $f : Y \to Z$ there is a homomorphism $U(f) : U(Y) \to U(Z)$ uniquely determined by $U(f)(u_Y(g)) = u_Z(f \circ g)$ for all objects g in $U(Y)$ because of the universal property. Define for an object X in $\mathcal{C}$ $u(X) \in U(X)$ by $u_X(id : X \to X)$.

Then (U,u) is the universal functorial additive invariant of $\mathcal{C}$.

Proof: We only explain the construction of the universal functorial additive invariant (U,u). Let $U(Y)$ be the quotient of the free abelian group generated by all morphisms $f : X \to Y$ with target Y and the subgroup generated by

$[f] - [g]$, if there is a weak equivalence h with $f \circ h = g$

$[f] - [f_1] - [f_2] + [f_o]$, if there is a commutative diagram with a push-out as square and i a cofibration

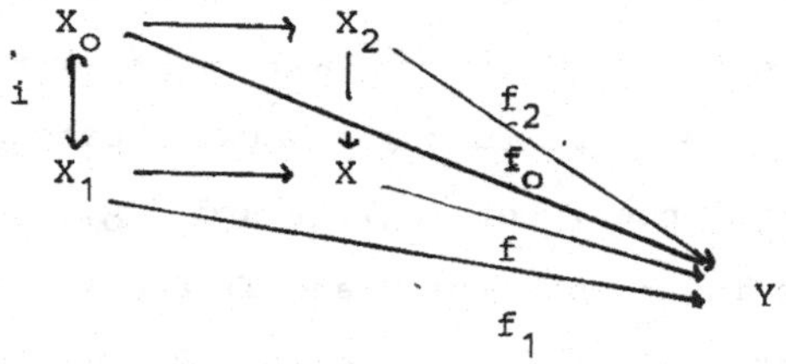

$[\emptyset \to Y]$

Define $U(g) : U(Y) \to U(Z)$ for a morphism $g : Y \to Z$ by composition and $u(Y) = [id : Y \to Y] \in U(Y)$. □

A functor $F : \mathcal{C} \to \mathcal{D}$ between categories with cofibrations and weak equivalences is a functor of pointed categories sending $co\,\mathcal{C}$ to $co\,\mathcal{D}$ and $w\mathcal{C}$ to $w\mathcal{D}$ such that F applied to a push-out with i a cofibration

$$\begin{array}{ccc} X_o & \longrightarrow & X_2 \\ {\scriptstyle i}\big\downarrow & & \big\downarrow \\ X_1 & \longrightarrow & X \end{array}$$

is again a push-out. If (A,a) is a (functorial) additive invariant for $\mathcal{D}$ we obtain a (functorial) additive invariant $F^*(A,a)$ for $\mathcal{C}$ by $A \circ F$ and $X \to a(F(X))$. Let $(U_{\mathcal{C}}, u_{\mathcal{C}})$ be the universal functorial additive invariant for $\mathcal{C}$.

The next result follows directly from the universal property and will be a useful tool for establishing various formulas.

Lemma 6.2. There is exactly one natural transformation $\Phi : U_{\mathcal{C}} \to A$ *with* $\Phi(X)(u_{\mathcal{C}}(X)) = a(F(X))$ *for all objects in* $\mathcal{C}$. □

Let $\mathcal{C}$, $\mathcal{D}$ and $\mathcal{E}$ be categories with cofibrations and weak equivalences. Consider a functor $\mathcal{C} \times \mathcal{D} \longrightarrow \mathcal{E}$ $X,Y \to X \times Y$ such that $X \times ? : \mathcal{D} \to \mathcal{E}$ and $? \times Y : \mathcal{C} \to \mathcal{E}$ are functors between categories with cofibrations and weak equivalences. Let $(U_{\mathcal{C}}, u_{\mathcal{C}})$ and $(U_{\mathcal{D}}, u_{\mathcal{D}})$ be the universal functorial additive invariants for $\mathcal{C}$ and $\mathcal{D}$ and (V,v) an arbitrary functorial additive invariant for $\mathcal{E}$. By Lemma 6.2. we obtain for any object Y in $\mathcal{D}$ a natural transformation $t(Y) : U_{\mathcal{C}}(?) \to V(? \times Y)$ uniquely determined by $t(Y)(u_{\mathcal{C}}(X)) = v(X \times Y)$ for all objects X in $\mathcal{C}$. Let $T(Y)$ be the abelian group of natural transformations $U_{\mathcal{C}}(?) \to V(? \times Y)$. Varying Y yields a functor $T : \mathcal{D} \to \{\text{abel. gr.}\}$. Since (T,t) is a functorial additive invariant for $\mathcal{D}$ there is exactly one natural transformation $\Phi : U_{\mathcal{D}} \to T$ such that $\Phi(Y)(u_{\mathcal{D}}(Y)) = t(Y)$ is valid for all Y in $\mathcal{D}$. We have shown:

Lemma 6.3. There is exactly one natural pairing

$$P(X,Y) : U_{\mathcal{C}}(X) \otimes U_{\mathcal{D}}(Y) \to V$$

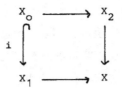

<u>satisfying</u> $P(X,Y)(u_C(X) \otimes u_{\mathcal{D}}(Y)) = v(X \times Y)$ <u>for all</u> X <u>in</u> C <u>and</u> Y <u>in</u> $\mathcal{D}$. □

<u>Corollary 6.4.</u> <u>Assume that</u> C <u>has an internal product such that</u> $? \times Y : C \to C$ <u>is a functor between categories with cofibrations and weak equivalences</u>. <u>Then there is a natural pairing</u>

$$P(X,Y) : U_C(X) \otimes U_C(Y) \to U_C(X \times Y)$$

<u>uniquely determined by the property that</u> $P(X,Y)(u_C(X) \otimes u_C(Y)) = u_C(X \times Y)$ <u>for all</u> X <u>and</u> Y <u>in</u> C. □

<u>Remark 6.5.</u> Under the conditions of Corollary 6.4. $U_C(*)$ becomes an associative ring with unit $u_C(*)$ and U_C a functor $C \to \{U_C(*)\text{-modules}\}$. □

Now we give some examples of categories with cofibrations and weak equivalences and compute some of the universal (functorial) additive invariants in this section or later. Let G be a topological group.

6.6. C = {G-spaces of the homotopy type of a finite G-CW-complex}

co C = {G-cofibrations}

w C = {G-homotopy equivalences}

This is a category with cofibrations and weak equivalences by Lemma 1.11. and Lemma 2.13.

<u>Theorem 6.7.</u> <u>The universal functorial additive invariant is</u> (U^G, χ^G).

<u>Proof:</u> (U^G, χ^G) is a functorial additive invariant by Theorem 5.4. It remains to verify the universal property. Given any functorial additive

invariant (B,b) we define a natural transformation $F : A \to B$ as follows. Given a G-space X and $\eta \in U^G(X)$ represented by a function $\mathrm{Is}\,\Pi_0(G,X) \to \mathbb{Z}$, the homomorphism $F(X) : U^G(X) \to B(X)$ sends η to $\Sigma\eta(x : G/H \to X) \cdot B(x)(b(G/H))$ where the sum runs over $\mathrm{Is}\,\pi_0(G,X)$. We must show that this is compatible with $\sim$. If $\sigma : G/K \to G/H$ is a G-homeomorphism and $h : G/K \times I \to X$ a G-homotopy between $y : G/K \to X$ and $x \circ \sigma$, this follows from homotopy invariance and the commutative diagram

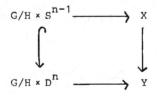

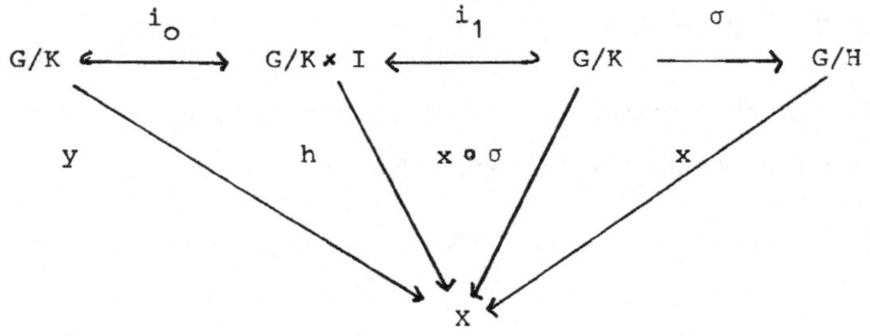

One should notice that for two G-homotopic G-maps f and g, $B(f) = B(g)$ must not necessarily be true.

Since each element in $U^G(X)$ is a linear $\mathbb{Z}$-combination of elements $U^G(x : G/H \to X)(\chi^G(G/H))$, the natural transformation F is the only one satisfying $F(G/H)(\chi^G(G/H)) = b(G/H)$ for all G/H. It remains to prove $F(X)(\chi^G(X)) = b(X)$ for any finite G-CW-complex. We do this inductively over the dimension n of X and subinduction over the number k of cells of maximal dimension. The induction begin $n = 0$ and $k = 0,1$ is already verified. Suppose that the assertion is true for X and Y is obtained from X by attaching a cell of dimension n

$$\begin{array}{ccc} G/H \times S^{n-1} & \longrightarrow & X \\ \downarrow & & \downarrow \\ G/H \times D^n & \longrightarrow & Y \end{array}$$

Apply additivity to this G-push-out, homotopy invariance to

$G/H \to G/H \times D^n$ and the induction hypotheses to G/H, $G/H \times S^{n-1}$ and Y. □

6.8. C = {G-spaces of the homotopy type of a finitely dominated G-CW-complex}

co C, w C as in 6.6.

<u>Theorem 6.9. The universal functorial additive invariant is</u> $(Wa^G \oplus U^G, (w^G, \chi^G))$.

<u>Proof:</u> $(Wa^G \oplus U^G, (w^G, \chi^G))$ is a functorial additive invariant by Theorem 3.2. and Theorem 5.4. Let (B,b) be any functorial additive invariant and suppose that $F : Wa^G \oplus U^G \to B$ is a natural transformation satisfying $F(Y)(w^G(Y), \chi^G(Y)) = b(Y)$. We show that then F is already determined. Namely, let $f : X \to Y$ represent $[f] \in Wa^G(Y)$ and $\eta : \{G/? \to Y\}/\sim$ be an element in $U^G(Y)$. Consider the computation where the sums run over Is $\Pi_o(G/Y)$ resp. Is $\Pi_o(G,X)$

$$F(Y)([f],\eta)$$

$$= F(Y)(0,\eta) - F(Y)(0,U^G(f)(\chi^G(X))) + F(Y)([f],U^G(f)(\chi^G(X)))$$

$$= \Sigma\eta(y) \cdot F(Y)(0,U^G(y)(\chi^G(G/H))) - F(Y)(0,U^G(f)(\Sigma\chi^G(X)(x) \cdot U^G(x)(\chi^G(G/H)))) + F(Y) \circ (Wa^G(f) \oplus U^G(f))(w^G(Y),\chi^G(Y))$$

$$= \Sigma\eta(y) \cdot B(y)(b(G/H)) - \Sigma\chi^G(X)(x) \cdot B(f \circ x)(b(G/H)) + B(f)(b(X)).$$

To prove the existence of F, we define F(Y) just by the formula above. The verification that this is well-defined is left to the reader. Obviously $F(Y)(w^G(Y),\chi^G(Y)) = b(Y)$ holds. □

6.10. C = {G-homotopy equivalences $f : X \to Y$ between finite G-CW-complexes}

Morphisms in $\mathcal{C}$ are given by commutative squares with g_o and g_1 cellular

$$\begin{array}{ccc} X_o & \xrightarrow{f_o} & Y_o \\ {\scriptstyle g_o}\downarrow & & \downarrow{\scriptstyle g_1} \\ X_1 & \xrightarrow{f_1} & Y_1 \end{array}$$

Such a morphism is a cofibration if g_o and g_1 are inclusions of pairs of G-CW-complexes and a weak equivalence if g_o and g_1 are simple G-homotopy equivalences.

<u>Theorem 6.11. The universal functorial additive invariant is</u> $(Wh^G \oplus U^G, (\tau^G, \chi^G))$.

<u>More precisely, it is given by the functor sending a morphism as above to</u> $Wh^G(g_1) \oplus U^G(g_1) : Wh^G(Y_o) \oplus U^G(Y_o) \to Wh^G(Y_1) \oplus U^G(Y_1)$ <u>and assigning to an object</u> $f : X \to Y$ <u>the element</u> $(\tau^G(f), \chi^G(Y)) \in Wh^G(Y) \oplus U^G(Y)$.

<u>Proof:</u> This is a functorial additive invariant by Theorem 4.8. and Theorem 5.4. Let (B,b) a functorial additive invariant. Suppose for the natural transformation $F : Wh^G \oplus U^G \to B$ that $F(f)(\tau^G(f), \chi^G(Y)) = b(f)$ holds for any object $f : X \to Y$. Then F is already determined by the following calculation for any object $f : X \to Y$. Let $r : Z \to Y$ be any strong G-deformation retraction for a pair of finite G-CW-complexes representing $[Z,Y] \in Wh^G(Y)$ and η be an element in $U^G(X)$. Since $U^G(f)$ is bijective, any element in $U^G(Y)$ can be written as $U^G(f)(\eta)$. The sums below run over $\mathrm{Is}\ \Pi_o(G,X)$. For $x : G/H \to X$ let $\hat{x}$ be the morphism in $\mathcal{C}$ from $id : G/H \to G/H$ to $f : X \to Y$ given by

$$\begin{array}{ccc} G/H & \overset{id}{\text{———}} & G/H \\ {\scriptstyle x}\downarrow & & \downarrow{\scriptstyle f \circ x} \\ X & \longrightarrow & Y \end{array}$$

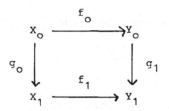

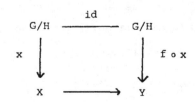

We get

$$F(f)([Z,Y],U^G(f)(\eta))$$

$$= F(f)(0,U^G(f)(\eta)) - F(f)(0,U^G(f)(\chi^G(X)))$$

$$+ F(f)(\tau^G(r),\chi^G(Y))$$

$$= \Sigma(\eta(x) - \chi^G(X)(x)) \cdot F(f)(0,U^G(f \circ x)(\chi^G(G/H)))$$

$$+ F(f)(\tau^G(r),\chi^G(Y))$$

$$= \Sigma(\eta(x) - \chi^G(X)(x)) \cdot B(\hat{x})(b(id : G/H \to G/H))$$

$$+ b(r)$$

Now define F just by this formula. □

6.12. $\mathcal{C}$ = {endomorphisms of G-spaces of the homotopy type of a finite G-CW-complex}

Morphisms are commutative squares

$$\begin{array}{ccc} X & \xrightarrow{f} & X \\ {\scriptstyle h}\downarrow & & \downarrow{\scriptstyle h} \\ Y & \xrightarrow{g} & Y \end{array}$$

co $\mathcal{C}$ = {morphisms with h a cofibration}

w $\mathcal{C}$ = {morphisms with h a G-homotopy equivalence}

The universal additive invariant is computed for G a finite abelian group in Okonek [1983] using Almkvist [1978] □

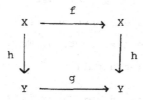

6.13. Let R be a ring. Let $\mathfrak{F}$ resp. $\mathfrak{P}$ be the category of finitely generated free resp. projective R-modules. They become categories with cofibrations and weak equivalences if split injections are the cofibrations and isomorphisms the weak equivalences. Suppose that there is a well defined notion of a rank of a finitely generated projective R-module such that rk R = 1 and rk $P \oplus Q$ = rk P + rk Q holds.

Then ($\mathbb{Z}$,rk) is the universal additive invariant for $\mathcal{F}$. The universal additive invariant for $\mathcal{P}$ is called the <u>projective class group</u> $K_o(R)$. Its restriction to $\mathcal{F}$ is an additive invariant for $\mathcal{F}$ and ($\mathbb{Z}$,rk) is also an additive invariant for $\mathcal{P}$. Hence there are homomorphisms $i : \mathbb{Z} \to K_o(R)$ and $r : K_o(R) \to \mathbb{Z}$ with $r \circ i = id$ such that $i(n) = [R^n]$ and $r([P]) =$ rk P. The cokernel of i is called the <u>reduced projective class group</u> $\tilde{K}_o(R)$.

The category CC($\mathcal{F}$) resp. CC($\mathcal{P}$) of finitely dimensional chain complexes over $\mathcal{F}$ resp. $\mathcal{P}$ is also a category with cofibrations and weak equivalences. Chain maps $i : C \to D$ such that each i_n is an split injection are cofibrations and homotopy equivalences are weak equivalences. Now one can also give chain complex versions of the examples above. □

<u>Comments 6.14.</u> The concept of a universal functorial additive invariant is introduced in Lück [1987b]. Its main use is that it characterizes the invariants above in a simple manner and leads directly to product and restriction formulas in the next section. We will see that it is much harder to describe them algebraically.

Additive invariants are used in tom Dieck [1987] IV. 8. to introduce induction categories which play a role in the axiomatic induction theory for compact Lie groups. Moreover, the Euler ring U(G) of a compact Lie group is defined as the universal additive invariant for finite G-CW-complexes (see tom Dieck [1979] 5.4.4.).

A similar characterization of Lefschetz indices is given in Laitinen-Lück [1987]. □

<u>Exercises 6.15.</u>

1. Show for a compact Lie group G that the Euler ring U(G) together with $X \to [X]$ is the universal additive invariant for finite G-CW-

complexes.

2. Prove that the universal additive invariant for the category of finite-dimensional G-CW-complexes is zero.

3. Let C be the category of G-homotopy projections $f : X \to X$ of finite G-CW-complexes for G a compact Lie group. Define equivariant Lefschetz index $\lambda^G(f) \in U^G(*)$ by the collection of integers $\Sigma(-1)^n$ trace $H_n((f^H, f^{>H})/WH) : H_n((X^H, X^{>H})/WH) \to H_n((X^H, X^{>H})/WH$ for $(H) \in \mathrm{Con}\, G$. Show that $U^G(*) \oplus U^G(*)$ with the function sending $f : X \to X$ to $\lambda^G(f), \lambda^G(id_X)$ is an additive invariant. Is it universal?

4. Show that $U^G(*)$ together with $f \to \lambda^G(f)$ (defined in exercise 3 above) is an additive invariant for the category of endomorphisms of finite G-CW-complexes (see 6.12.). Prove that it is not universal.

5. Let Gr(R) be the universal additive invariant for the category of finitely generated R-modules. Show for a principal domain that $Gr(R) \cong K_o(R) \cong \mathbb{Z}$ holds.

6. Give an example of a ring R with $\tilde{K}_o(R) \neq \{0\}$.

7. Considering a R-module M as a chain complex concentrated in dimension zero defines an homomorphism from $K_o(R)$ into the universal additive invariant $K_o(CC(\mathcal{P}))$ of $CC(\mathcal{P})$ (see 6.13.). Show that this is an isomorphism.

8. Let C be the category of free $\mathbb{Z}$-chain complexes such that $H_i(C)$ is finite for $i \geq 0$ and zero for large i. Compute the universal additive invariant.

7. Product and restriction formulas.

In this section we state product and restriction formulas. This requires to deal with the question whether the restriction of a finite G-CW-complex to a subgroup $H \subset G$ has the H-homotopy type of a finite H-CW-complex and, moreover, whether there is a (canonical) simple structure. We relate the finiteness obstruction and the Whitehead torsion by a geometric Bass-Heller-Swan-Homomorphism. We use it to define equivariant negative K-groups geometrically. We introduce an homomorphism $\omega : A(G)^* \to WH^G(*)$ using the diagonal product formula. Let G and H be topological groups.

Theorem 7.1. Product Formula

a) There is a natural pairing

$$P(X,Y) : (Wa^G(X) \oplus U^G(X)) \otimes (Wa^H(Y) \oplus U^H(Y)) \longrightarrow Wa^{G\times H}(X\times Y) \oplus U^{G\times H}(X\times Y)$$

uniquely determined by the property the $P(X,Y)$ sends $(w^G(X),\chi^G(X)) \otimes (w^H(Y),\chi^H(Y))$ to $(w^{G\times H}(X\times Y),\chi^G(X\times Y))$ for a finitely dominated G-space X and a finitely dominated H-space Y.

b) There is a natural pairing

$$P(X,Y) : (Wh^G(X) \oplus U^G(X)) \otimes (Wh^H(Y) \oplus U^H(Y)) \longrightarrow Wh^{G\times H}(X\times Y) \oplus U^{G\times H}(X\times Y)$$

uniquely determined by the property that it maps $(\tau^G(f),\chi^G(X)) \otimes (\tau^H(g),\chi^H(Y))$ to $(\tau^{G\times H}(f\times g),\chi^G(X\times Y))$ for a G- resp. H-homotopy equivalence $f : X' \to X$ resp. $g : Y' \longrightarrow Y$ between finite G-resp. H-CW-complexes.

c) Let the pairng

$$\otimes : Wh^G(X) \otimes U^H(Y) \longrightarrow Wh^{G\times H}(X\times Y)$$

send (u,v) to the component of $P(X,Y)((u,\chi^G(X)),(0,v))$ in $Wh^{G\times H}(X\times Y)$ and define analogously

$$\otimes : U^G(X) \otimes Wh^H(Y) \longrightarrow Wh^{G\times H}(X\times Y) .$$

Then we have

$$\tau^{G\times H}(f\times g) = \chi^G(X) \bullet \tau^H(g) + \tau^G(f) \bullet \chi^G(Y) .$$

Proof. Lemma 6.3. ,Theorem 6.11, Theorem 4.8. d. □

7.2. If X is a finitely dominated G-CW-complex and Y a finite H-CW-complex with $\chi^H(Y) = 0$ then $X\times Y$ is $G\times H$-homotopy equivalent to a finite $G\times H$-CW-complex. If $f : X' \longrightarrow X$ resp. $g : Y' \longrightarrow Y$ is a G- resp. H-homotopy equivalence between finite G- resp. H-CW-complexes and $\chi^G(X)$ and $\chi^H(Y)$ vanish, $f\times g$ is a simple $G\times H$-homotopy equivalence. □

Next we consider a (closed) subgroup H of G and want to deal with restriction to H . Of course we want to proceed as above. Therefore we must deal with the question whether the restriction of a finite G-CW-complex to H has a simple structure.

Assumption 7.3. For any (closed) $K \subset G$ the restriction res G/K of the homogeneous G-space G/K to H has the H-homotopy type of a finite H-CW-complex. □

We will get from the construction below or already from 4.31.

Lemma 7.4. If assumption 7.3. is fulfilled and X a G-space of the G-homotopy type of a finite, finitely dominated, skeletal finite resp. finite-dimensional G-CW-complex then res X is H-homotopy equivalent to a finite, finitely dominated, skeletal-finite resp. finite-dimensional H-CW-complex. □

We have introduced the component category $\Pi_0(G,X)$ in Definition 5.1. Given an object $x : G/K \longrightarrow X$ we want to define an homomorphism

7.5. $$\phi(x) : \mathrm{Aut}(x) \longrightarrow \mathrm{Wh}^H(\mathrm{res}\ X)$$

Let $\phi(x)$ send the automorphism σ of x given by the G-homeomorphism $\sigma : G/K \to G/K$ to $x_*\tau^H(\mathrm{res}\ \sigma : (\mathrm{res}\ G/K,\xi) \longrightarrow (\mathrm{res}\ G/K),\xi))$ for any simple H-structure ξ on res G/K . This is independent of the choice ξ because of Theorem 4.33.

If $\omega : x \longrightarrow y$ is any isomorphism in $\Pi_0(G,X)$ and $c(\omega) : \mathrm{Aut}(x) \longrightarrow \mathrm{Aut}(y)$ is conjugation with ω we have

7.6. $$\phi(y) \circ c(\omega) = \phi(x)$$

Definition 7.7. We call a G-space X H-simple if $\phi(x) : \mathrm{Aut}(x) \longrightarrow Wh^H(\mathrm{res}\ X)$ is zero for all $x \in \Pi_o(G,X)$. □

We discuss this definition later.

Consider a finite H-simple G-CW-complex X . Make the following choices.

7.8. For any $\bar{x} \in \mathrm{Is}\ \Pi_o(G,X)$ fix a representative $x : G/L \longrightarrow X$, a finite H-CW-complex $Z(x)$ and a H-homotopy equivalence $z(x) : Z(x) \longrightarrow \mathrm{res}\ G/L$. □

7.9.a) Choose for any n-cell a characteristic map. In other words, fix an explicite G-push out diagram where $\coprod$ runs over I_n

$$\begin{array}{ccc} \coprod G/K_i \times S^{n-1} & \xrightarrow{\ \coprod q_i^n\ } & X_{n-1} \\ \downarrow & & \downarrow k \\ \coprod G/K_i \times D^n & \xrightarrow{\ \coprod Q_i^n\ } & X_n \end{array}$$

b) $Q_i^n | G/K \times * : G/K_i \longrightarrow X$ for $* \in D^n$ is an object in $\Pi_o(G,X)$. Let $x_i^n : G/L_i \longrightarrow X$ be the representative of its isomorphism class choosen in a). Let $\sigma_i^n : G/L_i \longrightarrow G/K_i$ be an isomorphism in $\Pi_o(G,X)$ between these objects. □

7.10. With these choices we can equip $\mathrm{res}\ X$ inductively over the skeletons with a simple structure. Define a simple structure γ_i^n on $\mathrm{res}\ G/K_i$ for $i \in I_n$ by $\sigma_i^n \circ z(x_i^n) : Z(x_i^n) \longrightarrow \mathrm{res}\ G/K_i$. We obtain simple structures $\gamma_i^n \times S^{n-1}$ on $\mathrm{res}\ G/K_i \times S^{n-1}$ and $\gamma_i^n \times D^n$ on $\mathrm{res}\ G/K_i \times D^n$. Now suppose we have already a simple structure ξ_{n-1} on $\mathrm{res}\ X_{n-1}$. Then we equip $\mathrm{res}\ X_n$ with the simple structure ξ_n for which the G-push out in 7.9. becomes an H-push out of H-spaces with simple structure. □

7.11. If $f_{n-1} : Y_{n-1} \longrightarrow \mathrm{res}\ X_{n-1}$ is an explicite representative of the simple H-structure ξ_{n-1} on $\mathrm{res}\ X_{n-1}$ we obtain an explicite representative for the one of $\mathrm{res}\ X_n$ by the following diagram where $\coprod$ runs over I_n , f_{n-1}^{-1} is a H-homotopy inverse and h an H-homotopy with cellular h_1

$$\begin{array}{ccccc}
\coprod G/K_i \times D^n & \hookleftarrow & \coprod G/K_i \times S^{n-1} & \longrightarrow & \operatorname{res} X_{n-1} \\
{\scriptstyle \coprod (\sigma_i^n \circ z(x_i^n)) \times id} \uparrow & & {\scriptstyle \coprod (\sigma_i^n \circ z(x_i^n)) \times id} \uparrow & & {\scriptstyle id} \uparrow \\
\coprod Z(x_i^n) \times D^n & \hookleftarrow & \coprod Z(x_i^n) \times S^{n-1} & \longrightarrow & \operatorname{res} X_{n-1} \\
{\scriptstyle i_o} \downarrow & & {\scriptstyle i_o} \downarrow & & {\scriptstyle f_{n-1}^{-1}} \downarrow \\
\coprod Z(x_i^n) \times D^n \times I & \hookleftarrow & \coprod Z(x_i^n) \times S^{n-1} \times I & \xrightarrow{h} & Y_{n-1} \\
{\scriptstyle i_1} \uparrow & & {\scriptstyle i_1} \uparrow & & {\scriptstyle id} \uparrow \\
\coprod Z(x_i^n) \times D_n & \hookleftarrow & \coprod Z(x_i^n) \times S^{n-1} & \xrightarrow{h_1} & Y_{n-1}
\end{array}$$

□

Next we examine how this simple structure depends on the choices 7.8. and 7.9. We will see that only 7.8. is relevant. Suppose we have made a second choice 7.8'. and 7.9'. We want to define an homomorphism

7.12. $$\Theta_X : U^G(X) \longrightarrow Wh^H(\operatorname{res} X)$$

Given $u \in \operatorname{Is} \Pi_o(G,Y)$, choose an isomorphism $\sigma : x \longrightarrow x'$ in $\Pi_o(G,Y)$ between the representatives x and x' of 7.8. and 7.8'. Define $\Theta_X(u)$ by the image of the Whitehead torsion of the H-homotopy equivalence

$$Z(x) \xrightarrow{z(x)} \operatorname{res} G/L \xrightarrow{\sigma} \operatorname{res}(G/L') \xrightarrow{(z(x')')^{-1}} Z(x')'$$

under the homomorphism $(x' \circ z(x')')_* : Wh^H(Z(x')') \longrightarrow Wh^H(\operatorname{res} X)$. This is independent of the choice of σ as X is H-simple. The main technical result is

<u>Lemma 7.13.</u> <u>Let</u> X <u>be a finite</u> G-CW-<u>complex which is</u> H-<u>simple</u>. <u>Let</u> ξ <u>and</u> ξ' <u>be the simple structures on</u> res X <u>we get by 7.10. for the choices 7.8, 7.9, and 7.8', 7.9'. Then we have in</u> $Wh^H(\operatorname{res} X)$

$$\tau^H(id : (\operatorname{res} X,\xi) \longrightarrow (\operatorname{res} X,\xi')) = \Theta_X(\chi^G(X))$$

<u>Proof</u> Let ξ_n and ξ_n' be the simple structures we get on X_n . Let $k(n) : X_n \rightarrow X$ be the inclusion. We show inductively over $n \geq -1$

7.14. $$k(n)_*\tau^H(id : (\operatorname{res} X_n,\xi_n) \longrightarrow (\operatorname{res} X_n,\xi_n')) = \Theta_X \circ k(n)_*(\chi^G(X_n))$$

We prove the induction step from $(n-1)$ to n. In the sequel all sums run over I_n and we sometimes drop the index n. In the following process we thicken X_{n-1} into X_n. For $t \in [0,1]$ define $D^n(t) = \{y \in D^n \mid \|y\| \leq t\}$, $S^{n-1}(t) = \{y \in D^n \mid \|y\| = t\}$. Let $X_{n-1}(t)$ be $X_{n-1} \cup \coprod Q_i(G/K_i \times (D^n \setminus \mathrm{int}\, D^n(t)))$. Composing Q_i with the map $G/K_i \times D^n \longrightarrow G/K_i \times D^n$ and $G/K_i \times S^n \longrightarrow G/K_i \times D^n$ given by multiplication with the scalar t defines $Q_i(t) : G/H_i \times D^n \longrightarrow X_n \setminus \mathrm{int}\, X_{n-1}(t)$, $q_i(t) : G/H_i \times S^{n-1} \longrightarrow \partial X_{n-1}(t)$. Then we have the G-push out

7.15.
$$\begin{array}{ccc}
\coprod G/K_i \times S^{n-1} & \xrightarrow{\coprod q_i(t)} & X_{n-1}(t) \\
\downarrow & & \downarrow \\
\coprod G/K_i \times D^n & \xrightarrow{\coprod Q_i(t)} & X_n
\end{array}$$

Notice that 7.15. reduces for $t = 1$ to the given G-push out 7.9. We can do the same for the choices 7.8'., 7.9'. obtaining $q_i(t')'$, $Q_i(t')'$, $X_{n-1}(t')'$,... for $t' \in [0,1]$. We can assume without loss of generality that $Q_i(G/K_i \times \{0\}) = Q_i'(G/K_i' \times \{0\})$ for the origin $0 \in D^n$. Construct continuous strictly monoton increasing functions $\varepsilon : [0,1] \longrightarrow [0,1]$ and $\delta : [0,1] \longrightarrow [0,1]$ such that $X_{n-1}(\delta(t'))' \subset X_{n-1}(\varepsilon(t')) \subset X_{n-1}(t')'$ holds. Define $h : X_{n-1} \times [0,1] \longrightarrow X_{n-1}$ by requiring $h_o = id$, $h_{t'}|X_{n-1}(t')' = id$, $h_{t'} \circ Q_i'(gK_i',s) = Q_i'(gK_i',st'/\delta(t'))$ for $0 \leq s \leq \delta(t')$, $\delta(t') \neq 0$ and $h_{t'} \circ Q_i'(gK_i',s) = Q_i'(gK_i',t')$ for $0 < \delta(t') \leq s \leq t'$. Let $r(t') : X_{n-1}(\varepsilon(t')) \longrightarrow X_{n-1}(t')'$ be $h_{t'}|X_{n-1}(\varepsilon(t'))$. Define $(V_i(t'),v_i(t')) : G/K_i \times (D^n,S^{n-1}) \longrightarrow G/K_i' \times (D^n,S^{n-1})$ by $(Q_i(t')',q_i(t')') \circ (V_i(t'),v_i(t')) = h_{t'} \circ (Q_i(\varepsilon(t'))$, $q_i(\varepsilon(t'))$ for $t' \in [0,1[$. Define $f(t')$ by the G-push out for some $t' \in]0,1[$

7.16.

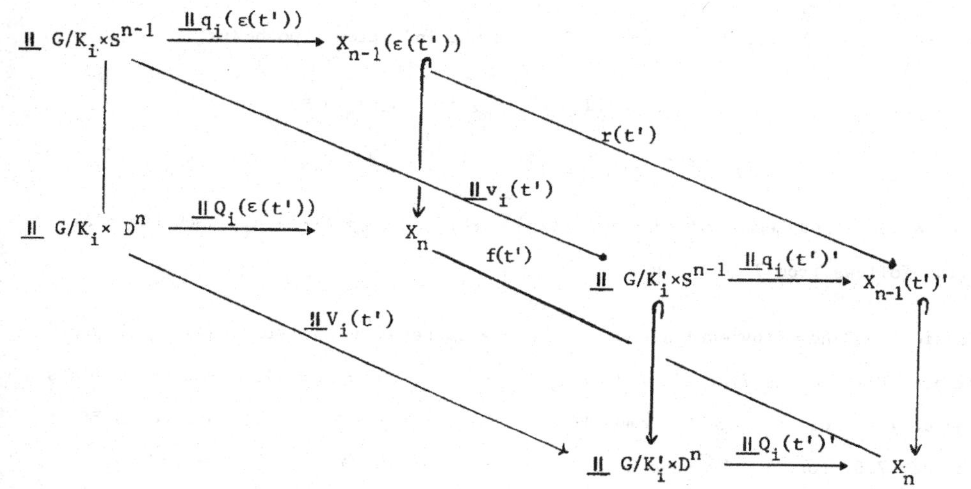

We already have defined simple H-structures γ_i and γ_i' on G/K_i and G/K_i'. Equip $X_{n-1}(\varepsilon(t'))$ with the simple structure $\xi_{n-1}(\varepsilon(t'))$ for which the inclusion $(X_{n-1},\xi_{n-1}) \longrightarrow (X_{n-1}(\varepsilon(t')),\xi_{n-1}(\varepsilon(t')))$ is simple. One easily checks that the left G-push out in 7.16. is a H-push out of H-spaces with simple structure if we use ξ_n on X_n. The analogous is true for the prime-version. Using the homotopy h above one shows that $f(t')$ is G-homotopic to the identity. If $k : X_{n-1} \longrightarrow X_n$ is the inclusion we get from Theorem 4.33. as $r(t')|X_{n-1} = \mathrm{id}$ is true.

7.17. $\tau^H(\mathrm{id} : (X_n,\xi_n) \longrightarrow (X_n,\xi_n')) =$

$$k_*\tau^H(\mathrm{id} : (X_{n-1},\xi_{n-1}) \longrightarrow (X_{n-1},\xi_{n-1}')) - \sum k_* q'_{i_*}\tau^H(v_i(t')) + \sum Q'_{i_*}\tau^H(V_i(t')) .$$

Next choose a homotopy equivalence $(W_i,w_i) : (D^n,S^{n-1}) \longrightarrow (D^n,S^{n-1})$ and a G-homomorphism $\omega_i : G/K_i \longrightarrow G/K_i'$ such that $(V_i(t'),v_i(t'))$ is G-homotopic to $\omega_i \times (W_i,w_i)$. Let $\overline{u}_i \in \mathrm{Is}\,\Pi_o(G,X_n)$ be given by $Q_i|G/K_i' \times \{0\}$. Since W_i and w_i are simple and $\chi(D^n) - \chi(S^{n-1})$ is $(-1)^n$ we derive from Theorem 4.33.

7.18. $$Q'_{i_*}\tau^H(V_i(t')) - k_* q'_{i_*}\tau^H(v_i(t')) = (-1)^n u_{i_*}\tau^H(\omega_i) .$$

Now ω_i is a morphism in $\Pi_o(G,X)$ from $Q_i|G/K_i \times \{0\}$ to $Q_i'|G/K_i' \times \{0\}$. Hence we get from the definitions if $k(n) : X_n \longrightarrow X$ is the inclusion

7.19. $$k(n)_* u_{i_*} \tau^H(\omega_i) = \Theta_X(k(n)_*(\bar{u}_i))$$

We derive from 7.17., 7.18., and 7.19., and the induction hypothesis

7.20. $$k(n)_* \tau^H(\mathrm{id} : (X_n,\xi_n) \longrightarrow (X_n,\xi_n')) =$$
$$\Theta_X \circ k(n-1)_* \chi^G(X_{n-1}) + (-1)^n \Sigma\, \Theta_X \circ k(n)_*(\bar{u}_i\ )$$

Theorem 5.4. and Lemma 5.6. imply $\chi^G(X_n) = k_* \chi^G(X_{n-1}) + \Sigma(-1)^n \bar{u}_i$. Now the claim 7.14. follows from 7.20. □

Consider a G-homotopy equivalence $f : X \longrightarrow Y$ between G-CW-complexes which are H-simple. Fix choices 7.8. and 7.9. for X and Y . The choice 7.8. for X defines a choice 7.8. for Y just by composition with $f : X \longrightarrow Y$. So we have two different choices 7.8. for Y . We have defined an homomorphism $\Theta_f : U^G(Y) \longrightarrow Wh^H(\mathrm{res}\ Y)$ in 7.12. regarding the one coming from X as the first. We define

7.22. $$\tau^H(\mathrm{res}\ f) \in Wh^H(\mathrm{res}\ X)$$

by $\tau^H(\mathrm{res}\ f : (\mathrm{res}\ X,\xi_X) - (\mathrm{res}\ Y,\xi_Y)) - \Theta_f(\chi^G(Y))$.

This element $\tau^H(\mathrm{res}\ f)$ is independent of the choices 7.8. and 7.9. Make different choices 7.8'. and 7.9'. for X and Y . We get ξ_X', ξ_Y' and Θ_f' . Let Θ_X and Θ_Y be the homomorphisms of 7.12. according to these different choices. One easily checks

$$\Theta_f'(\chi^G(Y)) - \Theta_f(\chi^G(Y)) = \Theta_Y(\chi^G(Y)) - f_*\Theta_X(\chi^G(X)).$$

Now we get from Lemma 7.13.

$$\tau^H(\mathrm{res}\ f : (\mathrm{res}\ X,\xi_X) \longrightarrow (\mathrm{res}\ Y,\xi_Y)) - \Theta_f(\chi^G(Y)) =$$
$$\tau^H(\mathrm{res}\ f : (\mathrm{res}\ X,\xi_X') \longrightarrow (\mathrm{res}\ Y,\xi_Y')) + f_*\Theta_X(\chi^G(Y))$$
$$- \Theta_Y(\chi^G(Y)) - \Theta_f(\chi^G(Y))$$

$$= \tau^H(\mathrm{res}\ f : (\mathrm{res}\ X,\xi_X') \longrightarrow (\mathrm{res}\ Y,\xi_Y')) - \Theta_f'(\chi^G(Y)).$$

Assumption 7.23. Any finite G-CW-complex X is H-simple. □

Lemma 7.24. If assumption 7.23. holds, we get a functorial additive invariant on the category $\mathcal{C}$ of 6.10. by $Wh^H(\mathrm{res}\ X), \tau^H(\mathrm{res}\ f)$.

Proof. Theorem 4.33. □

Theorem 7.25. Restriction formula.

a) If assumption 7.3. holds there is exactly one natural homomorphism

$$\mathrm{Res}\ X : Wa^G(X) \otimes U^G(X) \longrightarrow Wa^H(\mathrm{res}\ X) \otimes U^H(\mathrm{res}\ X)$$

sending $(w^G(X), \chi^G(X))$ to $(w^H(\mathrm{res}\ X), \chi^H(\mathrm{res}\ X))$ for any finitely dominated G-space X .

b) Suppose assumptions 7.3. and 7.23. Then there is exactly one natural homomorphism

$$\mathrm{Res}\ X : Wh^G(X) \otimes U^G(X) \longrightarrow Wh^H(\mathrm{res}\ X) \otimes U^H(\mathrm{res}\ X)$$

sending $(\tau^G(f), \chi^G(X))$ to $(\tau^H(\mathrm{res}\ f), \chi^H(\mathrm{res}\ X))$ for any G-homotopy equivalence $f : X' \longrightarrow X$ of finite G-CW-complexes.

Proof.

a) Let $\mathcal{C}(G)$ be the category defined in 6.8. Restriction defines a functor $\mathcal{C}(G) \longrightarrow \mathcal{C}(H)$ of categories with cofibrations and weak equivalences because of Lemmata 1.12., 1.13., and 7.4. Now apply Lemma 6.2. and Theorem 6.9.

b) Theorem 6.11. and Lemma 7.24. □

Combining the product and restriction formula yields a diagonal product formula. Its meaning lies in the fact that it is an internal formula. In the sequel $H = G \subset G \times G$ is the diagonal subgroup.

Theorem 7.26. Diagonal Product Formula

a) If assumption 7.3. holds for $G \subset G \times G$ there is exactly one natural pairing

$$P(X,Y) : (Wa^G(X) \otimes U^G(X)) \otimes (Wa^G(Y) \otimes U^G(Y)) \longrightarrow Wa^G(X\times Y) \otimes U^G(X\times Y)$$

sending $(w^G(X), \chi^G(X)) \otimes (w^G(Y), \chi^G(Y))$ to $w^G(X\times Y), \chi^G(X\times Y)$ for finitely dominated G-spaces X and Y .

b) Suppose assumptions 7.3 and 7.23. hold for $G \subset G\times G$. Then there is a natural pairing

$$P(X,Y) : (Wh^G(X) \circledast U^G(Y)) \circledast (Wh^G(Y) \circledast U^G(Y)) \longrightarrow Wh^G(X\times Y) \circledast U^G(X\times Y)$$

uniquely determined by the property that it sends $(\tau^G(f),\chi^G(X)) \circledast (\tau^G(g),\chi^G(Y))$ to $(\tau^G(f\times g),\chi^G(X\times Y))$ for G-homotopy equivalences between finite G-CW-complexes $f : X' \longrightarrow X$ and $g : Y' \longrightarrow Y$.

c) Suppose assumption 7.3. and 7.23. hold for $G \subset G\times G$. Let the pairing

$$\circledast : Wh^G(X) \circledast U^G(Y) \longrightarrow Wh^G(X\times Y)$$

send (u,v) to the component of $P(X,Y)((u,\chi^G(X)) \circledast (0,v))$ in $Wh^G(X\times Y)$. Define analogously

$$\circledast : U^G(X) \circledast Wh^G(Y) \longrightarrow Wh^G(X\times Y) .$$

Then we have:

$$\tau^G(f\times g) = \chi^G(X) \circledast \tau^G(g) + \tau^G(f) \circledast \chi^G(Y) \qquad \square$$

Now we come to the promised discussion of our assumptions.

Lemma 7.27. Let G be a compact Lie group. Then assumptions 7.3. and 7.23. are satisfied for any (closed) subgroup $H \subset G$.

Proof: This follows from 4.36. Namely, the restriction res G/K is a compact smooth H-manifold and has therefore a preferred simple structure ξ. This shows 7.3. If $\sigma : G/K \longrightarrow G/K$ is a G-homeomorphism, σ must be smooth so that $\tau^H(\sigma : (\text{res } G/K,\xi) \longrightarrow (\text{res } G/K,\xi))$ vanishes by 4.36. This verifies 7.23. $\square$

Remark 7.28. Of course we could have started in the case of a compact Lie group G in 7.8. always with the preferred simple H-structures of 4.36. Then res X has also a preferred simple H-structure ξ_X by 7.10. and Lemma 7.13. if X is a finite G-CW-complex and $\tau^H(\text{res } f)$ is just $\tau^H(\text{res } f : (X,\xi_X) \longrightarrow (Y,\xi_Y))$. However, the proofs would not become really simpler , one has to do the same type of calculations e.g. to prove that the choice of characteristic maps do not matter. It is remarkable that $\tau^H(\text{res } f)$ can be defined independently of the choice of 7.8. This observation will become important when we will regard fibrations and their transfer maps. Notice that no correction term Θ occurs when $\chi^G(X) = \chi^G(Y)$ vanish. In particular res X has a preferred simple H-structure if assumptions 7.3. is true, X is H-simple and $\chi^G(X)$ vanishes . Then the choice 7.8. does not matter. □

7.29. Let G be a compact Lie group. By the diagonal product formula $Wa^G(*) \otimes U^G(*)$ becomes an associative commutative ring with unit $u(*)$ and $Wa^G \otimes U^G$ a functor into the category of $Wa^G(*) \otimes U^G(*)$-modules. The analogue is true for $Wh^G \otimes U^G$ □

If X is a finitely dominated G-CW-complex and S^1 the one-dimensional circle with trivial G-action, $X \times S^1$ is G-homotopy equivalent to a finite G-CW-complex by 7.2. The geometric proof of Mather [1965] carries over to the equivariant case as follows.

Given a G-self map $f : Z \longrightarrow Z$, its <u>mapping torus</u> $T(f)$ is obtained from the mapping cylinder $Cyl(f)$ by identifying the top and bottom. In other words, $T(f)$ is the G-push out

7.30.
$$\begin{array}{ccc} Z \amalg Z & \xrightarrow{id \amalg id} & Z \\ \downarrow & & \downarrow \\ Cyl(f) & \longrightarrow & T(f) \end{array}$$

Let $f : X \longrightarrow Y$ and $g : Y \longrightarrow Z$ be G-maps. Then there is a G-homotopy equivalence $Cyl(g \circ f) \longrightarrow Cyl(f) \cup_Y Cyl(g)$ relative $X \amalg Z$ given in 4.17. It induces a G-homotopy equivalence

7.31.
$$T(f \circ g) \longrightarrow T(g \circ f)$$

If f and $g : X \longrightarrow Y$ are G-homotopic, Lemma 4.18. gives a G-homotopy equivalence $Cyl(f) \longrightarrow Cyl(g)$ rel $X \amalg Y$ and hence a G-homotopy equivalence

7.32.
$$T(f) \longrightarrow T(g) .$$

Consider a G-space X with a finite G-domination (Z,r,i) . We get from 7.31. and 7.32. a G-homotopy equivalence unique up to G-homotopy

7.33.
$$\Phi : T(i \circ r) \longrightarrow T(r \circ i) \longrightarrow T(id) \longrightarrow X \times S^1$$

Notice that $T(i \circ r)$ is a finite G-CW-complex. Given a G-space Y , we define an homomorphism

7.34. $$\phi(Y) : Wa^G(Y) \longrightarrow Wh^G(Y\times S^1)$$

as follows. Let X be a finitely dominated G-space and $f : X \longrightarrow Y$ be a G-map representing $[f] \in Wa^G(Y)$. Let $\Theta : S^1 \longrightarrow S^1$ send z to its complex conjugate $\bar{z}$ and choose $\Phi : T(i \circ r) \longrightarrow X\times S^1$ as constructed in 7.33. If Φ^{-1} is any G-homotopy inverse, define

$$\phi(Y)([f]) := (f \times id)_* \Phi_* (\tau^G(\Phi^{-1} \circ \Theta \circ \Phi))$$

We leave it to the reader to carry over the proof in Ferry [1981 a] that $\phi(Y)$ is well-defined. Moreover, a finitely dominated G-space X is G-homotopy equivalent to a finite G-CW-complex if and only if $\phi(X)(w^G(X))$ vanishes. This is equivalent to the next result

<u>Lemma 7.35.</u> $\phi(Y)$ <u>is injective</u>. □

Let $\sigma(j) : T^{n+1} \longrightarrow T^{n+1}$ be the permutation of coordinates

$$(z_1, z_2, \ldots, z_n, z) \longrightarrow (z_1, \ldots, z_{j-1}, z, z_j, \ldots, z_n)$$

for $j = 1, \ldots, n+1$.

<u>Definition 7.36.</u> <u>The equivariant negative</u> K-<u>groups are defined for</u> $n \geq 0$ <u>by</u>

$$\tilde{K}^G_{-n}(Y) = \bigcap_{j=1}^{n+1} \text{image}((id \times \sigma(j))_* \circ \phi(Y\times T^n)) \subset Wh^G(Y\times T^{n+1}) \quad \square$$

Let $(k_1, k_2, \ldots, k_{n+1})$ be the n+1-tuple of positive integers. If $p(k_1, \ldots, k_{n+1}) : T^{n+1} \longrightarrow T^{n+1}$ is the covering given by the products of the coverings $S^1 \longrightarrow S^1$ $z \longrightarrow z^{k_i}$, we obtain from the pull-back construction an homomorphism

7.37. $$p(k_1, \ldots, k_{n+1})^* : Wh^G(Y\times T^{n+1}) \longrightarrow Wh^G(Y\times T^{n+1}) .$$

This yields an operation of the monoid $\mathbb{N}^{n+1}$ on $Wh^G(Y\times T^{n+1})$. Later we prove

<u>Theorem 7.38.</u> <u>We have for</u> $n \geq 0$

$$\tilde{K}^G_{-n}(Y) = Wh^G(Y\times T^{n+1})^{\mathbb{N}^{n+1}} \quad \square$$

We can use the diagonal product formula to define for a compact Lie group G an homomorphism

7.39.
$$\omega : A(G)^{*} \longrightarrow Wh^{G}$$

where $A(G)^{*}$ are the units in the Burnside ring and Wh^{G} is the equivariant Whitehead group $Wh^{G}(*)$ of the trivial G-space consisting of one point. Some preparations are needed.

Let X be a G- and Y a H-space. Define the <u>join</u> $X * Y$ as the G×H-push out

7.40.
$$\begin{array}{ccc} X \times \{0\} \times Y \amalg X \times \{1\} \times Y & \longrightarrow & Y \amalg X \\ \downarrow & & \downarrow \\ X \times I \times Y & \longrightarrow & X * Y \end{array}$$

Up to G-homeomorphism this can also be defined as the G-push out

7.41.
$$\begin{array}{ccc} X \times Y & \longrightarrow & X \times \mathrm{Cone}(Y) \\ \downarrow & & \downarrow \\ \mathrm{Cone}(X) \times Y & \longrightarrow & X * Y \end{array}$$

If X and Y are G-CW-complexes equip $X * Y$ with the G-CW-complex structure such that 7.41. is cellular (see 4.1). If V and W are representations, $S(V \oplus W)$ and $SV * SW$ are G-homeomorphic. If X and Y are finite G-CW-complexes, we get from Theorem 5.4. and Theorem 7.26. in $U(G)$

7.42.
$$\chi^{G}(X*Y) = \chi^{G}(X) + \chi^{G}(Y) - \chi^{G}(X) \bullet \chi^{G}(Y) .$$

The projection $p : U(G) \longrightarrow A(G)$ and the character map $ch : A(G) \rightarrow \prod_{(H)} C(G)$ are ring homomorphisms. We get from 7.42.

7.43.
$$ch\circ p(1-\chi^{G}(SV))\cdot ch\circ p(1-\chi^{G}(SV))$$
$$= ch\circ p(1-\chi^{G}(SV*SV)) = ch\circ p(1-\chi^{G}(S(V\oplus V)) \equiv 1$$

since $\chi(SV \oplus V)^{H} = 0$ for all $H \subset G$. As ch is injective and the kernel of p is the nilradical of $U(G)$ (see tom Dieck [1979], Proposition 5.53), $1-\chi^{G}(SV)$ is a unit in $U(G)$ for any G-representation V. Next consider a G-self equivalence $f : SV \longrightarrow SV$ of a G-representation. Let $\hat{\tau}^G(f)$ denote the image of $\tau^G(f)$ under $pr_* : Wh^G(SV) \longrightarrow Wh^G(*)$. If $g : SV \longrightarrow SW$ is a G-homotopy equivalence, $\hat{\tau}^G(g\circ f\circ g^{-1}) = \hat{\tau}^G(f)$ by Theorem 4.8. d). This shows in particular that it does not matter which simple structure we use on SV. Consider two G-self equivalences $f : SV \to SV$ and $g : SW \longrightarrow SW$. We get from Theorem 4.8., Theorem 7.26. and 7.42.

7.44. $$(1-\chi^G(SV*SW))^{-1}\cdot \hat{\tau}^G(f*g) = (1-\chi^G(SV))^{-1}\hat{\tau}^G(f)+(1-\chi^G(SW))^{-1}\hat{\tau}^G(g)$$

Let $u \in A(G)^*$ be given. We can find a G-representation V and a G-self-map $f : SV \longrightarrow SV$ such that $ch(u)_H = \deg f^H$ holds for $H \subset G$, WH finite (see tom Dieck [1987], II.8). Then we define $\omega(u) \in Wh^G$ by $(1-\chi^G(SV))^{-1}\cdot\hat{\tau}^G(f))$. This gives a well defined homomorphism $\omega : A(G)^* \longrightarrow Wh^G$ by 7.44. since stable G-homotopy classes of G-maps $f : SV \longrightarrow SV$ are classified by the degrees $\deg f^H$ for $H \subset G$, WH finite.

Finally we examine for a compact Lie group G and a G-manifold M the restriction of the G-simple structure to $K \subset G$. Let $\xi^G(M)$ resp. ξ^K (res M) be the simple structure on M regarded as a G- resp. K-manifold which we have defined in 4.36. Using the simple structures $\xi^K(\text{res } G/H)$ we get a preferred choice 7.8. for M and thus by 7.10. a second K-simple structure on res M from $\xi^G(M)$ denoted by res $\xi^G(M)$.

<u>Lemma 7.4 5.</u> res $\xi^G(M) = \xi^K(\text{res } M)$

<u>Proof.</u> We use induction over the number r of orbit types. We start with the induction step from $r-1$ to r for $r \geq 2$. Choose $H \in Iso(M)$ to be maximal, i.e. $L \in Iso(M)$, $H \subset L$ implies $H = L$. In the sequel we use the notation of 4.36., e.g. $\nu = \nu(M_{(H)},M), \overline{M} = M\backslash \mathring{D}\nu$. Consider the G-push outs

7.46. $$\begin{array}{ccc} S\nu & \longrightarrow & \overline{M} \\ \downarrow & & \downarrow \\ D\nu & \longrightarrow & M \end{array} \qquad\qquad \begin{array}{ccc} S\nu & \longrightarrow & M_{(H)} \\ \downarrow & & \downarrow \\ S\nu\times I & \longrightarrow & D\nu \end{array}$$

These are G-push outs of G-spaces with simple structures with respect to $\xi^G(?)$

by definition. This is also true over K for $\xi^K(\mathrm{res}\ ?)$ by Proposition 4.60. One easily checks by inspecting 7.10. that these are K-push outs of K-spaces with simple structure with respect to $\mathrm{res}\ \xi^G(?)$. By induction hypothesis the claim is true for $S\nu$, $S\nu \times I$ and $\overline{M}$. Hence it holds for $D\nu$ and M by the K-push outs 7.46.

It remains to verify the induction begin. In other words, it suffices to prove $\mathrm{res}\ \xi^G(M) = \xi^K(\mathrm{res}\, M)$ under the assumption that M has only one orbit type, say G/H . Consider a fibre bundle $G/H \longrightarrow M \longrightarrow N$ with structure group WH and N a manifold. Recall that $\xi^G(M)$ arises from this situation using a non-equivariant triangulation of N . We have to show $\mathrm{res}\ \xi^G(M) = \xi^K(\mathrm{res}\, M)$ in this situation. We use induction over the dimension of N . The begin $\dim N = 0$ is trivial. In the induction step we use induction over the number of handle bodies of N . Suppose that N is obtained from $\overline{N}$ by attaching a handle $D^k \times D^{n-k}$

$$\begin{array}{ccc} S^{k-1} \times D^{n-k} & \longrightarrow & \overline{N} \\ \downarrow & & \downarrow \\ D^k \times D^{n-k} & \longrightarrow & N \end{array}$$

This push out induces a G-push out

7.47.
$$\begin{array}{ccc} \overline{M}_o & \longrightarrow & \overline{M} \\ \downarrow & & \downarrow \\ M_o & \longrightarrow & M \end{array}$$

by pulling back the fibre bundle. The claim holds for $\overline{M}_o$ and $\overline{M}$ by induction hypothesis and for M_o, as M_o is G-diffeomorphic to $G/H \times D^k \times D^{n-k}$. As 7.47. is a K-push out of K-spaces with simple structure with respect to both $\xi^K(\mathrm{res}\ ?)$ and $\mathrm{res}\ \xi^G(?)$, the claim for M follows. □

Corollary 7.48. *Let* $K \subset G$ *be compact Lie groups and* $f : M \longrightarrow N$ *be a G-homotopy equivalence of G-manifolds. Define* $\tau^G(f) \in Wh^G(N)$ *resp* $\tau^K(\mathrm{res}\, f) \in Wh^K(\mathrm{res}\, N)$ *with respect to the simple structure* $\xi^G(M), \xi^G(N)$ *resp .* $\xi^K(\mathrm{res}\, M), \xi^K(\mathrm{res}\, N)$ *of* 4.36. *Let* $\mathrm{Res}\, N : Wh^G(N) \longrightarrow Wh^K(\mathrm{res}\, N)$ *be induced from the map in Proposition* 7.25. *Then*

$$\operatorname{Res} N(\tau^G(f)) = \tau^K(\operatorname{res} f) \qquad \square$$

Comments 7.49. The non-equivariant sum and product formulas for the finiteness obstruction and the Whitehead torsion are established in Cohen [1973], Gersten [1966], Kwun-Szczarba [1965], Siebenmann [1965]. The product formula in the equivariant setting is treated in Andrzejewski [1986], Illman [1986] and Lück [1983]. A diagonal product formula for finite groups is established in tom Dieck [1981] using the language of modules over the orbit category. It is used for example in the theory of homotopy representations (see tom Dieck-Petrie [1982] § 8)

Mather's trick [1965] is the main ingredient in Ferry [1981a] to develop a simple homotopy approach to the finiteness obstruction. This is extended to the equivariant case in Kwasik [1983]. The invariant $o^G(X) \in Wh^G(X \times S^1)$ defined there is just $\phi(X)(w^G(X))$ in our notion. The statement in Kwasik [1983] 3.4 that $\tilde{o}^G(X)$ vanishes, if and only if X is finite, is equivalent to $\phi(Y)$ being always injective.

Negative K-groups are introduced in Bass [1968]. Algebraically equivariant negative K-groups are defined and splitted into ordinary ones (compare Theorem 3.11.) in Svensson [1985] provided that X is a point. Equivariant negative K-groups appear for example as obstruction groups for equivariant transversality in Madsen-Rothenberg [1985a].

Later we will deal with the various formulas, the geometric Bass-Heller-Swan-homomorphism and the groups $\tilde{K}^G_{-n}$ algebraically. $\square$

Exercises 7.50.

1. Let G be an abelian Lie group with finite $\pi_o(G)$. Do assumptions 7.3. and 7.24. hold for $H \subset G$?

2. Let G be a path-connected topological group. Suppose that G is homotopy equivalent to a finite CW-complex. Show that assumptions 7.3. and 7.24. hold if $H \subset G$ is the trivial subgroup. If $G \longrightarrow E \xrightarrow{p} B$ is a principal G-bundle define a transfer map $p^* : Wh^1(B) \longrightarrow Wh^1(E)$ and $p^* : Wa^1(B) \longrightarrow Wa^1(E)$ using the restriction formula.

3. Let G be a simply connected Lie group with $\dim G \geq 1$. Let $G \to E \xrightarrow{p} B$ be a principal G-bundle over a finitely dominated CW-complex. Show that E is homotopy equivalent to a finite CW-complex. (Hint: $p_* : Wa^1(E) \to Wa^1(B)$ is bijective and $p^*E \longrightarrow E$ has a section).

4. Let $SO(3)$ act diagonally on $\mathbb{R}^6 = \mathbb{R}^3 \times \mathbb{R}^3$. Compute $\chi^{SO(3)}(S(\mathbb{R}^6)) \in U(G)$ in terms of the base elements $[SO(3)/H]$ for $H \subset SO(3)$.

5. Let X be a finite G- and Y a finite H-CW-complex. Show that the two simple $G\times H$-structures on $X*Y$ defined by 7.40. and 7.41. agree.

6. Let V and W be G-representations. We get from Remark 7.28. and 7.40. or 7.41. a simple G-structure on $SV*SW$. Since $S(V\oplus W)$ is a G-manifold there is a simple G-structure on $S(V\oplus W)$ by 4.36. Show that the obvious G-homeomorphism $S(V\oplus W) \longrightarrow SV*SW$ is a simple G-homotopy equivalence.

7. Let G be the product of a finite 2-group H and a finite group K of odd order. Show that $\omega : A(G)^* \longrightarrow Wh^G$ is trivial (see Dovermann-Rothenberg [1988], p. 51a) (Hint: Use that $A(K)^* = \{\pm 1\}$ (see tom Dieck [1979] 1.5.) and any element in $A(H)^*$ can be represented by a self-map $SV \longrightarrow SV$ induced from an orthogonal map $V \longrightarrow V$ (see Tornhave [1984]). Show $A(G)^* = A(K)^* \otimes_{\mathbb{Z}} A(H)^*$)

8. Show $\tilde{K}_0^G(X) = Wa^G(X)$.

9. Let G be a compact Lie group and $f : M \longrightarrow N$ be a G-homotopy equivalence of G-manifolds. Define $\tau^H(f)$ for any $H \subset G$ using the simple structures of 4.36. Show

   a) If $\tau^G(f)$ is zero, $\tau^H(f)$ vanishes for all $H \subset G$ with $H \neq G$.

   b) The converse of a) is false.

10. Show the existence of two S^1-manifolds M and N which are not S^1-diffeomorphic but are H-diffeomorphic for any finite subgroup $H \subset S^1$.

8. Lift-extensions and the (discrete) fundamental category.

Given an appropriate G-space X with universal covering $p : \tilde{X} \longrightarrow X$, we want to lift the G-action to a $\tilde{G}$-action on $\tilde{X}$. More precisely, we construct an extension $1 \longrightarrow \pi_1(X) \longrightarrow \tilde{G} \longrightarrow G \longrightarrow 1$ and a $\tilde{G}$-action on $\tilde{X}$ extending the $\pi_1(X)$-action and covering the G-action. Then we will divide out the action of the component $\tilde{G}_o$ of the identity and obtain a $\tilde{G}/\tilde{G}_o$-space $\tilde{X}/\tilde{G}_o$. Later we will see that all the information we are interested in can be read off from the $\tilde{G}/\tilde{G}_o$-space $\tilde{X}/\tilde{G}_o$. The advantage in comparision with the G-space X is that $\tilde{X}/\tilde{G}_o$ is simply connected and $\pi_o(\tilde{G}) = \tilde{G}/\tilde{G}_o$ is discrete.

If we study a G-space X, we must deal with all WH-spaces X^H for $H \subset G$. We introduce the fundamental category and the discrete fundamental category. They are used to organize all the data like the component structure and fundamental groups of the various fixed point sets. It includes the lifting of the action of a certain subgroup $WH(C)$ of WH on a component C of X^H to its universal covering $\tilde{C}/WH(C)_o$ for all $H \subset G$.

8 A. Lift-extensions.

We start with lifting a G-action on X to a $\bar{G}$-action on $\bar{X}$ for any regular covering $p : \bar{X} \longrightarrow X$. In the sequel G is a locally path connected regular topological group and any G-space or space is required to be locally path-connected and regular. Recall that a topological space is regular if any neighbourhood of a point contains a closed neighbourhood. We need this condition to ensure that an open subset is again compactly generated (see Whitehead [1978] I.4.15). This guarantees for example that the pull-back (within the category of compactly generated spaces) of a covering is again a covering. Consider a regular covering $p : \bar{X} \longrightarrow X$ of the G-space X, i.e. a covering such that the group of deck-transformation acts transitively on the fibre or, equivalently, p is a $\Delta(p)$-principal bundle. A covering p is regular if and only if the image of $p_* : \pi_1(\bar{X}) \longrightarrow \pi_1(X)$ is normal in $\pi_1(X)$. Suppose the existence of an epimorphism $\bar{G} \longrightarrow G$ and a $\bar{G}$-action on $\bar{X}$ such that the following diagram commutes

$$\begin{array}{ccc} \bar{G} \times \bar{X} & \longrightarrow & \bar{X} \\ \downarrow & & \downarrow \\ G \times X & \longrightarrow & X \end{array}$$

Recall that [ ] denotes (free) homotopy classes of maps. Then G acts on $[S^1,X]$ and has to respect image $p_* : [S^1,\bar{X}] \longrightarrow [S^1,X]$. We will see that this is the only condition for lifting the G-action. If $p : \bar{X} \longrightarrow X$ is the universal covering, this condition is empty.

The category $\mathcal{C}(G)$ has as objects regular, connected, locally path-connected G-spaces X together with a regular covering $p : \bar{X} \longrightarrow X$ such that the G-action respects image $p_* : [S^1,\bar{X}] \longrightarrow [S^1,X]$. A morphism $(\bar{f},f) : p \longrightarrow q$ is a commutative diagram

$$\begin{array}{ccc} \bar{X} & \xrightarrow{\bar{f}} & \bar{Y} \\ {\scriptstyle p}\downarrow & & \downarrow{\scriptstyle q} \\ X & \xrightarrow{f} & Y \end{array}$$

such that f is G-equivariant. Let G also denote the constant functor $G : \mathcal{C}(G) \longrightarrow \{\text{top.gr.}\}$, whereas $\Delta : \mathcal{C}(G) \longrightarrow \{\text{top.gr.}\}$ sends p to the discrete group of deck -transformations $\Delta(p) = \{(\bar{f},\mathrm{id}) : p \longrightarrow p\}$.

A <u>lift-extension</u> $(\bar{G},\rho,i,q)$ consists of:

a) A functor $\bar{G} : \mathcal{C}(G) \longrightarrow \{\text{top.gr.}\}$.

b) An in p natural group action $\rho(p) : \bar{G}(p) \times \bar{X} \longrightarrow \bar{X}$.

c) Natural transformations $i : \Delta \longrightarrow \bar{G}$ and $q : \bar{G} \longrightarrow G$.

satisfying for any object p :

i) $1 \longrightarrow \Delta(p) \xrightarrow{i(p)} \bar{G}(p) \xrightarrow{q(p)} G \longrightarrow 1$ is an exact sequence of topological groups. Moreover $q(p)$ is a $\Delta(p)$-principal bundle.

ii) The following diagram commutes

$$\begin{array}{ccc} \Delta(p) \times \bar{X} & \longrightarrow & \bar{X} \\ \downarrow & i(p)\times \mathrm{id} & \downarrow \ \mathrm{id} \\ \bar{G}(p) \times \bar{X} & \xrightarrow{\rho(p)} & \bar{X} \\ \downarrow & q(p)\times \mathrm{id} & \downarrow \ p \\ G \times X & \longrightarrow & X \end{array}$$

Recall Convention 1.1. that we work in the category of compactly generated spaces. We frequently make use of the adjointness between the functors $X \times ?$ and $map(X,?)$ without mentioning it (Whitehead [1978], I 4.23).

<u>Theorem 8.1:</u> <u>Up to natural equivalence there is exactly one lift-extension for the locally path-connected regular topological group</u> G.

<u>Proof:</u> We construct for a given regular covering $p : \bar{X} \longrightarrow X$ in $\mathcal{C}(G)$ the desired quadruple $(G(p),\rho(p),i(p),q(p))$ as follows and leave the verification of functionality and uniqueness to the reader.

Define $\bar{G}(p)$ by the pull-back

8.2.
$$\begin{array}{ccccc} \bar{G}(p) & & \xrightarrow{\quad\rho(p)\quad} & & map(\bar{X},\bar{X}) \\ \downarrow{\scriptstyle q(p)} & & & & \downarrow{\scriptstyle p_*} \\ G & \xrightarrow{\rho} & map(X,X) & \xrightarrow{p^*} & map(\bar{X},X) \end{array}$$

Let $\Delta(p) \subset map(\bar{X},\bar{X})$ be the subset of deck-transformations $f : \bar{X} \longrightarrow \bar{X}$, i.e. maps covering the identity. We obtain $i(p) : \Delta(p) \longrightarrow \bar{G}(p)$ by 8.2. and the inclusion $\Delta(p) \longrightarrow map(\bar{X},\bar{X})$ and the constant map $\Delta(p) \longrightarrow G$ with unit $e \in G$ as value.

Choose $\bar{x} \in \bar{X}$ and $x \in X$ with $x = p\bar{x}$. In the following diagram $\overline{ev}$ and ev are given by evaluation at $\bar{x}$ and x.

8.3.
$$\begin{array}{ccc} \bar{G}(p) & \xrightarrow{\overline{ev}} & \bar{X} \\ \downarrow{\scriptstyle q(p)} & & \downarrow{\scriptstyle p} \\ G & \xrightarrow{ev} & X \end{array}$$

We claim that 8.3. is a pull-back. Choose a pull-back

8.4.
$$\begin{array}{ccc} ev^*\bar{X} & \xrightarrow{ev'} & \bar{X} \\ \downarrow{\scriptstyle \bar{p}} & & \downarrow{\scriptstyle p} \\ G & \xrightarrow{ev} & X \end{array}$$

We have to show that the map $f : \bar{G}(p) \longrightarrow ev^*\bar{X}$ induced by 8.3. is a homeomorphisms. We construct an inverse h . Notice that $ev^*\bar{X} \longrightarrow G$ is a covering so that $ev^*\bar{X}$ is locally path connected as G is. We want to construct in the diagram

8.5.
$$\begin{array}{ccccc} ev^*\bar{X} \times \bar{X} & \xrightarrow{\bar{p} \times p} & G \times X & \xrightarrow{\rho} & X \\ \uparrow{\scriptstyle i_{\bar{x}}} & & \searrow{\scriptstyle h'} & & \uparrow{\scriptstyle p} \\ ev^*\bar{X} & & \xrightarrow{ev'} & & \bar{X} \end{array}$$

a lift h' . Choose an arbitrary base point $(g,\bar{y}) \in ev^*\bar{X} \subset G \times \bar{X}$. The homomorphism $\rho_* : \pi_1(G,g) \times \pi_1(X,x) \longrightarrow \pi_1(X,gx)$ assigns $ev_*(v) \cdot l_{g*}(w)$ to (v,w) . The map $l_{g*} : \pi_1(X,x) \longrightarrow \pi_1(X,gx)$ sends the image of $p_* : \pi_1(\bar{X},\bar{x}) \longrightarrow \pi_1(X,x)$ into the image of $p_* : \pi_1(\bar{X},\bar{y}) \longrightarrow \pi_1(X,gx)$ because $l_{g*} : [S^1,X] \longrightarrow [S^1,X]$ respects image $[S^1,\bar{X}] \longrightarrow [S^1,X]$ (use Whitehead [1978] III 1.11). Hence the image of $(\rho \circ \bar{p} \times p)_* : \pi_1(ev^*\bar{X} \times \bar{X},((g,\bar{y}),\bar{x})) \longrightarrow \pi_1(X,gx)$ is contained in the image of $p_* : \pi_1(\bar{X},y) \longrightarrow \pi_1(X,gx)$. By Greenberg [1967] 6.1 such a lift h' exists.

By construction $\bar{p} : ev^*\bar{X} \to G$ and $h' : ev^*\bar{X} \to map(\bar{X},\bar{X})$ induce $h : ev^*\bar{X} \to \bar{G}(p)$ using 8.2. The composition $f \circ h : ev^*\bar{X} \to ev^*\bar{X}$ is the identity by the pull-back property of 8.4. and f is injective because of the Unique Lifting Theorem for coverings (Greenberg [1967] 5.1). Hence 8.3. is a pull back.

Composition defines a topological monoid structure on $map(\bar{X},\bar{X})$. By 8.2. we get the structure of a topological group on $\bar{G}(p)$. By 8.3. the sequence $1 \longrightarrow \Delta(p) \xrightarrow{i(p)} \bar{G}(p) \xrightarrow{q(p)} G \longrightarrow 1$ is an exact sequence of topological groups and a $\Delta(p)$-principal bundle. The operation of $\bar{G}(p)$ on $\bar{X}$ is given by $\rho(p)$ of 8.2. □

Remark 8.6. If G is a Lie group then $\bar{G}$ becomes a functor $\bar{G} : \mathcal{C}(G) \to \{\text{Lie groups}\}$. Namely, $\bar{G}(p) \longrightarrow G$ is a $\Delta(p)$-principal bundle for the discrete group $\Delta(p)$ and a topological group has at most one Lie group structure (see Bröcker-tom Dieck I.3.12)

□

Let $p : \tilde{X} \longrightarrow X$ be the universal covering. Write $\tilde{G} = \bar{G}(p)$, $\Delta = \Delta(p)$, $i = i(p)$ and $q = q(p)$. Consider the following diagram whose exact upper row is part of the long homotopy sequence of $\Delta \longrightarrow \tilde{G} \longrightarrow G$ and the lower exact row of $X \to EG \times_G X \longrightarrow BG$.

$$\begin{array}{ccccccccccc}
\pi_1(\tilde{G}) & \longrightarrow & \pi_1(G) & \longrightarrow & \Delta & \longrightarrow & \pi_o(\tilde{G}) & \longrightarrow & \pi_o(G) & \longrightarrow & 1 \\
\partial_2 \big\uparrow\!\!\uparrow & & \delta_2 \big\uparrow \cong & & \omega \big\uparrow \cong & & \partial_1 \big\uparrow \cong & & \delta_1 \big\uparrow \cong & & \\
\pi_2(EG \times_G X) & \longrightarrow & \pi_2(BG) & \longrightarrow & \pi_1(X) & \longrightarrow & \pi_1(EG \times_G X) & \longrightarrow & \pi_1(BG) & \longrightarrow & 1
\end{array}$$

The map ∂_i is the boundary map in the long homotopy sequence of $\tilde{G} \longrightarrow EG \times \tilde{X} \longrightarrow EG \times_{\tilde{G}} \tilde{X}$ if we regard EG as a $\tilde{G}$-space by $\tilde{G} \longrightarrow G$ and identify $EG \times_{\tilde{G}} \tilde{X}$ with $EG \times_G X$. Define δ_i similarly for $G \longrightarrow EG \longrightarrow BG$. The map ω sends $[w] \in \pi_1(X,x)$ to the deck transformation $\tilde{f} : \tilde{X} \longrightarrow \tilde{X}$ such that for any path $\tilde{w}$ from $\tilde{x}$ to $\tilde{f}\tilde{x}$ for a fixed $\tilde{x} \in \tilde{X}$ with $p\tilde{x} = x$ we have $[p \circ \tilde{w}] = [w]$.

<u>Lemma 8.7.</u> <u>The diagram above has exact rows and commutes. Moreover,</u> ∂_2 <u>is surjective and the other vertical maps are bijective.</u>

<u>Proof:</u> left to the reader. □

Consider a locally path connected regular topological group G and a connnected locally path connected regular G-space X possessing a universal covering $p : \tilde{X} \to X$. Let $\Delta = \Delta(p)$, $i = i(p)$, $q = q(p)$, $\tilde{G} = G(p)$ and $\tilde{\rho} = \tilde{\rho}(p)$ be given by the lift-extension (Theorem 8.1.). Since G is locally path connected, $\tilde{G}$ is locally path-connected. Let G_o and $\tilde{G}_o$ be the open closed normal subgroups given by the component of the unit. Notice that $\tilde{G}/\tilde{G}_o$ is discrete and isomorphic to $\pi_o(\tilde{G})$. We will later see that all the algebraic invariants like finiteness obstruction and White head torsion can be read off from the $\pi_o(\tilde{G})$-space $\tilde{X}/\tilde{G}_o$. Hence we study the square

8.8.
$$\begin{array}{ccc}
\tilde{X} & \xrightarrow{\ \tilde{o}\ } & \tilde{X}/\tilde{G}_o \\
p \big\downarrow & & \big\downarrow p/ \\
X & \xrightarrow{\ o\ } & X/G_o
\end{array}$$

<u>Lemma 8.9.</u>

a) <u>Suppose that</u> G <u>is a Lie group and</u> X <u>is proper and completely regular. Then</u> $\tilde{G}$ <u>is a Lie group and</u> $\tilde{X}$ <u>a proper completely regular</u> $\tilde{G}$ - <u>space and</u> $\tilde{X}/\tilde{G}_o$ <u>is a simply</u>

connected proper $\pi_o(\tilde{G})$-space.

b) Suppose that X is a proper G-CW-complex. Then $\tilde{X}$ is a proper $\tilde{G}$-CW-complex and $\tilde{X}/\tilde{G}_o$ is a simply connected proper $\pi_o(\tilde{G})$-CW-complex.

Proof:

a) One easily checks that $\tilde{X}$ is completely regular as X has this property. Consider the commutative diagram

$$\begin{array}{ccc} \tilde{G} \times \tilde{X} & \xrightarrow{\tilde{\Theta}} & \tilde{X} \times \tilde{X} \\ {\scriptstyle q\times p}\downarrow & & \downarrow{\scriptstyle p\times p} \\ G \times X & \xrightarrow{\Theta} & X \times X \end{array}$$

where $\tilde{\Theta}$ sends $(\tilde{g},\tilde{x})$ to $(\tilde{x},\tilde{g}\tilde{x})$ and Θ is defined analogously. Since $q \times p$ and $p \times p$ are normal coverings and $\tilde{\Theta}$ is bijective on each fibre this is a pull-back. As X is proper Θ is by definition proper. Then $\tilde{\Theta}$ is proper and hence $\tilde{X}$ is proper by Lemma 1.16. f. Now $\tilde{X}/\tilde{G}_o$ is simply connected by Corollary 1.40.

b) The n-skeleton and the open cells of the $\tilde{G}$-CW-complex $\tilde{X}$ are just the pre-images of the one of the G-CW-complex X. As in 1.25. one shows that the conditions of a $\tilde{G}$-CW-complex structure are fulfilled. This induces a $\pi_o(\tilde{G})$-CW-structure on $\tilde{X}/\tilde{G}_o$ by 1.4. Since X is proper and Δ acts freely on $\tilde{X}$ the $\tilde{G}$-action on $\tilde{X}$ is proper by Theorem 1.23. Because of Corollary 1.40. the orbit space $\tilde{X}/\tilde{G}_o$ is simply connected. □

One might think that $\tilde{p}/ : \tilde{X}/\tilde{G}_o \longrightarrow X/G_o$ is the universal covering of X/G but this is the not true in general. We determine the universal covering of X/G and $\pi_1(X/G)$ now.

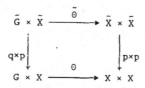

Suppose either that G is a Lie group and X is a connected locally path-connected completely regular G-space with a universal covering or that G is a locally path-connected regular topological group and X is a proper G-CW-complex with universal covering. Let $(\tilde{G},\tilde{\rho},\tilde{i},\tilde{q})$ be the associated lift-extension. Let K be the subgroup of $\tilde{G}$ generated by $\tilde{G}_o$ and all isotropy groups $G_{\tilde{x}}$ for $\tilde{x} \in \tilde{X}$. Since $\tilde{G}_o$ is normal in $\tilde{G}$ and $\tilde{G}_{\tilde{g}\tilde{x}} = \tilde{g}^{-1}\tilde{G}_{\tilde{x}}\tilde{g}$ holds, K is normal in $\tilde{G}$. Let $L \subset \pi_1(X,x)$ be the subset of homotopy classes of loops of the shape $w_1 * g_1\bar{w}_1 * g_1 w_2 * g_1 g_2 \bar{w}_2 * \ldots$ $* \ldots g_1 g_2 \ldots g_r \bar{w}_r * u g_1 g_2 \ldots g_r x$ where w_i is any path from x to some point $w_i(1)$ with inverse path $\bar{w}_i$ and u is a path in $\tilde{G}_o$ from the unit e to $u(1)$ such that $g_i \in G_{w_i(1)}$ and $u(1)g_1 g_2 \ldots g_r \in G_x$ holds. One easily checks that L is the image of $\Delta \cap K$ under the standard identification of Δ with $\pi_1(X,x)$ and hence is a normal subgroup in $\pi_1(X,x)$. Let $pr : X \longrightarrow X/G$ be the projection. Denote by $M \subset \pi_o(G)$ the normal subgroup generated by all $\pi_o(G_x)$ for all $x \in X$. Define a homomorphism $\partial : \pi_1(X/G,xG) \longrightarrow \pi_o(G)/M$ by assigning to $[w]$ the element represented by any $g \in G$ such that there exists a lift v of w with $v(0) = x$ and $v(1) = gx$.

<u>Proposition 8.10.</u> <u>Under the conditions above the universal covering of</u> X/G <u>is</u>

$$p_o : \tilde{X}/K \longrightarrow X/G$$

<u>Moreover, there is a well-defined exact sequence</u>

$$1 \longrightarrow L \longrightarrow \pi_1(X,x) \xrightarrow{p_*} \pi_1(X/G,xG) \xrightarrow{\partial} \pi_o(G)/M \longrightarrow 1$$

<u>Proof.</u> By Lemma 8.9. the $\tilde{G}$-space $\tilde{X}$ is proper. In the case of a Lie group G this is equivalent to the statement that for any two points $\tilde{x}$ and $\tilde{y}$ in $\tilde{X}$ there are neighbourhoods $V_{\tilde{x}}$ and $V_{\tilde{y}}$ such that $\operatorname{clos}\{g \in G \mid gV_x \cap V_y \neq \emptyset\}$ is compact in $\tilde{G}$ (tom Dieck [1987] I. 3.21.). Now it is easy to check that the $\tilde{G}/K$-space $\tilde{X}/K$ is proper. In the case of a G-CW-complex X this follows from Theorem 1.23. The action of $\tilde{G}/K$ on $\tilde{X}/K$ is free and $\tilde{G}/K$ is discrete by construction. Hence $p_o : \tilde{X}/K \rightarrow X/G$ is a $\tilde{G}/K$-principal bundle (tom Dieck [1987] I. 3.24.).

It remains to show that $\tilde{X}/K$ is simply connected. Consider any loop w in $\tilde{X}/K$ at $\tilde{x}K$ for $\tilde{x} \in \tilde{X}$. By Corollary 1.40. the loop w lifts to a path $\tilde{w}$ with $\tilde{w}(0)=\tilde{x}$. There is $k \in K$ with $\tilde{w}(1) = kx$. We can write $k = \tilde{g}_o\tilde{g}_1 \, \tilde{g}_r$ for $\tilde{g}_o \in \tilde{G}_o$ and $\tilde{g}_i \in \tilde{G}_{\tilde{x}_i}$ for $i = 1,\ldots,r$ and appropriate $\tilde{x}_i$. Choose a path $\tilde{w}_i$ from $\tilde{x}$ to $\tilde{x}_i$ in $\tilde{X}$ and a path u from 1 to $\tilde{g}_o$ in $\tilde{G}_o$. Let $\tilde{v}$ be the path

$$\tilde{w}_1 * \tilde{g}_1\bar{\tilde{w}}_1 * \tilde{g}_1\tilde{w}_2 * \tilde{g}_1\tilde{g}_2\tilde{w}_2 * \tilde{g}_1\tilde{g}_2\ldots\tilde{g}_r\bar{\tilde{w}}_r * u\cdot\tilde{g}_1\tilde{g}_2\ldots\tilde{g}_r\tilde{x} .$$

Then $\tilde{w} * \bar{\tilde{v}}$ is a loop in $\tilde{X}$ at $\tilde{x}$ and its projection is homotopic relative end points to w . Since $\tilde{X}$ is simply connected $\tilde{w} * \tilde{v}^-$ and hence w are nullhomotopic. We have shown that $\tilde{X}/K \longrightarrow X/G$ is the universal covering. This implies $\pi_1(X/G) = \tilde{G}/K$. We have the obvious exact sequence

$$1 \longrightarrow \Delta \cap K \longrightarrow \Delta \longrightarrow \tilde{G}/K \longrightarrow \tilde{G}/\Delta K \longrightarrow 1$$

We leave it to the reader to identify it with the sequence under consideration □

If one furthermore assumes that any G_x is path-connected, we obtain an exact sequence

8.11.
$$\pi_1(G,1) \xrightarrow{ev_*} \pi_1(X,x) \xrightarrow{p_*} \pi_1(X/G,xG) \longrightarrow \pi_o(G) \longrightarrow 1$$

The next result generalizes the observation above. It says that $p : X \longrightarrow X/G$ looks like a fibration in a certain range depending on the connectivity of $G \longrightarrow G/G_x$ for $x \in X$. For a Lie group G this is the connectivity of G_x increased by one. We omitt the proof as the result is not needed in this book.

Proposition 8.12. Let G be a topological group and X be a G-CW-complex. Let n be a non-negative integer such that $G \longrightarrow G/G_x$ is n-connected for any $x \in X$. Denote by $ev : G \longrightarrow X$ $g \longrightarrow gx$ the evaluation and by $p : X \longrightarrow X/G$ the projection.
Then there is an exact sequence

$$\pi_n(G,1) \xrightarrow{ev_*} \pi_n(X,x) \xrightarrow{p_*} \pi_n(X/G,xG) \xrightarrow{\partial} \pi_{n-1}(G,1) \xrightarrow{ev_*} \dots \xrightarrow{ev_*} \pi_o(X,x) \rightarrow \pi_o(X/G,xG) \rightarrow 1.$$ □

8 B. The (discrete) fundamental category and the (discrete) universal covering functor.

Now we come to the organization of all data like the component structure and fundamental groups of the various fixed point sets. It will be based on the following categories we will introduce now. The following assumptions are necessary and we suppose them for the remainder of this section.

Assumption 8.13.

i) G is a Lie group.

ii) If X is a G-space we suppose for any $H \subset G$ that X^H is locally path-connected and each component has a universal covering. Moreover, X is completely regular and a proper G-space. □

For the reader 's convenience we list the various conclusions out of this assumption we already have verified and need in the sequel.

8.14.

i) $WH^{op} \longrightarrow map(G/H,G/H)^G \quad gH \longrightarrow R_g$ is a homeomorphism for $H \in Iso(X)$ (Lemma 1.31.). Recall that R_g sends $g'H$ to $g'gK$.

ii) For $K \subset G$ and compact $H \subset K$ the space $G/K^H/WH$ is discrete (Theorem 1.33.)

iii) Let X be a G-CW-complex such that G_x is compact for any $x \in X$. Then assumption 8.13. is fulfilled (Theorem 1.23.). There is a canonical WH-CW-structure on X^H for any $H \subset G$. (1.36.). □

Notice that ii) and iii) can only be guaranteed under assumption 8.13.

Definition 8.15. The fundamental category $\Pi(G,X)$ of a G-space X is defined by

i) An object $x : G/H \longrightarrow X$ is a G-map. Sometimes we write $x(H)$ for shortness.

ii) A morphism $(\sigma,[w]) : x(H) \longrightarrow y(K)$ consists of a G-map $\sigma : G/H \longrightarrow G/K$ and a homotopy class $[w]$ relative $G/H \times \partial I$ of G-maps $w : G/H\times I \longrightarrow G/K$ with $w_1=x$ and $w_o = y\circ \sigma$. We often abbreviate $(\sigma,[w])$ by (σ,w) .

iii) The composition of the morphisms $(\sigma,w) : x(H) \longrightarrow y(K)$ and $(\tau,v) : y(K) \to z(L)$ is given by a "semi direct product formula" $(\tau\circ\sigma,\sigma^* v * w)$ with

$$\sigma^* v * w : G/H\times I \longrightarrow X$$

$$(gH,t) \longrightarrow \begin{cases} v(\sigma(gH),2t) & 0 \leq t \leq 1/2 \\ w(gH,2t-1) & 1/2 \leq t \leq 1 \end{cases}$$

An object $x : G/H \longrightarrow X$ is the same as a point x in X^H . A morphism $(w,\sigma) : x(H) \longrightarrow y(K)$ is a homotopy class of paths relative end points from $\sigma^* y(K)$ to $x(H)$ in X^H and $\sigma^* y(K)$ is $g\cdot y(K)$ for any g with $\sigma(eH) = g^{-1}K$. The composition uses the ordinary composition of paths. If G is compact and X a point, $\Pi(G,X)$ reduces to:

Definition 8.16. The orbit category OrG of a Lie group G has as objects proper homogeneous spaces G/H (i.e. H is compact) and as morphisms G-maps. If G is a compact topological group define OrG analogously. □

Remark 8.17. If G is trivial, $\Pi(G,X)$ is the fundamental groupoid $\Pi(X)$ of the space X . The fundamental category $\Pi(G,X)$ is the homotopy colimit (see Thomason [1982], p. 1624) of the covariant functor

$$\begin{aligned} \mathrm{Or}G^{op} &\longrightarrow \{\text{groupoids}\} \\ G/H &\longrightarrow \Pi(X^H) = \Pi(\mathrm{map}(G/H,X)^G) \end{aligned}$$

It is useful to topologize $\Pi(G,X)$. Recall that a category $\mathcal{C}$ is called topological if Ob $\mathcal{C}$ and Mor $\mathcal{C}$ are equipped with a topology such that all structure maps belonging to a category are continuous.

First we give the fundamental groupoid $\Pi(X)$ of a space X with a universal covering $p : \tilde{X} \longrightarrow X$ a topology. If Δ is the discrete group of deck transformations, let Δ act diagonally on $\tilde{X} \times \tilde{X}$. We require that the following bijections are homeomorphisms

8.18.
$$\begin{aligned} X &\longrightarrow \mathrm{Ob}\,\Pi(X) \\ \tilde{X} \times \tilde{X}/\Delta &\longrightarrow \mathrm{Mor}\,\Pi(X) \\ (\tilde{x},\tilde{y})\Delta &\longrightarrow [p\circ u] : p(\tilde{x}) \longrightarrow p(\tilde{y}) \end{aligned}$$

where u is any path from $\tilde{y}$ to $\tilde{x}$ in $\tilde{X}$.

We topologize $\mathrm{Ob}(\Pi/G,X))$ by the disjoint union

8.19.
$$\mathrm{Ob}(\Pi(G,X)) = \coprod_{H \subset G} \mathrm{map}(G/H,X) .$$

Let $M(G/H,G/K)$ be the pull-back

$$\begin{array}{ccc} M(G/H,G/K) & \longrightarrow & \mathrm{Mor}(\Pi(X^H)) \\ \downarrow & & \downarrow t \\ \mathrm{map}(G/H,G/K) \times \mathrm{map}(G/K,X) & \xrightarrow{\rho} & \mathrm{Ob}(\Pi(X^H)) \end{array}$$

where t sends a morphism to its target and ρ maps (σ,y) to $\sigma^* y$. Then define the topology on $\mathrm{Mor}(\Pi(G,X))$ by

8.20. $$\mathrm{Mor}(\Pi(G,X)) = \coprod_{H,K \subset G} M(G/H,G/K)$$

Notation 8.21. For an object $x : G/H \longrightarrow X$ define

a) $X^H(x)$ or briefly $X(x)$ is the component of X^H containing x .

b) $WH(x)$ is the isotropy group of $X^H(x)$ in $\pi_o(X^H)$ under the WH-action.

c) $\tilde{X}^H(x)$ (or briefly $\tilde{X}(x)$) is the set of all morphisms $(\sigma,w) : x \longrightarrow y$ with σ the identity $G/H \longrightarrow G/H$ equipped with the subspace topology of $\mathrm{Mor}(\Pi(G,X))$. Let $\tilde{x} \in \tilde{X}^H(x)$ be the identity $x \longrightarrow x$. The map $p(x) : \tilde{X}^H(x) \longrightarrow X^H(x)$ sends $(\sigma,w) : x \longrightarrow y$ to y .

d) $X^{>H}(x)$ (or briefly $X^>(x)$) is $X^H(x) \cap X^{>H}$ and $\tilde{X}^{>H}(x)$ (or $\tilde{X}^>(x)$) is $p(x)^{-1}(X^{>H}(x))$. □

By construction $p(x) : \tilde{X}(x) \longrightarrow X(x)$ is the standard model of the universal covering of $X(x)$.

Definition 8.22. The universal covering functor is the contravariant functor

$$\tilde{X} : \Pi(G,X) \longrightarrow \{\text{top. spaces}\}$$

sending $(\sigma,w) : x(H) \longrightarrow x(K)$ to $\tilde{X}(\sigma,w) : \tilde{X}^K(y) \longrightarrow \tilde{X}^H(x)$ induced by the composition of morphisms. □

8.23. Let $x = x(H)$ and $y = y(K)$ be objects in $\Pi(G,X)$ and $\mathrm{Mor}(x,y)$ be the space of morphisms. Let $\mathrm{map}(G/H,G/K)^G_{x,y}$ be the subspace of $\mathrm{map}(G/H,G/K)^G$ consisting of G-maps $\sigma : G/H \longrightarrow G/K$ for which $\pi_o(\sigma^*) : \pi_o(X^K) \longrightarrow \pi_o(X^H)$ sends $X^K(y)$ to $X^H(x)$. Denote by $q(x,y) : \mathrm{Mor}(x,y) \longrightarrow \mathrm{map}(G/H,G/K)^G_{x,y}$ the projection $(\sigma,w) \longrightarrow \sigma$. Let $\widetilde{ev} : \mathrm{Mor}(x,y) \longrightarrow \tilde{X}^H(x)$ send (σ,w) to $\tilde{X}(\sigma,w)(\tilde{y})$ for $\tilde{y} \in \tilde{X}^K(y)$ given by id_y and $ev : \mathrm{map}(G/H,G(K)^G_{x,y} \longrightarrow X^H(x)$ map σ to $\sigma^* y$. Similarly to the proof of 8.3. one checks that we obtain a pull back

$$
\begin{array}{ccc}
\mathrm{Mor}(x,y) & \xrightarrow{\widetilde{ev}} & \tilde{X}^H(x) \\
\Big\downarrow{\scriptstyle q(x,y)} & & \Big\downarrow{\scriptstyle \tilde{p}(x)} \\
\mathrm{map}(G/H,G/K)^G_{x,y} & \xrightarrow{ev} & X^H(x)
\end{array}
$$

Then $q(x,y)$ is a $\pi_1(X^H(x),x)$-principal bundle. Especially we obtain for $x = y$ using 8.14. i an exact sequence

$$1 \longrightarrow \pi_1(X^H(x),x) \xrightarrow{i(x)} \mathrm{Aut}(x) \xrightarrow{q(x)} WH(x)^{op} \longrightarrow 1$$

where $i(x)(w) = (id,w)$ and $q(x)=q(x,x)$. Since $q(x)$ is a $\pi_1(X^H(x),x)$-bundle, $\mathrm{Aut}(x)$ has a unique Lie group structure and the sequence above becomes a sequence of Lie groups (see Remark 8.6.). □

By functoriality we obtain an operation

$$\rho(x) : \mathrm{Aut}(x)^{op} \times \tilde{X}(x) \longrightarrow \tilde{X}(x)$$

Identify Δ and $\pi_1(X^H(x),x)^{op}$ in the usual way. Then we have

<u>Lemma 8.24.</u> <u>We obtain a lift extension for the</u> $WH(x)$-<u>space</u> $X^H(x)$ <u>with universal covering</u> $p(x) : \tilde{X}^H(x) \longrightarrow X^H(x)$ <u>by</u> $\mathrm{Aut}(x)^{op}, \rho(x), i(x), q(x)$. □

We have done the first step, going up to the universal covering. The second step, dividing out the action of the component of the unit needs some preparations. Let the maps in the diagram

8.25.

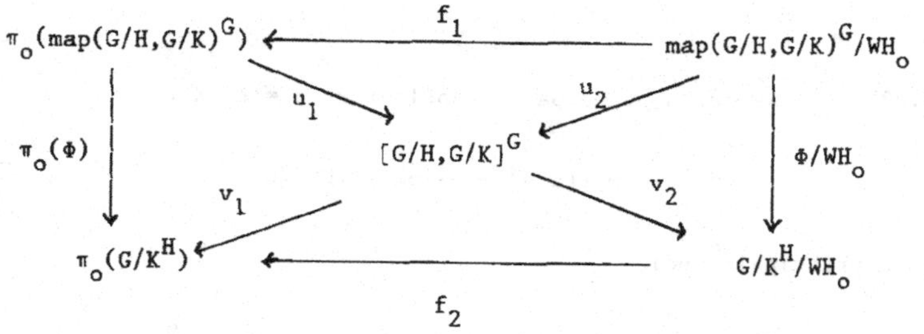

be defined by

$\Phi : \mathrm{map}(G/H,G/K)^G \longrightarrow G/K^H \qquad \phi \longrightarrow \phi(eH)$

$f_1(\phi \cdot WH_o) =$ component of ϕ

u_1 (component of ϕ) $= [\phi]$

$u_2(\phi \cdot WH_o) = [\phi]$

$v_1([\phi]) =$ component of $\Phi(\phi)$

$v_2([\phi]) = \Phi(\phi) \cdot WH_o$

<u>Lemma 8.26.</u> <u>All these maps are well defined and bijective and the diagram commutes</u>.

<u>Proof:</u> The homeomorphism Φ is introduced in 1.30. The only remaining difficulty is to show that $f_2 : G/K^H/WH_o \longrightarrow \pi_o(G/K^H)$ is bijective. This follows from 8.14.ii). □

<u>Lemma 8.27.</u> <u>If</u> x <u>and</u> y <u>are objects in</u> $\Pi(G,X)$, <u>the following diagram commutes and all maps are bijections</u>

$$\begin{array}{ccc} & \mathrm{Mor}(x,y)/\mathrm{Aut}(x)_o & \\ \swarrow & & \searrow \\ \mathrm{Aut}(y)_o\backslash \mathrm{Mor}(x,y)/\mathrm{Aut}(x)_o & \longrightarrow & \pi_o(\mathrm{Mor}(x,y)) \end{array}$$

<u>Proof:</u> The only non-trivial part is the verification that $\mathrm{Mor}(x,y)/\mathrm{Aut}(x)_o \longrightarrow \pi_o(\mathrm{Mor}(x,y))$ is injective. It suffices to show that $\mathrm{Mor}(x,y)/\mathrm{Aut}(x)_o$ is discrete.

From 8.23. we get a principal π-bundle for $\pi = \pi_1(X^H(x),x)$

$$\pi \longrightarrow \mathrm{Mor}(x,y) \longrightarrow \mathrm{map}(G/H,G/K)^G_{x,y}$$

This induces a principal bundle

$$\pi/\pi \cap \mathrm{Aut}(x)_o \longrightarrow \mathrm{Mor}(x,y)/\mathrm{Aut}(x)_o \longrightarrow \mathrm{map}(G/H,G/K)^G_{x,y}/WH(x)_o .$$

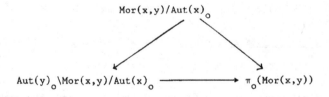

By Lemma 8.26. $\mathrm{map}(G/H,G/K)^G/WH_o$ is discrete so that the subspace $\mathrm{map}(G/H,G/K)^G_{x,y}/WH(x)_o$ is discrete. Since $\pi/\pi \cap \mathrm{Aut}(x)$ is discrete, the same is true for $\mathrm{Mor}(x,y)/\mathrm{Aut}(x)_o$.

If $\mathcal{C}$ is a topological category, its induced discrete category $\mathcal{C}/$ has the same objects as $\mathcal{C}$ and $\mathrm{Mor}_{\mathcal{C}/}(x,y)$ is $\pi_o(\mathrm{Mor}(x,y))$ for objects x,y .

Definition 8.28. The discrete fundamental category $\Pi/(G,X)$ is the discrete category induced by $\Pi(G,X)$. □

In the sequel we denote for $x,y \in \mathrm{Ob}\,\Pi(G,X) = \mathrm{Ob}\,\Pi/(G,X)$ the set of morphisms $\mathrm{Mor}_{\Pi(G,X)}(x,y)$ in $\Pi(G,X)$ by $\mathrm{Mor}(x,y)$ and $\mathrm{Mor}_{\Pi/(G,X)}(x,y)$ in $\Pi/(G,X)$ by $\mathrm{Mor}/(x,y)$. We have

8.29. $\mathrm{Mor}/(x,y) = \mathrm{Mor}(x,y)/\mathrm{Aut}(x)_o = \mathrm{Aut}(y)_o\backslash \mathrm{Mor}(x,y)/\mathrm{Aut}(x)_o = \pi_o(\mathrm{Mor}(x,y))$

The universal covering functor $\tilde{X} : \Pi(G,X) \longrightarrow \{\text{top. spaces}\}$ induces a contravariant functor $\tilde{X}/ : \Pi/(G,X) \longrightarrow \{\text{top. spaces}\}$ by $\tilde{X}/(x) := \tilde{X}(x)/\mathrm{Aut}(x_o)$. A morphism in $\Pi/(G,X)$ represented by the morphism $(\sigma,w) : x \longrightarrow y$ is sent to the map induced by $\tilde{X}(\sigma,w)$:

$$\begin{array}{ccc} \tilde{X}(y) & \xrightarrow{\tilde{X}(\sigma,w)} & \tilde{X}(x) \\ \downarrow & & \downarrow \\ \tilde{X}(y)/\mathrm{Aut}(y)_o & \xrightarrow{\tilde{X}/(\sigma,w)} & \tilde{X}(x)/\mathrm{Aut}(x)_o \end{array}$$

This is well-defined by Lemma 8.27.

Definition 8.30. The contravariant functor

$$\tilde{X}/ : \Pi/(G,X) \longrightarrow \{\text{top. spaces}\}$$

is called the discrete universal covering functor. □

Remark 8.31. The category $\Pi/(G,X)$ together with $\tilde{X}/$ are the basic objects we seek in this section. Of course their definition is complicated. This is not surprising since they contain such a lot of information about the G-space X . Their main advantage will be that in the applications one can forget the involved structure and has only to keep in mind that we are given a category Γ with a contravariant functor $\Gamma \longrightarrow$ {top. spaces} and that any endomorphism in Γ is an isomorphism (use 8.14. i). If G is discrete, $\Pi(G,X)$ and $\Pi/(G,X)$ coincide. □

Notation 8.32. If $x : G/H \longrightarrow X$ is an object define

a) $o(x) : X^H(x) \longrightarrow X^H/(x)$ is the projection under the $WH(x)_o$-action writing $X^H/(x) = X^H(x)/WH(x)_o$.

b) $\tilde{o}(x) : \tilde{X}^H(x) \longrightarrow \tilde{X}^H/(x)$ is the projection under the $Aut(x)_o$-action where $\tilde{X}^H/(x)$ is $\tilde{X}^H(x)/Aut(x)_o$.

c) Let $p/(x)$ be the projection such that we obtain a commutative diagram

$$\begin{array}{ccc} \tilde{X}^H(x) & \xrightarrow{\tilde{o}(x)} & \tilde{X}^H/(x) \\ {\scriptstyle p(x)}\downarrow & & \downarrow{\scriptstyle p/(x)} \\ X^H(x) & \xrightarrow{o(x)} & X^H/(x) \end{array}$$

d) Write $X^{>H}/(x) = o(x)(X^{>H}(x))$ and $\tilde{X}^{>H}/(x) = \tilde{o}(x)(\tilde{X}^{>H}(x))$.

e) We sometimes omitt the H, in expressions like $X^H(x), \tilde{X}^H(x), \ldots$ □

The main properties of these definitions are

Proposition 8.33. *Let G be a Lie group and X a proper G-CW-complex.*

a) *The discrete universal covering functor is a functor $\tilde{X}/ : \Pi/(G,X) \rightarrow$ {CW-compl} into the category of simply connected CW-complexes with cell preserving maps as morphisms.*

b) *Let $x : G/H \longrightarrow X$ be an object in $\Pi/(G,X)$. Then the following sequence is exact*

$$\pi_1(WH(x),e) \xrightarrow{ev_*} \pi_1(X^H(x),x) \xrightarrow{i/(x)} Aut/(x) \xrightarrow{q/(x)} \pi_o(WH(x)) \longrightarrow 1$$

if $ev : WH(x) \longrightarrow X^H(x)$ sends w to $w \cdot x$, $Aut/(x)$ is the group of automorphisms of x in $\Pi/(G,X)$, $i/(x)$ sends $[v]$ to $[id,[v])$ and $q/(x)$ maps $[\sigma,[v]]$ to the component of $\sigma \in WH(x) \subset WH = map(G/H,G/H)$.

c) *The sequence in b) is naturally isomorphic to the following part of the long exact sequence of the bundle $X^H(x) \longrightarrow EWH(x) \times_{WH(x)} X^H(x) \longrightarrow BWH(x)$.*

$$\pi_2(BWH(x)) \longrightarrow \pi_1(X^H(x)) \longrightarrow \pi_1(EWH(x) \times_{WH(x)} X^H(x)) \longrightarrow \pi_1(BWH(x)) \longrightarrow 1 .$$

Proof: Lemma 8.7., Lemma 8.9. and Lemma 8.24. □

Remark 8.34. We have introduced the component category $\Pi_o(G,X)$ in Definition 5.1. Recall that Is $\mathcal{C}$ denotes the set of isomorphism classes of objects for a category $\mathcal{C}$. Using 5.2. we get bijections

$$\text{Is } \Pi_o(G,X) = \text{Is } \Pi(G,X) = \text{Is } \Pi/(G,X) = \coprod_{(H) \in \text{Con } G} \pi_o(X^H)/WH \qquad □$$

If $f : X \longrightarrow Y$ is a G-map, we get functors $\Pi(G,X) : \Pi(G,X) \longrightarrow \Pi(G,Y)$ and $\Pi/(G,f)$ $\Pi/(G,f) : \Pi/(G,X) \longrightarrow \Pi/(G,Y)$ by composition. We later need

Lemma 8.35. Let $f : X \longrightarrow Y$ be a G-map between G-spaces such that $\pi_o(f^H) : \pi_o(X^H) \longrightarrow \pi_o(Y^H)$ and $\pi_1(f^H,x) : \pi_1(X^H,x) \longrightarrow \pi_1(Y^H,fx)$ for $x \in X^H$ and $H \subset G$ are bijective. Then $\Pi/(G,f)$ and $\Pi(G,f)$ are equivalences of categories.

Proof: A functor $F : \mathcal{C} \longrightarrow \mathcal{D}$ is an equivalence of categories if and only if F satisfies

i) F induces a bijection Is $\mathcal{C} \longrightarrow$ Is $\mathcal{D}$

ii) F induces a bijection $\text{Mor}_{\mathcal{C}}(x,y) \longrightarrow \text{Mor}_{\mathcal{D}}(Fx,Fy)$ $f \longrightarrow F(f)$ for any $x,y \in \mathcal{C}$.

(see MacLane [1971]). Because of Remark 8.34. both $\Pi(G,f)$ and $\Pi/(G,f)$ satisfy i). One easily checks that it suffices to prove ii) for $x = y$. Then ii) follows from the five lemma and Theorem 8.33. □

We mention the following variant

8.36. The discrete orbit category $Or/(G)$ of a Lie group G has as objects homogeneous spaces G/H with compact H and G-homotopy classes of G-maps as morphisms. Given a G-space X define contravariant functors

$$X : \text{Or } G \longrightarrow \{\text{top. spac.}\} \quad G/H \longrightarrow X^H$$

$$X/ : \text{Or}/G \longrightarrow \{\text{top. spac.}\} \quad G/H \longrightarrow X^H/WH_o$$

The verification that they are well-defined is completely analogous to the consideration above and is based on Lemma 8.26. The advantage of this situation is that the functors X and X/ have always the same source Or G and Or/G □

The decisive step from geometry and algebra lies in the following notion of the cellular chain complex functor. If Z is a CW-complex, its <u>cellular chain complex</u> $C^c(X;R)$ with R coefficients is defined by

$$\cdots \xrightarrow{\Delta_{n+1}} H_n(X_n,X_{n-1};R) \xrightarrow{\Delta_n} H_{n-1}(X_{n-1},X_{n-2};R) \xrightarrow{\Delta_{n-1}} \cdots$$

where Δ_n is the connecting homomorphism of the triple (X_n,X_{n-1},X_{n-2}). We obtain a functor

$$C^c : \{\text{CW-compl.}\} \longrightarrow \{\text{R-chain compl.}\}$$

<u>Definition 8.37.</u> <u>Let</u> G <u>be a Lie group and</u> X <u>a proper</u> G-CW-<u>complex. Define its cellular</u> $R\Pi/(G,X)$ <u>chain complex</u> $C^c(X;R)$ <u>as the contravariant functor given by the composition</u>

$$\Pi/(G,X) \xrightarrow{\tilde{X}/} \{\text{CW-compl.}\} \longrightarrow \{\text{R-chain compl.}\}$$

<u>Define the cellular</u> ROr/G-<u>chain complex analogously</u>. □

Here we can again see the use of the passage from $\tilde{X}$ to $\tilde{X}/$. It brings us back to the discrete case. Recall that a G-CW-complex has a canonical CW-complex structure if G is discrete but not in general. Hence it makes no sense to compose $\tilde{X}$ with C^c .

<u>Comments 8.38.</u> Lifting a G-action on X to a covering of X is treated also in Bredon [1972] I.9.. Proposition 8.10. generalizes a result of Armstrong [1982] where only discrete groups are considered.

The orbit category Or G is used in Bredon [1967] to define equivariant homology and cohomology with local coefficients. The notion of the cellular chain complex over the orbit category is developed to define equivariant finiteness obstructions for G-CW-complexes with simply connected fixed point sets in tom Dieck [1981]. Further examples of categories useful for transformation groups are given in tom Dieck [1987] I. 10.

We have already mentioned that the cellular $R\Pi/(G,X)$-chain complex $C^c(X):\Pi/(G,X) \longrightarrow$ {R-chain compl.} is an important link between geometry and algebra. Although its precise definition is complicated, its formal properties are very similar to the one of the ordinary cellular chain complex of the universal covering of a CW-complex. We recommend to the reader to think of $C^c(X)$ in this simple fashion.

The notion of the cellular $R\Pi/(G,X)$-chain complex initiates the study of the algebra of contravariant functors from a category Γ into the category of R-modules and R-chain complexes we begin in the next section. The only property of $\Pi/(G,X)$ we must keep in mind is the EI-property, i.e. endomorphisms are isomorphisms. □

Exercises 8.40.

1) Give an explicite example of a G-space X together with a regular covering $p : \bar{X} \longrightarrow X$ such that no lift extension exists.

2) Let $\mathbb{Z}/2$ act on S^1 by the antipodal map resp. complex-conjugation resp. trivially. Determine for any covering of S^1 the lift-extension.

3) Let G be a locally path-connected regular topological group and $p : \bar{X} \to X$ be a regular covering of the locally path-connected regular G-space X. Let $\bar{x} \in \bar{X}$ and $x \in X$ be points with $p\bar{x} = x$ such that G leaves x stationary. Suppose that the G-action on $\pi_1(X,x)$ respects the image of $p_* : \pi_1(\bar{X},\bar{x}) \longrightarrow \pi_1(X,x)$.

Show the unique existence of a G-action on $\bar{X}$ leaving $\bar{x}$ stationary and covering the given G-action on X.

4) Let G be a locally path-connected regular topological group acting on the non-orientable smooth manifold M. Then there is a unique G-action on the orientation covering $\hat{M}$ of M by orientation preserving homeomorphisms covering the given G-action.

5) Let G be a locally path-connected regular topological group and X be a connected locally path-connected regular G-space with universal covering $\tilde{X} \longrightarrow X$. Let $(\tilde{G},\tilde{\phi},\tilde{i},\tilde{q})$ be the associated lift-extension.

a) If $X \longrightarrow X/G$ is a G-principal bundle and X/G is regular, show $\pi_o(\tilde{G}) = \pi_1(X/G)$

b) Suppose that X has a fixed point x. Then $\pi_o(G)$ acts on $\pi_1(X,x)$ and

$\pi_o(\tilde{G})$ is the semi-direct product.

6) Give an explicite example of a compact Lie group G and a G-CW-complex X such that $\tilde{X}/\tilde{G}_o \longrightarrow X/G_o$ is not the universal covering.

7) Let G be a Lie group and X be a connected locally path-connected completely regular G-space possessing a universal covering. Suppose that X^H is path connected for any $H \subset G$ and there is $x \in X^G$ such that G acts trivially on $\pi_1(X,x)$. Then $p_* : \pi_1(X,x) \longrightarrow \pi_1(X/G,xG)$ is bijective.

8) Consider a locally path-connected simply-connected completely regular $\mathbf{Z}/p^n$-space X. Then X has a fixed point if and only if $X/(\mathbf{Z}/p^n)$ is simply-connected.

9) The Hawaiian earring H is the union of circles $\bigcup_{n=1}^{\infty} C_n$ with

$$C_n = \{(x,y) \in \mathbf{R}^2 \mid (x-1/n)^2 + y^2 = 1/n^2\}$$

equipped with the subspace topology $H \subset \mathbf{R}^2$. Let $\mathbf{Z}/2$ act on H by reflecting in the x-axis. Show that the sequence appearing in Proposition 8.10. is not exact for the $\mathbf{Z}/2$-space H. Which condition is not satisfied?

10) Try to generalize the exact sequence of Proposition 8.10. to the case where X might not be connected. One would expect a sequence for appropriate L and M of the shape

$$1 \longrightarrow L \longrightarrow \pi_1(X,x) \xrightarrow{p_*} \pi_1(X/G,xG) \xrightarrow{\partial} \pi_o(G)/M \xrightarrow{ev_*} \pi_o(X) \xrightarrow{p_*} \pi_o(X/G) \longrightarrow 1$$

11) Let G be a compact Lie group and X a G-space such that X^H is simply-connected and non-empty for $H \subset G$. Show that the obvious projections $\Pi(G,X) \longrightarrow \mathrm{Or}\, G$ and $\Pi/(G,X) \longrightarrow \mathrm{Or}/(G)$ are equivalences of categories. Is the converse true?

12) Let G be a compact Lie group and X be a completely regular G-space such that X^H is connected, locally path-connected and non-empty and has a universal covering for any $H \subset G$. Then the following assertions are equivalent:

i) For any object x in $\Pi(G,X)$ we have $\pi_o(\mathrm{Aut}(x)) = \mathrm{Aut}/(x) = \{id\}$.

ii) G is a torus and X^H is simply connected for any $H \subset G$.

13) Let $\mathbb{Z}/2$ act on S^n by reflecting in the equator $S^{n-1} \subset S^n$. Write down explicitely the fundamental category, the universal covering functor and the cellular $R\Pi(\mathbb{Z}/2,S^n)$-chain complex for $n \geq 0$.

14) Let G be a locally compact topological group and $G \longrightarrow X \xrightarrow{p} X/G$ be a G-principal bundle in the category of topological spaces. Assume that X/G is paracompact and compactly generated. Then $G \longrightarrow X \xrightarrow{p} X/G$ is a G-principal bundle in the category of compactly generated spaces.

Remark: A CW-complex is paracompact (see Michael [1956]).

15) A space X is semi-locally simply-connected if any point $x \in X$ possesses a neighbourhood U such that any loop in U is nullhomotopic in X. Show for a connected locally path-connected and semi-locally path-connected space X that it has a universal covering in the category of topological spaces. Prove furthermore that this universal covering belongs to the category of compactly generated spaces if X is compactly generated and paracompact.

16) Let $\exp:\mathbb{R} \longrightarrow S^1$ be the universal covering of $S^1 \subset \mathbb{R}^2$. Define $Z \subset \mathbb{R}^2$ as the union of $A = \{(s,t) \mid s^2+t^2 = 1,\ t \geq 0\}$, $B = \{(s,t) \mid -1 \leq s \leq 0,\ t = 0\}$ and $C = \{(s,1/2 \sin \pi/s) \mid 0 < s \leq 1\}$. Let $z \in Z$ be $(1,0)$. Let $f : Z \longrightarrow S^1$ be the map which is the identity on A and sends $(s,t) \in B \cup C$ to $(s',t') \in S^1$ with $s = s'$ and $t' \leq 0$. Show that f does not lift to a (continuous) map $(Z,z) \longrightarrow (\mathbb{R},0)$ and that Z is not locally path-connected. □

CHAPTER II

ALGEBRAICALLY DEFINED INVARIANTS

Summary

In section 9 we analyse the notion of a module over a category Γ, i.e. a contravariant functor $\Gamma \longrightarrow$ R-MOD into the category of R-modules for a commutative ring R (see 9.4.). This is a generalization of the algebra of modules over a group ring R[G] (see Example 9.5.). We keep all notions as close as possible to this special case. We define "based free" (Definition 9.17.), "finitely generated"(9.19.) such that a lot of well-known statements about R[G]-modules still make sense if one substitutes RG-module by $R\Gamma$-module. For example, the cellular $\mathbb{Z}\Pi/(G,X)$-chain complex is free with a bases whose elements are in bijective correspondence with the equivariant cells (Example 9.18.). The category of $R\Gamma$-modules MOD-$R\Gamma$ inherits the structure of an abelian category from R-MOD so that notions like "direct sum", "exact sequence", "projective", "chain complex", "homology", "projective resolution" are defined. In geometric applications Γ is the discrete fundamental category $\Pi/(G,X)$ (see Definition 8.28.). For the algebra we can forget its complicated structure, but consider any EI-category Γ (i.e. a small category such that all endomorphisms are isomorphisms.) The orbit category Or G of a finite group G has as objects homogeneous G-spaces G/H and as morphisms G-maps and is also an EI-category.

The general strategy is to extend notions well-known for R[G]-modules and R[G]-chain complexes to $R\Gamma$-modules and $R\Gamma$-chain complexes such that all of the familiar formal properties survive. This enables us to carry over definitions and constructions for spaces without group actions and their universal coverings to G-spaces and their discrete universal covering functor $\tilde{X}/ : \Pi/(G,X) \longrightarrow$ {top.spac.} (see Definition 8.30.). Then we compute or approximate these notions and constructions for $R\Gamma$-modules in terms of R[x]-modules for $x \in \mathrm{Ob}\,\Gamma$, where R[x] is the group ring R[Aut(x)].

The following functors relate $R\Gamma$-modules and R[x]-modules

9.26 splitting functor S_x : MOD-$R\Gamma \longrightarrow$ MOD-R[x]

9.27 restriction functor Res_x : MOD-$R\Gamma \longrightarrow$ MOD-R[x]

9.28 extension functor $E_x : \text{MOD-}R[x] \longrightarrow \text{MOD-}R\Gamma$

9.29 inclusion functor $I_x : \text{MOD-}R[x] \longrightarrow \text{MOD-}R\Gamma$

Given a $R\Gamma$-module M, let $\operatorname{Res}_x M$ be $M(x)$. Let $M(x)_s$ be the $R[x]$-submodule generated by all images $M(f) : M(y) \longrightarrow M(x)$ if $f : x \longrightarrow y$ runs over all non-isomorphisms f with x as source. Then $S_x M$ is $M(x)/M(x)_s$. If $R\operatorname{Hom}(?,x)$ is the free R-module generated by the set of morphisms $? \longrightarrow x$, let E_x send the $R[x]$-module N to $N \otimes_{R[x]} R\operatorname{Hom}(?,x)$. Define $I_x N$ to be the $R\Gamma$-module sending y to $N \otimes_{R[x]} R\operatorname{Hom}(y,x)$ if x and y are isomorphic and to $\{0\}$ otherwise. Then $(E_x, \operatorname{Res}_x)$ and (S_x, I_x) are pairs of adjoint functors. Given $x \in \operatorname{Ob}\Gamma$, let $\bar{x}$ be its isomorphism class. We write $\bar{x} \le \bar{y}$, if $\operatorname{Hom}(x,y) \neq \emptyset$, and $\bar{x} < \bar{y}$, if $\bar{x} \leqq \bar{y}$ and $\bar{x} \neq \bar{y}$. Denote by $\operatorname{Is}\Gamma$ the set of isomorphism classes of objects. We call a $R\Gamma$-module M of type T for $T \subset \operatorname{Is}\Gamma$ if $\operatorname{Iso} M = \{\bar{x} \in \operatorname{Is}\Gamma \mid S_x M \neq \{0\}\}$ is contained in T. The main result of section 9 is

<u>Theorem 9.39.</u> <u>The Cofiltration Theorem for projective</u> $R\Gamma$-<u>modules</u>. <u>Let</u> Γ <u>be a EI-category</u>. <u>Let</u> $\emptyset = T_o \subset T_1 \subset \ldots \subset T_\ell = T$ <u>be a filtration of</u> T <u>such that</u> $\bar{x} \in T_i$, $\bar{y} \in T_j$, $\bar{x} < \bar{y}$ <u>implies</u> $i > j$. <u>Consider a projective</u> $R\Gamma$-<u>module</u> P <u>of type</u> T. <u>Then there is a natural cofiltration</u>

$$P = P_o \xrightarrow{PR_o} P_1 \xrightarrow{PR_1} P_2 \xrightarrow{PR_2} \ldots \xrightarrow{PR_{\ell-1}} P_\ell = \{0\}$$

<u>satisfying</u>

a) P_i <u>is projective of type</u> $T \setminus T_i$. <u>If</u> P <u>is finitely generated</u>, <u>then</u> P_i <u>is finitely generated</u>.

b) <u>Let</u> $PR^i : P \longrightarrow P_i$ <u>be</u> $PR_{i-1} \circ PR_{i-2} \circ \ldots \circ PR_o$. <u>Then</u> $S_x PR^i : S_x P \rightarrow S_x P_i$ <u>is an isomorphism for</u> $\bar{x} \in T \setminus T_i$ <u>and</u> $S_x P_i = \{0\}$ <u>for</u> $\bar{x} \notin T \setminus T_i$.

c) <u>There is a natural exact sequence which splits (not naturally)</u>

$$0 \longrightarrow \bigoplus_{\bar{x} \in T_i \setminus T_{i-1}} E_x \circ S_x P \longrightarrow P_{i-1} \xrightarrow{PR_{i-1}} P_i \longrightarrow 0 \qquad \square$$

This implies

<u>Corollary 9.40.</u> <u>Let</u> P <u>be a</u> $R\Gamma$-<u>module with finite</u> $\operatorname{Iso} P$. <u>Then the following statements are equivalent:</u>

i) P <u>is projective</u>.

ii) $S_x P$ <u>is projective for all</u> $x \in \mathrm{Ob}\,\Gamma$ <u>and there is a (not natural) isomorphism</u>

$$P \cong \bigoplus_{\bar{x} \in \mathrm{Iso}\, P} E_x \circ S_x P .$$

In section 10. we introduce the K-theory $K_n(R\Gamma)$ (Def. 10.7.) and the Whitehead group $Wh(R\Gamma)$ (Def. 10.8). We derive from the Cofiltration Theorem using Waldhausen's Additivity Lemma

<u>Theorem 10.34.</u> <u>Splitting Theorem for algebraic</u> K-<u>theory of</u> $R\Gamma$-<u>modules</u>. <u>Let</u> Γ <u>be a</u> EI-<u>category</u>. <u>Then the splitting and extension functors induce a pair of natural inverse isomorphisms</u>

$$K_n(R\Gamma) \underset{E}{\overset{S}{\rightleftarrows}} \bigoplus_{\bar{x} \in \mathrm{Is}\,\Gamma} K_n(R[x])$$

<u>and</u>

$$Wh(R\Gamma) \underset{E}{\overset{S}{\rightleftarrows}} \bigoplus_{\bar{x} \in \mathrm{Is}\,\Gamma} Wh(R[x]) \qquad \square$$

In section 11 we define and study the <u>finiteness obstruction</u> $o(C)$ of a finitely dominated $R\Gamma$-chain complex C. If P is a finite projective $R\Gamma$-chain complex $R\Gamma$-chain homotopy equivalent to C, we put $o(C) = \Sigma(-1)^m[P_n] \in K_o(R\Gamma)$ (Definition 11.1.). We give various tools for computations, <u>additivity</u> (Theorem 11.2.), <u>homological computation</u> (Proposition 11.9.), <u>instant formula</u> (Proposition 11.12.), <u>computation by highly connected approximations</u> (Proposition 11.13.) and <u>product formulas</u> (11.18. and Theorem 11.24.). We examine its behaviour under the splitting of Theorem 10.34 and deal with the finiteness obstruction of a chain <u>homotopy projection</u>.

The subject of section 12. are torsion invariants of $R\Gamma$-chain complexes. Let $f : C \longrightarrow D$ be a $R\Gamma$-chain homotopy equivalence of finite projective $R\Gamma$-chain complexes and $\Phi : C_{odd} \oplus D_{ev} \longrightarrow C_{ev} \oplus D_{odd}$ be a stable isomorphism. Let c be the differential and γ be any chain contraction of $\mathrm{Cone}(f)$. If π denotes the obvious permutation map, we obtain a stable automorphism of a finitely generated projective $R\Gamma$-module

$$\mathrm{Cone}(f)_{odd} \xrightarrow{(c+\gamma)} \mathrm{Cone}(f)_{ev} \xrightarrow{\pi} C_{odd} \oplus D_{ev} \xrightarrow{\Phi} C_{ev} \oplus D_{odd} \xrightarrow{\pi} \mathrm{Cone}(f)_{odd} .$$

Its class in $K_1(R\Gamma)$ is the <u>torsion</u> $t(f,\Phi)$ (Definition 12.4.). In particular this applies to the case where $R\Gamma$ stands just for a group ring $R[G]$. Then the various notions of torsion like Whitehead torsion (see e.g. Cohen [1973]). Milnor's extension of Reidemeister torsion (see Milnor [1966]), self-torsion (see Gersten [1967]) and variations for additive categories (see Ranicki [1985a], [1987]) are special cases. Another advantage in comparison with other definitions will be crucial for our applications. We do not need bases, and can also deal with finite projective not necessarily free chain complexes and our invariants lie in unreduced K_1-groups. Hence we can also deal with the $\mathbb{Q}G$-chain complex given by the rational homology of a connected G-space X using the trivial differential. Notice that $H_o(X) = \mathbb{Q}$ is never free over $\mathbb{Q}G$ unless G is trivial.

In section 13 we introduce the singular and cellular $R\Pi/(G,X)$-chain complex $C^s(X)$ and $C^c(X)$ by composing the discrete universal covering functor with the functors "singular and cellular chain complex" (Definition 13.1.). They are based free (Lemma 13.2.) and are chain homotopy equivalent (Proposition 13.10.). We prove an <u>Equivariant Hurewicz Theorem</u> 13.15. We show the <u>Equivariant Realization Theorem</u> 13.19 whose non-equivariant version is due to Wall [1966]. It claims roughly the following. Let $h : Z \longrightarrow Y$ be a G-map of G-CW-complexes and $f : C_* \longrightarrow C^c_*(Y)$ a chain equivalence such that Z is 2-dimensional, h^H is 2-connected for all $H \subset G$ and the restrictions to dimension ≤ 2 of C_* and f are $C^c_*(Z)$ and $C^c_*(h)$. Then there is an extension $g : X \longrightarrow Y$ of $h : Z \longrightarrow Y$ with $X_2 = Z$ such that g is a G-homotopy equivalence and $C^c_*(X) = C_*$ and $C^c_*(g) = f$. This result allows us to switch between algebra and geometry.

In section 14 we apply the algebra of $R\Gamma$-modules to proper G-spaces for G a Lie group. We define invariants like <u>finiteness obstruction</u> $o^G(X) \in K^G_o(X)$, <u>reduced finiteness obstruction</u> $\tilde{o}^G(X) \in \tilde{K}^G_o(X)$, and <u>Euler characteristic</u> $\chi^G(X) \in U^G(X)$ of a finitely dominated G-space (Definition 14.4.) and <u>Whitehead torsion</u> $\tau^G(f) \in Wh^G(Y)$ of finite G-CW-complexes (Definition 14.13.). We show in an algebraic setting that the reduced finiteness obstruction is the obstruction for X to be G-homotopy equivalent to a finite G-CW-complex and that the Whitehead torsion is the obstruction for f to be simple and state fundamental properties like <u>homotopy invariance</u>, <u>additivity</u> and the <u>logarithmic property</u> (Theorem 14.6. and 14.14.). We establish

<u>product formulas</u> 14.19, <u>restriction formulas</u> 14.20. and 14.37. and <u>diagonal product formulas</u> 14.21 and 14.42. We identify the geometric approach of Chapter I and the algebraic approach of Chapter II.

<u>Theorem 14.12.</u> <u>There are natural isomorphisms</u>

$$\Phi(Y) : Wa^G(Y) \oplus U^G(Y) \longrightarrow K_o^G(Y)$$

<u>and</u>

$$\tilde{\Phi}(Y) : Wa^G(Y) \longrightarrow \tilde{K}_o^G(Y)$$

<u>such that</u> $\Phi(Y)((w^G(Y),\chi^G(Y))) = o^G(Y)$ <u>and</u> $\Phi(Y)(w^G(Y)) = \tilde{o}^G(Y)$ <u>holds for any finitely dominated</u> G-<u>space</u> Y . □

<u>Theorem 14.16.</u> <u>There is a natural isomorphism</u>

$$\Phi(Y) : Wh^G_{geo}(Y) \longrightarrow Wh^G(Y)$$

<u>such that</u> $\Phi(Y)(\tau^G_{geo}(f)) = \tau^G(f)$ <u>holds for any</u> G-<u>homotopy equivalence of finite</u> G-CW-<u>complexes</u> $f : X \longrightarrow Y$.

We also introduce negative algebraic K-groups and identify them with the geometric counterparts (Proposition 14.44.).

Although the geometric approach and the algebraic one turn out to be equivalent, they are interesting in their own right. The geometric view point is appealing because of its simplicity, whereas the algebraic treatment is important for calculations.

<u>Theorem 14.46.</u> <u>We have natural isomorphisms</u>

$$K_n^G(X) = \bigoplus_{(H)} \bigoplus_{C \in \pi_o(X^H)/WH} K_n(\mathbf{Z}[\pi_1(EWH(C) \times_{WH(C)} C)])$$

$$Wh^G(X) = \bigoplus_{(H)} \bigoplus_{C \in \pi_o(X^H)/WH} Wh(\mathbf{Z}[\pi_1(EWH(C) \times_{WH(C)} C)])$$

<u>where</u> $WH(C)$ <u>is the isotropy group of</u> $C \in \pi_o(X^H)$ <u>under the</u> WH-<u>action</u>.

The computation for the Whitehead group and lower K-groups can also be obtained by different methods (see e.g. Araki [1986], Dovermann-Rothenberg [1986], Illman [1985],

Hauschild [1977], Svenson [1985]). Our methods seem to be adequate for the treatment of higher K-groups, restriction formulas and diagonal formulas for compact Lie groups which seem not to be in the literature. We use the algebraic description in Theorem 14.49 to extend the Burnside ring congruences of a torus T^n for the Euler characteristics, i.e. $\chi(X^H) = \chi(X)$ for $H \subset T^n$, to the finiteness obstruction. Namely, we get $i_* o(X^H) = o(X) \in K_o(\mathbb{Z}\pi_1(X))$ for $i : X^H \longrightarrow X$ the inclusion and $H \subset G$ provided that X^H is connected and non-empty for each $H \subset G$ and X is a finitely dominated T^n-space. We will fully exploite the algebra of $R\Gamma$-modules in Chapter III.

9. Modules over a category and a splitting of projectives.

In this section we develop the elementary algebra of modules over a category Γ. We explain basic notions like submodule, quotient module, exact sequence, kernel, cokernel, finitely generated module, projective module, free module with a base, direct sum, product, tensor product, induction, restriction, ... We show how one can split a projective module over a EI-category Γ into a direct sum of projective modules living over the various group rings $R\mathrm{Aut}(x)$ for $x \in \mathrm{Ob}\,\Gamma$. This is the main ingredient in the splitting of the algebraic K-theory of Γ into the algebraic K-theories associated with all group rings $R\mathrm{Aut}(x)$ for $x \in \mathrm{Ob}\,\Gamma$.

9.A. Elementary facts about $R\Gamma$-modules.

Assumption 9.1. Let R be an associative commutative ring with unit. Suppose that we have specified the notion of a rank $\mathrm{rk}\, M$ for a finitely generated R-module such that $\mathrm{rk}(M \oplus N) = \mathrm{rk}\, M + \mathrm{rk}\, N$ and $\mathrm{rk}\, R = 1$ is valid. □

Definition 9.2. An EI-category Γ is a small category Γ (i.e. $\mathrm{Ob}\,\Gamma$ is a set) such that any endomorphism $f : x \longrightarrow x$ in Γ is an isomorphism. □

Example 9.3. If G is a Lie group and H a compact subgroup then any G-map $G/H \longrightarrow G/H$ is a G-diffeomorphism (Lemma 1.31). Hence the orbit category $\mathrm{Or}\, G$ (Definition 8.16.) and the discrete orbit category $\mathrm{Or}/(G)$ (see 8.36.) are EI-categories. Let X be a G-space satisfying Assumption 8.13. Then the fundamental category $\Pi(G,X)$ (Definition 8.15.) and the discrete fundamental category $\Pi/(G,X)$ (Definition 8.28.) are EI-categories. □

9.4. A $R\Gamma$-comodule resp. $R\Gamma$-contramodule is a covariant resp. contravariant functor

$$M : \Gamma \longrightarrow R\text{-MOD}$$

from Γ into the category R-MOD of R-modules. An homomorphism between $R\Gamma$-modules is a natural transformation. The functor category of covariant resp. contravariant functors $\Gamma \longrightarrow R\text{-MOD}$ is called the category of $R\Gamma$-comodules resp. $R\Gamma$-contramodules and denoted by

$$R\Gamma\text{-MOD} \quad \text{resp.} \quad \text{MOD-}R\Gamma \qquad \square$$

Since we are more interested in $R\Gamma$-contramodules, we mean with $R\Gamma$-<u>modules</u> always a $R\Gamma$-contramodule. Similarly a module over a ring is understood to be a right module. The notion of a $R\Gamma$-module generalizes the notion of a module over a group ring.

<u>Example 9.5.</u> Let G be a (discrete) group. The group ring RG is additively the free R-module generated by G. Multiplication is given by

$$(\sum_{g \in G} \lambda_g g) \cdot (\sum_{h \in G} \mu_h h) = \sum_{g \in G} (\sum_{h \in G} \lambda_{gh} \mu_{h^{-1}}) g$$

Let $\hat{G}$ be the groupoid with one object, namely G, and left translations $l_g : G \to G$ $h \to gh$ as morphisms. A left RG-module M is the same as a R-module together with a group homomorphism $\phi : G \to \mathrm{Aut}_R(M)$ sending g to $l_g : M \to M$ $m \to gm$. Similarly a right RG-module can be interpreted as a R-module together with a group homomorphism $\phi : G \to \mathrm{Aut}_R(M)^{op}$ mapping g to $r_g : M \to M$ $m \to mg$. Hence a left resp. right RG-module M determines uniquely a $R\hat{G}$-comodule resp. $R\hat{G}$-contra-module $\hat{M}$ and vice versa. We obtain identifications

$$RG\text{-MOD} = R\hat{G}\text{-MOD}$$

$$\text{MOD-}RG = \text{MOD-}R\hat{G} \qquad \square$$

<u>Example 9.6.</u> Let $G = \mathbf{Z}/p$ be the cyclic group of prime order p. A ROrG-module M consists of a RG-module $M(G) =: M_o$ and a R-module $M(G/G) =: M_1$ together with a R-map $M(pr) : M_1 \to M_o^G$ into the G-fixed point set of M_o. An homomorphism $f : M \to N$ between ROrG-modules is given by a RG-map $f_o : M_o \to N_o$ and a R-map $f_1 : M_1 \to N_1$ such that the following diagram commutes

$$\begin{array}{ccc} M_o & \xrightarrow{f_o} & N_o \\ {\scriptstyle M(pr)}\uparrow & & \uparrow{\scriptstyle N(pr)} \\ M_1 & \xrightarrow{f_1} & N_1 \end{array} \qquad \square$$

<u>Example 9.7.</u> Let $\hat{I}$ be the EI-category having two objects 0 and 1 and three morphisms $id : 0 \to 0$ and $id : 1 \to 1$ and $u : 0 \to 1$. The category of $R\hat{I}$-modules has as objects homomorphisms $f : M_o \to M_1$ of R-modules. Morphisms from

$f : M_o \longrightarrow M_1$ to $g : N_o \longrightarrow N_1$ are commutative squares of R-modules

$$\begin{array}{ccc} M_o & \xrightarrow{f} & M_1 \\ h_o \downarrow & & \downarrow h_1 \\ N_o & \xrightarrow{g} & N_1 \end{array}$$ □

<u>Example 9.8.</u> If M is a set, let RM or $R(M)$ be the free R-module generated by M . We obtain a $R\Gamma$-contramodule

$$R\Gamma(?,x) : \Gamma \longrightarrow R\text{-MOD} \qquad y \longrightarrow R\,\mathrm{Hom}(y,x)$$

for any $x \in \mathrm{Ob}\,\Gamma$. □

<u>Example 9.9.</u> Let G be a Lie group and X a proper G-CW-complex. Then the n-th chain module $C_n^c(X)$ of the cellular $R\Pi/(G,X)$-chain complex $C^c(X)$ (Definition 8.37) is a $R\Pi/(G,X)$-module.

Using 8.36 we obtain also a ROr/G-module $C^c(X)$. In particular we get for a finite group G and a finite G-set S a ROrG-module

$$\begin{array}{rcl} C^c(S) : \mathrm{Or}\,G & \longrightarrow & R\text{-MOD} \\ G/H & \longrightarrow & RS^H = R\,\mathrm{map}(G/H,S)^G \end{array}$$ □

<u>Example 9.10.</u> Consider a covariant resp. contravariant functor

$$F : \{\text{comp. Lie gr.}\} \longrightarrow R\text{-MOD} \quad .$$

Assume that F applied to any inner automorphism $c(k) : G \longrightarrow G$ $g \longrightarrow kgk^{-1}$ is the identity. Then we can define for any compact Lie group G a ROrG-co resp. ROrG-contramodule F_G resp. F^G as follows. It assigns to G/H the R-module $F(H)$. Given a G-map $\sigma : G/H \longrightarrow G/K$, choose $g \in G$ with $\sigma(eH) = gk$. Then $c(g) : H \longrightarrow K$ $h \longrightarrow g^{-1}hg$ is a well-defined group homomorphism. Let $F_G(\sigma)$ resp. $F^G(\sigma)$ be $F(c(g))$ This is independent of the choice of g because $c(gk)$ is $c(g) \circ c(k)$ and $F(c(k)) = \mathrm{id}$ for $k \in K$ holds. Examples for such functors are "real or complex representation ring", "Burnside ring" or the functor $G \longrightarrow K_n(\mathbb{Z}\pi_o(G))$, where functoriality is given either by induction or by restriction. □

The category of R-modules is an abelian category. Roughly speaking a category $\mathcal{C}$ is <u>abelian</u> (compare Schubert [1970 a], p. 103) if the following holds:

i) $\mathcal{C}$ possesses a zero object 0 .

ii) There are finite products and coproducts.

iii) Each morphism has a kernel and a cokernel.

iv) Each monomorphism is a kernel and each epimorphism is a cokernel.

If $\mathcal{C}$ is any category and $\mathcal{A}$ an abelian category then the functor category of covariant resp. contravariant functors $\mathcal{C} \longrightarrow \mathcal{A}$ has the structure of an abelian category which is more or less objectwise induced from the abelian category structure on $\mathcal{A}$ (see Schubert [1970 a] p. 104). Hence the category of RΓ-modules is an abelian category. For example, if $f : M \longrightarrow N$ is an homomorphism of RΓ-modules, define the kernel of f by

$$\ker f : \Gamma \longrightarrow R\text{-MOD} \qquad x \longrightarrow \ker(f(x) : M(x) \longrightarrow N(x)).$$

A sequence $0 \longrightarrow M_o \longrightarrow M_1 \longrightarrow M_2 \longrightarrow 0$ of RΓ-modules is exact if its evaluation at any object x in Γ is an exact sequence in R-MOD. Notions like injective and projective are defined as usual. Namely, P is <u>projective</u> if the following lifting problem has a solution.

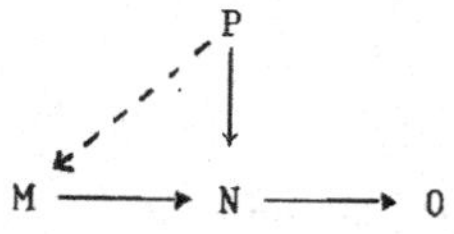

The definition of RΓ-chain complex, RΓ-chain map, RΓ-chain homotopy, homology of a RΓ-chain complex are obvious. One should notice that infinite products and coproducts exist in MOD-RΓ as they exist in R-MOD. There are some further notions of interest for us.

9.11. $\mathrm{Hom}_{R\Gamma}(M,N)$.

If M and N are RΓ-modules, let $\mathrm{Hom}_{R\Gamma}(M,N)$ be the R-module of RΓ-homomorphisms from M to N □

9.12. Tensorproduct $\otimes_{R\Gamma}$.

Let M be a $R\Gamma$-contramodule and N a $R\Gamma$-comodule. The R-module $M \otimes_{R\Gamma} N$ is the quotient P/Q of the R-module $P = \bigoplus_{x \in \Gamma} M(x) \otimes_R N(x)$ and the R-submodule Q generated by $\{mf \otimes n - m \otimes fn \mid m \in M(y), n \in M(x), f \in \mathrm{Mor}_\Gamma(x,y), x,y \in \Gamma\}$. Here we write $mf = M(f)(m)$ and $fn = N(f)(n)$ to make the definition more transparant . This is an explicit model for the tensorproduct of a contravariant and a covariant functor (see Schubert [1970 b], p. 45). □

9.13. Tensorproduct $\otimes_R$ over R .

If M is a $R\Gamma_1$-module and N a $R\Gamma_2$-module, let the $R\Gamma_1 \times \Gamma_2$-module $M \otimes_R N$ be the composition

$$\Gamma_1 \times \Gamma_2 \xrightarrow{M \times N} \text{R-MOD} \times \text{R-MOD} \xrightarrow{\otimes_R} \text{R-MOD} \quad \square$$

9.14. $R\Gamma_1$-$R\Gamma_2$-bimodule.

A $R\Gamma_1$-$R\Gamma_2$-bimodule M is a covariant functor

$$M : \Gamma_1 \times \Gamma_2^{op} \longrightarrow \text{R-MOD}$$

An example is $R\Gamma(??,?)$ given for $\Gamma_1 = \Gamma_2$ by

$$\Gamma_1 \times \Gamma_2^{op} \longrightarrow \text{R-MOD} \quad (x_1,x_2) \longrightarrow R\,\mathrm{Hom}(x_2,x_1)$$

9.15. Induction and restriction.

Let $F : \Gamma_1 \longrightarrow \Gamma_2$ be a covariant functor. Let $R\Gamma_2(??,F(?))$ be the $R\Gamma_1$-$R\Gamma_2$-bimodule $(x_1,x_2) \longrightarrow R\,\mathrm{Hom}_{\Gamma_2}(x_2,F(x_1))$ and define the $R\Gamma_2$-$R\Gamma_1$-bimodule $R\Gamma_2(F(?),??)$ analogously. Define the covariant functors induction and restriction with F by

$$\mathrm{ind}_F : \text{MOD-}R\Gamma_1 \longrightarrow \text{MOD-}R\Gamma_2 \qquad M \longrightarrow M \otimes_{R\Gamma_1} R\Gamma_2(??,F(?))$$

$$\mathrm{res}_F : \text{MOD-}R\Gamma_2 \longrightarrow \text{MOD-}R\Gamma_1 \qquad M \longrightarrow M \otimes_{R\Gamma_2} R\Gamma_2(F(?),??)$$

Of course res_F is just composition with F . It can also be described as

$$M \longrightarrow \mathrm{Hom}_{R\Gamma_2}(R(??,F(?)),M) \qquad \square$$

9.16. <u>Free</u> RΓ-<u>module with base</u>

Let I be any index set. An I-set B is a collection $\{B_i \mid i \in I\}$ of sets indexed by I. A map $f : B \longrightarrow C$ is a collection $\{f_i : B_i \longrightarrow C_i \mid i \in I\}$ of set maps. If we think of I as the category having I as set of objects and only the identities as morphisms then we can interprete an I-set as a functor $I \to \{\text{sets}\}$ and a map between I-sets as a natural transformation. Ocasionally we also think of an I-set B as a set B together with a map $\beta : B \longrightarrow I$ and of a map $f : B \to C$ between I-sets as a set map f making the following diagram commute

$$\begin{array}{ccc} B & \xrightarrow{\quad f \quad} & C \\ & \beta \searrow \quad \swarrow \gamma & \\ & I & \end{array}$$

Given $\beta : B \longrightarrow I$, we obtain an I-set B by $\{\beta^{-1}(i) \mid i \in I\}$. Given an I-set B, consider $\coprod_{i \in I} B_i$ with the obvious map into I. In the sequel I will often be the set $\mathrm{Ob}\,\Gamma$ of objects in an EI-category Γ.

Obviously any RΓ-module M has an underlying $\mathrm{Ob}\,\Gamma$-set, also denoted by M.

<u>Definition 9.17.</u> <u>A</u> RΓ-<u>module</u> M <u>is free with</u> $\mathrm{Ob}\,\Gamma$-<u>set</u> $B \subset M$ <u>as base if for any</u> RΓ-<u>module</u> N <u>and any map</u> $f : B \longrightarrow N$ <u>of</u> $\mathrm{Ob}\,\Gamma$-<u>sets there is exactly one</u> RΓ-<u>homomorphism</u> $F : M \longrightarrow N$ <u>extending</u> f. $\square$

The RΓ-module $R\Gamma(?,x)$ of Example 9.8. is free with the $\mathrm{Ob}\,\Gamma$-set B defined by $B(x) = \{id : x \longrightarrow x\} \subset R\mathrm{Hom}(x,x)$ and $B(y) = \emptyset$ for $y \neq x$ as base. This follows from the Yoneda-Lemma (see Schubert [1970 a], p. 23). Namely, if N is a RΓ-module and $n \in N(x)$, the desired homomorphism $F : R\mathrm{Hom}(?,x) \longrightarrow N(?)$ sends ϕ to $N(\phi)(n)$. Let B be any $\mathrm{Ob}\,\Gamma$-set, given by $\beta : B \longrightarrow \mathrm{Ob}\,\Gamma$. Then a model for the free RΓ-module with base B is

$$R\Gamma(B) = \bigoplus_{b \in B} R\Gamma(?,\beta(b)) \qquad \square$$

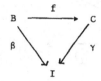

<u>Example 9.18.</u> Let G be a Lie group and X be a proper G-CW-complex. Let I_n be the set of n-dimensional cells of X and $C_n^c(X)$ the n-th chain module of the cellular $\mathbf{Z}\Pi/(G,X)$-chain complex of X (see Definition 8.37.). For each $i \in I_n$ choose a characteristic map

$$(Q_i,q_i) : G/H_i \times (D^n,S^{n-1}) \longrightarrow (X_n,X_{n-1})$$

Let $x_i : G/H_i \longrightarrow X$ be the restriction of Q_i to $G/H_i \times \{*\}$ for a base point $*$ in D^n. Choose a lift $(\tilde{Q}_i,\tilde{q}_i)$ of the restriction of (Q_i,q_i) to $eH_i \times (D^n,S^{n-1})$

$$\begin{array}{ccc} & & \tilde{X}_n^{H_i}(x_i)\ ,\ \tilde{X}_{n-1}^{H_i}(x_i) \\ & \nearrow^{(\tilde{Q}_i,\tilde{q}_i)} & \big\downarrow{\scriptstyle p(x_i)} \\ (D^n,S^{n-1}) & \xrightarrow[(Q_i,q_i)]{} & X_n^{H_i}(x_i)\ ,\ X_{n-1}^{H_i}(x_i) \end{array}$$

Consider the composition $(\tilde{Q}_i/,\tilde{q}_i/) : (D^n,S^{n-1}) \longrightarrow (\tilde{X}_n^{H_i}/(x_i),\tilde{X}_{n-1}^{H_i}/(x_i))$ of $(\tilde{Q}_i,\tilde{q}_i)$ and $\tilde{o}(x_i) : (\tilde{X}_n^{H_i}(x_i),\tilde{X}_{n-1}^{H_i}(x_i)) \longrightarrow (\tilde{X}_n^{H_i}/(x_i),\tilde{X}_{n-1}^{H_i}/(x_i))$. Fix a generator $\omega_n \in H_n(D^n,S^{n-1}) = \mathbf{Z}$ and define $\langle i\rangle \in C_n^c(X)(x_i)$ by its image under

$$(\tilde{Q}_i/,\tilde{g}_i/)_* : H_n(D^n,S^{n-1}) \longrightarrow H_n(\tilde{X}_n^{H_i}/(x_i),\tilde{X}_{n-1}^{H_i}/(x_i)) \quad .$$

Let $\langle I_n\rangle$ be the Ob Γ-set $\{\langle i\rangle \mid i \in I_n\}$ with $\gamma : \langle I_n\rangle \longrightarrow \mathrm{Ob}\,\Gamma$ sending $\langle i\rangle$ to x_i . We will later prove that $\langle I_n\rangle$ is a base for $C_n^c(X)$. Of course $\langle I_n\rangle$ depends on the various choices above, in particular the choice of the characteristic maps. We discuss this later. □

9.19. <u>Finitely generated</u> $R\Gamma$-<u>modules</u>.

If M is a $R\Gamma$-module and $S \subset M$ a Γ-subset, let span S be the smallest $R\Gamma$-submodule of M containing S .

$$\mathrm{span}\, S = \bigcap \{N \mid N\ R\Gamma\text{-submodule of } M,\ S \subset N\}$$

Notice that span S is the image of the unique $R\Gamma$-homomorphism $R\Gamma(S) \longrightarrow M$ extending $\mathrm{id} : S \longrightarrow S$. We call a Ob Γ-set $(B,\beta : B \longrightarrow \mathrm{Ob}\,\Gamma)$ <u>finite</u> if B is finite as a set. We say that the $R\Gamma$-module M is generated by the Ob Γ-set S if

M = span S holds and call it <u>finitely generated</u> if S is finite. □

We state some elementary properties of these notions and leave their verification to the reader.

9.20. The following statements for a $R\Gamma$-module P are equivalent:

i) P is projective.

ii) Each exact sequence $0 \longrightarrow M \longrightarrow N \longrightarrow P \longrightarrow 0$ splits.

iii) $\mathrm{Hom}_{R\Gamma}(P,?)$ is an exact function i.e. it preserves exact sequences.

iv) P is a direct summand in a free $R\Gamma$-module.

9.21. The tensorproduct and the Hom - functor are adjoint. More precisely, we have for any $R\Gamma_1$-$R\Gamma_2$-bimodule B, a $R\Gamma_1$-module M and a $R\Gamma_2$-module N a natural isomorphism of R-modules

$$\mathrm{Hom}_{R\Gamma_2}(M \otimes_{R\Gamma_1} B,N) \longrightarrow \mathrm{Hom}_{R\Gamma_1}(M,\mathrm{Hom}_{R\Gamma_2}(B,N))$$

It sends $\phi : M \otimes_{R\Gamma_1} B \longrightarrow N$ to the $R\Gamma_1$-homomorphism $M \longrightarrow \mathrm{Hom}_{R\Gamma_2}(B,N)$ assigning to $m \in M(x_1)$ the $R\Gamma_2$-homomorphism

$$B(x_1,??) \longrightarrow N(??) \qquad b \longrightarrow \phi(m \otimes b)$$ □

9.22. Given a covariant functor $F : \Gamma_1 \longrightarrow \Gamma_2$, ind_F and res_F are adjoint. Namely, there is a natural R-isomorphism for a $R\Gamma_1$-module M and a $R\Gamma_2$-module N

$$AD(M,N) : \mathrm{Hom}_{R\Gamma_2}(\mathrm{ind}_F M,N) \longrightarrow \mathrm{Hom}_{R\Gamma_1}(M,\mathrm{res}_F N)$$

This follows from 9.21. and the definition of ind_F and res_F given in 9.15. (see also Schubert [1970 b], § 17) □

9.23. The functor $? \otimes_{R\Gamma} N$ is right exact and the functor $\mathrm{Hom}_{R\Gamma}(?,N)$ is left exact. □

9.24. The induction functor respects "direct sum", "finitely generated", "free", and "projective" but is not exact in general. The restriction functor is exact but does not respect "finitely generated", "free" and "projective" in general. □

9.B. The structure of projective $R\Gamma$-modules

Now we show how one can built up a projective $R\Gamma$-module out of projective $R[\mathrm{Aut}(x)]$-modules over the group rings $R[\mathrm{Aut}(x)]$ for $x \in \mathrm{Ob}\,\Gamma$ provided that Γ is an EI-category.

Notation 9.25.

a) $R[x] := R[\mathrm{Aut}(x)]$ for $x \in \mathrm{Ob}\,\Gamma$

b) Define a partial ordering $\leq$ on the set $\mathrm{Is}\,\Gamma$ of isomorphism classes $\bar{x}$ of objects x in Γ by

$$\bar{x} \leq \bar{y} \iff \mathrm{Hom}(x,y) \neq \emptyset \qquad \square$$

The EI-property ensures $\bar{x} \leq \bar{y} \leq \bar{x} \Rightarrow \bar{x} = \bar{y}$. The partial ordering is crucial for the sequel. We introduce the following covariant functors for $x \in \mathrm{Ob}\,\Gamma$.

9.26. Splitting functor $S_x : \text{MOD-}R\Gamma \longrightarrow \text{MOD-}R[x]$

If M is a $R\Gamma$-module, let the singular $R[x]$-module at x be the R-submodule $M(x)_s$ of $M(x)$ generated by all images of R-homomorphisms $M(f) : M(y) \longrightarrow M(x)$ induced by all non isomorphisms $f : x \longrightarrow y$ with x as source. A priori $M(x)_s$ is only a R-submodule of $M(x)$ but it is even a $R[x]$-module, since for $f \in \mathrm{Aut}(x)$ the composition $g \circ f$ is an isomorphism if and only if g is. Define

$$S_x : \text{MOD-}R\Gamma \longrightarrow \text{MOD-}R[x] \qquad M \longrightarrow M(x)/M(x)_s \qquad \square$$

9.27. Restriction functor $\mathrm{RES}_x : \text{MOD-}R\Gamma \longrightarrow \text{MOD-}R[x]$

It sends M to $M(x)$

9.28. Extension functor $E_x : \text{MOD-}R[x] \longrightarrow \text{MOD-}R\Gamma$

It sends M to $M \otimes_{R[x]} R\,\mathrm{Hom}(?,x)$.

9.29. Inclusion functor $I_x : \text{MOD-}R[x] \longrightarrow \text{MOD-}R\Gamma$ $\square$

It assigns to the $R[x]$-module M the $R\Gamma$-module I_xM defined by

$$I_xM(y) = \begin{cases} M(x) \otimes_{R[x]} R\,\mathrm{Hom}(y,x) & \text{if } \bar{y} = \bar{x} \\ \{0\} & \text{if } \bar{y} \neq \bar{x} \end{cases}$$

This is well-defined because of the EI-property. □

9.30. Let $Aut(x)^{\wedge}$ be the groupoid defined by the group $Aut(x)$ for $x \in Ob\,\Gamma$ (see 9.5.). We have an obvious functor

$$i : Aut(x)^{\wedge} \longrightarrow \Gamma .$$

The restriction functor RES_x is the restriction with i and the extension functor E_x is the induction with i (see 9.15.). Let B be the $R\Gamma$-$R[x]$- bimodule given by the covariant functor

$$\Gamma_1 \longrightarrow MOD\text{-}R[x] \qquad y \longrightarrow \begin{cases} R\,Hom(x,y) & \text{if } \bar{x} = \bar{y} \\ \{0\} & \text{if } \bar{x} \neq \bar{y} \end{cases}$$

Then the splitting functor S_x is just $M \longrightarrow M \otimes_{R\Gamma_1} B$. Let C be the $R[x]$- $R\Gamma_1$-bimodule defined by the covariant functor

$$\Gamma_1 \longrightarrow R[x]\text{-}MOD \quad y \longrightarrow \begin{cases} R\,Hom(y,x) & \text{if } \bar{x} = \bar{y} \\ \{0\} & \text{if } \bar{x} \neq \bar{x} \end{cases}$$

The inclusion functor I_x sends M to $M \otimes_{R[x]} C$. It can also be described by assigning to M the $R\Gamma_1$-module $Hom_{R[x]}(B,M)$. □

Lemma 9.31.

a) <u>The functors</u> E_x <u>and</u> RES_x <u>and the functors</u> S_x <u>and</u> I_x <u>are adjoint</u>, i.e. <u>there are natural isomorphisms of</u> R-<u>modules</u>.

$$Hom_{R\Gamma}(E_xM,N) \longrightarrow Hom_{R[x]}(M,RES_xN)$$

<u>and</u>

$$Hom_{R[x]}(S_xM,N) \longrightarrow Hom_{R\Gamma}(M,I_xN)$$

b) $S_x(R\Gamma(?,y))$ <u>is</u> $R[x]$-<u>isomorphic to</u> $R[x]$ <u>for</u> $\bar{x} = \bar{y}$ <u>and zero otherwise</u>. <u>Moreover</u>, $E_x(R[x])$ <u>is</u> $R\Gamma(?,x)$.

c) S_x <u>and</u> E_x <u>respect</u> "<u>direct sum</u>", "<u>finitely generated</u>", "<u>free</u>" <u>and</u> "<u>projective</u>".

d) $S_x \circ E_x : MOD\text{-}R[x] \longrightarrow MOD\text{-}R[x]$ <u>is natural equivalent to the identity functor</u>. <u>The composition</u> $S_y \circ E_x$ <u>is zero for</u> $\bar{x} \neq \bar{y}$.

Proof: left to the reader. □

The role of the splitting and restriction functor for applications in geometry is explained by the next statement

Lemma 9.32. Let G be a Lie group and X be a proper G-CW-complex. Consider its cellular $R\Pi/(G,X)$-chain complex $C^c(X)$ and an object x in $\Pi/(G,X)$. Then

a) $S_x C^c(X)$ is naturally $R[x]$-isomorphic to the cellular $R[x]$-chain complex $C^c(\tilde{X}/(x), \tilde{X}^{>}/(x))$.

b) $RES_x C^c(X)$ is $C^c(\tilde{X}/(x))$. □

A direct consequence of Lemma 9.31. is that any free $R\Gamma$-module F can be built up by free $R[x]$-modules. More precisely, if we choose for any $\bar{x} \in \text{Is}\,\Gamma$ a representative $x \in \bar{X}$, there is an isomorphism

9.33. $$F \cong \bigoplus_{\bar{x} \in \text{Is}\,\Gamma} E_x \circ S_x F$$

We want to extend this to projective modules. This cannot be done directly as the isomorphism 9.33. is not natural. For arbitrary modules over an EI-category such a splitting is not true. Here is a counterexample.

Example 9.34. Let $G = \mathbb{Z}/p$ be the cyclic group of prime order p. Let Γ be $\text{Or}\,G$. Consider the $R\text{Or}G$-modules $M = I_{G/G}R$, $N = I_G R$ and $P = E_{G/G}R$. Then there is an obvious exact sequence which does not split

$$0 \longrightarrow N \longrightarrow P \longrightarrow M \longrightarrow 0 .$$

Obviously we have

$$M \neq P = E_G \circ S_G M \oplus E_{G/G} \circ S_{G/G} M$$

Hence 9.33. is not true for M. Applying S_G to the exact sequence above yields

$$0 \longrightarrow R \longrightarrow 0 \longrightarrow 0 \longrightarrow 0 .$$

Hence S_G is not an exact functor. The R-module R is of course free but $M = I_{G/G}R$ is not projective otherwise the exact sequence above would split. Hence $I_{G/G}$ does not respect "free" and "projective". The $R\Gamma$-module $P = R\,\mathrm{Hom}(?,G/G)$ is free. However, the R[G]-module $\mathrm{RES}_G(P) = R$ is not projective in general. Take for example $R = \mathbf{Z}$. Hence Res_G does not respect "free" and "projective". □

We introduce some notation. If $B = \{B_x \mid x \in \mathrm{Ob}\,\Gamma\}$ is a Ob Γ-set, let the <u>set of isomorphism classes of isotropy objects</u> $\mathrm{Iso}\,B \subset \mathrm{Is}\,\Gamma$ be $\{\bar{x} \in \mathrm{Is}\,\Gamma \mid$ there is a representative $x \in \bar{X}$ with $B_x \neq \emptyset\}$. The <u>set of isomorphism classes of isotropy objects</u> $\mathrm{Iso}\,M \subset \mathrm{Is}\,\Gamma$ of a $R\Gamma$-module is defined by $\{\bar{x} \in \mathrm{Is}\,\Gamma \mid S_x M \neq \{0\}\}$. If B is a Ob Γ-set and $R\Gamma(B)$ the free $R\Gamma$-module with base B then $\mathrm{Iso}\,R\Gamma(B) = \mathrm{Iso}\,B$. This is motivated by the following observation. Namely, let X be a proper G-CW-complex then

9.35.
$$\bigcup_{n \geq o} \mathrm{Iso}\, C_n^c(X) = \{\bar{x} \in \mathrm{Is}\,\Pi/(G,X) \mid X(x)\backslash X^{>}(x) \neq \emptyset\}$$

This is a consequence of Lemma 9.32.

Given a subset $T \subset \mathrm{Is}\,\Gamma$, call a $R\Gamma$-module <u>of type</u> T if there is an epimorphism $R\Gamma(B) \longrightarrow M$ for a free $R\Gamma$-module $R\Gamma(B)$ with $\mathrm{Iso}\,B \subset T$, or equivalently, if there is a Ob Γ-subset S of M with $\mathrm{Iso}\,S \subset T$ generating M. A quotient of a $R\Gamma$-module of type T is again of type T. This is not true for a submodule in general. If M is of type T for a finite Ob Γ-set T, we call M <u>of finite type</u>. If X is a proper G-CW-complex of finite orbit type such that $\pi_o(X^H)$ is finite for $H \in \mathrm{Iso}\,X$ then each $C_n^c(X)$ is a $\mathbf{Z}\Pi/(G,X)$-module of finite type.

Let T_o be a subset of $\mathrm{Is}\,\Gamma$ and M be a $R\Gamma$-module. Choose for any $\bar{x} \in T_o$ a representative $x \in X$. Recall that E_x and RES_x are adjoint (Lemma 9.31.). The following in M natural $R\Gamma$-homomorphism

9.36.
$$J_x M : E_x \circ \mathrm{RES}_x M \longrightarrow M$$

is the adjoint of the identity of $\mathrm{RES}_x M$:

$$J_xM : M(x) \otimes_{R[x]} R\,\mathrm{Hom}(?,x) \longrightarrow M(?) \quad m \otimes \phi \longrightarrow M(\phi)(m)\ .$$

The direct sum of these maps J_xM running over $\bar{x} \in T_o$ yields

$$9.37. \qquad J_{T_o}M : \bigoplus_{\bar{x} \in T_o} E_x \circ RES_x M \longrightarrow M$$

Define $COK_{T_o}M$ as the cokernel of $J_{T_o}M$. Let $PR_{T_o}M: M \longrightarrow COK_{T_o}M$ be the canonical projection. Then we have the in M natural exact sequence

$$9.38. \qquad \bigoplus_{\bar{x} \in T_o} E_x \circ RES_x M \xrightarrow{J_{T_o}M} M \xrightarrow{PR_{T_o}M} COK_{T_o}M \longrightarrow 0$$

Since E_x and RES_x are compatible with direct sums, the same is true for COK_{T_o}. If M is of type T for $T \subset \mathrm{Is}\,\Gamma$ then $COK_{T_o}M$ is of type $T \setminus (T \cap T_o)$. Namely, let B be a $\mathrm{Ob}\,\Gamma$-set with $\mathrm{Iso}\,B \subset T$ and $\phi : R\Gamma(B) \longrightarrow M$ be an epimorphism. Write B as the disjoint union $B = B' \coprod B''$ of $\mathrm{Ob}\,\Gamma$-sets with $\mathrm{Iso}\,B' \subset T\setminus(T \cap T_o)$ and $\mathrm{Iso}\,B'' \subset T \cap T_o$. The homomorphism $COK_{T_o}(\phi) : COK_{T_o}(R\Gamma(B)) \longrightarrow COK_{T_o}M$ is surjective and

$$COK_{T_o}(R\Gamma(B)) = COK_{T_o}(R\Gamma(B')) \oplus COK_{T_o}(R\Gamma(B''))$$

One checks directly that $COK_{T_o}(R\Gamma(?,x)) = 0$ for $\bar{x} \in T_o$ and hence $COK_{T_o}(R\Gamma(B''))$ vanish. Hence we have an epimorphism

$$R\Gamma(B') \longrightarrow COK_{T_o}(R\Gamma(B')) = COK_{T_o}(R\Gamma(B)) \longrightarrow COK_{T_o}M$$

so that $COK_{T_o}M$ is of type $T \setminus (T \cap T_o)$.

The main result of this section is

<u>Theorem 9.39.</u> <u>Cofiltration Theorem for projective</u> $R\Gamma$-<u>modules</u>. <u>Let</u> Γ <u>be a</u> EI-<u>category</u>. <u>Let</u> $\emptyset = T_o \subset T_1 \subset T_2 \subset T_\ell = T$ <u>be a filtration of</u> $T \subset \mathrm{Is}\,\Gamma$ <u>such that</u> $\bar{x} \in T_i$, $\bar{y} \in T_j$, $\bar{x} < \bar{y}$ <u>implies</u> $i > j$. <u>If</u> P <u>is a projective</u> $R\Gamma$-<u>module of type</u> T, <u>then there is a natural cofiltration</u>

$$P = P_o \xrightarrow{PR_o} P_1 \xrightarrow{PR_1} P_2 \longrightarrow \ldots \longrightarrow P_\ell = \{0\}$$

satisfying

a) $P_i = COK_{T_i} P_{i-1}$ and $PR_i = P_{i-1} \longrightarrow COK_{T_i} P_{i-1}$ is the projection.

b) P_i is projective of type $T \backslash T_i$. If P is finitely generated, P_i is finitely generated.

c) Let $PR^i : P \longrightarrow P_i$ be $PR_{i-1} \circ PR_{i-2} \circ \ldots \circ PR_1$. Then $S_x PR^i : S_x P \longrightarrow S_x P_i$ is an isomorphism for $\bar{x} \in T \backslash T_i$ and $S_x P_i = 0$ for $\bar{x} \notin T \backslash T_i$.

d) We obtain from 9.38. a natural exact sequence which splits (not naturally)

$$0 \longrightarrow \bigoplus_{\bar{x} \in T_i \backslash T_{i-1}} E_x \circ S_x P \longrightarrow P_{i-1} \xrightarrow{PR_{i-1}} P_i \longrightarrow 0 .$$

e) If $0 \longrightarrow N \longrightarrow P \longrightarrow Q \longrightarrow 0$ is an exact sequence of projective $R\Gamma$-modules of type T, then the induced sequence $0 \longrightarrow N_i \longrightarrow P_i \longrightarrow Q_i \longrightarrow 0$ is exact.

Proof. We use induction over $i = 0,1,\ldots,\ell$. The begin $i = 0$ is trivial. Consider $\bar{x} \in T_i \backslash T_{i-1}$ and $\bar{z} \in \operatorname{Is}\Gamma$ with $\bar{x} < \bar{z}$. Then $R\operatorname{Hom}(z,y) = \{0\}$ holds for any $\bar{y} \in T \backslash T_{i-1}$ as $\bar{x} < \bar{y}$, $\bar{y} \in T_j$ implies $i > j$ by assumption. As P_{i-1} is of type $T \backslash T_{i-1}$ by induction hypothesis, $P_{i-1}(z) = \{0\}$ for $\bar{x} < \bar{z}$ and hence $S_x P_{i-1} = \operatorname{Res}_x P_{i-1}$. We get from 9.38 the exact sequence

$$\bigoplus_{\bar{x} \in T_i \backslash T_{i-1}} E_x \circ S_x P_{i-1} \xrightarrow{J_{i-1}} P_{i-1} \xrightarrow{PR_{i-1}} P_i \longrightarrow 0$$

We want to show that J_{i-1} is split injective. It is compatible with direct sums. As any projective $R\Gamma$-module P of type T is a direct summand in a free module of type T which is a direct sum of $R\Gamma$-modules $R\operatorname{Hom}(?,y)$ for $\bar{y} \in T$, we may suppose $P = R\operatorname{Hom}(?,y)$ for $\bar{y} \in T$. One easily checks that

$$\bigoplus_{\bar{x} \in T_i \backslash T_{i-1}} E_x \circ S_x P_{i-1}$$

is zero for $\bar{y} \notin T_i \backslash T_{i-1}$ and J_{i-1} is an isomorphism for $\bar{y} \in T_i \backslash T_{i-1}$ using Lemma 9.31. Hence we get d) if we compose J_{i-1} with the homomophism

$$\bigoplus_{\bar{x} \in T_i \backslash T_{i-1}} E_x \circ S_x PR^{i-1} : \bigoplus_{\bar{x} \in T_i \backslash T_{i-1}} E_x \circ S_x P \longrightarrow \bigoplus_{\bar{x} \in T_i \backslash T_{i-1}} E_x \circ S_x P_{i-1}$$

which is an isomorphism by induction hypothesis. The other claims follow from d), Lemma 9.31. and the remarks above. □

Now we can extend 9.33. to projective modules.

Corollary 9.40. Let P be a projective $R\Gamma$-module of finite type. Then there is an (not natural) isomorphism

$$P \cong \bigoplus_{\bar{x} \in \mathrm{Is}\,\Gamma} E_x \circ S_x P$$

Proof. Use induction over Iso P and Theorem 9.39 □

Example 9.41. Let G be the cyclic group $\mathbb{Z}/p$ of prime order p. We claim:

A ROrG-module M is free resp. projective if and only if $S_G M$ is a free resp. projective RG-module, $S_{G/G}M$ is a free resp. projective R-module and $M(pr) : M(G/G) \longrightarrow M(G)$ for the projection $pr : G \longrightarrow G/G$ is injective.

The only if statement follows from Lemma 9.31. and Corollary 9.40. For the verification of the if-statemant consider the sequence 9.38

$$E_{G/G} \circ S_{G/G}M \xrightarrow{J_{G/G}M} M \xrightarrow{PR_{G/G}M} COK_{G/G}M \longrightarrow 0$$

Since $M(pr) : M(G/G) \longrightarrow M(G)$ is injective, $J_{G/G}M$ is injective. Moreover, $COK_{G/G}M$ is just $E_G \circ S_G M$ and hence projective. We obtain

$$M = E_{G/G} \circ S_{G/G}M \oplus E_G \circ S_G M$$

If $S_{G/G}M$ and $S_G M$ are free resp. projective, M must be free resp. projective. □

The following example shows that the EI-property is necessary for Theorem 9.39. and Corollary 9.40.

Example 9.42. Let Γ be the category having two objects and the following set of morphisms

$$\mathrm{Hom}(x,y) = \{u\} \ , \ \mathrm{Hom}(y,x) = \{v\} \ ,$$

$$\mathrm{Hom}(x,x) = \{id, vu\} \ , \ \mathrm{Hom}(y,y) = \{id, uv\}$$

Moreover we demand $vuv = v$ and $uvu = u$. Notice that Γ is no EI-category since $\mathrm{Aut}(x) = \{id\} \neq \mathrm{Hom}(x,x)$ and $\mathrm{Aut}(y) = \{id\} \neq \mathrm{Hom}(y,y)$ holds.

Everything of this section until to the definition of S_x and E_x does not require the EI-property. Also Lemma 9.31 b),c) and d) remains true so that we still have for a free $R\Gamma$-module F

$$F \cong \bigoplus_{\bar{x} \in \mathrm{Is}\,\Gamma} E_x \circ S_x F$$

The analogous statement for a projective $R\Gamma$-module is false for the category Γ we have defined above.

Let M be the $R\Gamma$-submodule of $R\,\mathrm{Hom}(?,y)$ given by $M(x) = \{0\}$, $M(y) = \{r\cdot(id_y - uv) \mid r \in R\} \subset R\,\mathrm{Hom}(y,y)$. Define $r : R\,\mathrm{Hom}(?,x) \longrightarrow M$ by $r(id_y) = id_y - uv$. Then $r|M$ is the identity. Hence M is projective and of finite type and $S_x M = 0$ and $S_y M = M(y) \cong R$. But M and $E_x \circ S_x M \oplus E_y \circ S_y M \cong R\,\mathrm{Hom}(?,y)$ are not isomorphic. □

Comment 9.43. Rothenberg [1978] gives a definition of the Whitehead group $Wh^G(*)$ using the following category $\mathcal{C}$ for a finite group G. A G-based R-module M is given by the free R-module $R(S)$ generated by a finite G-set S. Let M^H be the R-submodule $R(S^H)$. Notice that M^H is not the H-fixed point set of M under the G-action. A morphism $f : M \longrightarrow N$ between G-based modules is a R-homomorphism such that $f(M^H) \subset N^H$ for all $H \subset G$ holds. Let $\mathcal{C}$ be the category of G-based R-modules.

Define $\mathcal{B}$ as the full subcategory of MOD-ROrG consisting of all finitely generated free ROrG-modules. Any G-based R-module $R(S)$ defines a finitely generated free ROrG-module by $C^c(S)$ (see Example 9.9). A map $f : R(S) \longrightarrow R(T)$ between G-based R-modules induces an homomorphism $C^c(S) \longrightarrow C^c(T)$ between ROrG-modules, since $C^c(S)(G/H)$ is $R(S^H)$ by definition. We obtain a functor $F : \mathcal{C} \longrightarrow \mathcal{B}$. One easily constructs an inverse functor $F^{-1} : \mathcal{B} \longrightarrow \mathcal{C}$ such that both compositions are naturally equivalent to the identity. Hence $\mathcal{C}$ and $\mathcal{B}$ are equivalent categories.

The main advantage of $\mathcal{B}$ in comparision with $\mathcal{C}$ is that there is an obvious embedding of $\mathcal{B}$ into the abelian cateory of $R\Gamma$- modules. Since we have in MOD-$R\Gamma$ the notion of exact sequences, projectives, and so on, it will be fairly easy to introduce algebraic K-groups in the next section. Moreover, the notions about MOD-$R\Gamma$ are very close to the analogous ones for modules over group rings (see Example 9.5.). We recommend to the reader to have this analogy always in mind. This is not such obvious if one works with $\mathcal{C}$.

Modules over a category appear in obstruction theory, a local coefficient system for a space X is a $R\Pi(X)$-module over the fundamental category (see Whitehead [1978]). Modules over the orbit category are used for equivariant obstruction theory in Bredon [1967]. Further references for modules over EI-categories are tom Dieck [1981], tom Dieck [1987] I, 11, Lück [1988], Mitchell [1972], Triantafillou [1982]. Modules over EI-categories appear also in Anderson-Munkholm [1988], Jackowski-McClure [1987], Ranicki-Weiss [1987], Weiss-Williams [1987].

Exercises 9.44.

1) Show that all the notions we have defined for $R\Gamma$-modules like finitely generated, free, projective, base,... agree with the usual definitions for modules over a group ring under the identification RG-MOD = $R\hat{G}$-MOD and MOD-RG = MOD-$R\hat{G}$ of Example 9.5.

2) Let $\hat{I}$ be the EI-category having two objects 0 and 1 and three morphisms $id : 0 \longrightarrow 0$, $id : 1 \longrightarrow 1$ and $u : 0 \longrightarrow 1$. Prove that a $R\hat{I}$-module M is projective if and only if $M(u) : M(1) \longrightarrow M(0)$ is injective and both $M(1)$ and cokernel $M(u)$ are projective R-modules.

3) Show that a $R\Gamma$-module M has a projective resolution, i.e. there is an exact sequence possibly infinite to the left

$$\ldots \longrightarrow P_n \longrightarrow P_{n-1} \longrightarrow \ldots \longrightarrow P_1 \longrightarrow P_o \longrightarrow M \longrightarrow 0$$

such that P_n is projective for $n \geq 0$.

4) Let Γ be an EI-category such that Ob Γ is finite. Then the following statements are equivalent:

   i) A $R\Gamma$-module M is finitely generated if and only if $M(x)$ is a finitely generated R-module for any $x \in \text{Ob}\,\Gamma$.

   ii) $\text{Hom}(x,y)$ is finite for any $x,y \in \text{Ob}\,\Gamma$.

5) Let Γ be an EI-category with the property that a $R\Gamma$-module M is free resp. projective if and only if S_xM is free resp. projective for any $x \in \text{Ob}\,\Gamma$. Show that then Γ is a groupoid.

6) Give an example of an EI-category Γ and a $\mathbb{Q}\Gamma$-module M with a $\mathbb{Q}\Gamma$-submodule $N \subset M$ such that M is finitely generated but N is not finitely generated. Show for an EI-category Γ such that $\text{Ob}\,\Gamma$ and $\text{Hom}(x,y)$ for any $x,y \in \text{Ob}\,\Gamma$ are finite and a noetherian ring R that any submodule of a finitely generated $R\Gamma$-module is again finitely generated.

7) Let G be a finite group and M be a $\mathbb{Z}\text{Or}G$-module such that $M(G/H)$ is finitely generated over $\mathbb{Z}$ for any $H \subset G$. Then M has a finitely generated projective resolution, i.e. there is an exact sequence

$$\dots \longrightarrow P_2 \longrightarrow P_1 \longrightarrow P_o \longrightarrow M \longrightarrow 0$$

such that P_n is finitely generated and projective for $n \geq 0$.

8) Show that the extension functor is not exact in general. However, if R is $\mathbb{Q}$ and $\text{Ob}\,\Gamma$ and $\text{Aut}(x)$ for any $x \in \text{Ob}\,\Gamma$ is finite then the extension functor E_x is exact for any $x \in \text{Ob}\,\Gamma$.

9) Let G be the cyclic group $\mathbb{Z}/p^n$ of prime power order. Then a $R\text{Or}G$-module M is free resp. projective if and only if $S_{G/H}M$ is free resp. projective for any $H \subset G$ and $M(pr) : M(G/H) \longrightarrow M(G)$ for the projection $pr : G \longrightarrow G/H$ is injective for any $H \subset G$.

10) Let G be $\mathbb{Z}/pq$ for p and q prime numbers with $p \neq q$. Give for any R an example of a $R\text{Or}G$-module M such that $S_{G/H}M$ is projective and $M(pr) : M(G/H) \longrightarrow M(G)$ is injective for any $H \subset G$ but M is not projective.

11) Show that the functor $\{\text{Ob}\,\Gamma\text{-sets}\} \longrightarrow \text{MOD-}R\Gamma$ sending a $\text{Ob}\,\Gamma$-set B to the free $R\Gamma$-module $R\Gamma(B)$ with base B and the forgetful functor $\text{MOD-}R\Gamma \rightarrow \{\text{Ob}\,\Gamma\text{-sets}\}$ are adjoint.

12) Give an example of a EI-category Γ and $R\Gamma$-modules $M \subset N$ such that N has finite type but M does not have finite type.

13) Given a functor $F : \Gamma_1 \longrightarrow \Gamma_2$, define the coinduction $F_{\#}M$ of a $R\Gamma_1$-module M by $\text{Hom}_{R\Gamma_1}(R\Gamma(F(?),??),M)$. Show that F^* and $F_{\#}$ are adjoint.

10. Algebraic K-theory of modules over a category and its splitting.

We introduce the algebraic K-groups of an EI-category Γ with R-coefficients. The category of finitely generated projective $R\Gamma$-modules is an exact category in the sense of Quillen [1973], § 2 or a category with cofibrations and weak equivalences in the sense of Waldhausen [1985], 1.2. so that we can define $K_n(R\Gamma)$ by the homotopy groups of the spaces associated with such categories. We study the functorial properties under induction, restriction and tensor product. This global approach is useful for theoretic questions. Finally we split the algebraic K-groups $K_n(R\Gamma)$ into ordinary K-groups $K_n(R[x])$ of the group rings $R[x] := R[Aut(x)]$ for $\bar{x} \in Is\,\Gamma$ This is the main tool for explicit computations.

10.A. The algebraic K-theory of $R\Gamma$-modules.

We start with introducing "lower" and "middle" K-groups $K_n(\mathcal{A})$, $n \leq 1$ of an additive category. Given a ring R, a small category $\mathcal{A}$ is called R-additive if it satisfies (compare Bass [1968], § I.3 , Schubert [1970a] 1.5.)

i) Finite coproducts exists.

ii) For all objects $x,y \in \mathcal{A}$ the set $Mor(x,y)$ has the structure of an R-module.

iii) Composition of morphisms is a R-bilinear map

iv) Coproduct of morphisms is a R-bilinear map.

v) $\mathcal{A}$ has a zero-object 0, i.e. $Mor(0,x)$ and $Mor(x,0)$ have exactly one element for any $x \in Ob\,\mathcal{A}$.

If R is $\mathbb{Z}$, we call $\mathcal{A}$ just additive .

Example 10.1. Any abelian category and any exact category is an additive category. The homotopy category of R-chain complexes is an additive category but fails to be abelian. □

Consider an additive category $\mathcal{A}$. We call a sequence in $\mathcal{A}$

$$0 \longrightarrow A_o \xrightarrow{i} A \xrightarrow{p} A_1 \longrightarrow 0 .$$

exact if it is split exact, i.e. there is a morphism $s : A_1 \longrightarrow A$ with $p \circ s = id$ such that $i \amalg s : A_o \amalg A_1 \longrightarrow A$ is an isomorphism. Equivalently, there is

$r : A \longrightarrow A_o$ with $r \circ i$ such that $j_o \circ r + j_1 \circ p : A \longrightarrow A_o \coprod A_1$ is an isomorphism where $j_i : A_i \longrightarrow A_o \coprod A_1$ is the canonical inclusion. In other words, the sequence is isomorphic to the obvious sequence

$$0 \longrightarrow A_o \longrightarrow A_o \coprod A_1 \longrightarrow A_1 \longrightarrow 0.$$

Definition 10.2.

a) The Grothendieck group $K_o(\mathcal{A})$ is defined as the quotient of the free abelian group generated by the isomorphism classes of objects in $\mathcal{A}$ and the subgroup generated by all elements $[A_o] - [A] + [A_1]$ for which there is an exact sequence

$$0 \longrightarrow A_o \longrightarrow A \longrightarrow A_1 \longrightarrow 0$$

b) Let $K_1(\mathcal{A})$ be the quotient of the free abelian group generated by the automorphism $f : A \longrightarrow A$ of objects A in $\mathcal{A}$ and the subgroup generated by elements $[f \circ g] - [f] - [g]$ for automorphism f and g of the same object and elements $[f_o] - [f] + [f_1]$ for which there is a commutative diagram with exact rows

$$\begin{array}{ccccccccc} 0 & \longrightarrow & A_o & \xrightarrow{i} & A & \xrightarrow{p} & A_1 & \longrightarrow & 0 \\ & & \downarrow f_o & & \downarrow f & & \downarrow f_1 & & \\ 0 & \longrightarrow & A_o & \xrightarrow{i} & A & \xrightarrow{p} & A_1 & \longrightarrow & 0 \end{array}$$

Now we are going to define negative K-groups. For this purpose we introduce the category $G_n(\mathcal{A})$ of $\mathbf{Z}^n$-graded objects in $\mathcal{A}$ with bounded morphisms. An object A is a collection $A = \{A(i) \mid i \in \mathbf{Z}^n\}$ of objects $A(i)$ of $\mathcal{A}$. A morphism $f : A \longrightarrow B$ is a collection $\{f(i,j) : A(i) \longrightarrow B(j) \mid i,j \in \mathbf{Z}^n\}$ of morphisms in $\mathcal{A}$ such that there is an integer $\ell = \ell(f)$ with the property that $f(i,j)$ vanishes whenever $|i(m) - j(m)| > \ell$ holds for some $m \in \{1,2,\ldots,n\}$. The composition $h = g \circ f$ of $f : A \longrightarrow B$ and $g : B \longrightarrow C$ is defined by

$$h(i,k) = \sum_{j \in \mathbf{Z}^n} g(j,k) \circ f(i,j)$$

The structure of an additive category carries over from $\mathcal{A}$ to $G_n(\mathcal{A})$.

<u>Definition 10.3.</u> Define for $n \geq -1$

$$K_{-n}(\mathcal{A}) := K_1(G_{n+1}(\mathcal{A})) \qquad \square$$

Obviously the Definition 10.2. of $K_1(\mathcal{A})$ and the Definition 10.3. of $K_1(\mathcal{A})$ agree since $G_o(\mathcal{A})$ is just $\mathcal{A}$. In general $K_1(G_1(\mathcal{A}))$ is not $K_o(\mathcal{A})$ but $K_o(\hat{\mathcal{A}})$ if $\hat{\mathcal{A}}$ denotes the idempotent completion of $\mathcal{A}$ (see Pedersen-Weibel [1985]). Objects in $\hat{\mathcal{A}}$ are projection $p : A \longrightarrow A$ in $\mathcal{A}$, i.e. $p \circ p = p$. A morphism f from $p : A \longrightarrow A$ to $q : B \longrightarrow B$ in $\hat{\mathcal{A}}$ is a morphism $f : A \longrightarrow B$ satisfying $p \circ f \circ q = f$. Since for all categories $\mathcal{A}$, we will consider, the obvious inclusion $\mathcal{A} \longrightarrow \hat{\mathcal{A}}$ is an equivalence of categories, we can regard the Definitions 10.2. and 10.3. for $n = 0,1$ as equivalent.

Let $\mathcal{E}$ be an exact category. In Quillen [1973] § 2 a new category $Q\mathcal{E}$ is defined. If $|Q\mathcal{E}|$ is the classifying space associated with $Q\mathcal{E}$, the K-groups of $\mathcal{E}$ are defined by

10.4. $$K_i(\mathcal{E}) = \pi_{i+1}(|Q\mathcal{E}|) = \pi_i(\Omega|Q\mathcal{E}|) \quad \text{for} \quad i \geq 0 .$$

Similarly Waldhausen [1985] 1.3. assigns to a category $\mathcal{C}$ with cofibrations and weak equivalences a space $|wS.\mathcal{C}|$. We can also define

10.5. $$K_i(\mathcal{C}) = \pi_{i+1}(|wS.\mathcal{C}|) = \pi_i(\Omega|wS.\mathcal{C}|)$$

Any exact category $\mathcal{E}$ determines a category $\mathcal{C}(\mathcal{E})$ with cofibrations and weak equivalences such that $|Q\mathcal{E}|$ and $|wS.\mathcal{C}(\mathcal{E})|$ are naturally homotopy equivalent so that 10.4. and 10.5. agree (see Waldhausen [1985] 1.9.) .

If $\mathcal{E}$ is an exact category the Definition 10.4 and the Definition 10.3. of $K_o(\mathcal{E})$ agree (see Quillen [1973], § 2 Theorem 1). There is a natural map from $K_1(\mathcal{E})$ of Definition 10.2. to $K_1(\mathcal{E})$ defined in 10.4. but it is neither surjective nor injective in general (Quillen [1973] Remarks on p. 104). However, if $\mathcal{E}$ is the exact category FPMOD-Λ of finitely generated projective modules over a ring Λ then they agree. Moreover $K_i(\text{FPMOD-}\Lambda)$ is isomorphic to $\pi_i(BGL(\Lambda)^+)$ for $i \geq 1$ (see Gersten [1973]). We will use Definition 10.2. for $K_i(\mathcal{E})$ but it follows from the splitting theorem below that in all cases we consider in this book the Definition 10.2. and 10.4. agree for $K_1(\mathcal{E})$,

We have now defined functors for $n \in \mathbf{Z}$

$$K_n : \{\text{exact cat.}\} \longrightarrow \{\text{abel. gr.}\}$$

Proposition 10.6.

i) If $0 \longrightarrow F_1 \longrightarrow F_o \longrightarrow F_2 \longrightarrow 0$ is an exact sequence of functors between exact categories then

$$K_n(F_1) - K_n(F_o) + K_n(F_2) = 0$$

ii) If $F_o, F_1 : \mathcal{E} \longrightarrow \mathcal{E}'$ are naturally equivalent, then $K_n(F_o) = K_n(F_1)$.

iii) Let $i : I \longrightarrow \{\text{exact cat.}\}$ be a functor from a filtering category (see Schubert [1970 a] 9.2.4.). Then the natural map

$$\operatorname{colim}(K_n \circ i) \longrightarrow K(\operatorname{colim} i)$$

is an isomorphism.

Proof: For $n \leq 1$ this follows directly from the definitions. The proof for $n \geq 0$ is given in Quillen [1973] p. 103-106. See also the Additivity Lemma in Waldhausen [1985] 1.3.2.). □

Let Γ be an EI-category. The category FFMOD-RΓ resp. FPMOD-RΓ of finitely generated free resp. projective RΓ-modules is an exact category in the sense of Quillen [1973] § 2. It can also be viewed as a category with cofibrations given by split injections and weak equivalences given by isomorphisms and is in particular an additive category.

Definition 10.7. We define for $n \in \mathbf{Z}$

$$K_n(R\Gamma) = K_n(\text{FPMOD-}R\Gamma)$$

and

$$K_n^f(R\Gamma) := K_n(\text{FFMOD-}R\Gamma) \quad \square$$

We also need the following variant of $K_1(R\Gamma)$. A <u>trivial unit</u> in $R\Gamma$ is an automorphism of the free $R\Gamma$-module $R\Gamma(?,x) := R\,\text{Hom}(?,x)$ of the shape

$$uf_* : R\,\text{Hom}(?,x) \longrightarrow R\,\text{Hom}(?,x) \qquad g \longrightarrow uf \circ g$$

for any $x \in \text{Ob}\,\Gamma$, $f \in \text{Aut}(x)$ and $u \in \{\pm 1\}$.

<u>Definition 10.8.</u> <u>Let the Whitehead group of</u> $R\Gamma$

$$\text{Wh}(R\Gamma) := K_1(R\Gamma) \,/\, \langle\text{trivial units}\rangle$$

<u>be the quotient of</u> $K_1(R\Gamma)$ <u>and the subgroup generated by all trivial units</u> $\square$

Finally we introduce

<u>Definition 10.9.</u> <u>Let</u> $U(\Gamma)$ <u>be the free abelian group generated by the set</u> $\text{Is}\,\Gamma$ <u>of isomorphism classes of objects</u>

$$U(\Gamma) := \bigoplus_{\text{Is}\,\Gamma} \mathbf{Z} \qquad \square$$

We often write an element $\eta \in U(\Gamma)$ as a function $\eta : \text{Is}\,\Gamma \longrightarrow \mathbf{Z}$ which takes a value different from zero only for finitely many elements in $\text{Is}\,\Gamma$.

10.B. <u>Natural properties of the algebraic K-theory of RΓ-modules</u>.

Now we have introduced all algebraic K-groups we are interested in except the reduced K-groups $\tilde{K}_n(R\Gamma)$ whose definition needs some preparation. Next we study the functorial properties under induction and restriction and the pairings induced by the tensor product $\otimes_R$.

Given a covariant functor $F : \Gamma_1 \longrightarrow \Gamma_2$ the induction functor induces a functoı between exact categories

$$\text{ind}_F : \text{FPMOD-}R\Gamma_1 \longrightarrow \text{FPMOD-}R\Gamma_2$$

and

$$\text{ind}_F : \text{FFMOD-}R\Gamma_1 \longrightarrow \text{FFMOD-}R\Gamma_2$$

by 9.24. Moreover, it sends a trivial unit uf_* of $R\Gamma_1$ to the trivial unit $u\,F(f)_*$ of $R\Gamma_2$. Hence we obtain homomorphisms

10.10.
$$F_* : K_n(R\Gamma_1) \longrightarrow K_n(R\Gamma_2)$$
$$F_* : K_n^f(R\Gamma_1) \longrightarrow K_n^f(R\Gamma_2)$$
$$F_* : Wh(R\Gamma_1) \longrightarrow Wh(R\Gamma_2)$$

Define an homomorphism

10.11.
$$F_* : U(\Gamma_1) \longrightarrow U(\Gamma_2)$$

by

$$F_*(\eta) : \text{Is}\,\Gamma_2 \longrightarrow \mathbb{Z} \quad y \longrightarrow \sum_{\substack{\bar{x} \in \text{Is}\,\Gamma_1 \\ \overline{Fx} = \bar{y}}} \eta(x)$$

for $\eta \in U(\Gamma_1)$.

Let EI-CAT be the category of EI-categories. Then we have defined covariant functors

$$K_n, K_n^f, Wh, U : \text{EI-CAT} \longrightarrow \{\text{abel. gr.}\}$$

One easily checks

Lemma 10.12. *The functor*

$$S_y \circ \text{ind}_F \circ E_x : \text{MOD-}R[x] \longrightarrow \text{MOD-}R[y]$$

is zero for $\bar{y} \neq \overline{Fx}$. *It is natural equivalent to the induction with the group homomorphism* $\text{Aut}(x) \longrightarrow \text{Aut}(Fx)$ $f \longrightarrow F(f)$ *for* $y = Fx$. □

Remark 10.13. If G is a (discrete) group we have denoted by $\hat{G}$ the corresponding groupoid with one object (see Example 9.5.). Composing the functors K_n, K_n^f and Wh with the functor $\{\text{groups}\} \longrightarrow \text{EI-CAT}$ $G \longrightarrow \hat{G}$ defines functors

$$\{\text{groups}\} \longrightarrow \{\text{abel. gr.}\} ,$$

They coincide with the classical definitions of $K_n(RG)$, $K_n^f(RG)$ and $Wh(RG)$ since $\text{MOD-}RG = \text{MOD-}R\hat{G}$ holds (see Example 9.5.). □

Now we want to define contravariant functors K_n, K_n^f, Wh and U using restriction with a functor. Here already problems arise in the case of a group homomorphism $\phi : G \longrightarrow H$. Namely, it is not true in general that the restriction of a finitely generated free resp. projective RH-module to RG by ϕ is again finitely generated free resp. projective. This is obviously necessary to define by restriction with ϕ an homomorphism $K_n(RH) \longrightarrow K_n(RG)$. A necessary and sufficient condition is that ϕ is injective and H/image ϕ is finite.

Definition 10.14. We call a (covariant) functor $F : \Gamma_1 \longrightarrow \Gamma_2$ between EI-categories admissible if $\mathrm{RES}_F : \text{MOD-}R\Gamma_2 \longrightarrow \text{MOD-}R\Gamma_1$ sends finitely generated free resp. projective $R\Gamma_2$-modules to finitely generated free resp. projective $R\Gamma_1$-modules for any commutative ring R with unit. □

Since RES_F is compatible with direct sums and a finitely generated projective $R\Gamma_2$-module is a direct summand in a finitely generated free $R\Gamma_2$-module, F is admissible if and only if $\mathrm{RES}_F(R\,\mathrm{Hom}(?,y)) = R\,\mathrm{Hom}(F(?),y)$ is a finitely generated free $R\Gamma_1$-module for all $y \in \mathrm{Ob}\,\Gamma_2$.

Therefore we get a neccessary condition for F being admissible from Lemma 9.31. Namely $S_x \circ \mathrm{RES}_F \circ E_y(R[y])$ must be a finitely generated free R[x]-module for any $x \in \mathrm{Ob}\,\Gamma_1$ and $y \in \mathrm{Ob}\,\Gamma_2$. We compute $S_x \circ \mathrm{RES}_F \circ E_y$. If B is the $R\Gamma_1$-R[x]-module

$$B(?) = \begin{cases} R\,\mathrm{Hom}(x,?) & \bar{x} = \bar{?} \\ \{0\} & \bar{x} \neq \bar{?} \end{cases}$$

the functor $S_x \circ \mathrm{RES}_F \circ E_y$ is given by taking the tensorproduct over R[y] with the R[y]-R[x]-bimodule

$$R\,\mathrm{Hom}(??,y) \otimes_{R\Gamma_2} R\,\mathrm{Hom}(F(?),??) \otimes_{R\Gamma_1} B$$

(see 9.30.)

We call a morphism $f : F(x) \longrightarrow y$ in Γ_2 for $x \in \mathrm{Ob}\,\Gamma_1$ and $y \in \mathrm{Ob}\,\Gamma_2$ irreducible if for any factorization

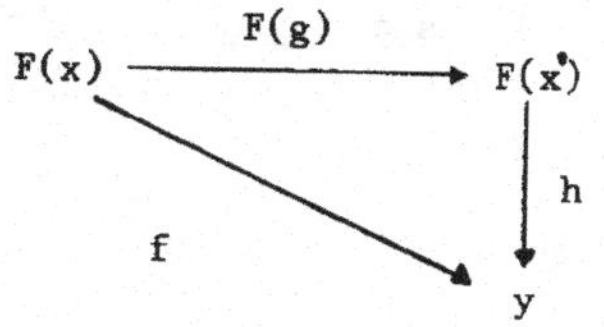

the morphism $g : x \longrightarrow x'$ in Γ_1 is an isomorphism. Let $Irr(x,y) \subset Mor_{\Gamma_2}(Fx,y)$ be the subset of irreducible morphisms. This is a Aut(y)-Aut(x)-set by composition so that $R\,Irr(x,y)$ is a $R[y]$-$R[x]$-bimodule.

We define inverse isomorphisms of $R[y]$-$R[x]$-bimodules.

$$R\,Hom(??,y) \otimes_{R\Gamma_2} R\,Hom(F(?),??) \otimes_{R\Gamma_1} B \longrightarrow R\,Irr(x,y)$$

$$f \otimes g \otimes h \longrightarrow \begin{cases} f\circ g\circ F(h), & \text{if } f\circ g\circ F(h) \text{ is irreducible} \\ 0 & \text{, otherwise} \end{cases}$$

$$R\,Irr(x,y) \longrightarrow R\,Hom(??,y) \otimes_{R\Gamma_2} R\,Hom(F(?),??) \otimes_{R\Gamma_1} B$$

$$g \longrightarrow id_y \otimes g \otimes id_x \quad .$$

Notice that $S_x R\,Hom(F(?),y)$ is by definition $R\,Irr(x,y)$. We get

Lemma 10.15. *The functor*

$$S_x \circ RES_F \circ E_y : \text{MOD-}R[y] \longrightarrow \text{MOD-}R[x]$$

is naturally equivalent to

$$M \longrightarrow M \otimes_{R[y]} R\,Irr(x,y) \quad \square$$

Proposition 10.16. *A functor* $F : \Gamma_1 \longrightarrow \Gamma_2$ *between EI-categories is admissible if and only if the following conditions hold for each* $y \in \Gamma_2$.

i) $Irr(x,y)$ *is non-empty only for finitely many* $\bar{x} \in Is\,\Gamma_1$ *and is a free* $Aut(x)$*-set with finitely many orbits for each* $x \in Ob\,\Gamma_1$.

ii) *Let* $h : F(z) \longrightarrow y$ *be a morphism in* Γ_2 *and* $z \in Ob\,\Gamma_1$. *There exist morphisms* $g : z \longrightarrow x$ *in* Γ_1 *and* $f \in Irr(x,y)$ *with* $f \circ F(g) = h$. *If* $g' : z \longrightarrow x'$ *and* $f' \in Irr(x',y)$ *also satisfy* $f' \circ F(g') = h$ *then there is*

<u>an isomorphism</u> $k : x \longrightarrow x'$ <u>in</u> Γ_1 <u>such that</u> $k \circ g = g'$ <u>and</u> $f = f' \circ F(k)$ <u>hold</u>.

<u>Proof.</u> Consider the homomorphism of $R\Gamma_1$-modules

$$T : \bigoplus_{\bar{x} \in \mathrm{Is}\,\Gamma_1} R\,\mathrm{Irr}(x,y) \otimes_{R[x]} R\,\mathrm{Hom}(?,x) \longrightarrow R\,\mathrm{Hom}(F(?),y)$$

$$f \otimes g \longrightarrow f \circ F(g)$$

One easily checks that T is bijective if and only if condition ii) is true. The source is finitely generated free if and only if $R\,\mathrm{Irr}(x,y)$ is a finitely generated free $R[x]$-module for any $x \in \mathrm{Ob}\,\Gamma_1$ and zero for almost all $\bar{x} \in \mathrm{Is}\,\Gamma_1$ or, equivalently, condition i) is satisfied (see Lemma 9.31.). Hence i) and ii) imply that RES_F is admissible.

Conversely, suppose that F is admissible. Then $R\,\mathrm{Hom}(F(?),y)$ is a finitely generated free $R\Gamma_1$-module. Applying S_x to T yields just the isomorphism of Lemma 10.15. By Lemma 9.31. the $R[x]$-module $R\,\mathrm{Irr}(x,y)$ is finitely generated free for any x and different from zero only for finitely many $x \in \mathrm{Ob}\,\Gamma_1$. This implies i). The source of T is a finitely generated free $R\Gamma_1$-module. Then condition ii) must be true as T is bijective by the Lemma below. □

<u>Lemma 10.17.</u> <u>Let</u> $F : P \longrightarrow Q$ <u>be an homomorphism of projective</u> $R\Gamma$-<u>modules of finite type</u>.

<u>Then</u> F <u>is bijective if and only if</u> S_xF <u>is bijective for any</u> $x \in \mathrm{Ob}\,\Gamma$.

<u>Proof.</u> Use induction over the type, the five lemma and Theorem 9.39. □

We will later use Proposition 10.16. to show for a Lie-group G, a G-space X satisfying Assumption 8.13. and a subgroup $H \subset G$ with finite G/H that restriction induces an admissible functor $\pi/(H,\mathrm{res}\,X) \longrightarrow \pi/(G,X)$ sending $x : H/L \longrightarrow \mathrm{res}\,X$ to its adjoint $\mathrm{ind}\,H/L = G/L \longrightarrow X$.

If $F : \Gamma_1 \longrightarrow \Gamma_2$ is admissible, it induces by definition functors between exact categories $\mathrm{RES}_F : \mathrm{FFMOD}\text{-}R\Gamma_2 \longrightarrow \mathrm{FFMOD}\text{-}R\Gamma_1$ and $\mathrm{RES}_F : \mathrm{FPMOD}\text{-}R\Gamma_2 \longrightarrow \mathrm{FPMOD}\text{-}R\Gamma_1$.

Hence we obtain homomorphisms

10.18.
$$F^* : K_n(R\Gamma_2) \longrightarrow K_n(R\Gamma_1)$$
$$F^* : K_n^f(R\Gamma_2) \longrightarrow K_n^f(R\Gamma_1) .$$

Next we examine what $F^* : K_1(R\Gamma_2) \longrightarrow K_1(R\Gamma_1)$ does with trivial units. Let $u\, f_* : R\, \mathrm{Hom}(?,y) \longrightarrow R\, \mathrm{Hom}(?,y) \quad g \longrightarrow u \cdot f \circ g$ be the trivial unit in Γ_2 given by $y \in \mathrm{Ob}\, \Gamma_2$, $f \in \mathrm{Aut}(y)$ and $u \in \{\pm 1\}$. For $x \in \mathrm{Ob}\, \Gamma_1$ let $v_x : R\, \mathrm{Irr}(x,y) \longrightarrow R\, \mathrm{Irr}(x,y)$ send g to $u \cdot f \circ g$. If T is the natural isomorphism appearing in the proof of Proposition 10.16. we have the commutative diagram of $R\Gamma_1$-modules

$$\begin{array}{ccc}
\bigoplus_{\bar{x} \in \mathrm{Is}\, \Gamma_1} E_x(R\, \mathrm{Irr}(x,y)) & \xrightarrow{T} & \mathrm{RES}_F(R\Gamma_2(?,y)) \\
\Big\downarrow \bigoplus_{\bar{x} \in \mathrm{Is}\, \Gamma_1} E_x(v_x) & & \Big\downarrow \mathrm{RES}_F(uf_*) \\
\bigoplus_{\bar{x} \in \mathrm{Is}\, \Gamma_1} E_x(R\, \mathrm{Irr}(x,y)) & \xrightarrow{T} & \mathrm{RES}_F(R\Gamma_2(?,y))
\end{array}$$

Hence we get in $K_1(R\Gamma_1)$

10.19.
$$F^*(uf_*) \quad = \quad \sum_{\bar{x} \in \mathrm{Is}\, \Gamma_1} [E_x(v_x)]$$

By Proposition 10.16. the $\mathrm{Aut}(x)$-set $\mathrm{Irr}(x,y)$ is a finite disjoint union of free $\mathrm{Aut}(x)$-orbits

$$\mathrm{Irr}(x,y) = \coprod_{i=1}^{r} g_i \cdot \mathrm{Aut}(x)$$

There is a unique permutation $\sigma \in \Sigma_r$ such that $f \circ g_i = g_{\sigma(i)} \circ f_i$ for appropriate $f_i \in \mathrm{Aut}(x)$ holds for $i = 1,2,\ldots,r$. Consider the $\mathrm{Aut}(x)$-isomorphisms

$$\Phi : \bigoplus_{i=1}^{r} R[x] \longrightarrow R\, \mathrm{Irr}(x,y)$$

sending $id_x \in R[x]$ of the i-th summand to g_i. Let

$$\sigma_* : \bigoplus_{i=1}^{r} R[x] \longrightarrow \bigoplus_{i=1}^{r} R[x]$$

be given by permuting the coordinates according to σ . Then the following diagram commutes

$$\begin{array}{ccccc} \bigoplus_{i=1}^{r} R[x] & \xrightarrow{\sigma_*} & \bigoplus_{i=1}^{r} R[x] & \xrightarrow{\bigoplus_{i=1}^{r} uf_i} & \bigoplus_{i=1}^{r} R[x] \\ \Phi \downarrow & & & & \downarrow \Phi \\ R\,\mathrm{Irr}(x,y) & & \xrightarrow{\quad v_x \quad} & & R\,\mathrm{Irr}(x,y) \end{array}$$

We get in $K_1(R[x])$ since $[\sigma_*]$ is $[(\text{sign }\sigma)\cdot \mathrm{id} : R[x] \longrightarrow R[x])$

$$[v_x] = \sum_{i=1}^{r} [\, u f_i] + [\text{sign } \sigma \cdot \mathrm{id}_{R[x]}]$$

Hence $[v_x]$ lies in the subgroup of $K_1(R[x])$ generated by the trivial units so that $F^* : K_1(R\Gamma_2) \longrightarrow K_1(R\Gamma_1)$ maps the subgroup generated by the trivial units in $K_1(R\Gamma_2)$ to the one in $K_1(R\Gamma_1)$ by 10.19. By definition we obtain an induced homomorphism

10.20. $$F^* : \mathrm{Wh}(R\Gamma_2) \longrightarrow \mathrm{Wh}(R\Gamma_1) .$$

Definean homomorphism

10.21. $$F^* : U(\Gamma_2) \longrightarrow U(\Gamma_1)$$

by the $\mathrm{Is}\,\Gamma_1 \times \mathrm{Is}\,\Gamma_2$ matrix $(d(\bar{x},\bar{y}))$ with

$$d(\bar{x},\bar{y}) = \mathrm{card}\ \mathrm{Irr}(x,y)/\mathrm{Aut}(x) .$$

This is well-defined by Proposition 10.16.

Let $\mathrm{EI\text{-}CAT}_A$ be the category of EI-categories with admissible functors as morphisms. All in all we have defined contravariant functors

10.22. $$K_n, K_n^f, \mathrm{Wh}, U : \mathrm{EI\text{-}CAT}_A \longrightarrow \{\text{abel. gr.}\}$$

where functionality is given by restriction. Obviously $(F \circ G)^* = G^* \circ F^*$ holds for K_n, K_n^f and Wh but is not clear for U . But this will follow from a result we prove later where U and K_o^f are identified. Of course one can verify this for U

also directly.

Now we come to products. The tensorproduct $\otimes_R$ over R is a functor

$$\otimes_R : \text{MOD-}R\Gamma_1 \times \text{MOD-}R\Gamma_2 \longrightarrow \text{MOD-}R\Gamma_1 \times \Gamma_2$$

(see 9.13.). Obviously $R\Gamma_1(?,x) \otimes R\Gamma_2(??,x_2)$ is $R\Gamma_1\times \Gamma_2((?,??),(x_1,x_2))$ and $M \otimes_R ?$ respects direct sums . If P is a finitely generated projective $R\Gamma_1$-module, $P \otimes_R ?$ induces a functor of exact categories $\text{FPMOD-}R\Gamma_2 \longrightarrow \text{FPMOD-}R\Gamma_1\times\Gamma_2$ and hence an homomorphism $K_n(R\Gamma_2) \rightarrow K_n(R\Gamma_1\times\Gamma_2)$ for $n \in \mathbf{Z}$. We obtain pairings by Proposition 10.6.

10.23.
$$\otimes_R : K_o(R\Gamma_1) \otimes K_n(R\Gamma_2) \longrightarrow K_n(R\Gamma_1 \times \Gamma_2)$$

and

$$\otimes_R : K^f_o(R\Gamma_1) \otimes K^f_n(R\Gamma_2) \longrightarrow K^f_n(R\Gamma_1 \times \Gamma_2) .$$

One easily checks that this induces

10.24.
$$\otimes_R : K^f_o(R\Gamma_1) \otimes Wh(R\Gamma_2) \longrightarrow Wh(R\Gamma_1\times \Gamma_2) .$$

The obvious identification $\text{Is }\Gamma_1 \times \text{Is }\Gamma_2 = \text{Is }\Gamma_1 \times \Gamma_2$ yields an isomorphism

10.25.
$$\otimes_R : U(\Gamma_1) \otimes U(\Gamma_2) \longrightarrow U(\Gamma_1 \times \Gamma_2).$$

<u>Lemma 10.26.</u> <u>We obtain natural transformations of covariant functors</u>

$$\text{EI-CAT} \times \text{EI-CAT} \longrightarrow \{\text{abel. gr.}\}$$

<u>and of contravariant functors</u>

$$\text{EI-CAT}_A \times \text{EI-CAT}_A \longrightarrow \{\text{abel. gr.}\}$$

<u>by the pairings</u> $\otimes_R$ <u>above</u>

$$K_o \otimes K_n \longrightarrow K_n$$
$$K^f_o \otimes K^f_n \longrightarrow K^f_n$$
$$K^f_o \otimes Wh \longrightarrow Wh$$
$$U \otimes U \longrightarrow U$$

Proof. For the first three pairings the desired compatibilities hold already between the functors between exact categories inducing the various maps on the K-groups The statement for the last paring can be verified directly or be viewed as a consequence of the identification $K_o^f = U$ we give later. □

If $\phi : R \longrightarrow S$ is an homomorphism of commutative associative rings with unit, composition with the induction functor R-MOD $\longrightarrow$ S-MOD yields change of coefficients homomorphisms

10.27.
$$\begin{aligned} \phi_* &: K_n(R\Gamma) \longrightarrow K_n(S\Gamma) \\ \phi_* &: K_n^f(R\Gamma) \longrightarrow K_n^f(S\Gamma) \\ \phi_* &: Wh(R\Gamma) \longrightarrow Wh(S\Gamma) . \end{aligned}$$

Roughly speaking, they are compatible with everything we have and will define.

Definition 10.28. Define the reduced K-groups $\tilde{K}_n(R\Gamma)$ as the cokernel of the homomorphism

$$K_o^f(R\Gamma) \otimes K_n(\mathbf{Z}) \xrightarrow{\mathrm{id} \otimes \phi_*} K_o^f(R\Gamma) \otimes K_n(R) \xrightarrow{\otimes_R} K_n(R\Gamma)$$

where $\phi : \mathbf{Z} \longrightarrow R$ is the canonical ring homomorphism □

$\tilde{K}_n$ is also a covariant functor EI-CAT $\longrightarrow$ {abel. gr.} by induction and a contravariant functor from $\text{EI-CAT}_A \rightarrow$ {abel.gr.} by restriction. We get a natural pairing

10.29.
$$\otimes_R : K_o^f(R\Gamma_1) \otimes \tilde{K}_n(R\Gamma_2) \longrightarrow \tilde{K}_n(R\Gamma_1 \times \Gamma_2) .$$

10.C. The splitting of the algebraic K-theory of RΓ-modules.

Now we have introduced all the algebraic K-groups $K_n(R\Gamma)$, $K_n^f(R\Gamma)$, $\tilde{K}_n(R\Gamma)$, $Wh(R\Gamma)$ and $U(\Gamma)$ we are interested in, including their behaviour under induction, restriction and tensor product $\otimes_R$. This global approach is adequate for theoretic questions. Now we will introduce a splitted version whose ingredients are K-groups of group rings and is useful for computations. Finally we identify them.

In the sequel let K be any of the functors K_n, K_n^f, $\tilde{K}_n$, Wh, EI-CAT $\longrightarrow$ {abel. gr.} defined above. If Γ is an EI-category and $u \in \text{Is}\,\Gamma$, let $\Gamma(u)$ be the full subcategory of Γ having all $x \in \text{Ob}\,\Gamma$ with $x \in u$ as objects. Notice that $\Gamma(u)$ is a groupoid and $\mathrm{Is}\,\Gamma(u)$ has exactly one object. Define

10.30. $$\mathrm{Split}\, K(R\Gamma) := \bigoplus_{u \in \mathrm{Is}\,\Gamma} K(R\Gamma(u))$$

Let $F : \Gamma_1 \longrightarrow \Gamma_2$ be a functor. If $\overline{F} : \mathrm{Is}\,\Gamma_1 \longrightarrow \mathrm{Is}\,\Gamma_2$ is the induced map, we obtain for any $u \in \mathrm{Is}\,\Gamma_1$ and $v \in \mathrm{Is}\,\Gamma_2$ with $v = \overline{F}(u)$ a functor $\Gamma_1(u) \rightarrow \Gamma_2(v)$ by restriction to $\Gamma_1(u)$ and hence an homomorphism

$$F(u,v)_* : K(R\Gamma(u)) \longrightarrow K(R\Gamma(v))$$

If $v \neq \overline{F}(u)$, define $F(u,v)_* : K(R\Gamma(u)) \longrightarrow K(R\Gamma(v))$ by the zero map. Let the homomorphism

10.31. $$F_* : \mathrm{Split}\, K(R\Gamma_1) \longrightarrow \mathrm{Split}\, K(R\Gamma_2)$$

be given by $(F(u,v)_*)_{u \in \mathrm{Is}\,\Gamma_1, v \in \mathrm{Is}\,\Gamma_2}$.

Thus we obtain a covariant functor

$$\mathrm{Split}\, K : \text{EI-CAT} \longrightarrow \{\text{abel. gr.}\}.$$

Consider an admissible functor $F : \Gamma_1 \longrightarrow \Gamma_2$. Let $R\,\mathrm{Irr}(u,v)$ be the $R\Gamma_1(u)$-$R\Gamma_2(v)$ bimodule for $u \in \mathrm{Is}\,\Gamma_1$, $v \in \mathrm{Is}\,\Gamma_2$

$$\Gamma_2(v) \times \Gamma_1(u)^{op} \longrightarrow R\text{-MOD} \qquad (x,y) \longrightarrow R\,\mathrm{Irr}(x,y).$$

We obtain an exact functor

$$\begin{array}{ccc} \text{FPMOD-}R\Gamma_2(v) & \longrightarrow & \text{FPMOD-}R\Gamma_1(u) \\ M & \longrightarrow & M \otimes_{R\Gamma_2(v)} R\,\mathrm{Irr}(u,v) \end{array}$$

since $R\,\mathrm{Irr}(x,y)$ is a finitely generated free $R[x]$-module by Proposition 10.16. It induces an homomorphism

$$F(v,u)^* : K(R\Gamma_2(v)) \longrightarrow K(R\Gamma_1(u)) .$$

By Proposition 10.16. we get a well-defined homomorphism

10.32. $$F^* : \mathrm{Split}\, K(R\Gamma_2) \longrightarrow \mathrm{Split}\, K(R\Gamma_1)$$

by $(F(v,u)^*)_{v \in \mathrm{Is}\,\Gamma_2,\ u \in \mathrm{Is}\,\Gamma_1}$.

We have introduced a contravariant functor

$$\mathrm{Split}\ K : \mathrm{EI\text{-}CAT}_A \longrightarrow \{\mathrm{abel.\ gr.}\} .$$

One can verify $(F \circ G)^* = G^* \circ F^*$ directly but it easily follows from the corresponding relation for K itself and the identification $K = \mathrm{Split}\ K$ we establish soon.

Let the pairings

10.33.
$$\begin{aligned}
\otimes_R &: \mathrm{Split}\ K_o(R\Gamma_1) \otimes \mathrm{Split}\ K_n(R\Gamma_2) \longrightarrow \mathrm{Split}\ K_n(R\Gamma_1 \times \Gamma_2)\\
\otimes_R &: \mathrm{Split}\ K_o^f(R\Gamma_1) \otimes \mathrm{Split}\ K_n^f(R\Gamma_2) \longrightarrow \mathrm{Split}\ K_n^f(R\Gamma_1 \times \Gamma_2)\\
\otimes_R &: \mathrm{Split}\ K_o^f(R\Gamma_1) \otimes \mathrm{Split}\ Wh(R\Gamma_2) \longrightarrow \mathrm{Split}\ Wh(R\Gamma_1 \times \Gamma_2)\\
\otimes_R &: \mathrm{Split}\ K_o^f(R\Gamma_1) \otimes \mathrm{Split}\ \tilde{K}_n(R\Gamma_2) \longrightarrow \mathrm{Split}\ \tilde{K}_n(R\Gamma_1 \times \Gamma_2)
\end{aligned}$$

be induced by the various pairings we get between $\Gamma_1(u)$ and $\Gamma_2(v)$ for all $u \in \mathrm{Is}\,\Gamma_1$ and $v \in \Gamma_2(v)$, e. g. all pairings

$$K_o(R\Gamma_1(u)) \otimes K_n(R\Gamma_2(v)) \longrightarrow K_n(R\Gamma_1(u) \times \Gamma_2(v)).$$

Notice that $\Gamma_1(u) \times \Gamma_2(v)$ is just $\Gamma_1 \times \Gamma_2(u,v)$.

Now we want to compare K and $\mathrm{Split}\,K$. For $u \in \mathrm{Ob}\,\Gamma$ define the <u>extension functor</u>

$$\begin{aligned}
E_u : \mathrm{MOD\text{-}R\Gamma}(u) &\longrightarrow \mathrm{MOD\text{-}R\Gamma}\\
M &\longrightarrow M \otimes_{R\Gamma(u)} R\,\mathrm{Hom}(?,??)
\end{aligned}$$

Analogously define the <u>splitting functor</u>

$$S_u : \mathrm{MOD\text{-}R\Gamma} \longrightarrow \mathrm{MOD\text{-}R\Gamma}(u)$$

by assigning to an $R\Gamma$-module the $R\Gamma(u)$-module

$$\Gamma(u) \longrightarrow R\text{-}\mathrm{MOD} \qquad x \longrightarrow S_x M$$

where $S_x M$ is the splitting functor of 9.26. We can S_u also define by $M \longrightarrow M \otimes_{R\Gamma} B$

for the $R\Gamma$-$R\Gamma(u)$-bimodule B given by $B(x,y) = R\,\mathrm{Hom}(x,y)$ for $y \in u$ and $B(x,y) = \{0\}$ for $y \notin u$, where x runs over $\mathrm{Ob}\,\Gamma(u)$ and y over $\mathrm{Ob}\,\Gamma$ (compare 9.30.).

Define the homomorphism

$$E(R\Gamma) : \mathrm{Split}\; K(R\Gamma) \longrightarrow K(R\Gamma)$$

by

$$\bigoplus_{u \in \mathrm{Is}\,\Gamma} K(E_u) : \bigoplus_{u \in \mathrm{Is}\,\Gamma} K(R\Gamma(u)) \longrightarrow K(R\Gamma)$$

Each S_u induces an homomorphism

$$K(S_u) : K(R\Gamma) \longrightarrow K(R\Gamma(u)) \quad .$$

Since finite coproducts and finite products agree in the category of abelian groups, we get for any finite subset $T \subset \mathrm{Is}\,\Gamma$ an homomorphism

$$S(R\Gamma)_T : K(R\Gamma) \longrightarrow \bigoplus_{u \in T} K(R\Gamma(u)) \quad .$$

Let I be the filtering category of finite subsets of $\mathrm{Is}\,\Gamma$ with inclusions as morphisms. In the sequel all colimits are taken over I.

Let $M(T)$ be the full subcategory of FPMOD-$R\Gamma$ consisting of finitely generated projective $R\Gamma$-modules of type T. Since each finitely generated $R\Gamma$-module is of finite type, we have

$$\text{FPMOD-}R\Gamma = \mathrm{colim}\; M(T) \; .$$

Proposition 10.6. implies

$$K(R\Gamma) := K(\text{FPMOD-}R\Gamma) = \mathrm{colim}\; K(M(T)).$$

Moreover we have

$$\mathrm{Split}\; K(R\Gamma) := \bigoplus_{u \in \mathrm{Is}\,\Gamma} K(R\Gamma(u)) = \mathrm{colim} \bigoplus_{u \in T} K(R\Gamma(u)) \; .$$

Hence we can define an homomorphism

$$S(R\Gamma) : K(R\Gamma) \longrightarrow \text{Split } K(R\Gamma)$$

by taking the colimit over all $S_T : K(M(T)) \longrightarrow \bigoplus_{u \in T} K(R\Gamma(u))$. The main result of this section is

<u>Theorem 10.34. Splitting Theorem for algebraic K-theory of $R\Gamma$-modules.</u>
<u>We have inverse pairs of natural equivalences E and S between the covariant functors</u>

$$K \text{ and Split } K : \text{EI-CAT} \longrightarrow \{\text{abel. gr.}\}$$

<u>and the contravariant functors</u>

$$K \text{ and Split } K : \text{EI-CAT}_A \longrightarrow \{\text{abel. gr.}\}$$

<u>where</u> K <u>stands for</u> K_n, K_n^f, $\tilde{K}_n$ <u>or</u> Wh . <u>Moreover</u> S <u>and</u> E <u>are compatible with the various pairings</u> $\otimes_R$.

<u>Proof.</u> We consider only the case $K = K_n$, the others follow easily. We start with proving for a fixed EI-category Γ that $E(R\Gamma)$ and $S(R\Gamma)$ are inverse isomorphisms .

For finite $T \subset \text{Is } \Gamma$ let

$$E_T : \bigoplus_{u \in T} K(R\Gamma(u)) \longrightarrow K(M(T))$$

be $\bigoplus_{u \in T} K(E_u)$ so that $E(R\Gamma)$ is colim E_T . Hence it suffices to show for any finite $T \subset \text{Is } \Gamma$ that $E_T \circ S_T = \text{id}$ and $S_T \circ E_T = \text{id}$ hold. We get $S_T \circ E_T = \text{id}$ directly from Lemma 9.31. We prove $E_T \circ S_T = \text{id}$ inductively over the cardinality of T. The begin $T = \emptyset$ is trivial.

In the induction step choose $u \in T$ maximal in T under $\leq$ and write $T' = T\setminus\{u\}$. We obtain from Theorem 9.39. a functor between exact categories

$$\text{COK}_u : M(T) \longrightarrow M(T')$$

Let $F : M(T') \longrightarrow M(T)$ be the inclusion. Theorem 9.39. yields an exact sequence

$$0 \longrightarrow E_u \circ S_u \longrightarrow ID \longrightarrow F \circ COK_u \longrightarrow 0$$

By Proposition 10.6. we have

$$K(E_u) \circ E(S_u) + K(F) \circ K(COK_u) = id$$

The following diagram commutes because of this equation, Theorem 9.39. and the induction hypothesis $E_{T'} \circ S_{T'} = id$.

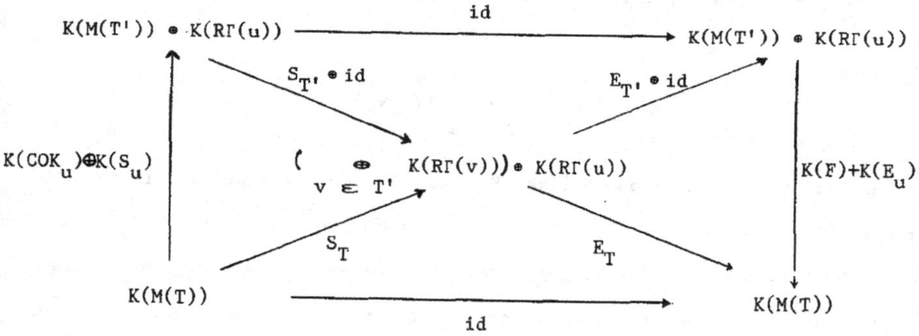

Hence $E_T \circ S_T$ is the identity. Therefore $E(R\Gamma)$ and $S(R\Gamma)$ are inverse isomorphisms.

It remains to show naturality. Given a covariant functor $F : \Gamma_1 \longrightarrow \Gamma_2$, we must show that the following diagram commutes

$$
\begin{array}{ccc}
K(R\Gamma_1) & \xrightarrow{\;F_*\;} & K(R\Gamma_2) \\[2mm]
{\scriptstyle E(R\Gamma_1)} \uparrow & & \downarrow {\scriptstyle S(R\Gamma_2)} \\[2mm]
\text{Split } K(R\Gamma_1) & \xrightarrow[\;F_*\;]{} & \text{Split } K(R\Gamma_2)
\end{array}
$$

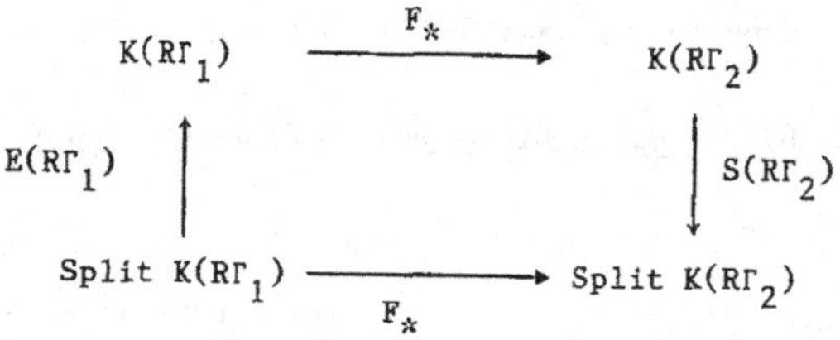

This follows from Lemma 10.12. In the contravariant case use Lemma 10.15. The easy verification that S and E are compatible with the various pairings $\bullet_R$ is left to the reader.   □

Remark 10.35. We emphasize that all the constructions about Split K do not involve choices of representatives $x \in u$ for the various $u \in \mathrm{Is}\,\Gamma$. For this reason we have always used the groupoid $\Gamma(u)$ and not the group $\mathrm{Aut}(x)$ for some representative x in u. They are related by the inclusion functor $i(x) : \mathrm{Aut}(x)^{\wedge} \longrightarrow \Gamma(u)$ and the identification of Remark 10.13. which gives an homomorphism

$$i(x)_* : K(R[x]) \longrightarrow K(R\Gamma(u)) .$$

This map $i(x)_*$ is an isomorphism by Proposition 10.6. because $i(x)$ is an equivalence of categories. Hence we have, indeed, expressed $K(R\Gamma)$ by algebraic K-groups of group rings. Namely, if we choose a representative x for any $\bar{x} \in \mathrm{Is}\,\Gamma$ we have

$$K(R\Gamma) \cong \bigoplus_{\bar{x} \in \mathrm{Is}\,\Gamma} K(R[x])$$

Also the natural homomorphism above like F_* and F^* can be expressed in terms of group rings. For example, let $F : \Gamma_1 \longrightarrow \Gamma_2$ be a functor and $x \in \mathrm{Ob}\,\Gamma_1$ and $y = F(x) \in \mathrm{Ob}\,\Gamma_2$. Let $\psi : \mathrm{Aut}(x) \longrightarrow \mathrm{Aut}(y)$ be the group homomorphism sending f to $F(f)$. The definition of F_* is based on the various homomorphisms $F(\bar{x},\bar{y})_*$. If ψ_* is the ordinary change of ring homomorphism we obtain a commutative diagram

$$\begin{array}{ccc} K(R\Gamma(\bar{x})) & \xrightarrow{F(\bar{x},\bar{y})_*} & K(R\Gamma(\bar{y})) \\ \uparrow{\scriptstyle i(x)_*} & & \uparrow{\scriptstyle i(y)_*} \\ K(R[x]) & \xrightarrow{\psi_*} & K(R[y]) \end{array}$$

Analogous statements are true for $F^*, \otimes_R, S(R\Gamma)$ and $E(R\Gamma)$. □

Theorem 10.36. Let G be a compact Lie group. Let K be K_n, K_n^f, $\tilde{K}_n$ or Wh. Then

$$K(R\mathrm{Or}/(G)) = \bigoplus_{(H) \in \mathrm{Con}\,G} K(R[\pi_0(WH)]) \qquad \square$$

Corollary 10.37. Let T^k be the k-dimensional torus. Then

$$\tilde{K}_n(\mathbf{Z}Or/(T^k)) = 0$$

and

$$Wh(\mathbf{Z}Or/(T^k)) = 0$$

Proof. For any $H \subset T^k$ we have $\pi_o(WH) = \{*\}$. By definition $\tilde{K}_n(\mathbf{Z}) = 0$ and $Wh(1)$ is zero by Cohen [1973] 11.1. Now apply Theorem 10.36. □

Recall assumption 9.1. that we have the notion of a rank $rk_R M$ of a finitely generated R-module M .

Definition 10.38. The rank of a finitely generated $R\Gamma$-module M $rk_{R\Gamma}M \in U(\Gamma)$ is given by $rk_{R\Gamma}M : Is\,\Gamma \longrightarrow \mathbf{Z}$ sending $\bar{x} \in Is\,\Gamma$ to $rk_R(S_x M \otimes_{R[x]} R)$. □

Remark 10.39. We have $M \cong N \implies rk_{R\Gamma}M = rk_{R\Gamma}N$ and $rk_{R\Gamma}(M \oplus N) = rk_{R\Gamma}M + rk_{R\Gamma}N$ since the splitting functor S_x is compatible with $\oplus$. However, it is not true in general for an exact sequence $0 \longrightarrow M_1 \longrightarrow M_o \longrightarrow M_2 \longrightarrow 0$ that $rk_{R\Gamma}M_1 - rk_{R\Gamma}M_o + rk_{R\Gamma}M_2 = 0$ holds as S_x fails to be an exact functor. Counterexamples are easily derived from Example 9.34. □

Lemma 10.40. Two finitely generated free $R\Gamma$-modules M and N are isomorphic if and only if they have the same rank.

Proof. We get for a finitely generated free $R\Gamma$-module M from Lemma 9.31. and 9.33.

$$M \cong \bigoplus_{\bar{x} \in Is\,\Gamma} (rk_{R\Gamma}(M)(\bar{x})) \cdot R\,Hom(?,x)$$

where $n \cdot N$ means $\bigoplus_{i=1}^{n} N$. □

Now we obtain homomorphisms

10.41. $$\begin{array}{llll} rk_{R\Gamma} : K_o(R\Gamma) \longrightarrow U(\Gamma) & [P] \longrightarrow rk_{R\Gamma}P \\ rk_{R\Gamma} : K_o^f(R\Gamma) \longrightarrow U(\Gamma) & [F] \longrightarrow rk_{R\Gamma}F \\ j_{R\Gamma} : U(\Gamma) \longrightarrow K_o^f(R\Gamma) & \eta \longrightarrow \sum_{\bar{x} \in \mathrm{Is}\,\Gamma} \eta(x)\cdot[R\,\mathrm{Hom}(?,x)] \\ i_{R\Gamma} : K_n^f(R\Gamma) \longrightarrow K_n(R\Gamma) & \\ p_{R\Gamma} : K_n(R\Gamma) \longrightarrow \tilde{K}_n(R\Gamma) & \end{array}$$

where $i_{R\Gamma}$ is induced by the inclusion functor FFMOD-RΓ ⟶ FPMOD-RΓ and $p_{R\Gamma}$ is the canonical projection.

<u>Proposition 10.42.</u>

a) <u>We have inverse isomorphisms</u>

$$K_o^f(R\Gamma) \underset{j_{R\Gamma}}{\overset{rk_{R\Gamma}}{\longleftrightarrow}} U(\Gamma)$$

b) <u>We get a split exact sequence</u>

$$0 \longrightarrow K_o^f(R\Gamma) \underset{i_{R\Gamma}}{\overset{j_{R\Gamma}\circ rk_{R\Gamma}}{\rightleftarrows}} K_o(R\Gamma) \xrightarrow{p_{R\Gamma}} \tilde{K}_o(R\Gamma) \longrightarrow 0$$

c) <u>For</u> $n \geq 1$ <u>the map</u>

$$i_{R\Gamma} : K_n^f(R\Gamma) \longrightarrow K_n(R\Gamma)$$

<u>is an isomorphism</u>.

<u>Proof.</u>

a) By Theorem 10.34. we have an isomorphism

$$S : K_o^f(R\Gamma) \longrightarrow \mathrm{Split}\ K_o^f(R\Gamma) = \bigoplus_{\bar{x} \in \mathrm{Is}\,\Gamma} K_o^f(R[x])$$

We obtain for any $x \in \mathrm{Ob}\,\Gamma$ an isomorphism $K_o^f(R[x]) \longrightarrow \mathbb{Z}$ $[F] \longrightarrow rk_R(F \otimes_{R[x]} R)$ Their direct sum over $\mathrm{Is}\,\Gamma$ yields an isomorphism $\mathrm{Split}\ K_o^f(R\Gamma) \longrightarrow U(\Gamma)$ whose composition with S is just $rk_{R\Gamma} : K_o^f(R\Gamma) \longrightarrow U(\Gamma)$. Obviously $rk_{R\Gamma} \circ j_{R\Gamma} = id$ holds because of Lemma 9.31.

b) Because of a) the map $j_{R\Gamma} \circ rk_{R\Gamma}$ is a retraction of $i_{R\Gamma}$. Since $\mathbb{Z}$ is a principle domain, $K_o(\mathbb{Z})$ is the infinite cyclic subgroup generated by $[\mathbb{Z}]$. Hence $i_{R\Gamma}$ can be identified with the map appearing in the Definition 10.28. of $\tilde{K}_o(R\Gamma)$ so that the sequence under consideration is exact.

c) This follows from the fact that each finitely generated projective RΓ-module is a direct summand in a finitely generated free one. □

Let $k_{R\Gamma} : K_o^f(R\Gamma) \otimes K_n(\mathbb{Z}) \longrightarrow K_n(R\Gamma)$ be the map appearing in the Definition 10.28 of $\tilde{K}_n(R\Gamma) := \operatorname{cok} k_{R\Gamma}$. Define the homomorphism

10.43. $$q_{R\Gamma} : K_n(R\Gamma) \longrightarrow U(\Gamma) \otimes K_n(R) = \bigoplus_{\bar{x} \in \operatorname{Is}\Gamma} K_n(R)$$

by the composition

$$K_n(R\Gamma) \xrightarrow{S} \operatorname{Split} K_n(R\Gamma) = \bigoplus_{\bar{x} \in \operatorname{Is}\Gamma} K_n(R[x]) \longrightarrow \bigoplus_{\bar{x} \in \operatorname{Is}\Gamma} K_n(R)$$

If $\phi : \mathbb{Z} \longrightarrow R$ is the canonical ring homomorphism the composition $q_{R\Gamma} \circ k_{R\Gamma}$ agrees with $rk_{R\Gamma} \otimes \phi_* : K_o^f(R\Gamma) \otimes K_n(\mathbb{Z}) \longrightarrow U(\Gamma) \otimes K_n(R)$. Expecially for $R = \mathbb{Z}$ we get

<u>Proposition 10.43.</u> <u>There is a split exact sequence</u>

$$0 \longrightarrow K_o^f(\mathbb{Z}\Gamma) \otimes K_n(\mathbb{Z}) \underset{k_{R\Gamma}}{\overset{(j_{R\Gamma} \otimes id)\circ q_{R\Gamma}}{\rightleftarrows}} K_n(\mathbb{Z}\Gamma) \longrightarrow \tilde{K}_n(\mathbb{Z}\Gamma) \longrightarrow 0$$

Now we examine the functorial properties of these homomorphisms.

<u>Proposition 10.44.</u>

a) <u>We have natural transformations of covariant functors</u> EI-CAT → {abel.gr.} <u>and of contravariant functors</u> $\text{EI-CAT}_A \longrightarrow$ {abel.gr.}

$$rk : K_o^f \longrightarrow U$$

$$j : U \longrightarrow K_o^f$$

$$i : K_n^f \longrightarrow K_n$$

$$p : K_n \longrightarrow \tilde{K}_n$$

$$k : K_o^f \bullet K_n(\mathbf{Z}) \longrightarrow \tilde{K}_n$$

b) <u>We get a natural transformation of covariant functors</u> EI-CAT $\longrightarrow$ {abel.gr.}

$$rk : K_o \longrightarrow U$$

c) <u>Suppose that</u> R <u>is a Dedekind domain with</u> ch R = 0 <u>and quotient field</u> F . <u>Let</u> $rk_R M$ <u>be defined by</u> $rk_F\ F \otimes_R M$. <u>Suppose for any</u> $y \in \mathrm{Ob}\Gamma$ <u>that</u> Aut(y) <u>is finite and no prime number</u> p <u>dividing</u> $|\mathrm{Aut}(y)|$ <u>is a unit in</u> R. <u>Then we obtain a natural transformation of contravariant functors</u> $\text{EI-CAT}_A \longrightarrow$ {abel.gr.}

$$rk : K_o \longrightarrow U .$$

<u>Proof.</u>

a) is left to the reader.

b+c) Let $r : \mathrm{Split}\ K_o(R\Gamma) \longrightarrow U(\Gamma)$ be the direct sum over Is Γ of the homomorphisms $K_o(R[x]) \longrightarrow \mathbf{Z}$ $[P] \longrightarrow rk_R P \otimes_{R[x]} R$. Then $rk_{R\Gamma} : K_o(R\Gamma) \to U$ is the composition of r with $S : K_o(R\Gamma) \longrightarrow \mathrm{Split}(R\Gamma)$. We have already seen that S is natural so that it suffices to show that r is natural.

Consider a functor $F : \Gamma_1 \longrightarrow \Gamma_2$. It induces an homomorphism $\psi : \mathrm{Aut}(x) \to \mathrm{Aut}(Fx)$ $f \longrightarrow F(f)$. Then the claim follows in the covariant case b) from $P \otimes_{R[x]} R \cong \mathrm{ind}_\psi P \otimes_{R[Fx]} R$. In the contravariant case c) we have to check for a finitely generated projective R[y]-module P and $x \in \mathrm{Ob}\ \Gamma_1$, $y \in \mathrm{Ob}\ \Gamma_2$

$$rk_F(P \otimes_{R[y]} R\ \mathrm{Irr}(x,y) \otimes_{R[x]} R \otimes_R F) = rk_F(P \otimes_{R[x]} R \otimes_R F) \cdot |\mathrm{Irr}(x,y)/\mathrm{Aut}(x)|$$

The assumptions in c) guarantee that $P \otimes_R F$ is a finitely generated free F[y]-module by Swan [1960a], Theorem 8.1. We get

$$rk_F((P \otimes_R F) \otimes_{F(y)} F(\mathrm{Irr}(x,y)/\mathrm{Aut}(x))) = rk_F((P \otimes_R F) \otimes_{F(y)} F) \cdot |\mathrm{Irr}(x,y)/\mathrm{Aut}(x)|$$

We have

$$(P \otimes_R F) \otimes_{F[y]} F(\mathrm{Irr}(x,y)/\mathrm{Aut}(x)) \cong_F P \otimes_{R[y]} R\ \mathrm{Irr}(x,y) \otimes_{R[x]} R \otimes_R F .$$

Now the claim follows. □

The following example shows that $rk : K_o \longrightarrow U$ is not compatible with restriction in general.

Example 10.45. Let $F : \{1\} \longrightarrow G$ be the inclusion of the trivial group into a finite group G. This can be interpreted as an admissible functor $F : \{1\}^\wedge \longrightarrow G^\wedge$ (Example 9.5.). The relation $F^* \circ rk_{QG^\wedge} = rk_{Q\{1\}^\wedge} \circ F^*$ is equivalent to the commutativity of the diagram

$$\begin{array}{ccccc} K_o(\mathbb{Q}G) & \xrightarrow{\otimes_{\mathbb{Q}G}\mathbb{Q}} & K_o(\mathbb{Q}) & \xrightarrow{rk_{\mathbb{Q}}} & \mathbb{Z} \\ \downarrow res & & & & \downarrow \cdot |G| \\ K_o(\mathbb{Q}) & & \xrightarrow{rk_{\mathbb{Q}}} & & \mathbb{Z} \end{array}$$

Inspecting it for $[\mathbb{Q}] \in K_o(\mathbb{Q}G)$ one recognizes that it commutes if and only if G is trivial. Notice that for non-trivial G the group order $|G| \neq 1$ is a unit in $\mathbb{Q}$ so that the conditions in Proposition 10.44. c) are not satisfied. □

Finally we show that $rk_{R\Gamma}$ is compatible with $\otimes_R$. Namely, one easily verifies

Proposition 10.46. The following diagram commutes

$$\begin{array}{ccc} K_o(R\Gamma_1) \otimes K_o(R\Gamma_2) & \longrightarrow & K_o(R\Gamma_1 \times \Gamma_2) \\ \downarrow rk_{R\Gamma_1} \otimes rk_{R\Gamma_2} & & \downarrow rk_{R\Gamma_1 \times \Gamma_2} \\ U(\Gamma_1) \otimes U(\Gamma_2) & \longrightarrow & U(\Gamma_1 \times \Gamma_2) \end{array}$$

□

Example 10.47. As an illustration we consider briefly $K_o^G := K_o(\mathbb{Z}OrG)$, $U^G = U(\mathbb{Z}OrG)$, and $\tilde{K}_o^G = \tilde{K}_o(G)$ for a finite group G. We later treat the case of a compact Lie group extensively. We want to work out the meaning of $\otimes_R$ and restriction.

The main observation is that the diagonal functor $\Delta : Or\,G \longrightarrow Or\,G \times Or\,G$ is admissible. We prove this later in a more general context using Proposition 10.16. Moreover $Irr(G/H, G/K \times G/L)$ is given by the set $Mono(G/H, G/K \times G/L)^G$ of injective G-maps $G/H \longrightarrow G/K \times G/L$.

Hence we get the structure of a commutative associative ring with unit on U^G by

$$U(\mathbf{Z}\mathrm{Or}G) \otimes U(\mathbf{Z}\mathrm{Or}G) \xrightarrow{\otimes_{\mathbf{Z}}} U(\mathbf{Z}\mathrm{Or}G \times \mathrm{Or}\ G) \xrightarrow{\Delta^*} U(\mathbf{Z}\mathrm{Or}G)$$

and analogously on K_o^G. The map $U(G) \cong K_o^f(\mathbf{Z}\mathrm{Or}G) \longrightarrow K_o(\mathbf{Z}\mathrm{Or}G) = K_o^G$ is a ring homomorphism. Since $\mathbf{Z}$ is a Dedekind domain with $\mathbf{Z}^* = \{\pm 1\}$, we get from Proposition 10.44. and Proposition 10.46. a ring homomorphism

$$rk : K_o^G \longrightarrow U^G$$

By Proposition 10.42. we have an isomorphism of abelian groups

$$rk \oplus p \qquad K_o^G \longrightarrow U^G \oplus \tilde{K}_o^G$$

We have a U^G-module structure on $\tilde{K}_o^G$

$$U^G \otimes \tilde{K}_o^G \longrightarrow K_o^f(\mathbf{Z}\mathrm{Or}G) \otimes \tilde{K}_o(\mathbf{Z}\mathrm{Or}G) \xrightarrow{\otimes_R} \tilde{K}_o(\mathbf{Z}\mathrm{Or}G \times \mathrm{Or}G) \xrightarrow{\Delta^*} \tilde{K}_o(\mathbf{Z}\mathrm{Or}G) = \tilde{K}_o^G .$$

We will later show using some algebra that $U^G \oplus \tilde{K}_o^G$ carries the structure of a commutative associative ring with unit given by $(u,v)\cdot(u',v') = (uu',uv'+u'v)$ and that $rk \oplus p : K_o^G \longrightarrow U^G \oplus \tilde{K}_o^G$ is a ring isomorphism.

Additively U^G is the free abelian group generated by Con G . We obtain an isomorphism between abelian group from U^G to the Euler ring $U(G)$ by $(H) \longrightarrow G/H$. This is even a ring isomorphism by Theorem 10.34. since $|\mathrm{MONO}(G/H,G/K\times G/L)^G/\mathrm{Aut}(G/H)|$ is just the number of orbits of type G/H in $G/K \times G/L$ equipped with the diagonal G-action. In other words $U(G) = A(G)$ is isomorphic to the Grothendieck ring $K_o^f(\mathbf{Z}\mathrm{Or}G)$ of the exact category of finitely generated free $\mathbf{Z}\mathrm{Or}G$-modules. □

10.D. The Bass-Heller-Swan decomposition.

Given an additive category $\mathcal{A}$, define $\mathrm{Nil}(\mathcal{A})$ as the quotient of the free abelian group generated by all nilpotent endomorphism $f : A \longrightarrow A$ in $\mathcal{A}$ (i.e. $f^n = 0$ for appropriate n) and the subgroup generated by all elements $[f_1]-[f_o]+[f_2]$ for which there is a commutative diagram with exact rows

$$\begin{array}{ccccccccc} 0 & \longrightarrow & A_1 & \xrightarrow{i} & A_o & \xrightarrow{p} & A_2 & \longrightarrow & 0 \\ & & \downarrow f_1 & & \downarrow f_o & & \downarrow f_2 & & \\ 0 & \longrightarrow & A_1 & \xrightarrow[i]{} & A_o & \xrightarrow[p]{} & A_2 & \longrightarrow & 0 \end{array}$$

Notice that conjugated nilpotent endomorphism define the same element in $\mathrm{Nil}(\mathcal{A})$. There is an inclusion $i : K_o(\mathcal{A}) \longrightarrow \mathrm{Nil}(\mathcal{A})$ $[A] \longrightarrow [0 : A \to A]$ and a retraction $r : \mathrm{Nil}(\mathcal{A}) \longrightarrow K_o(\mathcal{A})$ $[f : A \to A] \longrightarrow [A]$. Define $\widetilde{\mathrm{Nil}}(\mathcal{A})$ as the cokernel of i so that we obtain a splitting $\mathrm{Nil}(\mathcal{A}) = \widetilde{\mathrm{Nil}}(\mathcal{A}) \oplus K_o(\mathcal{A})$.

We define

10.48.
$$\mathrm{Nil}(R\Gamma) = \mathrm{Nil}(\text{FPMOD-}R\Gamma)$$
$$\widetilde{\mathrm{Nil}}(R\Gamma) = \widetilde{\mathrm{Nil}}(\text{FPMOD-}R\Gamma)$$

One easily checks that $\widetilde{\mathrm{Nil}}(R\Gamma)$ is $\widetilde{\mathrm{Nil}}(\text{FFMOD-}R\Gamma)$. Recall that $\hat{\mathbf{Z}}$ is the groupoid having only one object, namely $\mathbf{Z}$, and left translations $\ell(n) : \mathbf{Z} \longrightarrow \mathbf{Z}$ as morphisms. Let $\ell(1)_* : R\,\mathrm{Hom}(?,\mathbf{Z}) \longrightarrow R\,\mathrm{Hom}(?,\mathbf{Z})$ be given by composition with $\ell(1)$. Define injective homomorphisms

10.49.
$$B : K_o(R\Gamma) \longrightarrow K_1(R\Gamma \times \hat{\mathbf{Z}})$$
$$\tilde{B} : \tilde{K}_o(R\Gamma) \longrightarrow \mathrm{Wh}(R\Gamma \times \hat{\mathbf{Z}})$$

by sending $[P]$ to the class of $\mathrm{id} \otimes_R (-\ell(1)_*) : P \otimes_R R\,\mathrm{Hom}(?,\mathbf{Z}) \to P \otimes_R R\,\mathrm{Hom}(?,\mathbf{Z})$. These maps B and $\tilde{B}$ do not agree with the one defined in Bass, Heller, and Swan [1964], Bass [1968], XII where $\ell(1)_*$ instead of $-\ell(1)_*$ is taken. The definition 10.49. has the advantage that it corresponds to the geometrically defined map $\phi : \mathrm{Wa}^G(Y) \longrightarrow \mathrm{Wh}^G(Y \times S^1)$ of 7.34. and that its image can be characterized as follows

For $k \geq 2$ let $p(k) : \hat{\mathbf{Z}} \longrightarrow \hat{\mathbf{Z}}$ be the functor sending $\ell(n)$ to $\ell(kn)$. Then $\mathrm{id} \times p(k) : \Gamma \times \hat{\mathbf{Z}} \longrightarrow \Gamma \times \hat{\mathbf{Z}}$ is admissible and we obtain by restriction $(\mathrm{id} \times p(k))^* : \mathrm{Wh}(R\Gamma \times \hat{\mathbf{Z}}) \longrightarrow \mathrm{Wh}(R\Gamma \times \hat{\mathbf{Z}})$. Since $p(k_1) \circ p(k_2) = p(k_1 \cdot k_2)$ holds we get an operation of the multiplicative monoid $\mathbb{N}$ of positive integers on

$Wh(R\Gamma \times \hat{\mathbf{Z}})$ and analogously on $K_1(\mathbf{Z}\Gamma \times \hat{\mathbf{Z}})$

Proposition 10.50.

$$\text{image } B = K_1(R\Gamma \times \hat{\mathbf{Z}})^{\mathbb{N}}$$

$$\text{image } \tilde{B} = K_1(R\Gamma \times \hat{\mathbf{Z}})^{\mathbb{N}} \qquad \square$$

This is proved in Ranicki [1987a] Proposition 7.5. even for additive categories. In Ranicki [1987a], Proposition 7.4 there is defined a geometrically significant splitting (see also Svensson [1985])

Proposition 10.51.

$$K_1(R\Gamma \times \hat{\mathbf{Z}}) = K_1(R\Gamma) \oplus K_o(R\Gamma) \oplus \widetilde{Nil}(R\Gamma) \oplus \widetilde{Nil}(R\Gamma)$$

and

$$Wh(R\Gamma \times \hat{\mathbf{Z}}) = Wh(R\Gamma) \oplus \tilde{K}_o(R\Gamma) \oplus \widetilde{Nil}(R\Gamma) \oplus \widetilde{Nil}(R\Gamma)$$

generalizing the results for modules over a ring in Bass, Heller, and Swan [1964] and Bass [1968], XII to additive categories. For a discussion of the geometrically and algebraically significant splittings we refer to Ranicki [1985]. We mention two further results which can be derived from Ranicki [1987] § 8.

There are also injective Bass-Heller-Swan-homomorphisms $B : K_{-i}(R\Gamma) \to K_{-i+1}(R\Gamma \times \hat{\mathbf{Z}})$ for $i \geq 0$ such that image B is $K_{-i+1}(R\Gamma \times \mathbf{Z})^{\mathbb{N}}$. The compositions of these maps define an embedding $K_{-i}(R\Gamma) \subset K_1(R\Gamma \times \hat{\mathbf{Z}}^{i+1})$. Let $\sigma(j) : \hat{\mathbf{Z}}^{i+1} \to \hat{\mathbf{Z}}^{i+1}$ be the permutation functor $(z_1, z_2, \dots, z_{i+1}) \longrightarrow (z_1, \dots, z_{j-1}, z_{i+1}, z_j, \dots, z_i)$ for $j = 1, \dots i+1$. The various maps $(id \times p(k_1) \times \dots \times p(k_{i+1}))$ define an operation of the monoid $\mathbb{N}^{i+1}$ on $K_1(R\Gamma \times \hat{\mathbf{Z}}^{i+1})$

Proposition 10.52.

a) $$K_{-i}(R\Gamma) = \bigcap_{j=1}^{i+1} \text{image}((id \times \sigma(j))_* \circ B : K_o(R\Gamma \times \hat{\mathbf{Z}}^i) \longrightarrow K_1(R\Gamma \times \hat{\mathbf{Z}}^{i+1}))$$

if $B : K_o(R\Gamma \times \hat{\mathbf{Z}}^i) \longrightarrow K_1((R\Gamma \times \hat{\mathbf{Z}}^i) \times \hat{\mathbf{Z}})$ is the homomorphism 10.49.

b) $$K_{-i}(R\Gamma) = K_1(R\Gamma \times \hat{\mathbf{Z}}^{i+1})^{\mathbb{N}^{i+1}} \qquad \square$$

Define reduced negative K-groups by

10.53. $\tilde{K}_{-i}(R\Gamma) = \bigcap_{j=1}^{i+1} \text{image}((\text{id} \times \sigma(j))_* \circ \tilde{B} : \tilde{K}_o(R\Gamma \times \hat{\mathbf{Z}}^i) \longrightarrow Wh(R\Gamma \times \hat{\mathbf{Z}}^{i+1}))$.

<u>Proposition 10.54.</u>

a) $\tilde{K}_{-i}(R\Gamma) = Wh(R\Gamma \times \hat{\mathbf{Z}}^{i+1})^{\mathbb{N}^{i+1}}$

b) $\tilde{K}_{-i}(R\Gamma) = K_{-i}(R\Gamma)$ <u>for</u> $i \geq 1$ □

<u>Remark 10.55.</u> The Propositions 10.50., 10.51., 10.52. and 10.54. can also be proved by reducing them to the case of group rings by Theorem 10.34. □

<u>Comments 10.56.</u> Negative K-groups for a ring are introduced in Bass [1968]. We use the definition for additive categories due to Pedersen [1984]. Algebraic definitions of $K_{-n} = K_{-n}(\mathbf{Z}\text{Or}G)$ and $Wh^G = Wh(\mathbf{Z}\text{Or}G)$ using different methods and a splitting can be found in Svensson [1985] and Rothenberg [1978] for finite groups G . The algebraic K-groups $K_o(R\Gamma)$, $K_1(R\Gamma)$ and $Wh(R\Gamma)$ are treated in tom Dieck [1987, I.11.] and Lück [1983]. The details of Example 10.47. are carried out in tom Dieck [1981].

The main result of this section is the Splitting Theorem 10.34. We emphasize that it includes all the natural properties. It is very useful to have both the elegant but abstract global version and the concrete but very complicated splitted version. Especially we are interested in restriction where the formulas for Split K involving all these bimodules $R\,\text{Irr}(x,y)$ are massy.

We need the restriction to get a ring structure on K_o^G and $U^G = U(G)$ which is a very important part of the structure. For example, the Euler ring $U(G)$ is easy as an abelian group but becomes interesting and difficult if one examines the ring structure. Several product formulas stated later depend on this ring structure and have important applications to geometry.

Text books on algebraic K-theory are Bass [1968], Curtis-Reiner [1981], [1987], Milnor [1971], Oliver [1988], Swan [1968], Silvester [1981]

The following remark may illuminate the Splitting Theorem 10.34. and the EI-property. Assume that Γ is an EI-category with finite $\mathrm{Is}\,\Gamma$. We can numerate $\mathrm{Is}\,\Gamma = \{\bar{x}_1, \bar{x}_2, \dots, \bar{x}_n\}$ such that $\bar{x}_i \leq \bar{x}_j$ implies $i \geq j$. Then the set valued matrix $(\mathrm{Hom}(x_i, x_j))_{i,j}$ is triangular as $\mathrm{Hom}(x_i, x_j) = \emptyset$ for $i < j$. But it is a general phenomenon in algebraic K-theory that in a triangular matrix only the entries on the diagonal matter. These are the groups $\mathrm{Aut}(x)$ for $\bar{x} \in \mathrm{Is}\,\Gamma$. □

<u>Exercises 10.57.</u>

1) Consider pairs (P,Q) of finitely generated $R\Gamma$-modules. Call (P,Q) and (P',Q') equivalent if there is a finitely generated free $R\Gamma$-module F with $P \oplus Q' \oplus F \cong P' \oplus Q \oplus F$. Show that this is an equivalence relation and denote by A the set of equivalence classes. Prove that we get a well-defined structure of an abelian group on A by $(P,Q) + (P',Q') = (P \oplus P', Q \oplus Q')$ and that $(P,Q) \to [P] - [Q]$ defines an isomorphism $A \longrightarrow K_o(R\Gamma)$.

2) Show that $P \longrightarrow [P] \in K_o(R\Gamma)$ is the universal additive invariant for the category FPMOD-$R\Gamma$ of finitely generated projective $R\Gamma$-modules with split injections as cofibrations and isomorphisms as weak equivalences.

3) Let Γ be an EI-category such that $\mathrm{Is}\,\Gamma$ and $\mathrm{Hom}(x,y)$ for $x,y \in \mathrm{Ob}\,\Gamma$ is finite. Consider

$$\mathrm{ch} : U(\Gamma) = \bigoplus_{v \in \mathrm{Is}\,\Gamma} \mathbf{Z} \longrightarrow \prod_{u \in \mathrm{Is}\,\Gamma} \mathbf{Z}$$

sending $\eta : \mathrm{Is}\,\Gamma \longrightarrow \mathbf{Z}$ to the $\mathrm{Is}\,\Gamma$-tuple

$$\Big(\sum_{\bar{y} \in \mathrm{Is}\,\Gamma} |\mathrm{Hom}(x,y)| \cdot \eta(y) \,\Big|\, \bar{x} \in \mathrm{Is}\,\Gamma \Big) .$$

a) For any finitely generated free $R\Gamma$-module F we have:

$$\mathrm{ch}(\mathrm{rk}_{R\Gamma} F) = (\mathrm{rk}_R F(x) \mid \bar{x} \in \mathrm{Is}\,\Gamma)$$

b) The map ch is injective and has a finite cokernel of order $\prod_{\bar{x} \in \mathrm{Is}\,\Gamma} |\mathrm{Aut}(x)|$

4) Let Γ be a EI-category such that $\mathrm{Is}\,\Gamma$ and $\mathrm{Hom}(x,y)$ for $x,y \in \mathrm{Ob}\,\Gamma$ are finite. Define an homomorphism

$$\rho : K_o(\mathbb{Z}\Gamma) \longrightarrow \prod_{Is\,\Gamma} \mathbb{Z}$$

by $[P] \longrightarrow (rk_{\mathbb{Z}}(P(x)) \mid \overline{x} \in Is\,\Gamma)$.

Show that ρ factorizes in

$$\begin{array}{ccc} K_o(\mathbb{Z}\Gamma) & & \\ \Big\downarrow{\scriptstyle rK_{\mathbb{Z}\Gamma}} & \searrow{\scriptstyle \rho} & \\ U(\Gamma) & \xrightarrow[ch]{} & \prod_{Is\,\Gamma} \mathbb{Z} \end{array}$$

Prove by given a counterexample that this is not true if one considers $\mathbb{Q}\Gamma$ instead of $\mathbb{Z}\Gamma$.

<u>Hint:</u> Use Swan's result that for a finitely generated projective $\mathbb{Z}G$-module M the $\mathbb{Q}G$-module $M \otimes_{\mathbb{Z}} \mathbb{Q}$ is $\mathbb{Q}G$-free provided that G is a finite group.

5) Consider the cyclic group $\mathbb{Z}/m$ of order m . Compute $K_n(\mathbb{C}Or\mathbb{Z}/m)$ for $n=0,1$.

6) Show that we can define the structure of a commutative associative ring with unit on $\mathbb{Z} \oplus \mathbb{Z} \oplus \tilde{K}_o(\mathbb{Z}[\mathbb{Z}/p])$ by

$$(u,v,w) \cdot (u',v',w') := (uu', uv' + u'v + pvv', uw' + u'w)$$

and that it is isomorphic to the ring $K_o(\mathbb{Z}Or\mathbb{Z}/p)$.

7) Show for a finite group G that $\tilde{K}_o(\mathbb{Z}OrG)$ is finite. Consider two finitely generated projective $\mathbb{Z}OrG$-modules P and Q such that $rk_{\mathbb{Z}}\, P(G/H) = rk_{\mathbb{Z}} Q(G/H)$ holds for any $H \subset G$. Prove the existence of an integer n and a finitely generated free $\mathbb{Z}OrG$-module F satisfying

$$(\oplus_n P) \oplus F \cong (\oplus_n Q) \oplus F$$

<u>Hint:</u> Use the result due to Swan that $\tilde{K}_o(\mathbb{Z}G)$ is finite for a finite group G .

8) Prove for a finite group G that the diagonal functor $\Delta : OrG \to OrG \times OrG$ is admissible and $Irr(G/H, G/K \times G/L)$ is the set $Mono(G/H, G/K \times G/L)^G$ of injective G-maps $G/H \longrightarrow G/K \times G/L$.

9) Let H be a (closed) subgroup of the compact Lie group G. Let $i : Or/H \to Or/G$ be the functor induced by the induction. Namely, the class of $f : H/K \to H/L$ is sent to the class of $G \times_H f : G/K \to G/L$. Show that i is admissible if H and G have the same dimension. Verify that i for $H = \mathbf{Z}/m \subset S^1$ is not admissible.

10) Let Γ be a small category which is not necessarily an EI-category. Assume that $\operatorname{Is}\Gamma$ is finite. Show that we obtain well-defined homomorphisms

$$E : \bigoplus_{\bar{x} \in \operatorname{Is}\Gamma} K_n(R[x]) \xrightarrow{\bigoplus_{\bar{x} \in \operatorname{Is}\Gamma} K_n(E_x)} K_n(R\Gamma)$$

$$S : K_n(R\Gamma) \xrightarrow{\prod_{\bar{x} \in \operatorname{Is}\Gamma} K_n(S_x)} \bigoplus_{\bar{x} \in \operatorname{Is}\Gamma} K_n(R[x])$$

such that $S \circ E = ID$. Is for the category of Example 9.42. $E \circ S = ID$ true?

11) Show for a field F that

$$K_1(F[\mathbf{Z}]) \cong \mathbf{Z} \oplus F^*$$

holds.

12) Show $K_n(\mathbf{Z}Or/G) = \{0\}$ for G a compact Lie group and $n \leq -2$.
Hint: Use the result of Carter [1980] that $K_n(\mathbf{Z}\Gamma) = \{0\}$ for Γ a finite group and $n \leq -2$.

13) Show that the definitions 10.28. and 10.53. give the same groups $\tilde{K}_{-n}(R\Gamma)$ for $n \geq 1$

11. The algebraic finiteness obstruction

We assign to a finitely dominated $R\Gamma$-chain complex C its algebraic finiteness obstruction $o(C) \in K_0(R\Gamma)$. It vanishes if and only if C is $R\Gamma$-homotopy equivalent to a finite free $R\Gamma$-chain complex. We collect its main properties like homotopy invariance and additivity. Moreover, we extend its definition to arbitrary $R\Gamma$-chain complexes having a finite resolution. We examine the behaviour of the finiteness obstruction under induction, restriction and the tensor product and under the splitting of the algebraic K-theory of $R\Gamma$.

A $R\Gamma$-chain complex $C = (C_*, c_*)$ consists of $R\Gamma$-modules C_n and $R\Gamma$-maps $c_n : C_n \to C_{n-1}$ for $n \in \mathbf{Z}$ satisfying $c_{n-1} \circ c_n = 0$ for $n \in \mathbf{Z}$. Equivalently, this is a covariant functor $\Gamma \longrightarrow$ {R-chain compl.} . Extend the notions of a chain map, chain homotopy, homology ... for R-chain complexes to $R\Gamma$-chain complexes in the obvious way .

A $R\Gamma$-chain complex C is always assumed to be positive i.e. $C_n = 0$ for $n < 0$. We call C free, projective finitely generated resp. of type T for $T \subset \mathrm{Is}\,\Gamma$ if each C_n has this property. A $R\Gamma$-chain complex C is of finite type if there is a finite $T \subset \mathrm{Is}\,\Gamma$ such that C is of type T . Notice that this implies that each C_n is of finite type but the converse is false in general. In the literature the notion of finite type sometimes means what we have called finitely generated. We will always use the notion introduced above. We say that C is d-dimensional if $C_n = 0$ for $n > d$ holds. A $R\Gamma$-chain complex C is finite-dimensional if C is d-dimensional for some d . A $R\Gamma$-chain complex C is finite if it is finitely generated and finite-dimensional, i.e. C_n is finitely generated for $n \geq 0$ and zero for large n.

An approximation (P,f) of a $R\Gamma$-chain complex C is a projective $R\Gamma$-chain complex P together with a weak homology equivalence $f : P \to C$ (i.e. $H_n(f)$ is bijective for any $n \geq 0$). We call (P,f) finitely generated, finite-dimensional, finite,... if P has this property.

Definition 11.1. *If C is a $R\Gamma$-chain complex possessing a finite approximation define its finiteness obstruction*

$$o(C) \in K_0(R\Gamma)$$

by $o(C) = \sum_{n=o}^{\infty} (-1)^n [P_n]$ for any finite approximation (P,f) . Let $\tilde{o}(C) \in \tilde{K}_o(R\Gamma)$ be its reduction under $K_o(R\Gamma) \longrightarrow \tilde{K}_o(R\Gamma)$ □

We want to show that $o(C)$ is well-defined and the following result.

Theorem 11.2.

a) Obstruction property.

Let C be a $R\Gamma$-chain complex possessing a finite approximation. Then $\tilde{o}(C)$ in $\tilde{K}_o(R\Gamma)$ vanishes if and only if C has a finite free approximation. Provided that C is projective this is equivalent to C being homotopy equivalent to a finite free $R\Gamma$-chain complex.

b) Weak homology invariance.

Let $f : C \longrightarrow D$ be a weak homology equivalence between $R\Gamma$-chain complexes possessing a finite appproximation. Then: $o(C) = o(D)$.

c) Additivity.

Consider the exact sequence of $R\Gamma$-chain complexes

$$0 \longrightarrow C^1 \xrightarrow{i} C^o \xrightarrow{p} C^2 \longrightarrow 0$$

If two of them have a finite approximation then so the third and we have

$$o(C^1) - o(C^o) + o(C^2) = 0 \qquad □$$

We need for its proof and later applications some preparations. A chain map $f: C \to D$ is n-connected if $H_i(f)$ is bijective for $i < n$ and surjective for $i = n$. It is a weak homology equivalence or briefly homology equivalence if $H_i(f)$ is bijective for all i . One easily checks for a chain map $f : C \longrightarrow D$ between (not neccessarily projective, but positive) $R\Gamma$-chain complexes (compare Proposition 2.3.).

Lemma 11.3.

a) f is n-connected if and only if $f_* : [P,C] \longrightarrow [P,D]$ between the sets of homotopy classes of chain maps is bijective for any (n-1)-dimensional projective $R\Gamma$-chain complex P and is surjective for any n-dimensional projective $R\Gamma$-chain complex P .

b) f is a weak homology equivalence if and only if $f_* : [P,C] \longrightarrow [P,D]$ is bijective for any projective RΓ-chain complex P .

c) A weak homology equivalence $f : P \longrightarrow Q$ between projective RΓ-chain complexes is a homotopy equivalence. □

11.4. We set some sign-conventions.

Consider a chain map $f : C \longrightarrow D$. Define its mapping cylinder $Cyl(f)$ by

$$\begin{array}{ccc} Cyl(f)_n & & Cyl(f)_{n-1} \\ \| & & \| \\ C_{n-1} \oplus C_n \oplus D_n & \xrightarrow[\begin{pmatrix} -c_{n-1} & 0 & 0 \\ -id & c_n & 0 \\ f_{n-1} & 0 & d_n \end{pmatrix}]{} & C_{n-2} \oplus C_{n-1} \oplus D_{n-1} \end{array}$$

The obvious inclusions yields chain maps

$$i : C \longrightarrow Cyl(f)$$

$$j : D \longrightarrow Cyl(f)$$

Define the projection

$$p : Cyl(f) \longrightarrow D$$

by $(0,f_n,id) : C_{n-1} \oplus C_n \oplus D_n \longrightarrow D_n$. One has $p \circ i = f$ and $p \circ j = id$. Define the mapping cone $Cone(f)$ as the cokernel of i .

$$\begin{array}{ccc} Cone(f)_n & & Cone(f)_{n-1} \\ \| & & \| \\ C_{n-1} \oplus D_n & \xrightarrow[\begin{pmatrix} -c_{n-1} & 0 \\ f_{n-1} & d_n \end{pmatrix}]{} & C_{n-2} \oplus D_{n-1} \end{array}$$

The suspension ΣC of C is the mapping cone of $C \longrightarrow 0$

$$(\Sigma C)_n = C_{n+1} \xrightarrow{-c_{n-1}} (\Sigma C)_{n-1} = C_{n-2}$$

If $f : M \longrightarrow N$ is a RΓ-map and $n \geq 0$, let $n(M)$ be the RΓ-chain complex concentrated in dimension n given by $n(M)_i = \{0\}$ for $i \neq n$ and $n(M)_n = M$ and $n(f) : n(M) \longrightarrow n(N)$ the obvious chain map. We have $\Sigma^n 0(f) = n(f)$. Let $e\ell_n(f)$ be the mapping cone of $(n-1)(f)$

$$\begin{array}{ccccccc} e\ell_n(f)_{n+1} & & e\ell_n(f)_n & & e\ell_n(f)_{n-1} & & e\ell_n(f)_{n-2} \\ \| & & \| & & \| & & \| \\ 0 & \longrightarrow & M & \xrightarrow{f} & N & \longrightarrow & 0 \end{array}$$

We write $e\ell_n(M) = e\ell_n(id : M \longrightarrow M)$. We call $e\ell_n(f)$ or $e\ell_n(M)$ the n-<u>dimensional elementary chain complex</u> of f or M.

Frequently we make use of the exact sequences of RΓ-chain complexes

$$0 \longrightarrow C \xrightarrow{i} Cyl(f) \longrightarrow Cone(f) \longrightarrow 0$$

$$0 \longrightarrow D \xrightarrow{j} Cyl(f) \longrightarrow Cone(C) \longrightarrow 0$$

$$0 \longrightarrow D \longrightarrow Cone(f) \longrightarrow \Sigma C \longrightarrow 0$$

$$0 \longrightarrow (n-1)(M) \longrightarrow e\ell_n(M) \longrightarrow n(M) \longrightarrow 0$$

We get long exact homology sequences. $Cone(C) := Cone(id_C)$ is contractible, i.e. chain homotopy equivalent to the zero-chain complex, and $j : D \longrightarrow Cyl(f)$ and $p : Cyl(f) \longrightarrow D$ are chain homotopy equivalences. □

If C is a RΓ-chain complex we write

$$C_{odd} = C_1 \oplus C_3 \oplus C_5 \oplus \ldots$$

$$C_{ev} = C_o \oplus C_2 \oplus C_4 \oplus \ldots$$

<u>Lemma 11.5.</u>

a) <u>The chain map</u> $f : C \longrightarrow D$ <u>is a homotopy equivalence if and only if</u> $Cone(f)$ <u>is contractible</u>.

b) <u>Let</u> C <u>be a contractible</u> RΓ-<u>chain complex</u>. <u>Let</u> c <u>be its differential and</u> γ

and $\bar{\gamma}$ *two chain contractions*, i.e. *chain homotopies between* id *and* 0 . *Then*:

i) *The following maps are isomorphisms*

$$(c+\gamma) : C_{odd} \longrightarrow C_{ev}$$

$$(c+\bar{\gamma}) : C_{ev} \longrightarrow C_{odd}$$

ii) *If* C *is finite and projective the composition* $(c+\bar{\gamma}) \circ (c+\gamma) \quad C_{odd} \to C_{odd}$ *is an automorphism of a finitely generated projective* $R\Gamma$-*module*. *Its class in* $K_1(R\Gamma)$ *is zero*.

a) Let $\gamma : \mathrm{Cone}(f) \longrightarrow \mathrm{Cone}(f)$ be the chain contraction

$$\begin{pmatrix} h_{n-1} & g_n \\ \ell_{n-1} & k_n \end{pmatrix} : C_{n-1} \oplus D_n \longrightarrow C_{n} \oplus D_{n+1}$$

Then $g : D \longrightarrow C$ is a chain map and $h : \mathrm{id} - gf$ and $k : \mathrm{id} - fg$ are chain homotopies. Conversely, given g,h,k define γ as above with $\ell = o$. Then γ is a chain homotopy between a chain isomorphism $F : \mathrm{Cone}(f) \longrightarrow \mathrm{Cone}(f)$ of the shape $\begin{pmatrix} \mathrm{id} & 0 \\ * & \mathrm{id} \end{pmatrix}$ and the zero-map . We obtain a chain contraction by $F^{-1} \circ \gamma$.

b) Let $\Delta : C \longrightarrow C$ be the map of degree two given by $(\bar{\gamma}-\gamma) \circ \gamma$. Let the isomorphism $f : C_{ev} \longrightarrow C_{ev}$ be determined by the triangular matrix

$$\begin{pmatrix} \mathrm{id} & 0 & 0 & \cdots \\ \Delta & \mathrm{id} & 0 & \cdots \\ 0 & \Delta & \mathrm{id} & \cdots \\ \vdots & \vdots & \vdots & \end{pmatrix} : C_o \oplus C_2 \oplus C_4 \oplus \ldots \longrightarrow C_o \oplus C_2 \oplus C_4 \oplus$$

The composition

$$g : C_{odd} \xrightarrow{(c+\gamma)} C_{ev} \xrightarrow{f} C_{ev} \xrightarrow{(c+\bar{\gamma})} C_{odd}$$

is given by the triangular matrix

$$\begin{pmatrix} \bar{\gamma}c + c\Delta c + c\gamma & 0 & 0 & \cdots \\ * & \bar{\gamma}c + c\Delta c + c\gamma & 0 & \cdots \\ * & * & \bar{\gamma}c + c\Delta c + c\gamma & \cdots \\ \vdots & \vdots & \vdots & \end{pmatrix}$$

Now Δ is a homotopy of homotopies. Namely:

$$c\Delta - \Delta c = \gamma - \bar{\gamma} .$$

Hence $\bar{\gamma}c + c\Delta c + c\gamma$ is the identity, so that g is an isomorphism.

Obviously f and g represent zero in $K_1(R\Gamma)$ □

<u>Lemma 11.6.</u> <u>Consider the following commutative diagram of (positive but not necessarily projective) chain complexes with exact rows</u>

$$\begin{array}{ccccccccc} 0 & \longrightarrow & C^1 & \xrightarrow{i} & C^o & \xrightarrow{p} & C^2 & \longrightarrow & 0 \\ & & \downarrow{\scriptstyle g^1} & & \downarrow{\scriptstyle g^o} & & \downarrow{\scriptstyle g^2} & & \\ 0 & \longrightarrow & C^1 & \xrightarrow{i} & C^o & \xrightarrow{p} & C^2 & \longrightarrow & 0 \end{array}$$

<u>Assume that two of the chain complexes satisfy the assumption that there is a positive chain complex</u> $\bar{P}^i$ <u>and a weak homology equivalence</u> $k^i : \bar{P}^i \longrightarrow C^i$ <u>with one of the properties</u>

a) $\bar{P}^i$ <u>is projective</u>.

b) $\bar{P}^i$ <u>is projective and finite-dimensional</u>.

c) $\bar{P}^i$ <u>is projective and finitely generated</u>.

d) $\bar{P}^i$ <u>is projective and finite</u> .

Then there is a commutative diagram with exact rows of chain complexes with the property a) resp. b) resp. c) resp. d).

$$\begin{array}{ccccccccc} 0 & \longrightarrow & P^1 & \xrightarrow{j} & P^o & \xrightarrow{q} & P^2 & \longrightarrow & 0 \\ & & \downarrow{\scriptstyle f^1} & & \downarrow{\scriptstyle f^o} & & \downarrow{\scriptstyle f^2} & & \\ 0 & \longrightarrow & P^1 & \xrightarrow[j]{} & P^o & \xrightarrow[q]{} & P^2 & \longrightarrow & 0 \end{array}$$

and weak homology equivalences $k^i : P^i \longrightarrow C^i$ for $i = 0,1,2$ such that $g^i \circ k^i$ is homotopic to $k^i \circ f^i$ for $i = 0,1,2$ and $i \circ k^1 = k^o \circ j$ and $p \circ k^o = k^2 \circ q$ holds.

<u>Proof:</u> The proof is divided into two steps.

<u>Step 1.</u> Construct everything except the

$$f^i : P^i \longrightarrow P^i .$$

We start with the case, where we have weak homology equivalences $k^i : \bar{P}^i \longrightarrow C^i$ for $i = 0,1$. By Lemma 11.3. there is a chain map $j' : \bar{P}^1 \longrightarrow \bar{P}^o$ with $k^o \circ j' \simeq i \circ k^1$. Let $j : \bar{P}^1 \longrightarrow Cyl(j')$ be the inclusion and $pr : Cyl(j') \longrightarrow \bar{P}^o$ be the projection Since j is a cofibration, that is, all j_n are split injective, we can change $k^o \circ pr$ homotopically into $x : Cyl(j') \longrightarrow C^o$ satisfying $x \circ j = i \circ k^1$. If $y : Cone(j') \longrightarrow C^2$ is the induced map we have the commutative diagram with exact rows

$$\begin{array}{ccccccccc} 0 & \longrightarrow & \bar{P}^1 & \longrightarrow & Cyl(j') & \longrightarrow & Cone(j') & \longrightarrow & 0 \\ & & \downarrow{\scriptstyle k^o} & & \downarrow{\scriptstyle x} & & \downarrow{\scriptstyle y} & & \\ 0 & \longrightarrow & C^1 & \longrightarrow & C^o & \longrightarrow & C^2 & \longrightarrow & 0 \end{array}$$

Obviously $Cyl(j')$ and $Cone(j')$ have the property a) resp. b) resp. c) resp. d) as $\bar{P}^1$ and $\bar{P}^o$ have. By construction k^o and x are weak homology equivalences. The same holds for y by the long homology sequence and the five lemma.

Now we deal with the case where C^o and C^2 fulfill the assumption. We have the exact sequence

$$0 \longrightarrow C^o \longrightarrow Cone(i) \longrightarrow \Sigma C^1 \longrightarrow 0$$

and a weak homology equivalence $Cone(i) \longrightarrow C^2$. By the first case we also get a weak homology equivalence with the desired property for C^1. A further application of the first case finishes the proof of the second case.

The last case where C^1 and C^2 satisfy the assumption is reduced to second case by the sequence above.

<u>Step 2</u>. Constructing the $f^i : P^i \longrightarrow P^i$.

We have already constructed $0 \longrightarrow P^1 \xrightarrow{i} P^o \xrightarrow{q} P^2 \longrightarrow 0$ and weak homology equivalences $k^i : P^i \longrightarrow C^i$ satisfying $i \circ k^1 = k^o \circ j$ and $p \circ k^1 = k^2 \circ q$. Now one easily checks that a chain map $f^i : P^i \longrightarrow P^i$ together with a chain homotopy $\phi^i : P^i \longrightarrow C^i$ between $g^i \circ k^i$ and $k^i \circ f^i$ is the same as a commutative diagram of chain complexes

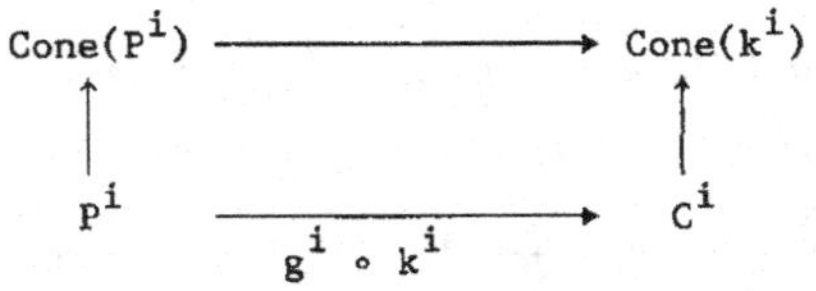

where the vertical arrows are the canonical inclusions. Since they are cofibrations and $\text{Cone}(k^i)$ acyclic we get the existence of f^1 and $\phi^1 : g^1 \circ k^1 = k^1 \circ f^1$.

In the following commutative diagram we want to construct the dotted arrow such that it remains commutative

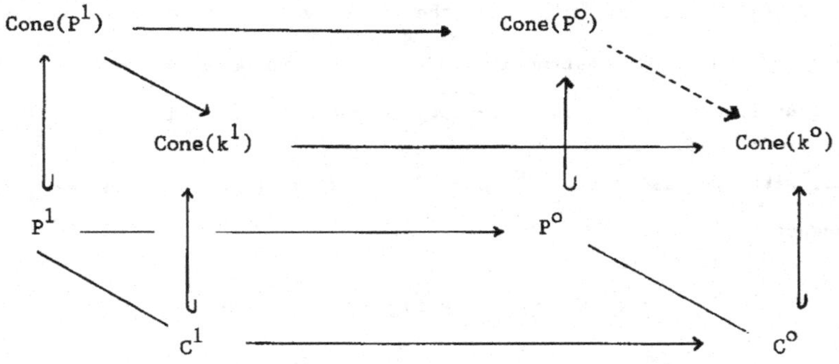

This can be done as $\mathrm{Cone}(k^o)$ is acyclic, and the inclusion of the image of $\mathrm{Cone}(P^1) \oplus P^o \longrightarrow \mathrm{Cone}(P^o)$ into $\mathrm{Cone}(P^o)$ is a cofibration.

Hence we have constructed a chain map $f^o : P^o \longrightarrow P^o$ and a chain homotopy $\phi^o : P^o \to C^o$ between $g^o \circ k^o$ and $k^o \circ f$ such that $i \circ f^1 = f^o \circ j$ and $i \circ \phi^1 = \phi^o \circ j$ holds. Let $f^2 : P^2 \longrightarrow P^2$ and $\phi^2 : P^2 \longrightarrow C^2$ be the induced map and chain homotopy. This finishes the proof of Step 2 and Lemma 11.6. □

Proof of Theorem 11.2.

b) We show homotopy invariance and that Definition 11.1. makes sense simultaneously. Namely, by Lemma 11.3. it suffices to show for a homotopy equivalence $f : P \to Q$ between finite projective $R\Gamma$-chain complexes that $o(P) = o(Q)$ is true. This follows from Lemma 11.5. telling us that $P_{ev} \oplus Q_{odd}$ and $P_{odd} \oplus Q_{ev}$ are isomorphic.

c) If $0 \longrightarrow P^1 \longrightarrow P^o \longrightarrow P^2 \longrightarrow 0$ is an exact sequence of finite projective $R\Gamma$-chain complexes we obviously have $o(P^1) - o(P^o) + o(P^2) = 0$. Now apply Lemma 11.6.

a) Recall the split exact sequence of Proposition 10.42.

$$0 \longrightarrow K_o^f(R\Gamma) \longrightarrow K_o(R\Gamma) \longrightarrow \tilde{K}_o(R\Gamma) \longrightarrow 0$$

If C has a finite free approximation (F,f), then $o(C) = o(F)$ holds and $o(F)$ lies in $K_o^f(R\Gamma)$ so that $\tilde{o}(C)$ is zero. Conversely, consider the finite projective $R\Gamma$-chain complex P with $\tilde{o}(P) = 0$. Choose d so big such that P is d-dimensional. Any finitely generated projective $R\Gamma$-module is a direct summand in a finitely generated free $R\Gamma$-module. For any finitely generated projective $R\Gamma$-module Q and $n \geq 0$ $el_n(Q)$ is contractible. By adding appropiate $R\Gamma$-chain complexes $el_n(Q)$ we can change P up to homotopy such that P_n is finitely generated free for $n < d$. Then $\tilde{o}(P) = [P_d]$ is zero in $\tilde{K}_o(R\Gamma)$ so that there is a finitely generated free $R\Gamma$-module F such that $P_d \oplus F$ is finitely generated free. Adding $el_d(F)$ yields a finite free $R\Gamma$-chain complex homotopy equivalent to P. □

Sometimes the finiteness obstruction can be read off from homology.

Let M be a $R\Gamma$-module. A <u>resolution</u> (P,f) of M is a projective $R\Gamma$-chain complex P together with an isomorphism $f : H_o(P) \longrightarrow M$ such that $H_i(P) = 0$ for $i > 0$. A resolution is <u>finitely generated</u>, <u>finite</u>, ... if P is. We can interprete a resolution (P,f) of M as an approximation of the $R\Gamma$-chain complex $O(M)$. In the sequel we identify $H_o(P)$ and M. We give now a version of the fundamental Lemma of homological algebra. Its proof is omitted.

<u>Lemma 11.7.</u> <u>Let</u> n <u>be an integer and</u> C <u>be a</u> $R\Gamma$-<u>chain complex with</u> $H_i(C)=0$ for $i > n$. <u>If</u> P <u>is a projective</u> $R\Gamma$-<u>chain complex with</u> $P_i = 0$ <u>for</u> $i < n$, we get a bijection

$$[P,C] \longrightarrow \mathrm{Hom}_{R\Gamma}(H_n(P),H_n(C)) \qquad [f] \longrightarrow H_n(f) \qquad \square$$

If M is a $R\Gamma$-module possessing a finite resolution define

11.8. $$[M] \in K_o(R\Gamma)$$

by $\Sigma(-1)^n[P_n]$ for any finite resolution. This is the finiteness obstruction of $O(M)$ and hence well-defined.

<u>Proposition 11.9.</u> <u>Let</u> C <u>be a</u> $R\Gamma$-<u>chain complex such that</u> $H_i(C)$ <u>has a finite resolution for</u> $i \geq 0$ <u>and is zero for large</u> i.

<u>Then</u> C <u>has a finite approximation and</u>

$$o(C) = \sum_{i=o}^{\infty} (-1)^i \, [H_i(C)]$$

<u>Proof.</u> We use induction over an integer n with the property $H_i(C) = 0$ for $i > n$. The begin $n = -1$ is trivial. In the induction step from $n-1$ to n choose a finite resolution P of $H_n(C)$. By Lemma 11.7. there is a $R\Gamma$-chain map $f : \Sigma^n P \longrightarrow C$ with $H_n(f) = \mathrm{id}$. Consider the exact sequence

$$0 \longrightarrow C \longrightarrow \mathrm{Cone}(f) \longrightarrow \Sigma^{n+1}P \longrightarrow 0 .$$

By Theorem 11.2. Cone(f) has a finite approximation and

$$o(C) = o(\mathrm{Cone}(f)) - o(\Sigma^{n+1}P) .$$

Using the exact sequence

$$0 \longrightarrow \Sigma^{i}P \longrightarrow \mathrm{Cone}(\Sigma^{i}P) \longrightarrow \Sigma^{i+1}P \longrightarrow 0$$

one shows $-o(\Sigma^{n+1}P) = (-1)^{n}[H_n(C)]$.

By the long homology sequence we get $H_i(C) = H_i(\mathrm{Cone}\ f)$ for $i < n$ and $H_i(\mathrm{Cone}(f)) = 0$ for $i \geq n$. Now apply the induction hypothesis to $\mathrm{Cone}(f)$. □

A <u>domination</u> (D,r,i,h) for a chain complex C consists of a $R\Gamma$-chain complex D together with chain maps $r : D \longrightarrow C$, $i : C \longrightarrow D$ and a chain homotopy $h : r \circ i \simeq id$. We call C <u>finitely dominated</u> if there is a domination (D,r,i,h) by a finite projective $R\Gamma$-chain complex D .

Now we give some criterions when a projective $R\Gamma$-chain complex is homotopy equivalent to a finite resp. finite-dimensional one.

<u>Proposition 11.10.</u> *The following statements for a projective* $R\Gamma$*-chain complex* C *are equivalent.*

i) *Let* $B_n(C)$ *be the image of* $c_{n+1} : C_{n+1} \longrightarrow C_n$ *and* $j : B_n(C) \longrightarrow C_n$ *be the inclusion. There is* $C'_n \subset C_n$ *such that for the inclusion* $i : C'_n \longrightarrow C_n$ *the map* $i \oplus j : C'_n \oplus B_n(C) \longrightarrow C_n$ *is an isomorphism. Moreover, the following chain map from a n-dimensional projective* $R\Gamma$*-chain complex into* C *is a chain equivalence*

$$\begin{array}{ccccccccccc}
\cdots & 0 & \longrightarrow & 0 & \longrightarrow & C'_n & \xrightarrow{c_n \circ i} & C_{n-1} & \xrightarrow{c_{n-1}} & C_{n-2} & \xrightarrow{c_{n-2}} \\
 & \downarrow & & \downarrow & & \downarrow{\scriptstyle i} & & \downarrow{\scriptstyle id} & & \downarrow{\scriptstyle id} & \\
\cdots & C_{n+2} & \xrightarrow{c_n} & C_{n+1} & \xrightarrow{c_{n+1}} & C_n & \xrightarrow{c_n} & C_{n-1} & \xrightarrow{c_{n-1}} & C_{n-2} & \xrightarrow{c_{n-2}} \cdots
\end{array}$$

ii) C *is homotopy equivalent to a n-dimensional projective* $R\Gamma$*-chain complex*

iii) C *is dominated by a n-dimensional* $R\Gamma$*-chain complex.*

iv) $B_n(C) \subset C_n$ *is a direct summand and* $H_i(C) = 0$ *for* $i > n$.

v) $H^{n+1}(C;M) := H^{n+1}(\mathrm{Hom}_{R\Gamma}(C,M)) = 0$ *for any* $R\Gamma$*-module* M *and* $H_i(C)$ *for* $i > n$ *vanishes.*

Proof. i) ==> ii) and ii) ==> iii) are obvious.

iii) ==> iv). Let (D,r,i,h) be a domination of C by a n-dimensional $R\Gamma$-chain complex D. Then $H_i(C)$ is a direct summand in $H_i(D)$ and obviously $H_i(D) = 0$ for $i > n$. A retraction of $B_n(C) \subset C_n$ is given by $c_{n+1} \circ h_n : C_n \longrightarrow B_n(C)$. Namely

$$c_{n+1} \circ h_n \circ c_{n+1} = c_{n+1} \circ h_n \circ c_{n+1} + c_{n+1} \circ c_{n+2} \circ h_{n+1} = c_{n+1} \circ id - c_{n+1} \circ r_{n+1} \circ i_{n+1} = c_{n+1}$$

iv ==> i). Obviously the chain map in i) is a weak homology equivalence and hence a homotopy equivalence by Lemma 11.3.

ii) ==> v) is obvious.

v) ==> iv). Let M be $B_n(C)$. Then the following sequence is exact

$$\mathrm{Hom}_{R\Gamma}(C_n, B_n(C)) \xrightarrow{c_{n+1}^*} \mathrm{Hom}_{R\Gamma}(C_{n+1}, B_n(C)) \xrightarrow{c_{n+2}^*} \mathrm{Hom}_{R\Gamma}(C_{n+2}, B_n(C))$$

Consider $c_{n+1} \in \mathrm{Hom}_{R\Gamma}(C_{n+1}, B_n(C))$. Since $c_{n+2}^*(c_{n+1}) = c_{n+1} \circ c_{n+2}$ is zero there is $r : C_n \longrightarrow B_n(C)$ with $r \circ c_{n+1} = c_{n+1}$ so that r is a retraction.

Proposition 11.11. Consider a $R\Gamma$-chain complex C

C is homotopy equivalent to a projective resp. finitely generated projective resp. n-dimensional projective resp. n-dimensional finitely generated projective $R\Gamma$-chain complex if and only if it is dominated by such a $R\Gamma$-chain complex.

Proof. Let (D,r,i,h) be a domination of C. Define a $R\Gamma$-chain complex (D',d') by

$$D'_n = \bigoplus_{j=o}^{n} D_j \xrightarrow{d'_n} D'_{n-1} = \bigoplus_{j=o}^{n-1} D_j$$

with $d'_n(j,k) : D_j \longrightarrow D_k$ given by

$$d_n'(j,k) = \begin{cases} 0 & j \geq k+2 \\ (-1)^{n+k} d & j = k+1 \\ id-ir & j = k,\ j \equiv n \quad (2) \\ ir & j = k,\ j \equiv n+1\ (2) \\ (-1)^{n+k+1} ih^{k-j} r & j \leq k-1 \end{cases}$$

Written as a matrix we get for d_n'

$$\begin{pmatrix} ir & & 0 & \cdots \\ -ihr & id-ir & d & \cdots \\ -ih^2r & ihr & ir & \cdots \\ \vdots & \vdots & \vdots & \end{pmatrix} \qquad n \equiv 0 \quad (2)$$

$$\begin{pmatrix} id-ir & d & 0 & \cdots \\ ihr & ir & -d & \cdots \\ ih^2r & -ihr & id-ir & \cdots \\ \vdots & \vdots & \vdots & \end{pmatrix} \qquad n \equiv 1 \quad (2)$$

Define chain maps

$$\bar{r} : C \longrightarrow D' \quad \text{by} \quad (0,0,\ldots,r)^{tr}$$

and

$$\bar{i} : D' \longrightarrow C \quad \text{by} \quad \bar{i}_n = (h^n i, h^{n-1} i, \ldots, hi, i)$$

Then $\bar{r} \circ \bar{i}$ is the identity and a chain homotopy $k : D' \longrightarrow D'$ between $i \circ \bar{r}$ and the identity is given by the obvious inclusions

$$\bigoplus_{j=o}^{n} D_j \longrightarrow \bigoplus_{j=o}^{n+1} D_j$$

Hence D' and C are homotopy equivalent. Now the claim follows using Proposition 11.10. □

The explicite construction in the proof of Proposition 11.11. leads to the following explicite formula.

Proposition 11.12. *Let* (D,r,i,h) *be a domination of the* $R\Gamma$*-chain complex* C *such that* D *is finite projective*.

Then the map

$$p : \bigoplus_{j=o}^{\infty} D_j \longrightarrow \bigoplus_{j=o}^{\infty} D_j$$

$$\begin{pmatrix} i\circ r & -d & 0 & 0 & \cdots \\ -ihr & id-ir & d & 0 & \cdots \\ -ih^2r & ihr & ir & -d & \cdots \\ -ih^3r & ih^2r & ihr & id-r & \cdots \\ \vdots & \vdots & \vdots & \vdots & \end{pmatrix}$$

is a projection (i.e. $p^2 = p$) *of a finitely generated projective* $R\Gamma$*-module and we have in* $K_o(R\Gamma)$

$$o(C) = [\text{image } p] - [D_{odd}]$$

Proof: In the proof of Proposition 11.11. we have constructed a $R\Gamma$-chain complex (D',d') with $D' \simeq C$. By Proposition 11.10. we obtain a finitely generated n-dimensional projective $R\Gamma$-chain complex D'' with $D'' \simeq D'$ for an even integer $n \geq \dim D$. Namely, D'' looks like

$$0 \longrightarrow \text{image } d'_{n+1} \longrightarrow D'_{n-1} \longrightarrow D'_{n-2} \longrightarrow \cdots$$

Hence we get

$$o(C) = [\text{image } d'_{n+1}] + \sum_{i=o}^{n-1} (-1)^i [D'_i] \quad .$$

But d'_{n+1} is just p and $[D_i'] = \sum_{j=o}^{i} [D_j]$. □

<u>Proposition 11.13.</u> <u>Let</u> (D,r,i) <u>be a domination of</u> C <u>by a finitely generated</u> <u>n-dimensional projective</u> $R\Gamma$-<u>chain complex</u> D . <u>Assume that</u> r <u>is</u> n-<u>connected</u>.

<u>Then</u> $H_{n+1}(r)$ <u>is finitely generated projective and we have in</u> $K_o(R\Gamma)$

$$o(C) = o(D) + (-1)^n[H_{n+1}(r)] .$$

<u>Proof</u>. Consider the exact sequence

$$0 \longrightarrow D \longrightarrow \mathrm{Cone}(r) \longrightarrow \Sigma C \longrightarrow 0 .$$

By Proposition 11.11. D and C are homotopy equivalent to finitely generated projective n-dimensional $R\Gamma$-chain complexes. Hence $\mathrm{Cone}(r)$ is homotopy equivalent to a finitely generated projective $R\Gamma$-chain complex P of dimension $n+1$. Since $H_i(r) = H_i(P)$ is zero for $i \leq n$. $H_{n+1}(r) = H_{n+1}(P)$ is finitely generated projective. Theorem 11.2. and Proposition 11.9. imply

$$o(C) = -o(\Sigma C) = o(D) - o(\mathrm{Cone}(r)) = o(D) - (-1)^{n+1}[H_{n+1}(r)] = o(D) + (-1)^n[H_{n+1}(r)] \qquad \square$$

There is also a splitted version of the finiteness obstruction. Let C be a $R\Gamma$-chain complex of the homotopy type of a finite projective $R\Gamma$-chain complex. Then S_xC is of the homotopy type of a finite projective $R[x]$-chain complex for all $x \in \mathrm{Is}\,\Gamma$. Hence we can define

11.14.
$$\mathrm{Split}\ o(C) \in \mathrm{Split}\ K_o(R\Gamma)$$
$$\mathrm{Split}\ \tilde{o}(C) \in \mathrm{Split}\ \tilde{K}_o(R\Gamma)$$

by $(o(S_xC) \in K_o(R[x]) \mid \bar{x} \in \mathrm{Is}\,\Gamma)$ and $(\tilde{o}(S_xC) \in \tilde{K}_o(R[x]) \mid \bar{x} \in \mathrm{Is}\,\Gamma)$.

Recall the isomorphism of Theorem 10.34. $S : K_o(R\Gamma) \longrightarrow \mathrm{Split}\ K_o(R\Gamma)$.

<u>Proposition 11.15.</u> <u>Let</u> C <u>be a</u> $R\Gamma$-<u>chain complex of the homotopy type of a finite projective</u> $R\Gamma$-<u>chain complex</u> . <u>Then</u>

$$S(o(C)) = \mathrm{Split}\ o(C) .$$

Proof. Without loss of generality we can assume that C itself is a finite projective $R\Gamma$-chain complex. Then there is a finite subset $T \subset \operatorname{Is}\Gamma$ such that C is of type T. We use induction over the cardinality of T. In the induction step choose $\bar{x} \in T$ maximal in T under $\leq$. From Theorem 9.39. we obtain an exact sequence of finite projective $R\Gamma$-chain complexes

$$0 \longrightarrow E_x \circ S_x C \longrightarrow C \longrightarrow COK_x C \longrightarrow 0$$

such that $COK_x C$ is of type $T' = T \setminus \{\bar{x}\}$ and $S_y COK_x C = S_y C$ for $\bar{y} \in T'$. We get from Theorem 11.2. in $K_o(R\Gamma)$

$$o(C) = o(COK_x C) + o(E_x \circ S_x C)$$

and in Split $K_o(R\Gamma)$ by Lemma 9.31.

$$\text{Split } o(C) = \text{Split } o(COK_x C) + \text{Split } o(E_x \circ S_x C)$$

Moreover $S(o(E_x \circ S_x C)) = \text{Split } o(E_x \circ S_x C)$ follows from Lemma 9.31. and $S(o(COK_x C)) = \text{Split } o(COK_x C)$ from the induction hypothesis. This implies $S(o(C)) =$ Split $o(C)$ □

Proposition 11.15. is not true if C is only a $R\Gamma$-chain complex possessing a finite approximation.

If $F : \Gamma_1 \longrightarrow \Gamma_2$ is a functor and C a $R\Gamma_1$-chain complex of the homotopy tpye of a finite projective $R\Gamma_1$-chain complex then $F_* C := \operatorname{ind}_F C$ is a $R\Gamma_2$-chain complex of the homotopy type of a finite projective $R\Gamma_2$-chain complex. Let $F_* : K_o(R\Gamma_1) \longrightarrow K_o(R\Gamma_2)$ be induced by induction (10.10.). Then we have

11.16. $$F_*(o(C)) = o(F_* C) \quad .$$

Consider an admissible functor $F : \Gamma_1 \longrightarrow \Gamma_2$ and a $R\Gamma_2$-chain complex C of the homotopy type of a finite projective $R\Gamma_2$-chain complex. Then $F^* C := \operatorname{res}_F C$ is a $R\Gamma_1$-chain complex of the homotopy type of a finite projective $R\Gamma_1$-chain complex. Let $F^* : K_o(R\Gamma_2) \longrightarrow K_o(R\Gamma_1)$ be induced by the restriction (10.18.). We get

11.17. $$F^*(o(C)) = o(F^*C) .$$

Let C be a $R\Gamma_1$- and D be a $R\Gamma_2$-chain complex. Let <u>their tensor product over</u> R $C \otimes_R D$ be defined by

$$\partial_n \mid C_i \otimes D_j = c_i \otimes id + (-1)^i id \otimes d_j ,$$

where c, d and ∂ are the differentials of C, D and $C \otimes_R D$. If C and D have the homotopy type of a finite projective chain complex, then also $C \otimes_R D$. Recall the pairing $\otimes_R : K_o(R\Gamma_1) \otimes K_o(R\Gamma_2) \longrightarrow K_o(R\Gamma_1 \times \Gamma_2)$ (10.23.). We claim

11.18. $$o(C) \otimes_R o(D) = o(C \otimes_R D) .$$

In the proof of 11.18. we can assume that C and D are finite projective. Then one computes

$$\begin{aligned} o(C \otimes_R D) &= \sum_{n=o}^{\infty} (-1)^n [(C \otimes_R D)_n] \\ &= \sum_{n=o}^{\infty} (-1)^n \sum_{i+j=n} [C_i \otimes_R D_j] \\ &= \sum_{n=o}^{\infty} \sum_{i+j=n} (-1)^i [C_i] \otimes_R (-1)^j [D_j] \\ &= (\sum_{i=o}^{\infty} (-1)^i [C_i]) \otimes_R (\sum_{j=o}^{\infty} (-1)^j [D_j]) = o(C) \otimes_R o(D) . \end{aligned}$$

Related to the finiteness obstruction is the Euler characteristic.

<u>Definition 11.19.</u> <u>Let</u> C <u>be a</u> $R\Gamma$-<u>chain complex possessing a finite approximation</u> <u>Define its Euler characteristic</u>

$$\chi(C) \in U(\Gamma)$$

<u>by the image of its finiteness obstruction</u> $o(C)$ <u>under the homomorphism</u> $rk_{R\Gamma} : K_o(R\Gamma) \longrightarrow U(\Gamma)$ (10.41.) . □

If (P,f) is a finite approximation of C then

11.20. $$\chi(C) = \Sigma(-1)^n rk_{R\Gamma}(P)$$

If C possesses even a finite free approximation (F,f) and $rk_{R\Gamma} : K_o^f(R\Gamma) \to U(\Gamma)$ is the isomorphism of Proposition 10.42. we get

11.21. $$rk_{R\Gamma}(\Sigma(-1)^n[F_n]) = \chi(C)$$

Let $F : \Gamma_1 \longrightarrow \Gamma_2$ be a functor and C a $R\Gamma_1$-chain complex of the homotopy type of a finite projective $R\Gamma_1$-chain complex. Then Proposition 10.44. and 11.16. imply for the homomorphism $F_* : U(\Gamma_1) \longrightarrow U(\Gamma_2)$ (10.11.)

11.22. $$F_*(\chi(C)) = \chi(F_*C)$$

Let $F : \Gamma_1 \longrightarrow \Gamma_2$ be an admissible functor and C a $R\Gamma_2$-chain complex. Suppose either that C has the homotopy type of a finite projective $R\Gamma_2$-chain complex and the conditions of Proposition 10.44. c) are satisfied for $R\Gamma_1$ and $R\Gamma_2$, or that C has the homotopy type of a finite free $R\Gamma_2$-chain complex. Then we get from Proposition 10.44. and 11.17. for the homomorphism $F^* : U(\Gamma_2) \longrightarrow U(\Gamma_1)$ (10.21.)

11.23. $$F^*(\chi(C)) = \chi(F^*C)$$

We can improve the product formula 11.18. under certain condition for Γ_1 and Γ_2 which are for example satisfied for the discrete orbit category Or/G of a compact Lie group. Recall the pairings

$$\otimes_R : U(\Gamma_1) \otimes U(\Gamma_2) \longrightarrow U(\Gamma_1 \times \Gamma_2)$$

$$\otimes_R : U(\Gamma_1) \otimes K_o(R\Gamma_2) \longrightarrow K_o(R\Gamma_1 \times \Gamma_2)$$

$$\otimes_R : K_o(R\Gamma_1) \otimes U(\Gamma_2) \longrightarrow K_o(R\Gamma_1 \times \Gamma_2)$$

the isomorphism

$$U(\Gamma) \cong K_o^f(R\Gamma)$$

and the split injection

$$K_o^f(R\Gamma) \longrightarrow K_o(R\Gamma)$$

(see 10.23., 10.25., Proposition 10.42.).

Theorem 11.24. Let R be a Dedekind domain with $\mathrm{ch}\, R = 0$ and quotient field F. Let Γ_1 and Γ_2 be EI-categories. Suppose for $i = 1,2$ and $x \in \mathrm{Ob}\,\Gamma_i$ that $\mathrm{Aut}(x)$ is finite and no prime p dividing $|\mathrm{Aut}(x)|$ is a unit in R. Consider a $R\Gamma_1$-chain complex C and a $R\Gamma_2$-chain complex D such that both have the homotopy tpye of a finite projective chain complex. Define $rk_R M$ by $rk_F(M \otimes_R F)$. Then we have in $K_o(R\Gamma)$

$$o(C \otimes_R D) = \chi(C) \otimes_R o(D) + o(C) \otimes_R \chi(D) - \chi(C) \otimes_R \chi(D)$$

Proof. We get from 11.18.

$$o(C) \otimes_R o(D) \quad = \quad o(C \otimes_R D)$$

Hence it suffices to prove for a finitely generated projective $R\Gamma_1$-module P and a finitely generated $R\Gamma_2$-module Q that in $K_o(R\Gamma_1 \times \Gamma_2)$ the relation

$$[P] \otimes_R [Q] \quad = rk_{R\Gamma_1}(P) \otimes_R Q + P \otimes_R rk_{R\Gamma_2} Q - rk_{R\Gamma_1}(P) \otimes_R rk_{R\Gamma_2}(Q)$$

holds. By Theorem 10.34. this reduces to the case of group rings, i.e. Γ_1 is a finite group G and Γ_2 a finite group H and P and Q are RG- and RH-modules. By assumption no prime p dividing $|G|$ or $|H|$ is a unit in R.

By Swan [1960 a] Theorem 8.1. $F \otimes_R P$ is a finitely generated free $F[G]$-module and $F \otimes_R Q$ a finitely generated free $F[H]$-module if F is the quotient field of R. Now we can apply Swan [1960 a] Theorem 7.1. saying that for any non-zero ideal I in R we can find a finitely generated free RG- resp. RH-submodule $M \subset P$ resp. $N \subset Q$ such that $\mathrm{Ann}_R(P/M) = \{r \in R \mid rx = 0 \text{ for } x \in P/M\}$ resp. $\mathrm{Ann}_R(Q/M)$ is prime to I. Hence we can assume that $\mathrm{Ann}_R(P/M) \neq \{0\}$ and $\mathrm{Ann}_R(Q/M) \neq \{0\}$ are prime. This implies that $P/M \otimes_R Q/N = 0$ and $\mathrm{Tor}_R(P/M),Q/N) = 0$ holds so that $P/M \otimes_R N \longrightarrow P/M \otimes_R Q$ is an isomorphism. We have the exact sequences

$$0 \longrightarrow M \otimes_R Q \longrightarrow P \otimes_R Q \longrightarrow P/M \otimes_R Q \longrightarrow 0$$

and

$$0 \longrightarrow M \otimes_R N \longrightarrow P \otimes_R N \longrightarrow P/M \otimes_R N \longrightarrow 0$$

so that $P/M \otimes_R Q$ and $P/M \otimes_R N$ have finite resolutions. This implies

$$[P \otimes_R Q] - [M \otimes_R Q] = [P/M \otimes_R Q] = [P/M \otimes_R N] = [P \otimes_R N] - [M \otimes_R N] \quad .$$

Since $\mathrm{Ann}_R(P/M)$ and $\mathrm{Ann}_R(Q/M)$ are non-zero we get $M \otimes_R F \cong_{FG} P \otimes_R F$ and $N \otimes_R F \cong_{FH} Q \otimes_R F$. This implies $r(P) := rk_F((M \otimes_R F) \otimes_{FG} F) = rk_F((P \otimes_R F) \otimes_{FG} F)$ and $r(Q) := rk_F((N \otimes_R F) \otimes_{FH} F) = rk_F((Q \otimes_{FH} F)$. We have shown in $K_o(RG \times H)$

$$[P] \otimes_R [Q] = r(P) \cdot [RG] \otimes_R Q + P \otimes r(Q) \cdot [RH] - r(P) \cdot r(Q) \cdot [RG] \otimes_R [RH].$$

This finishes the proof. □

Notice that the assumption of Theorem 11.24. are satisfied if $\mathrm{Aut}(x)$ is finite for $x \in \mathrm{Ob}\,\Gamma_i$ and $i = 1,2$ and R is $\mathbf{Z}$.

Given a projection $p : P \longrightarrow P$ of a finitely generated projective $R\Gamma$-module, image p is again a finitely generated projective $R\Gamma$-module and we can assign to p an element $[p] \in K_o(R\Gamma)$ by [image p] . We want to generalize this to homotopy projections $p : C \longrightarrow C$ (i.e. $p \circ p \simeq p$) of finitely dominated $R\Gamma$-chain complexes.

A _splitting_ (D,r,i) for a homotopy projection $p : C \longrightarrow C$ consists of a projective $R\Gamma$-chain complex D and chain maps $r : C \longrightarrow D$ and $i : D \longrightarrow C$ such that the following diagram commutes up to homotopy

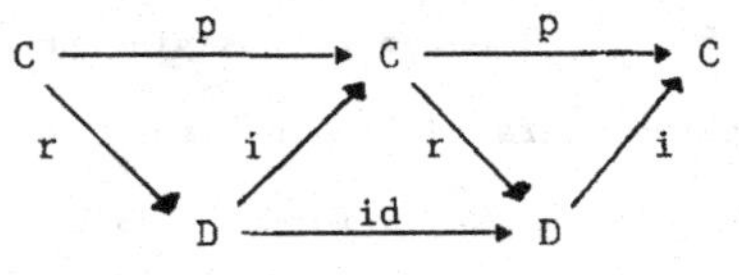

Lemma 11.25. Any homotopy projection $p : C \longrightarrow C$ _of a projective_ $R\Gamma$-_chain complex_ C _has a splitting_ (D,r,i) . _If_ (D',r',i') _is another splitting_ $r' \circ i : D \longrightarrow D'$

<u>and</u> $r \circ i' : D' \longrightarrow D$ <u>are inverse homotopy equivalences</u>

<u>Proof.</u> The existence is proved by a kind of Eilenberg-swindle. By Proposition 11.11. we can assume that C is projective. Let E^1 and E^o be copies of $\bigoplus_{n=o}^{\infty} C$ and $q : E^1 \longrightarrow E^o$ be the chain map

$$\begin{pmatrix} id-p & 0 & 0 & 0 & \cdots \\ p & id-p & 0 & 0 & \cdots \\ 0 & p & id-p & 0 & \cdots \\ 0 & 0 & p & id-p & \cdots \\ \vdots & \vdots & \vdots & \vdots & \end{pmatrix}$$

The chain map $s : E^o \longrightarrow E^1$ is given by the transposed matrix. Define $u : E^o \to C$ by $(p,0,0,0,\ldots)$ and $v : C \longrightarrow E^o$ by $(id,0,0,\ldots,0)^{tr}$. Choose a homotopy $h : p \circ p \simeq p$ and define chain homotopies $\phi^o : E^o \longrightarrow E^1$, $\phi^1 : E^1 \longrightarrow E^1$ and $\psi : E^1 \longrightarrow C^1$ by

$$\phi^o = \begin{pmatrix} h & -h & 0 & 0 & \cdots \\ -h & 2h & -h & 0 & \cdots \\ 0 & -h & 2h & -h & \cdots \\ 0 & 0 & -h & 2h & \cdots \\ \vdots & \vdots & \vdots & \vdots & \end{pmatrix}$$

$$\phi^1 : \begin{pmatrix} 2h & -h & 0 & 0 & \cdots \\ -h & 2h & -h & 0 & \cdots \\ 0 & -h & 2h & -h & \cdots \\ 0 & 0 & -h & 2h & \cdots \\ \vdots & \vdots & \vdots & \vdots & \end{pmatrix}$$

$$\psi : (-h,0,0,\ldots)$$

One easily checks $\phi^o : id \simeq v \circ u + q \circ s$ and $\phi^1 : s \circ q \simeq id$ and $\psi : u \circ q = 0$. We obtain chain maps

$$r : C \xrightarrow{(0,v)^{tr}} \mathrm{Cone}(q)$$

and

$$i : C \xrightarrow{(\psi,u)} C$$

Let Ω : Cone(a) $\longrightarrow$ Cone(q) be the chain homotopy

$$\begin{pmatrix} id & 0 \\ \varepsilon & id \end{pmatrix} \cdot \begin{pmatrix} \phi^1 & s \\ 0 & \phi^o \end{pmatrix}$$

for $\varepsilon = -v \circ \psi - q \circ \phi^1 - \phi^o \circ q$. Then Ω is an homotopy between id and $r \circ i$ and we have $i \circ r = p$. Hence (Cone(q),r,i) is a splitting for $p : C \longrightarrow C$. □

Because of Lemma 11.25. we can define

<u>Definition 11.26.</u> <u>Let</u> $p : C \longrightarrow C$ <u>be a homotopy projection of a</u> RΓ-<u>chain complex</u> C <u>possessing a finite approximation</u>. <u>Choose a finite approximation</u> $h : P \longrightarrow C$ <u>and a homotopy projection</u> $q : P \longrightarrow P$ <u>satisfying</u> $h \circ q \simeq p \circ h$. <u>Let</u> (D,r,i) <u>be any splitting of</u> q. <u>Define</u>

$$o(p) \in K_o(R\Gamma)$$

<u>by the finiteness obstruction</u> o(D). □

<u>Theorem 11.27.</u>

a) <u>Homology invariance</u>

<u>Let</u> $p : C \longrightarrow C$ <u>and</u> $q : D \longrightarrow D$ <u>be homotopy projections of</u> RΓ-<u>chain complexes having a finite approximation</u>. <u>Suppose the existence of a weak homology equivalence</u> $f : C \longrightarrow D$ <u>satisfying</u> $q \circ f \simeq f \circ p$.

<u>Then</u>

$$o(p) = o(q) .$$

b) <u>Additivity</u>

<u>Consider the commutative diagram of</u> RΓ-<u>chain complexes possessing a finite approxi-</u>

mation such that the rows are exact and the vertical arrows homotopy projections

$$\begin{array}{ccccccccc} 0 & \longrightarrow & C^1 & \xrightarrow{j} & C^o & \xrightarrow{q} & C^2 & \longrightarrow & 0 \\ & & \downarrow{\scriptstyle p^1} & & \downarrow{\scriptstyle p^o} & & \downarrow{\scriptstyle p^2} & & \\ 0 & \longrightarrow & C^1 & \xrightarrow{j} & C^o & \xrightarrow{q} & C^2 & \longrightarrow & 0 \end{array}$$

Then:

$$o(p^1) - o(p^o) + o(p^2) = 0 .$$

Proof:

b) use Lemma 11.6., the explicite construction of a splitting in Lemma 11.25. and Theorem 11.2. □

Remark 11.28. If $p : C \longrightarrow C$ is a homotopy projection of a finite projective $R\Gamma$-chain complex, we have constructed an explicite splitting (D,r,i) in Lemma 11.25. Since (C,r,i) is a domination of D and $o(p) = o(D)$, we obtain an explicite formula for $o(p)$ by Proposition 11.12. involving only C, p and an homotopy $h : p \circ p \simeq p$. □

Lemma 11.29. Let $p : C \longrightarrow C$ be a homotopy projection of a $R\Gamma$-chain complex such that $H_n(C)$ has a finite resolution for $n \geq 0$ and is zero for large n. Then C has a finite approximation and $o(p) \in K_o(R\Gamma)$ is defined. Moreover, image $H_n(p) : H_n(C) \longrightarrow H_n(C)$ has a finite resolution for $n \geq 0$ and is zero for large n and we get in $K_o(R\Gamma)$

$$o(p) = \Sigma(-1)^n \,[\text{image } H_n(p)]$$

Proof. Since image $H_n(p)$ is a direct summand in $H_n(C)$ and isomorphic to $H_n(D)$ for any splitting (D,r,i) of C the result follows from Proposition 11.9.

Comments 11.30. The finiteness obstruction is introduced in Wall [1965] in a form close to Proposition 11.13. Wall's paper seems to be inspired by Swan [1960 b]. Our approach is similar to Wall [1966] where also Proposition 11.10. is proved. Most of the material of this section except the homological computations can be generalized to additive categories. This is done in Ranicki [1985] where the formulas appearing in Proposition 11.11. and Proposition 11.12. are taken from. The finiteness obstruction and its splitting for chain complexes over the orbit category Or G of a finite group is treated in tom Dieck [1981] including a product formula like Theorem 11.24. The finiteness obstruction of a homotopy projection and the unique existence of splittings can be found in Lück [1986]. See also Lück-Ranicki [1986] where among other things the second part of Remark 11.28. is worked out. For a splitting of idempotents see Freyd [1966], Hastings-Heller [1981], [1982]. □

<u>Exercises 11.31.</u>

1. Let G be a Lie group and X be a proper G-CW-complex. Show that X is finite resp. skeletal finite resp. n-dimensional resp. of finite orbit type if and only if its cellular $\mathbb{Z}\pi/(G,X)$-chain complex $C^c(X)$ is.

2. Let G be a finite group and C be a projective $\mathbb{Z}$OrG-chain complex. Suppose the existence of a number n such that $H_i(C)$ is finitely generated for $i \geq 0$, $H_i(C)$ is zero for $i \geq n$ and $H^{n+1}(C,M) = 0$ for any $\mathbb{Z}$OrG-module M. Prove that C is homotopy equivalent to a n-dimensional finite projective $\mathbb{Z}$OrG-chain complex.

3. Let C be a finitely dominated $\mathbb{Z}Or/T^n$-chain complex for T^n the n-dimensional torus. Show that C is homotopy equivalent to a finite free $\mathbb{Z}Or/T^n$-chain complex.

4. Let G be a finite group and C be a $\mathbb{Z}$OrG-chain complex possessing a finite approximation. Show that $\chi_{\mathbb{Z}}C(G/H) \in \mathbb{Z}$ is defined for any $H \subset G$ and that $(\chi_{\mathbb{Z}}C(G/H) \mid (H) \in \mathrm{Con}\, G)$ lies in the image of the character map

$$ch : A(G) \longrightarrow \prod_{(H) \in \mathrm{Con}\, G} \mathbb{Z}$$

5. Let $G \neq \{1\}$ be finite. Give an example of a finite $\mathbb{Z}$OrG-chain complex possessing no finite approximation

6. Let G be $\mathbb{Z}/p$. Consider the $\mathbb{Q}$OrG-module $M = I_{G/G}(\mathbb{Q})$. Show that M has a finite $\mathbb{Q}$OrG-approximation. Obviously Split $o(M) \in$ Split $K_o(\mathbb{Q}OrG)$ given by $\{S_{G/H}M \in K_o(\mathbb{Q}WH) \mid (H) \in \mathrm{Con}(G)\}$ is defined since $S_{G/H}M$ is a finitely ge-

nerated projective $\mathbb{Q}WH$-module. Show that $S : K_o(\mathbb{Q}\mathrm{Or}G) \longrightarrow \mathrm{Split}\ K_o(\mathbb{Q}\mathrm{Or}G)$ does not send $o(M)$ to $\mathrm{Split}\ o(M)$.

7. Let C^i be a finitely dominated $\mathbb{Z}\Gamma_i$-chain complex with $\chi_{\mathbb{Z}\Gamma_i}(C^i) = 0$ for $i=1,2$. Assume for $i = 1,2$ that $\mathrm{Aut}(x)$ is finite for any $x \in \mathrm{Ob}\,\Gamma$. Show that the $\mathbb{Z}\Gamma_1 \times \Gamma_2$-chain complex $C^1 \otimes C^2$ is homotopy equivalent to a finite free $\mathbb{Z}\Gamma_1 \times \Gamma_2$-chain complex F with $\chi_{\mathbb{Z}\Gamma_1 \times \Gamma_2}(F) = 0$.

8. Let G be $\mathbb{Z}/p$. Then $\tilde{K}_o(\mathbb{Z}\mathrm{Or}G) \cong \tilde{K}_o(\mathbb{Z}G)$. Consider finitely dominated $\mathbb{Z}\mathrm{Or}G$-chain complexes C and D . Regard $C \otimes_{\mathbb{Z}} D$ as a $\mathbb{Z}\mathrm{Or}G$-chain complex by restriction with $\mathrm{Or}G \longrightarrow \mathrm{Or}G \times \mathrm{Or}G$. Show that $C \otimes_{\mathbb{Z}} D$ is again finitely dominated, the ordinary Euler-characteristics $\chi_{\mathbb{Z}}(C(G/G))$ and $\chi_{\mathbb{Z}}(D(G/G)) \in \mathbb{Z}$ are defined and that one has in $\tilde{K}_o(\mathbb{Z}\mathrm{Or}G) = \tilde{K}_o(\mathbb{Z}G)$

$$\tilde{o}(C \otimes_{\mathbb{Z}} D) = \chi_{\mathbb{Z}}(C(G/G)) \cdot \tilde{o}(D) + \chi_{\mathbb{Z}}(D(G/G)) \cdot \tilde{o}(C)$$

9. Let $p : C \longrightarrow C$ be a homotopy projection of a projective $R\Gamma$-chain complex and (D,r,i) a splitting. Show the existence of a $R\Gamma$-chain complex D' and an homotopy equivalence $f : D \oplus D' \longrightarrow C$ such that the following diagram commutes up to homotopy

$$\begin{array}{ccc} C & \xrightarrow{\quad p \quad} & C \\ f \uparrow & & \uparrow f \\ D \oplus D' & \xrightarrow[\begin{pmatrix} 1 & 0 \\ 0 & 0 \end{pmatrix}]{} & D \oplus D' \end{array}$$

10. Let $\mathcal{C}$ be the category having as objects homotopy projections $p : C \longrightarrow C$ of finitely dominated $R\Gamma$-chain complexes and as morphisms from $p : C \longrightarrow C$ to $q : D \longrightarrow D$ chain maps $f : C \longrightarrow C$ satisfying $q \circ f = f \circ p$. Weak cofibrations are given by morphism such that each f_n is a split injection and weak equivalences are morphisms with f an homotopy equivalence. Show that $K_o(R\mathrm{Or}G) \oplus K_o(R\mathrm{Or}G)$ together with function sending an object $p : C \longrightarrow C$ to $(o(C),o(p))$ is the universal additive invariant.

12. The algebraic torsion

In this section we want to define the algebraic torsion of a homotopy equivalence between appropriate RΓ-chain complexes. We want to treat various kinds of torsion like Whitehead-torsion, Reidemeister torsion, self-torsion, round torsion, ... simultaineously unifying the various approaches. The main difference in our treatment compared with the usual ones is that we substitute the notion of a base for a RΓ-chain complex C by a stable isomorphism $C_{odd} \longrightarrow C_{ev}$. The main advantage will be that then one can treat also the case where C is projective but not necessarily free. This will enable us in particular to define equivariant Reidemeister torsion in a transparent way.

To motivate our approach we recall briefly the definition of the Whitehead torsion of a homotopy equivalence $f : C \longrightarrow D$ between finite based free R-chain complexes. By Lemma 11.5. we obtain an isomorphism

$$(c+\gamma) : \mathrm{Cone}(f)_{odd} \longrightarrow \mathrm{Cone}(f)_{ev}$$

if c is the differential and γ a chain contraction of $\mathrm{Cone}(f)$. The basis determine an isomorphism

$$\phi : \mathrm{Cone}(f)_{ev} \longrightarrow \mathrm{Cone}(f)_{odd}$$

Then the Whitehead torsion is defined as the element $\tau(f)$ in $\tilde{K}_1(R) = K_1(R) / \{\pm 1\}$ represented by the automorphism $\phi\circ(c+\gamma)$ of the finitely generated free R-module $\mathrm{Cone}(f)_{odd}$. The Whitehead torsion could also be defined as an element in $K_1(R)$. However, in $\tilde{K}_1(R)$ the important formulas $t(f \oplus g) = t(f) + t(g)$ and $t(f \circ g) = t(f) + t(g)$ are true but not in $K_1(R)$. Here is a counterexample.

<u>Example 12.1.</u> Let $u_1 : F_1 \longrightarrow F_1$ and $u_2 : F_2 \longrightarrow F_2$ be automorphisms of the based free R-module of rank 1. Consider the i-dimensional elementary RΓ-chain complexes $e\ell_i(u_i)$ (see 11.4.) and the homotopy equivalences $f_i : 0 \longrightarrow e\ell_i(u_i)$ for $i = 1,2$. One easily checks that $\tau(f_1)$ is $[u_1] \in K_1(R)$ and $\tau(f_2)$ is $-[u_2] \in K_1(R)$. The mapping cone of $f_2 \oplus f_1$ together with a chain contradiction looks like

$$0 \longrightarrow F_2 \underset{(u_2^{-1},0)}{\overset{\binom{u_2}{0}}{\rightleftarrows}} F_2 \oplus F_1 \underset{\binom{0}{u_1^{-1}}}{\overset{(0,u_1)}{\rightleftarrows}} F_1 \longrightarrow 0$$

Then $t(f_2 \oplus f_1)$ is given by

$$\begin{pmatrix} 0 & u_1 \\ u_2^{-1} & 0 \end{pmatrix} : F_2 \oplus F_1 \longrightarrow F_1 \oplus F_2 .$$

Hence $t(f_2) + t(f_1)$ and $t(f_2 \oplus f_1)$ are not the same element in $K_1(R)$, they differ by

$$\begin{pmatrix} 0 & 1 \\ 1 & 0 \end{pmatrix} : R \oplus R \longrightarrow R \oplus R \qquad \square$$

Now we develop a theory of torsion not using bases but a stable map $\phi : \mathrm{Cone}(f) \longrightarrow \mathrm{Cone}(f)_{odd}$. This is precisely what the bases give in the usual treatment.

Fix a subgroup $U \subset K_1(R\Gamma)$. It will play the role of the trivial units. Given a not necessarily commutative square of isomorphisms of finitely generated projective $R\Gamma$-modules

$$\begin{array}{ccc} X_1 & \xrightarrow{u_1} & X_2 \\ \uparrow u_4 & & \downarrow u_2 \\ X_4 & \xleftarrow{u_3} & X_3 \end{array}$$

define its torsion $t(S) \in K_1(R)$ by $[u_4 \circ u_3 \circ u_2 \circ u_1]$. We get the same if we start at any corner and run around in the clockwise direction. If S is commutative $t(S)$ vanishes. Define the sum $S_1 \oplus S_2$ of two such squares in the obvious way. Then $t(S_1 \oplus S_2) = t(S_1) + t(S_2)$ holds. If S_1 and S_2 are squares with a common edge we can built a composed square $S_2 \circ S_1$. Namely, the outer square in

$$
\begin{array}{ccc}
X_1 & \xrightarrow{u_1} & X_2 \\
{\scriptstyle u_4}\uparrow & S_1 & \downarrow{\scriptstyle u_2} \\
X_4 & \xleftarrow{u_3} & X_3 \\
{\scriptstyle u_7}\uparrow & S_2 & \downarrow{\scriptstyle u_5} \\
X_6 & \xleftarrow{u_6} & X_5
\end{array}
$$

We get $t(S_2 \circ S_1) = t(S_1) + t(S_2)$. Isomorphic squares have the same torsion.

Given two finitely generated projective $R\Gamma$-modules M and N , a <u>stable map</u> from $M \longrightarrow N$ is a $R\Gamma$-homomorphism $f : M \bullet X \longrightarrow N \bullet X$ for some finitely generated projective $R\Gamma$-module X . In the sequel π always denotes the permutation map which is clear from the context by specifying the source and the target. If $g : M \bullet Y \to N \bullet Y$ is another stable map from M to N, we call f and g U-<u>stably equivalent</u> if there is a finitely generated projective $R\Gamma$-module P such that the torsion of the following square lies in $U \subset K_1(R\Gamma)$

12.2.
$$
\begin{array}{ccc}
M \bullet X \bullet Y \bullet P & \xrightarrow{f \bullet id} & N \bullet X \bullet Y \bullet P \\
\downarrow{\scriptstyle \pi} & & \downarrow{\scriptstyle \pi} \\
M \bullet Y \bullet X \bullet P & \xrightarrow{g \bullet id} & N \bullet Y \bullet X \bullet P
\end{array}
$$

This relation is clearly symmetric and reflexive. We next prove transitivity. If g is U-stably equivalent to $h : M \bullet Z \longrightarrow N \bullet Z$ by a diagram of the shape 12.2. with Q instead of P then f is U-stably equivalent to h by

$$
\begin{array}{ccc}
M \bullet X \bullet Z \bullet Y \bullet P \bullet Q & \xrightarrow{f \bullet id} & N \bullet X \bullet Z \bullet Y \bullet P \bullet Q \\
\downarrow{\scriptstyle \pi} & & \downarrow{\scriptstyle \pi} \\
M \bullet X \bullet Y \bullet P \bullet Z \bullet Q & \xrightarrow{f \bullet id} & N \bullet X \bullet Y \bullet P \bullet Z \bullet Q \\
\downarrow{\scriptstyle \pi} & & \downarrow{\scriptstyle \pi} \\
M \bullet Y \bullet X \bullet P \bullet Z \bullet Q & \xrightarrow{g \bullet id} & N \bullet Y \bullet X \bullet P \bullet Z \bullet Q \\
\downarrow{\scriptstyle \pi} & & \downarrow{\scriptstyle \pi} \\
M \bullet Y \bullet Z \bullet Q \bullet X \bullet P & \xrightarrow{g \bullet id} & N \bullet Y \bullet Z \bullet Q \bullet X \bullet P \\
\downarrow{\scriptstyle \pi} & & \downarrow{\scriptstyle \pi} \\
M \bullet Z \bullet Y \bullet Q \bullet X \bullet P & \xrightarrow{h \bullet id} & N \bullet Z \bullet Y \bullet Q \bullet X \bullet P \\
\downarrow{\scriptstyle \pi} & & \downarrow{\scriptstyle \pi} \\
M \bullet Z \bullet X \bullet Y \bullet P \bullet Q & \xrightarrow{h \bullet id} & N \bullet Z \bullet X \bullet Y \bullet P \bullet Q
\end{array}
$$

We denote by $\{f\} : M \longrightarrow N$ the U-stable equivalence class of a stable map $f : M \oplus X \longrightarrow N \oplus Y$. Given two stable maps $f : M \oplus X \longrightarrow N \oplus X$ and $f' : M' \oplus X' \rightarrow N' \oplus X'$, let $f \oplus_s f'$ be the stable map from $M \oplus M' \longrightarrow N \oplus N'$ making the following diagram commute

$$\begin{array}{ccc} M \oplus M' \oplus X \oplus X' & \xrightarrow{f \oplus_s f'} & N \oplus N' \oplus X \oplus X' \\ \downarrow \pi & & \downarrow \pi \\ M \oplus X \oplus M' \oplus X' & \xrightarrow[f \oplus f']{} & N \oplus X \oplus N' \oplus X' \end{array}$$

It is left to the reader to show that $(f \oplus_s f') \oplus_s f'' \sim_U f \oplus_s (f' \oplus_s f'')$ holds and $f \sim_U g$ and $f' \sim_U g'$ implies $f \oplus_s f' \sim_U g \oplus_s g'$ where $\sim_U$ means U-stably equivalent. Hence we get a well-defined notion of a direct sum $\{f\} \oplus \{g\} := \{f \oplus_s g\}$ of U-stable equivalence classes of maps .

If $f : L \oplus X \longrightarrow M \oplus X$ and $g : M \oplus Y \longrightarrow N \oplus Y$ are stable maps let $g \circ_s f$ be the stable map from L to N making the following diagram commute

$$\begin{array}{ccccc} L \oplus X \oplus Y & & \xrightarrow{g \circ_s f} & & N \oplus X \oplus Y \\ \downarrow f \oplus id & & & & \uparrow \pi \\ M \oplus X \oplus Y & \xrightarrow{\pi} & M \oplus Y \oplus X & \xrightarrow{g \oplus id} & N \oplus Y \oplus X \end{array}$$

Let f be U-stably equivalent to f' . We leave it to the reader to verify that then $g \circ_s f$ and $g \circ_s f'$ are U-stably equivalent. Analogously $g \sim_U g' \Rightarrow g \circ_s f \sim g' \circ_s f$. Hence we can define the composition $\{g\} \circ \{f\}$ of U-stably equivalence classes of stable maps $\{f\} : L \longrightarrow M$ and $\{g\} : M \longrightarrow N$ by $\{g \circ_s f\}$. One easily checks associativity and that $\{f\} \circ \{f^{-1}\} = \{id\}$ and $\{f\} \circ \{id\} = \{id\} \circ \{f\} = \{f\}$ holds.

An U-stable equivalence class of automorphisms $\{f\} : M \longrightarrow M$ determines an element $t(\{f\}) \in K_1(R\Gamma)/U$. We have $t(\{f\} \circ \{g\}) = t(\{f\}) + t(\{g\})$ and $t(\{f\} \oplus \{g\}) = t(\{f\}) \oplus t(\{g\})$. Hence we can also assign to a square of U-stable equivalence classes of isomorphisms

$$\begin{array}{ccc} X_1 & \xrightarrow{\{u_1\}} & X_2 \\ \{u_4\} \uparrow & S & \downarrow \{u_2\} \\ X_4 & \xleftarrow{\{u_3\}} & X_3 \end{array}$$

and element

12.3. $$t(S) \in K_1(R\Gamma)/U$$

by $t(\{u_4\} \circ \{u_3\} \circ \{u_2\} \circ \{u_1\})$. We still have $t(S_1 \oplus S_2) = t(S_1) \oplus t(S_2)$ and $t(S_2 \circ S_1) = t(S_1) + t(S_2)$

Now we can define the torsion we seek. Let $(f,\{\phi\}) : C \longrightarrow D$ be a homotopy equivalence between finite projective $R\Gamma$-chain complexes together with a U-stable equivalence class of isomorphisms $\{\phi\} : C_{odd} \oplus D_{ev} \longrightarrow D_{odd} \oplus C_{ev}$. If c is the differential and γ any chain contraction of $\mathrm{Cone}(f)$, we get by Lemma 11.5. an isomorphism $(c+\gamma) : \mathrm{Cone}(f)_{odd} \longrightarrow \mathrm{Cone}(f)_{ev}$.

<u>Definition 12.4.</u> <u>Define the torsion</u>

$$t(f,\{\phi\}) \in K_1(R\Gamma)/U$$

<u>by the torsion of the square</u>

$$\begin{array}{ccc}
\mathrm{Cone}(f)_{odd} & \xrightarrow{\{c+\gamma\}} & \mathrm{Cone}(f)_{ev} \\
\uparrow \{\pi\} & & \downarrow \{\pi\} \\
D_{odd} \oplus C_{ev} & \xleftarrow{\{\phi\}} & C_{odd} \oplus D_{ev}
\end{array}$$

□

This is independent of the choice of γ by Lemma 11.5.

<u>Example 12.5.</u> Let Γ be trivial and $U = \{\pm 1\} \subset K_1(R)$. Consider an homotopy equivalence $f : C \longrightarrow D$ of finite based free R-chain complexes. Let $\phi : C_{odd} \oplus D_{ev} \longrightarrow D_{odd} \oplus C_{ev}$ be a base preserving isomorphism. Then the Whitehead torsion of f in $\tilde{K}_1(R)$ defined in Cohen [1973] coincides with $t(f,\{\phi\})$. □

Let $f : C \longrightarrow D$ and $f' : C' \longrightarrow D'$ be chain equivalences of finite projective $R\Gamma$-chain complexes and $\{\phi\} : C_{odd} \oplus D_{ev} \longrightarrow D_{odd} \oplus C_{ev}$ and $\{\phi'\} : C'_{odd} \oplus D'_{ev} \longrightarrow D'_{odd} \oplus C'_{ev}$ be U-stable equivalence classes of isomorphisms. Define

$$\{\phi\} \oplus \{\phi'\} : (C \oplus C')_{odd} \oplus (D \oplus D'_{ev}) \longrightarrow (D \oplus D')_{odd} \oplus (C \oplus C')_{ev}$$

as the U-stable equivalence class of stable isomorphisms for which the following square has torsion zero in $K_1(R\Gamma)/U$

$$\begin{array}{ccc} (C \oplus C')_{odd} \oplus (D \oplus D')_{ev} & \xrightarrow{\{\phi\} \oplus \{\phi'\}} & (D \oplus D')_{odd} \oplus (C \oplus C')_{ev} \\ \downarrow \{\pi\} & & \downarrow \{\pi\} \\ (C_{odd} \oplus D_{ev}) \oplus (C'_{odd} \oplus D'_{ev}) & \xrightarrow{\{\phi \oplus_s \phi'\}} & (D_{odd} \oplus C_{ev}) \oplus (D'_{odd} \oplus C'_{ev}) \end{array}$$

Then we have:

12.6. $$t(f \oplus f', \{\phi\} \oplus \{\phi'\}) = t(f,\{\phi\}) + t(f',\{\phi'\})$$

Remark 12.7. The reader should check this in the case described in Example 12.1. In the notion used there let $\{\phi_i\} : e\ell_i(u_i)_{odd} \longrightarrow e\ell_i(u_i)_{ev}$ be $\{id\}$ for $i=1,2$. Then $\{\phi_2\} \oplus \{\phi_1\}$ is also given by the identity and one easily verifies 12.6. The difference between our approach and the one in Example 12.2. using bases is that

$$\begin{pmatrix} 0 & u_1 \\ u_2^{-1} & 0 \end{pmatrix} : F_2 \oplus F_1 \longrightarrow F_1 \oplus F_2$$

must be composed in our approach by the flip, in the approach using bases by the identity. □

We want to extend 12.6. to exact sequences.

Lemma 12.8. Let $0 \longrightarrow C \xrightarrow{i} D \xrightarrow{p} E \longrightarrow 0$ be an exact sequence of $R\Gamma$-chain complexes. If E is projective and acyclic then there is a $R\Gamma$-chain map $s : E \rightarrow D$ satisfying $p \circ s = id$.

Proof. Choose for any $n \geq 0$ a map $\sigma_n : E_n \longrightarrow D_n$ with $p_n \circ \sigma_n = id$. If γ is a chain contraction of E define $s_n : E_n \longrightarrow D_n$ by $d_{n+1} \circ \gamma_n \circ \sigma_n + \sigma_n \circ \gamma_{n-1} \circ e_n$ if d and e are the differentials of D and E. □

Now suppose that we are given a commutative diagram of finite projective $R\Gamma$-chain complexes with exact rows and homotopy equivalences as vertical arrows

12.9.

$$\begin{array}{ccccccccc}
0 & \longrightarrow & C^1 & \xrightarrow{i} & C^o & \xrightarrow{p} & C^2 & \longrightarrow & 0 \\
 & & \downarrow{\scriptstyle f^1} & & \downarrow{\scriptstyle f^o} & & \downarrow{\scriptstyle f^2} & & \\
0 & \longrightarrow & D^1 & \xrightarrow[j]{} & D^o & \xrightarrow[q]{} & D^2 & \longrightarrow & 0
\end{array}$$

and U-stable equivalence classes of isomorphisms $\{\phi_i\} : C^i_{odd} \oplus D^i_{ev} \longrightarrow D^i_{odd} \oplus C^i_{ev}$ for $i = 0,1,2$. Choose sections s and t for the exact sequences

$$0 \longrightarrow C^1_{odd} \oplus D^1_{ev} \xrightarrow{i_{odd} \oplus j_{ev}} C^o_{odd} \oplus D^o_{ev} \underset{s}{\overset{p_{odd} \oplus q_{ev}}{\rightleftarrows}} C^2_{odd} \oplus D^2_{ev} \longrightarrow 0$$

and

$$0 \longrightarrow D^1_{odd} \oplus C^1_{ev} \xrightarrow{j_{odd} \oplus i_{ev}} D^o_{odd} \oplus C^o_{ev} \underset{t}{\overset{q_{odd} \oplus q_{ev}}{\rightleftarrows}} D^2_{odd} \oplus C^2_{ev} \longrightarrow 0$$

Define $t(\{\phi^1\},\{\phi^o\},\{\phi^2\}) \in K_1(R\Gamma)/U$ by the torsion of the square

$$\begin{array}{ccc}
D^o_{odd} \oplus C^o_{ev} & \xleftarrow{\{\phi_o\}} & C^o_{odd} \oplus D^o_{ev} \\
\uparrow{\scriptstyle (j_{odd} \oplus i_{ev}) \oplus t} & & \uparrow{\scriptstyle (i_{odd} \oplus j_{ev}) \oplus s} \\
D^1_{odd} \oplus C^1_{ev} \oplus D^2_{odd} \oplus C^2_{ev} & \xleftarrow[\{\phi^1\} \oplus \{\phi^2\}]{} & C^1_{odd} \oplus D^1_{ev} \oplus C^2_{odd} \oplus D^2_{ev}
\end{array}$$

One easily checks that this is independent of the choice of s and t . Requiring $t(\{\phi^1\}, \{\phi^o\}, \{\phi^2\})$ to be zero corresponds to demanding that the rows in 12.9. are based exact if the chain complexes were based.

<u>Proposition 12.10.</u> <u>Additivity</u>

<u>We have in</u> $K_1(R\Gamma)/U$

$$t(f^1,\{\phi^1\}) - t(f^o,\{\phi^o\}) + t(f^2,\{\phi^2\}) = t(\{\phi^1\},\{\phi^o\},\{\phi^2\}) .$$

<u>Proof.</u> We get from 12.9. and Lemma 12.8. a split exact sequence of finite projective $R\Gamma$-chain complexes

$$0 \longrightarrow \mathrm{Cone}(f^1) \xrightarrow{i \oplus j} \mathrm{Cone}(f^o) \underset{s}{\overset{p \oplus q}{\rightleftarrows}} \mathrm{Cone}\ (f^2) \longrightarrow 0$$

yielding an isomorphism

$$(i \oplus j) \oplus s : \mathrm{Cone}(f^1) \oplus \mathrm{Cone}(f^2) \longrightarrow \mathrm{Cone}(f^o)$$

Let γ^1 and γ^2 be chain contractions of $\mathrm{Cone}(f^1)$ and $\mathrm{Cone}(f^2)$. Conjugating them with the isomorphism above yields a chain contraction γ^o of $\mathrm{Cone}(f^o)$. We have the commutative diagram

$$\begin{array}{ccc}
\mathrm{Cone}(f^1)_{odd} \oplus \mathrm{Cone}(f^2)_{odd} & \xrightarrow{(c^1+\gamma^1) \oplus (c^2+\gamma^2)} & \mathrm{Cone}(f^1)_{ev} \oplus \mathrm{Cone}(f^2)_{ev} \\
\downarrow \pi & & \downarrow \pi \\
(\mathrm{Cone}(f^1) \oplus \mathrm{Cone}(f^2))_{odd} & & (\mathrm{Cone}(f^1) \oplus \mathrm{Cone}(f^2))_{ev} \\
\downarrow ((i \oplus j) \oplus s)_{odd} & & \downarrow ((i \oplus j) \oplus s)_{ev} \\
\mathrm{Cone}(f^o)_{odd} & \xrightarrow[(c^o+\gamma^o)]{} & \mathrm{Cone}(f^o)_{ev}
\end{array}$$

The torsion of the following square is $t(f^1,\{\phi^1\}) + t(f^2,\{\phi^2\})$

$$\begin{array}{ccc}
\mathrm{Cone}(f^1)_{odd} \oplus \mathrm{Cone}(f^2)_{ev} & \xrightarrow{\{c^1+\gamma^1\} \oplus \{c^2+\gamma^2\}} & \mathrm{Cone}(f^1)_{ev} \oplus \mathrm{Cone}(f^2)_{ev} \\
\downarrow \{\pi\} & & \downarrow \{\pi\} \\
(D^1_{odd} \oplus C^1_{ev}) \oplus (D^2_{odd} \oplus C^2_{ev}) & \xleftarrow{\{\phi^1\} \oplus \{\phi^2\}} & (C^1_{odd} \oplus D^1_{ev}) \oplus (C^2_{odd} \oplus D^2_{ev})
\end{array}$$

By definition $t(f^o,\{\phi^o\})$ is the torsion of

$$\begin{array}{ccc}
\mathrm{Cone}(f^o)_{odd} & \xrightarrow{\{c^o+\gamma^o\}} & \mathrm{Cone}(f^o)_{ev} \\
\{\pi\} \downarrow & & \downarrow \{\pi\} \\
D^o_{odd} \oplus C^o_{ev} & \xleftarrow[\{\phi^o\}]{} & C^o_{odd} \oplus D^o_{ev}
\end{array}$$

Because of the last three squares $t(f^1,\{\phi^1\}) - t(f^o,\{\phi^o\}) + t(f^1,\{\phi^1\})$ is the torsion of the square

$$\begin{array}{ccc}
D^o_{odd} \oplus C^o_{ev} & \xleftarrow{\{\phi^o\}} & C^o_{odd} \oplus D^o_{ev} \\
\uparrow j_{odd} \oplus i_{ev} \oplus (\pi \circ s \circ \pi) & & \uparrow i_{odd} \oplus j_{ev} \oplus (\pi \circ s \circ \pi) \\
(D^1_{odd} \oplus C^1_{ev}) \oplus (D^2_{odd} \oplus C^2_{ev}) & \xleftarrow[\{\phi^1\} \oplus \{\phi^2\}]{} & (C^1_{odd} \oplus D^1_{ev}) \oplus (C^2_{odd} \oplus D^2_{ev})
\end{array}$$

But this is by definition $t(\{\phi^1\},\{\phi^o\},\{\phi^2\})$. □

In particular Proposition 12.10. implies

12.11. $$t(f,\{\phi\}) - t(f,\{\psi\}) = t(\{\phi\} \circ \{\psi^{-1}\})$$

Proposition 12.12. Homotopy invariance

Consider two homotopic chain equivalences f *and* $g : C \longrightarrow D$ *between finite projective* $R\Gamma$-*chain complexes and a* U-*stable class of stable isomorphisms* $\{\phi\} : C_{odd} \oplus D_{ev} \longrightarrow D_{odd} \oplus C_{ev}$. *Then*

$$t(f,\{\phi\}) = t(g,\{\phi\}) .$$

Proof: If $h : f \simeq g$ is an homotopy, we have an isomorphism $I : \mathrm{Cone}(f) \rightarrow \mathrm{Cone}(g)$ given by

$$\begin{pmatrix} id & 0 \\ h & id \end{pmatrix} : C_{*-1} \oplus D_* \longrightarrow C_{*-1} \oplus D_*$$

If γ is a chain contraction for $\mathrm{Cone}(f)$ $\delta = I \circ \gamma \circ I^{-1}$ is one for $\mathrm{Cone}(g)$. The following diagram commutes and I_{odd} and I_{ev} represent zero in $K_1(R\Gamma)$

$$\begin{array}{ccc} \mathrm{Cone}(f)_{odd} & \xrightarrow{(c+\gamma)} & \mathrm{Cone}(f)_{ev} \\ \downarrow I_{odd} & & \downarrow I_{ev} \\ \mathrm{Cone}(g)_{odd} & \xrightarrow[(c+\delta)]{} & \mathrm{Cone}(g)_{ev} \end{array}$$ □

The next result is a consequence of additivity and homotopy invariance. Consider homotopy equivalences $f : C \longrightarrow D$ and $g : D \longrightarrow E$ between finite projective $R\Gamma$-chain complexes and U-stable equivalence classes of stable isomorphisms $\{\phi\} : C_{odd} \oplus D_{ev} \longrightarrow D_{odd} \oplus E_{ev}$ and $\{\psi\} : D_{odd} \oplus E_{ev} \longrightarrow E_{odd} \oplus D_{ev}$.

Define a U-stable equivalence class of stable isomorphisms $\{\psi\}\#\{\phi\} : C_{odd} \oplus E_{ev} \rightarrow E_{odd} \oplus C_{ev}$ by requiring that for the following square S $t(S) = 0$ in $K_1(R\Gamma)/U$ holds

$$\begin{array}{ccc}
(C_{odd} \oplus D_{ev}) \oplus (D_{odd} \oplus E_{ev}) & \xrightarrow{\{\phi\} \oplus \{\psi\}} & (D_{odd} \oplus C_{ev}) \oplus (E_{odd} \oplus D_{ev}) \\
\{\pi\} \downarrow & & \downarrow \{\pi\} \\
(C_{odd} \oplus E_{ev}) \oplus (D_{odd} \oplus D_{ev}) & \xrightarrow[(\{\psi\}\#\{\phi\}) \oplus \{id\}]{} & (E_{odd} \oplus C_{ev}) \oplus (D_{odd} \oplus D_{ev})
\end{array}$$

<u>Proposition 12.13.</u> <u>Logarithmic property.</u>

<u>We have</u>

$$t(g \circ f, \{\psi\}\#\{\phi\}) = t(f,\{\phi\}) + t(g,\{\psi\}) .$$

<u>Proof:</u> Let $\{\bar{\phi}\} : \mathrm{Cone}(f)_{ev} \longrightarrow \mathrm{Cone}(f)_{odd}$ be the U-stable class of stable isomorphisms for which the torsion of the following square vanishes in $K_1(R\Gamma)/U$

$$\begin{array}{ccc}
\mathrm{Cone}(f)_{ev} & \xrightarrow{\{\bar{\phi}\}} & \mathrm{Cone}(f)_{odd} \\
\{\pi\} \downarrow & & \downarrow \{\pi\} \\
C_{odd} \oplus D_{ev} & \xrightarrow{\{\phi\}} & D_{odd} \oplus C_{ev}
\end{array}$$

Define $\{\bar{\psi}\} : \mathrm{Cone}(g)_{ev} \longrightarrow \mathrm{Cone}(f)_{odd}$ and $\overline{\{\psi\}\#\{\phi\}} : \mathrm{Cone}(g \circ f)_{ev} \longrightarrow \mathrm{Cone}(f)_{odd}$ analogously. Then

12.14. $$t(f,\{\phi\}) = t(\mathrm{Cone}(f),\{\bar{\phi}\}) := t(0 \longrightarrow \mathrm{Cone}(f),\{\bar{\phi}\})$$

$$t(g,\{\psi\}) = t(\mathrm{Cone}(g),\{\bar{\psi}\})$$

$$t(g \circ f,\{\psi\}\#\{\phi\}) = t(\mathrm{Cone}(g \circ f),\overline{\{\psi\}\#\{\phi\}})$$

Let $h : \Sigma^{-1}\mathrm{Cone}(g) \longrightarrow \mathrm{Cone}(f)$ be the chain map given by

$$\begin{pmatrix} 0 & 0 \\ -1 & 0 \end{pmatrix} : D_n \oplus E_{n+1} \longrightarrow C_{n-1} \oplus D_n$$

Let $\{\rho\} : \mathrm{Cone}(h)_{ev} \longrightarrow \mathrm{Cone}(h)_{odd}$ be determined by requiring $t(S) = 0$ in $K_1(R\Gamma)/U$ for the diagram S

$$\begin{array}{ccc}
\mathrm{Cone}(h)_{ev} & \xrightarrow{\{\rho\}} & \mathrm{Cone}(h)_{odd} \\
\{\pi\} \downarrow & & \downarrow \{\pi\} \\
\mathrm{Cone}(f)_{ev} \oplus \mathrm{Cone}(g)_{ev} & \xrightarrow{\{\bar{\phi}\} \oplus \{\bar{\psi}\}} & \mathrm{Cone}(f)_{odd} \oplus \mathrm{Cone}(g)_{odd}
\end{array}$$

Consider the commutative diagram with the exact sequence of h as upper row

$$\begin{array}{ccccccccc} 0 & \longrightarrow & \mathrm{Cone}(f) & \longrightarrow & \mathrm{Cone}(h) & \longrightarrow & \mathrm{Cone}(g) & \longrightarrow & 0 \\ & & \uparrow & & \uparrow & & \uparrow & & \\ 0 & \longrightarrow & 0 & \longrightarrow & 0 & \longrightarrow & 0 & \longrightarrow & 0 \end{array}$$

Proposition 12.10. implies

12.15. $$t(\mathrm{Cone}(h),\{\rho\}) = t(\mathrm{Cone}(f),\{\bar{\phi}\}) + t(\mathrm{Cone}(g),\{\bar{\psi}\})$$

We have also the exact sequence

$$0 \longrightarrow \mathrm{Cone}(g\circ f) \xrightarrow{\ i\ } \mathrm{Cone}(h) \longrightarrow \mathrm{Cone}(D) \longrightarrow 0$$

with i given by

$$\begin{pmatrix} f & 0 & 1 & 0 \\ 0 & 1 & 0 & 0 \end{pmatrix}^{tr} : C_{*-1} \oplus D_* \longrightarrow D_{*-1} \oplus E_* \oplus C_{*-1} \oplus D_*$$

Let $\{id\} : \mathrm{Cone}(D)_{ev} \longrightarrow \mathrm{Cone}(D)_{odd}$ be the class of the identity. One verifies directly that $t(\overline{\{\psi\}\#\{\phi\}},\{\rho\},\{id\})$ vanishes. Proposition 12.10. implies

12.16. $$t(\mathrm{Cone}(h),\{\rho\}) = t(\mathrm{Cone}(g\circ f),\overline{\{\psi\}\#\{\phi\}}) + t(\mathrm{Cone}(D),\{id\})$$

Now the claim follows from 12.14., 12.15. and 12.16. since $t(\mathrm{Cone}(D),\{id\})$ vanishes. □

<u>Example 12.17.</u> Let P be a finite projective $R\Gamma$-chain complex and $f : P \longrightarrow P$ a self-equivalence. There is a canonical $\{1\}$-stable class of stable isomorphisms $P_{odd} \oplus P_{ev} \longrightarrow P_{odd} \oplus P_{ev}$, namely $\{id\}$. Define the <u>self-torsion</u> of f

$$t(f) \in K_1(R\Gamma)$$

by $t(f,\{id\})$. We get from Proposition 12.10., 12.12. and 12.13.

a) <u>Additivity</u>

Consider the commutative diagram with exact rows and self-equivalences as vertical arrows

$$\begin{array}{ccccccccc} 0 & \longrightarrow & C^1 & \xrightarrow{i} & C^o & \xrightarrow{p} & C^2 & \longrightarrow & 0 \\ & & \downarrow f^1 & & \downarrow f^o & & \downarrow f^2 & & \\ 0 & \longrightarrow & C^1 & \xrightarrow{i} & C^o & \xrightarrow{p} & C^2 & \longrightarrow & 0 \end{array}$$

Then $t(f^1) - t(f^o) + t(f^2) = 0$.

b) Homotopy and Conjugation invariance

i) $f \simeq f' \Rightarrow t(f) = t(f')$

ii) If $f : C \longrightarrow C$, $g : D \longrightarrow D$ and $h : C \longrightarrow D$ are homotopy equivalences and $h \circ f \simeq g \circ h$, then $t(f) = t(g)$.

c) Logarithmic property

$$t(f \circ g) = t(f) + t(g)$$

12.18. Let $f : C \longrightarrow C$ be a weak homology equivalence of a $R\Gamma$-chain complex possessing a finite approximation. Define

12.19. $$t(f) \in K_1(R\Gamma)$$

by $t(g)$ for any self-equivalence $g : P \longrightarrow P$ of a finite approximation (P,h) of C satisfying $h \circ g \simeq f \circ h$. Because of Lemma 11.3. and Lemma 11.6. this is well defined and still a), b) and c) holds. Analogously to Lemma 11.7. one proves:

Let $f : C \longrightarrow C$ be a weak homology equivalence of a $R\Gamma$-chain complex such that $H_n(C)$ has a finite resolution for $n \geq 0$ and is zero for large n . Then C has a finite approximation and we have in $K_1(R\Gamma)$

$$t(f) = \Sigma(-1)^n[H_n(f)] .$$

If $g : M \longrightarrow M$ is an automorphism of a $R\Gamma$-module possessing a finite resolution let $[g]$ be the torsion of $O(g) : O(M) \longrightarrow O(M)$. ∎

Example 12.20. Let C be a finite projective $R\Gamma$-chain complex. A U-round structure $\{\alpha\}$ on C is a U-stably equivalence class of stable isomorphisms $\{\alpha\} : C_{odd} \rightarrow C_{ev}$.

We call $(C,\{\alpha\})$ a <u>round</u> $R\Gamma$-<u>chain complex.</u> Consider a $R\Gamma$-homotopy equivalence $f : (C,\{\alpha\}) \longrightarrow (D,\{\beta\})$ between round $R\Gamma$-chain complexes. Let $\{\phi\} : C_{odd} \oplus D_{ev} \longrightarrow D_{odd} \oplus C_{ev}$ be the U-stable equivalence class of stable isomorphisms for which the following square has torsion zero

$$\begin{array}{ccc} C_{odd} \oplus D_{ev} & \xrightarrow{\{\phi\}} & D_{odd} \oplus C_{ev} \\ \{\pi\} \downarrow & & \downarrow \{\pi\} \\ C_{odd} \oplus D_{ev} & \xrightarrow[\{\alpha\} \oplus \{\beta^{-1}\}]{} & C_{ev} \oplus D_{odd} \end{array}$$

Define the <u>round torsion</u>

$$t(f : (C,\{\alpha\}) \longrightarrow (D,\{\beta\})) \in K_1(R\Gamma)/U$$

by $t(f,\{\phi\})$. Propositions 12.10., 12.12. and 12.13. imply

a) <u>Additivity</u>

Consider the exact sequence of U-round $R\Gamma$-chain complexes

$$0 \longrightarrow C^1,\{\alpha^1\} \xrightarrow{i} C^o,\{\alpha^o\} \xrightarrow{p} C^2,\{\alpha^2\} \longrightarrow 0 .$$

Choose splittings s_{odd} and s_{ev} of

$$0 \longrightarrow C^1_{odd} \xrightarrow{i_{odd}} C^o_{odd} \underset{s_{odd}}{\overset{p_{odd}}{\rightleftarrows}} C^2_{odd} \longrightarrow 0$$

and

$$0 \longrightarrow C^1_{ev} \xrightarrow{i_{ev}} C^o_{ev} \underset{s_{ev}}{\overset{p_{ev}}{\rightleftarrows}} C^2_{ev} \longrightarrow 0$$

Define $t(\{\alpha^1\},\{\alpha^o\},\{\alpha^2\})$ as the torsion of

$$\begin{array}{ccc} C^o_{ev} & \xleftarrow{\{\alpha^o\}} & C^o_{odd} \\ i_{ev} \oplus s_{ev} \uparrow & & \uparrow i_{odd} \oplus s_{odd} \\ C^1_{ev} \oplus C^2_{ev} & \xleftarrow{\{\alpha^1\} \oplus \{\alpha^2\}} & C^1_{odd} \oplus C^2_{odd} \end{array}$$

Now consider the commutative diagram with exact sequences of round $R\Gamma$-chain complexes as rows and homotopy equivalences as vertical arrows

$$\begin{array}{ccccccccc} 0 & \longrightarrow & (C^1,\{\alpha^1\}) & \longrightarrow & (C^o\{\alpha^o\}) & \longrightarrow & (C^2,\{\alpha^2\}) & \longrightarrow & 0 \\ & & \downarrow f^1 & & \downarrow f^o & & \downarrow f^2 & & \\ 0 & \longrightarrow & (D^1,\{\beta^1\} & \longrightarrow & (D^o,\{\beta^o\}) & \longrightarrow & (D^2,\{\beta^2\}) & \longrightarrow & 0 \end{array}$$

Then we have in $K_1(R\Gamma)/U$

$$t(f^1) - t(f^o) + t(f^2) = t(\{\alpha^1\},\{\alpha^o\},\{\alpha^2\}) - t(\{\beta^1\},\{\beta^o\},\{\beta^2\}) .$$

b) <u>Homotopy invariance</u>

$$f \simeq g \Longrightarrow t(f) = t(g)$$

c) <u>Logarithmic property</u>

$$t(f\circ g) = t(f) + t(g)$$

Notice that a round structure on a finite projective $R\Gamma$-chain complex C exists if and only if $o(C) \in K_o(R\Gamma)$ vanishes. □

<u>Example 12.21.</u> Let $(C,\{\alpha\})$ be a round $R\Gamma$-chain complex. Denote by $\mathfrak{X}(C)$ the trivial $R\Gamma$-chain complex given by the homology of C

$$\cdots \longrightarrow H_{n+2}(C) \xrightarrow{o} H_{n+1}(C) \xrightarrow{o} H_n(C) \xrightarrow{o} H_{n-1}(C) \xrightarrow{o} \cdots$$

Suppose that each $H_n(C)$ is finitely generated projective. Notice that $o(\mathfrak{X}(C))=o(C)$ holds by Proposition 11.9. so that $\mathfrak{X}(C)$ has some round structure. Choose a round structure $\{\beta\} : \mathfrak{X}(C)_{odd} \longrightarrow \mathfrak{X}(C)_{ev}$. Now there is up to homotopy exactly one chain map

$$h(C) : \mathfrak{X}(C) \longrightarrow C$$

such that $H_n(h)$ is the identity for $n \geq 0$. Define the <u>absolute torsion</u>

$$t(C,\{\alpha\},\{\beta\}) \in K_1(R\Gamma)/U$$

by the round torsion of $h(C) : (\mathfrak{X}(C),\{\beta\}) \longrightarrow (C,\{\alpha\})$.

a) <u>Additivity</u>

Let $0 \longrightarrow (C^1,\{\alpha^1\}) \xrightarrow{i} (C^o,\{\alpha^o\}) \xrightarrow{p} (C^2,\{\alpha^2\}) \longrightarrow 0$ be an exact sequence of round $R\Gamma$-chain complexes. Suppose we are given round structures $\{\beta^i\}$ on $\mathcal{H}(C^i)$ for $i = 0,1,2$. We have defined $t(\{\alpha^1\},\{\alpha^o\},\{\alpha^2\})$ in Example 12.20. The long homology sequence

$$\cdots \longrightarrow H_{n+1}(C^2) \xrightarrow{\partial} H_n(C^1) \xrightarrow{i_n} H_n(C^o) \xrightarrow{p_n} H_n(C^2) \longrightarrow \cdots$$

can be viewed as an acyclic finite projective chain complex D. Equip D with the round structure $\{\delta\}$ such that the following square has torsion zero in $K_1(R\Gamma)/U$

$$\begin{array}{ccc}
D_{odd} & \xrightarrow{\quad\{\delta\}\quad} & D_{ev} \\
\downarrow{\scriptstyle\{\pi\}} & & \downarrow{\scriptstyle\{\pi\}} \\
\mathcal{H}(C^1)_{odd} \oplus \mathcal{H}(C^2)_{odd} \oplus \mathcal{H}(C^o)_{ev} & \xrightarrow[\{\beta^1\}\oplus\{\beta^2\}\oplus\{\beta^o\}^{-1}]{} & \mathcal{H}(C^1)_{ev} \oplus \mathcal{H}(C^2)_{ev} \oplus \mathcal{H}(C^o)_{odd}
\end{array}$$

Then we have in $K_1(R\Gamma)/U$ if $t(D,\{\delta\})$ is $t(0 \rightarrow D,\{\delta\})$.

$$t(C^1,\{\alpha^1\},\{\beta^1\}) - t(C^o,\{\alpha^o\},\{\beta^o\}) + t(C^2,\{\alpha^2\},\{\beta^2\}) = t(\{\alpha^1\},\{\alpha^o\},\{\alpha^2\}) \cdot t(D,\{\delta\})$$

We give the proof only under the assumption that the boundary ∂ in the long homology sequence is zero. Then we have the diagram with exact sequences of round chain complexes as rows

$$\begin{array}{ccccccccc}
0 & \longrightarrow & C^1 & \xrightarrow{i} & C^o & \xrightarrow{p} & C^2 & \longrightarrow & 0 \\
 & & \uparrow{\scriptstyle h(C^1)} & & \uparrow{\scriptstyle h(C^o)} & & \uparrow{\scriptstyle h(C^1)} & & \\
0 & \longrightarrow & \mathcal{H}(C^1) & \xrightarrow{\mathcal{H}(i)} & \mathcal{H}(C^o) & \xrightarrow{\mathcal{H}(p)} & \mathcal{H}(C^2) & \longrightarrow & 0
\end{array}$$

We can choose $h(C^1), h(C^o)$ and $h(C^1)$ such that the diagram commutes since $\mathcal{H}(i)$ is a cofibration. Now apply the additivity of the round torsion using that $t(\beta^1,\beta^o,\beta^2) = t(D,\{\delta\})$ holds.

b) <u>Transformation under homotopy equivalences</u>

Let $f : (C,\{\alpha\}) \longrightarrow (C',\{\alpha'\})$ be a homotopy equivalence of round chain complexes.

Suppose that we are given round structures $\{\beta\}$ on $\mathfrak{X}(C)$ and $\{\beta'\}$ on $\mathfrak{X}(C')$ so that $\mathfrak{X}(f) : (\mathfrak{X}(C),\{\beta\}) \longrightarrow (\mathfrak{X}(C'),\{\beta'\})$ is a homotopy equivalence between round chain complexes. The logarithmic property of the round torsion implies

$$t(f) - t(\mathfrak{X}(f)) = t(C',\{\alpha'\},\{\beta'\}) - t(C,\{\alpha\},\{\beta\})$$

If in particular C and C' are acyclic we can compute $t(f)$ by the absolute invariants $t(C',\{\alpha'\})$ and $t(C,\{\alpha\})$. □

Example 12.22. Suppose that Γ is trivial. Consider a finite based free R-chain complex C . Suppose that $H_n(C)$ is stably free and has a stable base for each $n \geq 0$, i.e. we have choosen a base for $H_n(C) \oplus R^k$ for some k . Consider $h(C) : \mathfrak{X}(C) \longrightarrow C$. The bases determine a U-stable equivalence class of stable isomorphisms $\{\phi\} : \mathfrak{X}(C)_{odd} \oplus C_{ev} \longrightarrow C_{odd} \oplus \mathfrak{X}(C)_{ev}$ for $U = \{\pm 1\}$. Then we obtain by Definition 12.4. an element $t(h(C) : \mathfrak{X}(C) \longrightarrow C,\{\phi\}) \in \tilde{K}_1(R)$. We leave it to the reader to verify that this agrees with the torsion defined in Milnor [1966], p. 365. For a complete proof of the additivity of the absolute torsion in this special case see Milnor [1966], p. 367 + 368. □

Now we introduce the Whitehead torsion of a homotopy equivalence between finite based free $R\Gamma$-chain complexes. Some remarks about bases are needed.

Let F be a free $R\Gamma$-module and $(B,\beta : B \longrightarrow \mathrm{Ob}\,\Gamma)$ and $(B',\beta' : B' \longrightarrow \mathrm{Ob}\,\Gamma)$ be two bases (see 9.16.). We call them equivalent if there is a bijection of sets $\psi : B \longrightarrow B'$ such that for any $b \in B$ there is an isomorphism $f : \beta(b) \to \beta'(\psi(b))$ in Γ and a sign $u \in \{\pm 1\}$ with the property that $F(f) : F(\beta(b)) \to F(\beta'(\psi(b)))$ sends b to $u \cdot \psi(b)$. This defines an equivalence relation on the set of bases for F . If we have choosen such an equivalence class $[B]$, we call F a <u>free $R\Gamma$-module with preferred equivalence class of bases</u>. A free $R\Gamma$-chain complex C has <u>a preferred equivalence class of bases</u> if each C_n for $n \geq 0$ has.

Let $(F,[B])$ and $(F',[B'])$ be finitely generated free $R\Gamma$-modules with equivalence classes of bases such that $rk_{R\Gamma}F = rk_{R\Gamma}F'$ holds. Hence we can choose representatives B and B' such that there is an isomorphism $\psi : B \longrightarrow B'$ of $\mathrm{Ob}\,\Gamma$-sets. Let

$\phi : F \longrightarrow F'$ be the isomorphism determined by ψ. Let $U \subset K_1(R\Gamma)$ be the subgroup generated by all trivial units (see section 10.).

One easily checks

Lemma 12.23. The U-stable equivalence classes of stable isomorphism

$$\{\phi\} : F \longrightarrow F'$$

depends only on the equivalence classes $[B]$ and $[B']$. □

Definition 12.24. Let $f : C \longrightarrow D$ be a homotopy equivalence between finite free $R\Gamma$-chain complexes with preferred equivalence classes of bases. By Lemma 12.23. we obtain a U-stable equivalence class of stable isomorphisms $\{\phi\} : C_{odd} \oplus D_{ev} \to D_{odd} \oplus C_{ev}$ Define the Whitehead torsion

$$\tau(f) \in Wh(R\Gamma) = K_1(R\Gamma)/U$$

by the torsion $t(f,\{\phi\})$ (see Definition 12.4.) □

Consider the exact sequence of finite free $R\Gamma$-chain complexes with preferred equivalence classes of bases $[B^1]$, $[B^o]$ and $[B^2]$ $0 \to C^1 \xrightarrow{i} C^o \xrightarrow{p} C^2 \longrightarrow 0$. We call it based exact if there are representatives B^1, B^o and B^2 with $i(B^1) \subset B^o$ and $p(B^o) = B^2$ such that $p|B^o \setminus i(B^1) : B^o \setminus i(B^1) \longrightarrow B^2$ is a bijection of Ob Γ-sets. We get from Proposition 12.10., 12.12. and 12.13. the following Theorem. All chain complexes appearing in it are finite free and have a preferred equivalence class of bases.

Theorem 12.25.

a) Homotopy invariance

Let f and $g : C \longrightarrow D$ be homotopy equivalences. Then $f \simeq g \Longrightarrow \tau(f) = \tau(g)$.

b) Additivity

Consider the diagram with based exact rows and homotopy equivalences as vertical arrows

$$\begin{array}{ccccccccc} 0 & \longrightarrow & C^1 & \xrightarrow{i} & C^o & \xrightarrow{p} & C^2 & \longrightarrow & 0 \\ & & \downarrow f^1 & & \downarrow f^o & & \downarrow f^2 & & \\ 0 & \longrightarrow & D^1 & \xrightarrow{j} & D^o & \xrightarrow{q} & D^2 & \longrightarrow & 0 \end{array}$$

Then $\tau(f^1) - \tau(f^o) + \tau(f^2) = 0$.

c) Logarithmic property

$$\tau(f \circ g) = \tau(f) + \tau(g) \qquad \square$$

Let C be a finite free $R\Gamma$-chain complex with a preferred equivalence class of bases $[B]$. Given $x \in \mathrm{Ob}\,\Gamma$, choose a representative B such that for $b \in B$ $\overline{\beta(b)} = \bar{x}$ in $\mathrm{Is}\,\Gamma$ already implies $\beta(b) = x$ in $\mathrm{Ob}\,\Gamma$. Let $B_x \subset S_x M$ be

$$\{bM(x)_s \in S_x M = M(x)/M(x)_s \mid b \in B, \beta(b) = x\} .$$

Then $S_x[B] := [B_x]$ is a well-defined equivalence class of bases for the finite free $R[x]$-chain complex $S_x C$ depending only on $[B]$. If $f : C \longrightarrow D$ is a homotopy equivalence between finite free $R\Gamma$-chain complexes with preferred equivalence class of bases then $S_x f : S_x C \longrightarrow S_x D$ is a homotopy equivalence between finite free $R[x]$-chain complexes with preferred equivalence class of bases. Hence we can define

12.26. $$\mathrm{Split}\,\tau(f) \in \mathrm{Split}\,\mathrm{Wh}(R\Gamma)$$

by $\{\tau(S_x f) \in \mathrm{Wh}(R[x]) \mid \bar{x} \in \mathrm{Is}\,\Gamma\}$. Recall the isomorphism $S : \mathrm{Wh}(R\Gamma) \to \mathrm{Split}\,\mathrm{Wh}(R\Gamma)$ (Theorem 10.34.). We get from Theorem 9.39. and Theorem 12.25. (analogously to Proposition 11.15.)

Proposition 12.27. Let $f : C \longrightarrow D$ be a homotopy equivalence between finite free $R\Gamma$-chain complexes with preferred equivalence class of bases . Then

$$S(\tau(f)) = \mathrm{Split}\,\tau(f) \qquad \square$$

Consider a functor $F : \Gamma_1 \longrightarrow \Gamma_2$ and a finitely generated free $R\Gamma_1$-module M with a preferred equivalence class of bases $[B]$. Then $F_* M = \mathrm{ind}_F M$ is a finitely generated free $R\Gamma_2$-module with a preferred equivalence class of bases $F_*[B]$ defined as follows. Recall that F_*M is a quotient of $\bigoplus_{x \in \mathrm{Ob}\,\Gamma_1} M(x) \otimes_R R\,\mathrm{Hom}(?,Fx)$. If $B = (B, \beta : B \longrightarrow \mathrm{Ob}\,\Gamma_1)$ is a representative for $[B]$, a representative for F_*B is given by $\{b \otimes_R \mathrm{id}_{F(\beta(b))} \in F_*M(F(\beta(b)))\}$ and the obvious map $\gamma : b \otimes_R \mathrm{id}_{F(\beta(b))} \longmapsto F(\beta(b)) \in \mathrm{Ob}\,\Gamma_2$. Hence $F_*f : F_*C \longrightarrow F_*D$ is a homotopy equivalence between finite free chain complexes with preferred equivalence classes of bases if f is. We have defined $F_* : Wh(R\Gamma_1) \longrightarrow Wh(R\Gamma_2)$ in 10.10. We get from the definitions

12.28. $$F_*(\tau(f)) = \tau(F_*f)$$

Consider an admissible functor $F : \Gamma_1 \longrightarrow \Gamma_2$. Given a finitely generated free $R\Gamma_2$-module M, its restriction F^*M is finitely generated free by definition of "admissible". If $[B]$ is a preferred equivalence class of bases for M, we want to define a preferred equivalence class of bases $F^*[B]$ for F^*M. It suffices to treat the special case $M = R\,\mathrm{Hom}(?,y)$ with the $\mathrm{Ob}\,\Gamma_2$-base $B = \{\mathrm{id}_y \in R\,\mathrm{Hom}(y,y)\}$. Since F is admissible $\mathrm{Irr}(x,y)$ is a free $\mathrm{Aut}(x)$-set with finitely many orbits for $x \in \mathrm{Ob}\,\Gamma_1$ by Proposition 10.16. Choose

$$C_x = \{f_{x,i} \mid i=1,2,\ldots,\mathrm{card}\,\mathrm{Irr}(x,y)/\mathrm{Aut}(x)\} \subset \mathrm{Irr}(x,y)$$

such that each $\mathrm{Aut}(x)$-orbit contains exactly one $f_{x,i}$. If we fix for any $\bar{x} \in \mathrm{Is}\,\Gamma$ a representative x, consider the $\mathrm{Ob}\,\Gamma_1$-set $C = \coprod_{\bar{x} \in \mathrm{Is}\,\Gamma} C_x$. It is a base for $F^*\mathrm{Hom}(?,y)$. Define $F^*[B]$ to be $[C]$. The verification that there is a well-defined homomorphism $F^* : Wh(R\Gamma_2) \longrightarrow Wh(R\Gamma_1)$ (see 10.20.) also shows that $F^*[B]$ is well-defined. We get for a homotopy equivalence $f : C \longrightarrow D$ of finite free $R\Gamma_2$-chain complexes with preferred equivalence classes of bases that $F^*f : F^*C \longrightarrow F^*D$ is a homotopy equivalence of finite free $R\Gamma_1$-chain complexes with preferred equivalence classes of bases and

12.29. $$F^*(\tau(f)) = \tau(F^*f)$$

If $f_i : C^i \longrightarrow D^i$ is a homotopy equivalence of finite free $R\Gamma_i$-chain complexes with preferred equivalence classes of bases for $i = 1,2$ then $f_1 \otimes_R f_2 : C^1 \otimes_R C^2 \longrightarrow D^1 \otimes_R D^2$ is a $R\Gamma_i$-homotopy equivalence of finite free $R\Gamma_1 \times \Gamma_2$-chain complexes with preferred equivalence classes of bases. We get from 10.23. and Proposition 10.42. pairings $U(\Gamma_1) \times Wh(R\Gamma_2) \longrightarrow Wh(R\Gamma_1 \times \Gamma_2)$ and $Wh(R\Gamma_1) \times U(\Gamma_2) \rightarrow Wh(R\Gamma_1 \times \Gamma_2)$

12.30. $$\tau_{R\Gamma_1 \times \Gamma_2}(f_1 \otimes_R f_2) = \chi_{R\Gamma_1}(D^1) \otimes \tau_{R\Gamma_2}(f_2) + \tau_{R\Gamma_1}(f_1) \otimes \chi_{R\Gamma_2}(D^2) .$$

The proof of 12.30. is reduced by Theorem 12.25. c using $\tau(f_1 \otimes_R f_2) = \tau(id \otimes f_2) + \tau(f_1 \otimes id)$ to the case where $f_1 = id$ holds. Because of Theorem 12.25. b it suffices to treat $f_1 = id : R\,Hom(?,x) \longrightarrow R\,Hom(?,x)$ for $x \in Ob\,\Gamma_1$. Now the claim 12.30. follows from the definitions.

We call a $R\Gamma$-chain map $i : C \longrightarrow D$ of finite free $R\Gamma$-chain complexes with preferred equivalence class of bases an <u>elementary expansion</u> if there is a finitely generated free $R\Gamma$-module F with preferred equivalence class of bases and a based exact sequence $0 \longrightarrow C \xrightarrow{i} D \longrightarrow el_n(F) \longrightarrow 0$. Any homotopy inverse $r : D \longrightarrow C$ of i is an <u>elementary collapse</u>. A composition of elementary expansions and collapses is a <u>formal deformation</u>. An $R\Gamma$-chain map homotopic to a formal deformation is called a <u>simple homotopy equivalence</u>.

If $f : C \longrightarrow D$ is a simple homotopy equivalence $\tau(f)$ vanishes by Theorem 12.25. The proof of the converse is left as an exercise to the reader.

<u>Comments 12.31.</u> For finite based free R-chain complexes the Whitehead torsion is treated in Milnor [1966] and Cohen [1973] and the self-torsion in Gersten [1967]. One can easily extend our notion of torsion except the homological computations to additive categories. Notice that the notion of a stable map makes sense in any additive category whereas the notion of a base does not carry over directly. Torsion for additive categories is dealt with in Ranicki [1985 a], [1987] but using a substitute for the notion of a bases such that problems explained as in Example 12.1. occur . We have already mentioned that the approach using stable maps has the advantage that one can also work with finite projective R-chain complexes (see Example 12.21.). This will be crucial for defining Reidemeister torsion in section 18.

Let $\mathcal{C}$ be the category of finite projective $R\Gamma$-chain complexes with cofibrations and weak equivalences. The inclusion FPMOD-$R\Gamma \longrightarrow \mathcal{C}$ induces isomorphisms

$K_n(R\Gamma) \longrightarrow K_n(\mathbb{C})$ for $n \geq 0$. The finiteness obstruction and the self-torsion define explicite inverse maps for $n = 0$ and 1. □

Exercises 12.32.

1) Let $f : R \longrightarrow S$ be a ring homomorphism such that the R-module f^*S has a finite resolution. Show that we get a well-defined transfer homomorphism for $n = 0,1$

$$f^* : K_n(S) \longrightarrow K_n(R)$$

by sending $\eta \in K_o(S)$ represented by the image of the projection $p : S^n \to S^n$ to $o(f^*p : f^*S^n \longrightarrow f^*S^n)$ resp. $\eta \in K_1(S)$ represented by the automorphism $g : S^n \longrightarrow S^n$ to $t(f^*g : f^*S^n \longrightarrow f^*S^n)$.

2) Let C be a finite free $\mathbb{Z}$-chain complex. Choose a $\mathbb{Z}$-base for C and for $\mathcal{H}(C)/\text{Tors}\,\mathcal{H}(C)$. This induces $\mathbb{Q}$-bases for $C \otimes_{\mathbb{Z}} \mathbb{Q}$ and $\mathcal{H}(C \otimes_{\mathbb{Z}} \mathbb{Q}) = \mathcal{H}(C) \otimes_{\mathbb{Z}} \mathbb{Q} = \mathcal{H}(C)/\text{Tors}\,\mathcal{H}(C) \otimes_{\mathbb{Z}} \mathbb{Q}$. Show that

$$m\chi'(C) \in \tilde{K}_1(\mathbb{Q}) = \mathbb{Q}^*/\mathbb{Z}^*$$

given by $t(h(C \otimes_{\mathbb{Z}} \mathbb{Q}) : \mathcal{H}(C \otimes_{\mathbb{Z}} \mathbb{Q}) \longrightarrow C \otimes_{\mathbb{Z}} \mathbb{Q})$ is independent of the choices of the $\mathbb{Z}$-bases. Define the multiplicative Euler characteristic

$$m\chi(C) \in \mathbb{Q}^*/\mathbb{Z}^* \cong \{r \in \mathbb{Q} \mid r > 0\}$$

by $m\chi(C) = \prod_{n=o}^{\infty} |\text{Tors}\, H_n(C)|^{(-1)^n}$. Prove $m\chi'(c) = m\chi(C)$.

3) Let $f : C \longrightarrow D$ be a chain map between $\mathbb{Z}$-chain complexes such that $H_n(C)$ resp $H_n(D)$ is finitely generated for $n \geq 0$ and zero for large n. Provided that $H_n(f)$ is finite for $n \geq 0$ define

$$m\chi(f) = \prod_{n=o}^{\infty} |H_n(f)|^{(-1)^n} \in \mathbb{Q}^*/\mathbb{Z}^*$$

Prove.

i) $f \simeq g \Longrightarrow m\chi(f) = m\chi(g)$

ii) Consider the commutative diagram with exact rows

$$\begin{array}{ccccccccc}
0 & \longrightarrow & C^1 & \longrightarrow & C^o & \longrightarrow & C^2 & \longrightarrow & 0 \\
 & & \downarrow f^1 & & \downarrow f^o & & \downarrow f^2 & & \\
0 & \longrightarrow & D^1 & \longrightarrow & D^o & \longrightarrow & D^2 & \longrightarrow & 0
\end{array}$$

Then $m\chi(f_1) \cdot m\chi(f^o)^{-1} \cdot m\chi(f^2) = 1$.

iii) $m\chi(f \circ g) = m\chi(f) \cdot m\chi(g)$

4) Let $f : C \longrightarrow D$ be a homotopy equivalence between finite free $R\Gamma$-chain complexes with preferred equivalence classes of bases. Prove that f is simple if and only if $\tau(f)$ vanishes.

5) Let $f_1 : C^1 \longrightarrow D^1$ and $f_2 : C^2 \longrightarrow D^2$ be homotopy equivalences between finite free $R\Gamma_1$- and $R\Gamma_2$-chain complexes with preferred equivalence classes of bases. Show that the $R\Gamma_1 \times \Gamma_2$-homotopy equivalence $f_1 \otimes_R f_2$ is simple if $\chi_{R\Gamma_1}(C^1)$ and $\chi_{R\Gamma_2}(C^2)$ vanish.

6) Let $f : C \longrightarrow D$ be a $\mathbf{Z}Or/T^n$-homotopy equivalence between finite free $\mathbf{Z}Or/T^n$-chain complexes with preferred equivalence classes of bases. Show $\tau(f) = 0$.

7) Let $f : C \longrightarrow D$ be an isomorphism of finite free $R\Gamma$-chain complexes with preferred equivalence classes of bases. Show

$$\tau(f) = \sum_{n=o}^{\infty} (-1)^n \quad \tau(f_n : C_n \longrightarrow D_n)$$

8) Show that the function assigning to a homotopy equivalence $f : C \longrightarrow D$ between finite free $R\Gamma$-chain complexes with preferred equivalence classes of bases $(\chi(D);\tau(f)) \in U(\Gamma) \oplus Wh(R\Gamma)$ is universal for all functions $f \longrightarrow a(f) \in A$ into some abelian group A satisfying

i) $f \simeq g \Longrightarrow a(f) = a(g)$.

ii) The following diagram commutes and has based exact sequences rows

$$\begin{array}{ccccccccc} 0 & \longrightarrow & C^1 & \longrightarrow & C^o & \longrightarrow & C^2 & \longrightarrow & 0 \\ & & \downarrow{\scriptstyle f^1} & & \downarrow{\scriptstyle f^o} & & \downarrow{\scriptstyle f^2} & & \\ 0 & \longrightarrow & D^1 & \longrightarrow & D^o & \longrightarrow & D^2 & \longrightarrow & 0 \end{array}$$

Then $a(f^1) - a(f^o) + a(f^2) = 0$.

iii) If F is a finitely generated free $R\Gamma$-module with preferred equivalence classes of bases then $a(0 \longrightarrow e\ell_n(F)) = a(e\ell_n(F) \longrightarrow 0)$ is zero. □

.3. The cellular chain complex

n this section we treat the cellular $R\Pi/(G,X)$-chain complex $C^c(X)$ of a G-CW-omplex. We equip $C^c(X)$ with a cellular equivalence class of bases. We introduce ingular and cellular $R\Pi(G,X)$-homology and compare them. We define a $Z\Pi/(G,X)$-Hure-icz map and prove an Equivariant Hurewicz Theorem. It is used for the verification f the Realization Theorem of a $\mathbf{Z}\Pi/(G,X)$-chain complex as a cellular $\mathbf{Z}\Pi/(G,X)$-chain omplex. Finally we study the behaviour of the cellular $R\Pi(G,X)$-chain complex under estriction and tensor product. In this section G is always a Lie group and all G-paces are supposed to satisfy assumption 8.13.

e have defined the discrete fundamental category $\Pi/(G,X)$ and the discrete universa overing functor $\tilde{X}/:\Pi/(G,X) \longrightarrow \{top.sp.\}$ in Definition 8.28. and 8.30. Let ^{s} : $\{top.sp.\} \longrightarrow \{R\text{-ch.compl.}\}$ be the functor singular chain complex.

efinition 13.1. <u>The singular</u> $R\Pi/(G,X)$-<u>chain complex</u> $C^s(X)$ <u>is the composition</u>

$$C^s : \Pi/(G,X) \xrightarrow{\tilde{X}/} \{top.sp.\} \xrightarrow{C^s} \{R-ch.compl.\} \qquad \square$$

f X is a G-CW-complex we obtain by Theorem 8.33. a functor $\tilde{X}/ : \Pi/(G,X) \longrightarrow$ CW-compl.} . In Definition 8.37. we have introduced the cellular $R\Pi/(G,X)$-chain omplex $C^c(X)$ by the composition

$$C^c(X) : \Pi/(G,X) \xrightarrow{\tilde{X}/} \{CW\text{-compl.}\} \xrightarrow{C^c} \{R\text{-ch.compl.}\}$$

here C^c is the functor "cellular chain complex with R-coefficients".

here is of course a relative version of $C^s(X,A)$ and $C^c(X,A)$. If (X,A) is a air of G-spaces resp. a relative G-CW-complex we obtain a functor

$$\widetilde{(X,A)}/ : \Pi/(G,X) \longrightarrow \{\text{pairs of sp.}\} \text{ resp. } \{\text{rel. CW-compl.}\}$$

y sending $x : G/H \longrightarrow X$ to the pair $(\tilde{X}(x), p(x)^{-1}(X(x) \cap A))/Aut(x)_o$ using no -ation 8.32. Recall that $X(x)$ is the component of X^H containing x with univer-al covering $p(x) : \tilde{X}(x) \longrightarrow X(x)$ and $Aut(x)$ is the automorphism group of x in (G,X) operating on $\tilde{X}(x)$. Now define $C^s(X,A)$ and $C^c(X,A)$ by the composition f $\widetilde{(X,A)}/$ with C^s and C^c .

Sometimes it suffices to consider everything over the discrete orbit category $Or/(G)$. Then the singular $ROr/(G)$-chain complex $C^s(X)$ is the composition of C^s with

$$X/ : Or/G \longrightarrow \{\text{top.sp.}\} \qquad G/H \longrightarrow X^H/WH_o$$

and $C^c(X)$ is defined analogously (see Definition 8.37.).

Next we deal with the cellular base of $C^c(X,A)$ for a relative proper G-CW-complex (X,A). If $\{e_i^n \mid i \in I_n\}$ is the set of n-cells, we have defined in Example 9.18 a $Ob\Pi/(G,X)$ subset $\langle I_n \rangle \subset C_n^c(X,A)$. This subset depends on various choices like the one of a characteristic map. However, we have

<u>Lemma 13.2.</u>

a) <u>We obtain by</u> $\langle I_n \rangle$ <u>an equivalence class of bases</u> (see section 12) <u>for</u> $C_n^c(X,A)$ <u>depending only on the</u> G-CW-<u>complex structure</u>.

b) <u>The singular</u> $R\Pi/(G,X)$-<u>chain complex</u> $C^s(X,A)$ <u>is free</u>.

<u>Proof</u>. Suppose that X_n is the G-push out

$$\begin{array}{ccc} \coprod_{i \in I_n} G/H_i \times S^{n-1} & \xrightarrow{\coprod_{i \in I_n} q_i} & X_{n-1} \\ \downarrow & & \downarrow \\ \coprod_{i \in I_n} G/H_i \times D^n & \xrightarrow{\coprod_{i \in I_n} Q_i} & X_n \end{array}$$

Let $y_i : G/H_i \longrightarrow X_{n-1}$ be $Q_i | G/H_i \times \{*\}$. By 8.23. we obtain a push out for $x : G/H \longrightarrow X$

$$\begin{array}{ccc} \coprod_{i \in I_n} Mor(x,x_i) \times S^{n-1} & \longrightarrow & \tilde{X}(x)_{n-1} \\ \downarrow & & \downarrow \\ \coprod_{i \in I_n} Mor(x,x_i) \times D^n & \longrightarrow & \tilde{X}(x)_n \end{array}$$

Dividing out the $\mathrm{Aut}(x)_o$ action yields a push-out

$$\begin{array}{ccc} \coprod_{i \in I_n} \mathrm{Mor}/(x,x_i) \times S^{n-1} & \longrightarrow & \tilde{X}/(x)_{n-1} \\ \downarrow & & \downarrow \\ \coprod_{i \in I_n} \mathrm{Mor}/(x,x_i) \times D^n & \longrightarrow & \tilde{X}/(x)_n \end{array}$$

Since $\mathrm{Mor}/(x,x_i)$ is discrete we obtain $C_n^c(X)(x) = H_n(\tilde{X}/(x)_n, \tilde{X}/(x)_{n-1}) = \bigoplus_{i \in I_n} R\,\mathrm{Hom}(x,x_i)$ so that $C_n^c(X) = \bigoplus R\,\mathrm{Hom}(?,x_i)$ is free.

In section 12 we have assigned to $\langle I_n \rangle$ an equivalence class of $R[x]$-bases $\langle I_n \rangle_x$ for $S_x C_n^c(X,A)$. Hence we obtain an equivalence class of $R[x]$-bases for $C_n^c((\tilde{X}(x), \tilde{X}^>(x) \cup p(x)^{-1}(X(x) \cap A))/\mathrm{Aut}(x)_o)$ under the natural isomorphism of Lemma 9.32. This is precisely the one constructed in Cohen [1973] 19.1 where it is shown that it depends only on the relative $\mathrm{Aut}/(x)$-structure on $(\tilde{X}(x), \tilde{X}^>(x) \cup p(x)^{-1}(X(x) \cap A))/\mathrm{Aut}(x)_o)$. Since $\langle I_n \rangle_x$ depends only on the G-CW-complex structure of (X,A), the same is true for $\langle I_n \rangle$. □

b) For simplicity we show b) only for the singular ROr/G-chain complex $C_n^s(X)$. Consider $H \subset G$ and a singular simplex $s : \Delta_n \longrightarrow X^H/WH_o$ in $C_n^s(X)(G/H)$. Let $K = K(s,H)$ be $\cap \{G_y \mid y \in X^H,\ y \cdot WH_o \in \text{image } s\}$. We have $H \subset K$ and the projection $\sigma = \sigma(s,H) : G/H \longrightarrow G/K$. Since $WH_o \cdot K \cdot WH_o^{-1} = K$ is true there is for any $\tau \in WH_o$ given by $\tau : G/H \longrightarrow G/H$ an element $\tau' \in WK_o$ given by $\tau' : G/K \to G/K$ such that $\sigma \circ \tau = \tau' \circ \sigma$ holds. Hence $\sigma^* : X^K/WK_o \longrightarrow X^H/WH_o$ is injective. As image $s \subset$ image σ^* holds by construction, there is a singular simplex $\hat{s} : \hat{s}(H,s) : \Delta_n \to X^K/WK_o$ satisfying $\sigma^* \circ \hat{s} = s$. Now define $C_n^s(X)(G/H)_r$ as the free R-submodule of $C_n^s(X)(G/H)$ generated by all those singular n-simplices s with $K(s,H) = H$. Recall that $C_n^s(X)(G/H)_s$ is the R-submodule generated by all images $C_n^s(X)(f)$ where f runs over all non-isomorphisms in Or/G with G/H as source. Then we have a decomposition of $R\pi_o(WH)$-modules $C_n^s(X)(G/H) = C_n^s(X)(G/H)_r \oplus C_n^s(X)(G/H)_s$.
Now one easily checks that the natural map $\bigoplus_{(H)} E_{G/H} C_n^s(X)(G/H)_r \longrightarrow C^s(X)$ is an isomorphism. □

<u>Definition 13.3.</u> <u>We call the equivalence class of bases given by</u> $\langle I_n \rangle$ <u>for</u> $n \geq 0$ <u>the cellular equivalence class of bases for</u> $C^c(X,A)$. □

Let $f : (X,A) \longrightarrow (Y,B)$ be a cellular G-map. It induces a functor

$$\Pi/(G,f) : \Pi/(G,X) \longrightarrow \Pi/(G,Y)$$

by composition. Moreover, we obtain a natural transformation $\tilde{f}/ : \tilde{X}/ \longrightarrow \tilde{Y}/ \circ \Pi/(G,f)$. If f_* and f^* denote induction and restriction with $\Pi/(G,f)$, we obtain chain maps (use 9.22.)

$$13.4. \qquad C^c(f) : C^c(X,A) \longrightarrow f^* C^c(Y,B)$$

$$C^c(f) : f_* C^c(X,A) \longrightarrow C^c(Y,B)$$

Let $h : (X,A) \times I \longrightarrow (Y,B)$ be a cellular G-homotopy between f and g . It induces a natural equivalence $\Pi/(G,h) : \Pi/(G,f) \longrightarrow \Pi/(G,g)$. By composition we get a natural equivalence $\hat{h}/ : \tilde{Y}/ \circ \Pi/(G,f) \longrightarrow \tilde{Y}/ \circ \Pi/(G,g)$ and thus a $R\Pi/(G,X)$-isomorphism $C^c(\hat{h}/) : f^* c^c(Y,B) \longrightarrow g^* C^c(Y,B)$. Moreover, we obtain a natural transformation $\tilde{h}/ : \tilde{X}/ \times I \longrightarrow \tilde{Y}/ \circ \Pi/(G,g)$ with $\tilde{h}_o = \hat{h}/ \circ \tilde{f}/$ and $\tilde{h}_1 = \tilde{g}/$. This induces a $\mathbf{Z}\Pi/(G,X)$-chain homotopy

$$13.5. \qquad C^c(h) : C^c(\hat{h}/) \circ C^c(f) \simeq C^c(g)$$

Analogously we obtain a base preserving $R\Pi/(G,X)$-isomorphism $C^c(\hat{h}/) : f_* C^c(X,A) \longrightarrow g_* C^c(X,A)$ and a $R\Pi/(G,Y)$-chain homotopy

$$13.6. \qquad C^c(h) : C^c(g) \circ C^c(\hat{h}/) \simeq C^c(f)$$

Consider the cellular G-push out with i_2 a G-cofibration (see 4.1.)

$$\begin{array}{ccc} X_o & \xrightarrow{\ i_1\ } & X_1 \\ {\scriptstyle i_2}\downarrow & \searrow^{j_o} & \downarrow{\scriptstyle j_1} \\ X_2 & \xrightarrow{\ j_2\ } & X \end{array}$$

<u>Lemma 13.7.</u> <u>There is a based exact sequence of</u> $R\Pi/(G,X)$-<u>chain complexes</u>

$$0 \longrightarrow j_{o*}C^c(X_o) \xrightarrow{j_{1*}C^c(i_1) \oplus j_{2*}C^c(i_2)} j_{1*}C^c(X_1) \oplus j_{2*}C^c(X_2) \xrightarrow{C^c(j_1)-C^c(j_2)} C^c(X) \longrightarrow 0$$

□

Lemma 13.8. Let $i : A \longrightarrow X$ be the inclusion of a pair of G-CW-complexes Then we have the based exact sequence of $\mathbf{Z}\Pi/(G,X)$-chain complexes

$$0 \longrightarrow i_*C^c(A) \xrightarrow{C^c(i)} C^c(X) \longrightarrow C^c(X,A) \longrightarrow 0$$

□

The proof of Lemmata 13.7. and 13.8. is easily reduced to the well known non-equivariant case by evaluating at any $x : G/H \longrightarrow X \in \mathrm{Ob}\ \Pi/(G,X)$.

Definition 13.9. Let the singular resp. cellular $R\Pi/(G,X)$-homology $H^s(X)$ resp. $H^c(X)$ be the homology of the singular resp. cellular $R\Pi/(G,X)$-chain complex $C^s(X)$ resp. $C^c(X)$. □

Consider the functors singular and cellular chain complex C^s and C^c: {CW-compl.} $\longrightarrow$ {R-ch.compl.} . In order to compare them in a natural way we introduce an intermediate functor C^{in} : {CW-compl.} $\longrightarrow$ {R-ch.compl.} . It assigns to a CW-complex the R-subcomplex $C^{in}(X) \subset C^s(X)$ with $C^{in}_n(X)$ the kernel of

$$C^s_n(X_n) \xrightarrow{c^s_n} C^s_{n-1}(X_n) \longrightarrow C^s_{n-1}(X_n,X_{n-1}) := C^s_{n-1}(X_n)/C^s_{n-1}(X_{n-1}).$$

Let $i(X) : C^{in}(X) \longrightarrow C^s(X)$ be the inclusion. We show that $i(X)$ is a weak homology equivalence by verifying that $C^s(X)/C^{in}(X)$ is acyclic.

An n-cycle u in $C^s(X)/C^{in}(X)$ is given by $u \in C^s_n(X)$ satisfying $c^s_n(u) \in C^{in}_{n-1}(X)$. The class of u in $C^s_n(X,X_n)$ is a cycle and hence a boundary as $H^s_n(X,X_n)$ vanishes. Therefore we can find $v \in C^s_{n+1}(X)$ and $u' \in C^s_n(X_n)$ with $c^s_{n+1}(v) = u-u'$. Since $c^s_n(u') = c^s_n(u)$ lies in $C^{in}_{n-1}(X) \subset C_{n-1}(X_{n-1})$ we have $u' \in C^{in}_n(X)$. Hence the cycle u in $C^s_n(X)/C^{in}_n(X)$ is a boundary.

The canonical map $C^s_n(X_n) \longrightarrow C^s_n(X_n,X_{n-1})$ induces an epimorphism of R-chain complexes $p(X) : C^{in}(X) \longrightarrow C^c(X)$. We show that $\ker p(X)$ is acyclic so that $p(X)$ is a weak homology equivalence. A cycle $u \in \ker p(X)$ can be written as $u = u_1 + u_2$ for $u_1 \in \ker c^s_n : C^s_n(X_{n-1}) \longrightarrow C^s_{n-1}(X_{n-1})$ and $u_2 \in \mathrm{image}\ c^s_{n+1}: C^s_{n+1}(X_n) \to C^s_n(X_n)$.

Since $H_n(X_{n-1})$ vanishes, u_1 is contained in the image of $c^s_{n+1}: C^s_{n+1}(X_{n-1}) \longrightarrow C^s_n(X_{n-1})$. Therefore u lies in the image of $c^s_{n+1} : C_{n+1}(X_n) \longrightarrow C_n(X_n)$ and is a boundary in $\ker p(X)$.

Now we have natural weak homology equivalences $C^s \xleftarrow{i} C^{in} \xrightarrow{p} C^c$ between functors {CW-compl.} $\longrightarrow$ {R-ch.compl.} . Composing with the discrete universal covering functor yields

<u>Proposition 13.10.</u> <u>Let</u> X <u>be a</u> G-<u>CW-complex</u>.

a) <u>There are in</u> X <u>natural weak</u> $R\Pi/(G,X)$-<u>homology equivalence</u>

$$C^s(X) \xleftarrow{i} C^{in}(X) \xrightarrow{p} C^c(X)$$

b) <u>There are homotopy inverse and up to homotopy natural</u> $R\Pi/(G,X)$-<u>homotopy equivalences</u>

$$C^s(X) \longleftrightarrow C^c(X)$$

c) <u>The singular and cellular homology</u> $R\Pi/(G,X)$-<u>modules</u> $H^s(X)$ <u>and</u> $H^c(X)$ <u>are naturally isomorphic</u>. □

Because of Proposition 13.10. we often write $H(X)$ instead of $H^s(X)$ or $H^c(X)$. Given a G-pair (X,A) , define a contravariant functor for $n \geq 2$

13.11. $$\pi_n(X,A) : \Pi/(G,A) \longrightarrow \{\text{groups}\}$$

as follows. It sends $x : G/H \longrightarrow A$ to $\pi_n(X^H,X^H \cap A,x) = \pi_n(X^H(x),X^H(x) \cap A,x)$. Let $(\sigma,w) : x(H) \longrightarrow y(K)$ be a morphism in the fundamental category $\Pi(G,A)$ representing a morphism $[\sigma,w] : x(H) \longrightarrow y(K)$ in the discrete fundamental category $\Pi/(G,A)$. Then $\sigma : G/H \longrightarrow G/K$ induces a map $\sigma^* : (X^K,X^K \cap A,y) \to (X^H,X^H \cap A,\sigma^*y)$ and a homomorphism $\pi_n(\sigma^*) : \pi_n(X^K,X^K \cap A,y) \longrightarrow \pi_n(X^H,X^H \cap A,\sigma^*,y)$. Recall that w is a homotopy class of paths from σ^*y to x in A^H . Let

$$t_w : \pi_n(X^H,X^H \cap A,\sigma^*y) \longrightarrow \pi_n(X^H,X^H \cap A,x)$$

be the homomorphism defined by homotopies along w (see Whitehead [1978], p. 257). Define $\pi_n(X,A)([\sigma,w])$ by the composition $t_w \circ \pi_n(\sigma^*) : \pi_n(X^K,X^K \cap A,y) \longrightarrow$

$\pi_n(X^H, X^H \cap A, x)$. We leave it to the reader to check that the choice of $(\sigma,w) \in [\sigma,w]$ does not matter. Analogously we get for $n \geq 1$

13.12. $$\pi_n(X) : \Pi/(G,X) \longrightarrow \{\text{groups}\} .$$

If $n \geq 3$ in 13.11. we obtain a $\mathbf{Z}\Pi/(G,A)$-module since $\pi_n(X^H, X^H \cap A, x)$ is abelian for $n \geq 3$. Similarly we have a $\mathbf{Z}\Pi/(G,X)$-module in 13.12. for $n \geq 2$.

Consider the inclusion $i : A \longrightarrow X$ of G-spaces. For each $x : G/H \to A \in \mathrm{Ob}\,\Pi/(G,A)$ we obtain an homomorphism for $n \geq 2$

$$h(X,A)(x) : \pi_n(X^H, X^H \cap A, x) \longrightarrow i^*H_n(X,A)(x)$$

by the composition

$$\pi_n(X(x), X(x) \cap A, x) \xleftarrow[\cong]{p(x)_*} \pi_n(\tilde{X}(x), p(x)^{-1}(X(x) \cap A), x)$$
$$\xrightarrow{o(x)_*} \Pi_n((\bar{X}(x), p(x)^{-1}(X(x) \cap A))/\mathrm{Aut}(x)_o, x/)$$
$$\xrightarrow{h} H_n^s((\tilde{X}(x), p(x)^{-1}(X(x) \cap A))/\mathrm{Aut}(x)_o) ,$$

where h is the ordinary Hurewicz map (see Whitehead [1978] IV 4.1.). If $\pi_n(X,A)$ is a $\mathbf{Z}\Pi(G,A)$-module, this induces a $\mathbf{Z}\Pi/(G,A)$-Hurewicz-homomorphism

13.13. $$h(X,A) : \pi_n(X,A) \longrightarrow i^*H_n(X,A)$$

Analogously we get a $\mathbf{Z}\Pi/(G,X)$-Hurewicz-homomorphism

13.14. $$h(X) : \pi_n(X) \longrightarrow H_n(X)$$

We call a map $f : X \longrightarrow Y$ a Π-<u>isomorphism</u> if $\pi_o(f^H) : \pi_o(X^H) \to \pi_o(Y^H)$ and $\pi_1(f^H, x) : \pi_1(X^H, x) \longrightarrow \pi_1(Y^H, f^H x)$ are bijections for all $H \subset G$ and $x \in X^H$. Recall that f is (G,n)-connected if f^H is n-connected for $H \subset G$. A pair (X,A) is Π-<u>isomorphic</u> resp. (G,n)-connected if the inclusion $i : A \longrightarrow X$ is. Notice that $\Pi/(G,i) : \Pi/(G,A) \longrightarrow \Pi/(G,X)$ is an equivalence of categories (Lemma 8.35.) and in particular admissible if (X,A) is Π-isomorphic.

Theorem 13.15. The Equivariant Hurewicz Theorem .

Let (X,A) be a proper relative G-CW-complex. Suppose that (X,A) is Π-isomorphic and $(G,n-1)$-connected for $n \geq 2$. Then the $\mathbf{Z}\Pi/(G,A)$ Hurewicz-homomorphism

$$h(X,A) : \pi_n(X,A) \longrightarrow H_n(X,A)$$

is an isomorphism.

Proof. As (X,A) is $(G,n-1)$-connected, we can assume that X is obtained from A by attaching cells of dimension $\geq n$ (Proposition 2.3.). Consider $x : G/H \to A$. We abbreviate $Y = \tilde{X}(x)$, $B = p(x)^{-1}(X(x) \cap A)$, $K = \mathrm{Aut}(x)$ and surpress base points in the sequel. If p denotes the projection and h the non-equivariant Hurewicz-map the following diagram commutes

$$\begin{array}{ccc} \Pi_n(Y,B) & \xrightarrow{\ h\ } & H_n(Y,B) \\ \big\downarrow p_* & & \big\downarrow p_* \\ \Pi_n(Y/K_o,B/K_o) & \xrightarrow{\ h\ } & H_n(Y/K_o,B/K_o) \end{array}$$

It suffices to show that all maps above are bijections. We have $Y_{n-1} = Y_{-1} = B$ for the relative K-CW-complex (Y,B) . Hence (Y,B) and $(Y/K_o,B/K_o)$ are $(n-1)$-connected. By Proposition 8.33. Y,B , Y/K_o and B/K_o are simply connected so that both Hurewicz-maps h are bijective by the ordinary Hurewicz-theorem (see Whitehead [1978], IV, 7.9.). It remains to show that $p_* : H_n(Y,B) \longrightarrow H_n(Y/K_o,B/K_o)$ is an isomorphism.

Let K_o be the component of the identity of K . Consider (Y,B) as a relative K_o-CW -complex. We have $H_k(Y_m,Y_{m-1}) \cong \bigoplus_{i \in I_m} H_k(K_o/L_i \times (D^m,S^{m-1})) = \bigoplus_{i \in I_m} H_{k-m}(K_o/L_i)$

Since K_o/L_i is path connected, the projection $p_* : H_k(Y_m,Y_{m-1}) \to H_k(Y_m/K_o,Y_{m-1}/K_o)$ is bijective for $k \leq m$ and $H_k(Y_m,Y_{m-1}) = H_k(Y_m/K_o,Y_m/K_o) = 0$ for $k < m$. We show inductively for $m = n$, $n+1$, $n+2$,... that $p_* : H_n(Y_m,Y_{n-1}) \to H_n(Y_m/K_o,Y_{n-1}/K_o)$ is bijective. The induction step follows from the five-lemma (Dold [1980], I.2.9.) and the commutative diagram whose rows are exact sequence of triples

$$\begin{array}{ccccccc}
H_{n+1}(Y_{m+1},Y_m) & \longrightarrow & H_n(Y_m,Y_{n-1}) & \longrightarrow & H_n(Y_{m+1},Y_{n-1}) & \longrightarrow & 0 \\
\downarrow p_* & & \downarrow p_* & & \downarrow p_* & & \\
H_{n+1}(Y_{m+1}/K_o,Y_m/K_o) & \longrightarrow & H_n(Y_m/K_o,Y_{n-1}/K_o) & \longrightarrow & H_n(Y_{m+1}/K_o,Y_{n-1}/K_o) & \longrightarrow & 0
\end{array}$$

Now $p_* : H_n(Y,B) \longrightarrow H_n(Y/K_o,B/K_o)$ is an isomorphism since $H_*(Y,B)$ is lim $H_*(Y_m,Y_{n-1})$ and analogously for $(Y/K_o,B/K_o)$ by Milnor [1962] or Whitehead [1978], XIII.1.1. □

Let $f : X \longrightarrow Y$ be a G-map between G-spaces. Define the contravariant functor for $n \geq 2$

13.16.
$$\pi_n(f) : \Pi/(G,X) \longrightarrow \{\text{groups}\}$$

by $\pi_n(\mathrm{Cyl}(f),X)$ and the $\mathbf{Z}\Pi/(G,X)$-module $H_n(f)$ by $i^*H_n(\mathrm{Cyl}(f),X)$ for the inclusion $i : X \longrightarrow \mathrm{Cyl}(f)$.

<u>Corollary 13.17.</u> <u>Let</u> $f : X \longrightarrow Y$ <u>be a</u> Π-<u>isomorphism between</u> G-CW-<u>complexes</u> .

a) f <u>is</u> (G,n)-<u>connected for</u> $n \geq 2$ <u>if and only if</u> $H_m(f)$ <u>vanishes for</u> $m \leq n$.

b) f <u>is a</u> G-<u>homotopy equivalence if and only if</u> $H_m(f)$ <u>vanishes for</u> $m \geq 0$.

<u>Proof.</u> a) follows from Theorem 13.15. and implies b) using Theorem 2.4. Notice that for a Π-isomorphism $\pi_n(f)$ is a $\mathbf{Z}\Pi/(G,X)$-module also for $n = 2$ as it is a quotient of $\pi_n(X)$. □

13.18. Let (X,A) be a relative G-CW-complex with $X^2 = A$. Consider the $\mathbf{Z}\Pi/(G,A)$ chain complex $C^{\Pi}(X)$

$$\longrightarrow \dots\ i_n^*\pi_n(X_n,X_{n-1}) \xrightarrow{\Delta_n} i_{n-1}^*\pi_{n-1}(X_{n-1},X_{n-2}) \xrightarrow{\Delta_{n-1}} \dots$$

where $i_n : A \longrightarrow X_n$ is the inclusion and Δ_n is the boundary of the triple (X_n,X_{n-1},X_{n-2}) . The Hurewicz-maps define a chain map $C^{\Pi}(X) \longrightarrow i^*C^c(X)$ for the inclusion $i : A \longrightarrow X$ since $i^*C^c(X)$ is given by

$$\longrightarrow \dots\ i_n^*H_n(X_n,X_{n-1}) \xrightarrow{\Delta_n} i_{n-1}^*H_{n-1}(X_{n-1},X_{n-2}) \longrightarrow \dots$$

By Theorem 13.15. we obtain an isomorphism of $\mathbf{Z}\Pi/(G,A)$-chain complexes

$$C^{\Pi}(X) \xrightarrow{\cong} C^{c}(X)$$

Now we can prove the main result of this section, the equivariant version of the realization theorem of chain complexes as cellular chain complexes. It is the decisive link between geometry and algebra, or more precisely, between G-CW-complexes and chain complexes over a category.

Theorem 13.19. The Equivariant Realization Theorem.

Let $h : (Z,A) \longrightarrow (Y,B)$ be a G-map between proper relative G-CW-complexes such that $h|A : A \longrightarrow B$ is a G-homotopy equivalence and $h : Z \longrightarrow Y$ a Π-isomorphism. Consider an integer $r \geq 2$ such that $r \geq \dim(Z,A)$ holds for the relative dimension of (Z,A). Assume the existence of a free $\mathbf{Z}\Pi/(G,A)$-chain complex C with preferred equivalence class of bases and a homotopy equivalence $f : C \longrightarrow i^* h^* C^c(Y,B)$ for $i : A \longrightarrow Z$ the inclusion such that the restriction $C|r$ of C to dimension r is $i^* C^c(Z,A)$ and $f|r$ is $i^* C^c(h)$.

Then we can construct a pair of proper G-CW-complexes (X,A) containing (Z,A) as r-skeleton and a cellular G-homotopy equivalence $g : (X,A) \rightarrow (Y,B)$ satisfying

i) *g extends h.*

ii) $C = j^* C^c(X,A)$ for the inclusion $j : A \longrightarrow X$ (*including the equivalence class of bases*).

iii) $j^* C^c(g) = f$.

Proof. We construct inductively for $n = r, r+1, \ldots$ a pair of G-CW-complexes $X_n = (X_n, A)$, a cellular G-map $g_n : X_n \longrightarrow Y_{n+1}$ for $Y_{n+1} = (Y,A)_{n+1}$ satisfying

a) $X_r = (Z,A)$ and $g_r = h$.

b) The (n-1)-skeleton of (X_n, A) is X_{n-1} and $g_n|X_{n-1} = g_{n-1}$.

c) $C|_n = i_n^* C^c(X_n)$ for the inclusion $i_n : A \longrightarrow X_n$

d) $i_n^* C^c(g_n) = f|_n$.

:he induction begin $n = r$ is given by a). In the induction step from n to $n+1$ ye show firstly

.3.20. $\qquad g_n : X_n \longrightarrow Y_{n+1}$ is (G,n)-connected

ince f is a chain homotopy equivalence, the chain map $i_n^* C^c(g_n) : i_n^* C_n^c(X,A) \longrightarrow$ ${}_n^* g_n^* C^c(Y_{n+1},B)$ is n-connected. By Proposition 13.10. we get $H_m(g_n) = 0$ for $m \leq n$ ince $r \geq 2$ and $g_r = h$, the map g_n is a Π-isomorphism. Now 13.20. follows from orollary 13.17. Recall that $\Pi/(G,i_n) : \Pi/(G,A) \longrightarrow \Pi/(G,X_n)$ is an equivalence of ategories. Notice that 13.20. is not true for $g_n : X_n \longrightarrow Y_n$ in general.

f B_{n+1} is a base for C_{n+1} we construct X_{n+1} out of X_n by attaching one cell $/H_b \times D^{n+1}$ for each $b \in B_{n+1}$. Since we also want to extend g_n to g_{n+1} over he cell, we must specify for each $b \in B_{n+1}$ a G-diagram

$$\begin{array}{ccc} G/H_b \times S^n & \xrightarrow{\bar{q}_b} & X_n \\ \downarrow & & \downarrow g_n \\ G/H_b \times D^{n+1} & \xrightarrow{\bar{Q}_b} & Y_{n+1} \end{array}$$

et $x_b : G/H_b \longrightarrow A$ be the object of $\Pi/(G,A)$ belonging to $b \in B_{n+1}$. Then a G-iagram above is the same as a non-equivariant diagram

$$\begin{array}{ccc} S^n & \xrightarrow{q_b} & X_n(x_b) \\ \downarrow & & \downarrow g_n|X_n(x_b) \\ D^{n+1} & \xrightarrow{Q_b} & Y_{n+1}(g_n \circ x_b) \end{array}$$

ıt such a diagram represents an element $\omega_b' \in \pi_{n+1}(g_n|X_n(x_b))$. Since g_n is (G,n)-onnected by 13.20. we have by Theorem 13.15. the $\mathbb{Z}\Pi/(G,A)$-isomorphism $i_n^* \pi_{n+1}(g_n) \rightarrow$ ${}_n^* H_{n+1}(g_n)$

ıese considerations show

3.21. Each choice of elements $\omega_b \in i_n^* H_{n+1}^c(g_n)(x_b)$ for $b \in B_{n+1}$ determines a -extension

$$\begin{array}{ccc} X_n & \xrightarrow{g_n} & Y_n \\ \downarrow & & \downarrow \\ X_{n+1} & \xrightarrow{g_{n+1}} & Y_{n+1} \end{array}$$

such that $i_{n+1}^{*}C^c(X_{n+1})_{n+1} = C_{n+1}$.

The following diagram commutes

$$\begin{array}{ccc} C_{n+1} & \xrightarrow{f_{n+1}} & i_n^{*}g_n^{*}C_{n+1}^c(Y_{n+1},B) \\ \downarrow & & \downarrow \\ C_n = i_n^{*}C_n^c(X_n,A) & \xrightarrow{i_n^{*}C_n^c(g_n)} & i_n^{*}g_n^{*}C_n^c(Y_{n+1},B) \end{array}$$

Let $(u_b,v_b) \in \operatorname{Cone}(i_n^{*}C^c(g_n) : i_n^{*}C_n^c(X_n,A) \longrightarrow i_n^{*}g_n^{*}C_{n+1}^c(Y_{n+1},B))_{n+1}$ be the image of $b \in B_{n+1} \subset C_{n+1}$ under $c_{n+1} \bullet f_{n+1} : C_{n+1} \longrightarrow i_n^{*}C_n^c(X_n,A) \bullet i_n^{*}g_n^{*}C_{n+1}^c(Y_{n+1},B)$. Because of the diagram above , (u_b,v_b) is a cycle It determines an element $\omega_b \in H_{n+1}(g_n)(x_b)$ under the identification of Proposition 13.10. It is straight forward to check that under these choices ω_b we obtain by 13.21 an extension $g_{n+1} : X_{n+1} \longrightarrow Y_{n+1}$ satsisfying a), b), c) and d). This finishes the induction step.

Let $g : X \longrightarrow Y$ be the limit of the maps $g_n : X_n \longrightarrow Y$. It is a G-homotopy equivalence by Corollary 13.17. and fulfills i), ii), and iii) by construction. □

Next we deal with the natural properties of the cellular $R\Pi/(G,X)$-chain complex. We start with restriction for an inclusion $i : H \longrightarrow G$ of Lie groups. Given a G-space X , the adjunction between induction and restriction defines a homeomorphism $\mathrm{ad} : \mathrm{map}(G/K,X)^G \longrightarrow \mathrm{map}(H/K,\mathrm{res}\ X)^H$ for $K \subset H$. Consider objects $x : H/K \rightarrow \mathrm{res}\ X$ and $y : H/L \longrightarrow \mathrm{res}\ X$ and a morphism $(\sigma,w) : x \longrightarrow y$ in $\Pi(H,\mathrm{res}\ X)$. Then we obtain a morphism $(\mathrm{ind}(\sigma),\mathrm{ad}^{-1}(w)) : \mathrm{ad}^{-1}(x) \longrightarrow \mathrm{ad}^{-1}(y)$ in $\Pi(G,X)$. One easily verifies that we obtain a functor between topological categories

13.22. $\qquad \Pi(i,X) : \Pi(H,\mathrm{res}\ X) \longrightarrow \Pi(G,X)$.

It induces a functor

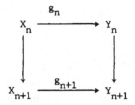

13.23. $$\Pi/(i,X) : \Pi/(H,\mathrm{res}\ X) \longrightarrow \Pi/(G,X) .$$

Lemma 13.24. Let $i : H \longrightarrow G$ be an inclusion of Lie groups such that G/H is finite. Let X be a proper G-CW-complex. Then

a) $\Pi/(i,X) : \Pi/(H,\mathrm{res}\ X) \longrightarrow \Pi/(G,X)$ is admissible.

b) Let $x : H/K \longrightarrow \mathrm{res}\ X$ be an object in $\Pi/(H,\mathrm{res}\ X)$ and $y : G/L \longrightarrow X$ an object in $\Pi/(G,X)$. Let $d(\bar{x},\bar{y})$ be the cardinality of $\mathrm{Irr}(x,y)/\mathrm{Aut}(x)$. Then $d(\bar{x},\bar{y})$ is the number of orbits $O \subset \mathrm{res}\ G/L$ of type H/K such that the H-orbit $y(O)$ meets $X^K(x)$.

c) There is a canonical H-CW-complex structure on $\mathrm{res}\ X$. We have a based isomorphism of $R\Pi/(H,\mathrm{res}\ X)$-chain complexes $C^c(\mathrm{res}\ X) \longrightarrow \Pi/(i,X)^* C^c(X)$.

Proof. The proof is based on Proposition 10.16. Consider objects $x : H/K \to \mathrm{res}\ X$ in $\Pi/(H,\mathrm{res}\ X)$ and $y : G/L \longrightarrow X$ in $\Pi/(G,X)$. We want to compute $\mathrm{Irr}(x,y)$. Regard any morphism $[\sigma,w] : \Pi/(i,X)(x) \longrightarrow y$ in $\Pi/(G,Y)$ represented by a morphism $(\sigma ,w) : \Pi(i,X)(x) \longrightarrow y$ in $\Pi(G,X)$. Let $i : \mathrm{im}(\mathrm{ad}\ \sigma) \longrightarrow \mathrm{res}\ G/L$ be the inclusion of the image of the adjoint $\mathrm{ad}\ \sigma : H/K \longrightarrow \mathrm{res}\ G/L$ of $\sigma : G/K = \mathrm{ind}\ H/K \longrightarrow G/L$. Write $x_o : \mathrm{im}(\mathrm{ad}\ \sigma) \longrightarrow \mathrm{res}\ X$ for $\mathrm{res}\ y \circ i$. Notice that $\mathrm{im}(\mathrm{ad}\ \sigma)$ is a homogeneous H-space again. We denote by $\mathrm{ad}\ \sigma$ also the obvious epimorphism $H/K \to \mathrm{im}(\mathrm{ad}\ \sigma)$ Let $c(x_o)$ be the constant path at x_o . We have the factorization in $\Pi/(G,X)$

13.25.
$$\begin{array}{ccc} \Pi/(i,X)(x) & \xrightarrow{\quad [\sigma,w] \quad} & y \\ {\scriptstyle \Pi/(i,X)([\mathrm{ad}\ \sigma,\mathrm{ad}\ w])} \searrow & & \nearrow {\scriptstyle [\mathrm{ad}\ i,c(x_o)]} \\ & \Pi/(i,X)(x_o) & \end{array}$$

Now consider an object $z : H/M \longrightarrow \mathrm{res}\ X$ and a morphism $[\tau,v] : x \longrightarrow z$ in $\Pi/(H,\mathrm{res}\ X)$ such that we have in $\Pi/(G,X)$ the factorization

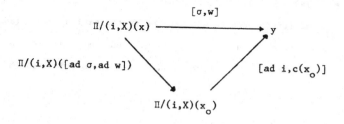

13.26.
$$\begin{array}{ccc} \Pi/(i,X)(x) & \xrightarrow{[\sigma,w]} & y \\ & {\scriptstyle \Pi/(i,X)([\tau,v])} \searrow \quad \nearrow {\scriptstyle [\omega,u]} & \\ & \Pi/(i,X)(Z) & \end{array}$$

Since $\omega \circ \operatorname{ind} \tau$ and σ are G-homotopic, $\operatorname{ad} \omega \circ \tau$ and $\operatorname{ad} \sigma$ are H-homotopic. As G/H is finite, $\operatorname{res} G/L$ is a finite disjoint union of homogeneous H-spaces. Hence the homogeneous H-spaces $\operatorname{im} \operatorname{ad} \omega$ and $\operatorname{im} \operatorname{ad} \sigma$ agree. Denote the epimorphism $H/M \longrightarrow \operatorname{im}(\operatorname{ad} \sigma)$ induced by $\operatorname{ad} \omega : H/M \longrightarrow \operatorname{res} G/L$ also by $\operatorname{ad} \omega$. Hence we obtain a morphism $[\operatorname{ad} \omega, \operatorname{ad} u] : z \longrightarrow x_o$ in $\Pi/(H, \operatorname{res} X)$. We leave it to the reader to show

13.27. $$[\operatorname{ad} \omega, \operatorname{ad} u] \circ [\tau,v] = [\operatorname{ad} \sigma, \operatorname{ad} w] \quad \text{in } \Pi/(H,\operatorname{res} X)$$

13.28. $$[\omega,u] = [\operatorname{ad} i, c(x_o)] \circ \Pi/(i,X)[\operatorname{ad} \omega, \operatorname{ad} u] \quad \text{in } \Pi/(G,X)$$

Now we can show

13.29. $\operatorname{Irr}(x,y)$ consists of all the morphisms $[\sigma,w] : \Pi/(i,X)(x) \longrightarrow y$ for which $\operatorname{ad} \sigma : H/K \longrightarrow \operatorname{res} G/L$ is injective.

Notice that $\operatorname{res} G/L$ is a finite union of H-orbits so that $\operatorname{ad} \sigma$ is injective for one representative (σ,w) of $[\sigma,w]$ if and only if it is for all. If $\operatorname{ad} \sigma$ is not injective $[\sigma,w]$ is not irreducible because in the factorization 13.25. $[\operatorname{ad} \sigma, \operatorname{ad} w]$ is no isomorphism. Suppose that $\operatorname{ad} \sigma$ is injective. Then $\operatorname{image} \operatorname{ad} \sigma = \operatorname{image} \operatorname{ad} \omega = \operatorname{image}(\operatorname{ad} \omega \circ \tau)$ implies that τ and hence $[\tau,v]$ are isomorphisms in 13.26.

Next we prove

13.30. $\operatorname{Irr}(x,y)$ is a free $\operatorname{Aut}/(x)$-set.

Let $[\sigma,w] : \Pi/(i,X)(x) \longrightarrow y$ be irreducible. Consider an automorphism $[\tau,v]$ of x in $\Pi/(H,\operatorname{res} X)$ such that $[\sigma,w] \circ \Pi/(i,X)([\tau,v]) = [\sigma,w]$ holds. We must show that $[\tau,v]$ is the identity under the hypothesis that $\operatorname{ad} \sigma$ is injective. By definition there is a path (ω_t,u_t) in $\operatorname{Mor}(\Pi(i,X)(x),y)$ from

$(\sigma \circ \mathrm{ind}\ \tau, (\mathrm{ind}\ \tau)^* w * \mathrm{ad}\ v)$ to (σ, w). We have $\mathrm{image}(\mathrm{ad}\ \omega_t) = \mathrm{image}(\mathrm{ad}\ \sigma)$ for all $t \in [0,1]$ so that we can define an automorphism of H/K by $\mathrm{ad}\ \sigma^{-1} \circ \mathrm{ad}\ \omega_t$. We get a path from (τ, v) to $(\mathrm{id}, c(x))$ in $\mathrm{Mor}(x,x)$ by $(\mathrm{ad}\sigma^{-1} \circ \mathrm{ad}\omega_t, (\mathrm{ad}\sigma^{-1} \circ \mathrm{ad}\omega_t)^* w^- * u_t)$. Hence $[\tau, v] : x \longrightarrow x$ in $\Pi/(H, \mathrm{res}\ X)$ is the identity. This shows 13.30.

Let M be the set of orbits O in $\mathrm{res}\ G/L$ of type H/K such that the H-orbit $\gamma(O)$ meets $X^K(x)$. Then

13.31. We have a bijection $\mathrm{Irr}(x,y)/\mathrm{Aut}(x) \longrightarrow M$ sending $[\sigma, w]$ to image ad σ.

We leave it to the reader to check that 13.30. and 13.31. imply condition i) and 13.27. and 13.28. condition ii) in Proposition 10.16. so that a) holds. We get b) from 13.31. whereas c) follows from 1.35. and the definitions. □

Next we treat products. Consider a proper G-CW-complex X and a proper H-CW-complex Y. The cartesian product defines a functor

13.32.
$$I : \Pi/(G,X) \times \Pi/(H,Y) \longrightarrow \Pi/(G \times H, X \times Y).$$

Using the Künneth formula we obtain a natural equivalence (see Whitehead [1978] II. 2.22.) of $R\Pi/(G,X) \times \Pi/(H,Y)$-chain complexes

13.33.
$$V : C^c(X) \otimes_R C^c(Y) \longrightarrow I^* C^c(X \times Y)$$

The adjoint is denoted in the same way.

<u>Lemma 13.34.</u> <u>We have a base preserving isomorphism of</u> $R\Pi/(G \times H, X \times Y)$-<u>chain complexes with preferred equivalence class of bases</u>

$$V : I_* C^c(X) \otimes_R C^c(Y) \longrightarrow C^c(X \times Y) \qquad \square$$

<u>Remark 13.35.</u> The statements above have analogues for the cellular ROr/G-chain complex $C^c(X)$ of a proper G-CW-complex X. Let $F : \Pi/(G,X) \longrightarrow Or/G$ be the obvious forgetful functor. Then induction with F applied to the cellular $R\Pi/(G,X)$-chain complex gives just the cellular ROr/G-chain complex. Hence we have also a cellular equivalence class of bases for the cellular ROr/G-chain complex. Given a pair of proper G-CW-complexes (X,A) there is an exact sequence of ROr/G-chain

complexes $0 \longrightarrow C^c(A) \longrightarrow C^c(X) \longrightarrow C^c(X,A)$ and analogously for G-push outs. Moreover, we have by induction an admissible functor $Or/i : Or/H \longrightarrow Or/G$ if $i : H \to G$ is an inclusion of Lie groups with finite G/H. The advantage of the cellular ROr/G-chain complex $C^c(X)$ is that it lives over a category independent of X. If X is a proper G-CW-complex with simply connected fixed point sets X^H for $H \subset G$ then $F : \Pi/(G,X) \longrightarrow Or/G$ is an equivalence of categories and it does not matter whether we work over $\Pi/(G,X)$ or Or/G. □

Comments 13.36. The cellular chain complex translates geometric information, given by a G-CW-complex structure, into algebraic information, given by a $\mathbf{Z}\Pi/(G,X)$-chain complex. A lot of geometric statements about G-CW-complexes have algebraic analogous for chain complexes. The way back from algebra to geometry is given by the Equivariant Realization Theorem. It is originally proved in Wall [1966] for $G = 1$. Other versions can be found in Smith [1986], [1987].

We emphasize again the strong analogy between our notions and results over $\mathbf{Z}\Pi/(G,X)$ and their non-equivariant analogues over $\mathbf{Z}\pi_1(X)$. Therefore we denote the $\mathbf{Z}\Pi/(G,X)$-chain complexes $C^c(X)$ and $C^s(X)$, their homology $H^c(X)$ and $H^s(X)$, the Hurewicz map $h(X) : \pi_n(X) \longrightarrow H^s(X)$ as in the non-equivariant situation. It is sometimes useful to forget their complicated structure but think of them as if one has no group action.

Because of this analogy there are no big conceptual difficulties to extend the ordinary obstruction theory as developed in Whitehead [1978] V.5 to the equivariant case (see Bredon [1967]). We indicate in the exercises how to deal with equivariant homology and cohomology theories (see also tom Dieck [1987], II.6, Illman [1975], Willson [1975].

Exercises 13.37.

1) Let $O(2)$ act on $\mathbf{R}^2$ in the obvious way. Let the $O(2)$-space S^2 be the one-point compactification of $\mathbf{R}^2$. Compute the cellular $\mathbf{Z}\Pi/(O(2);S^2)$-chain complex $C^c(S^2)$.

2) Compute the cellular $\mathbf{Z}Or\mathbf{Z}/2$-chain complex of SV for any real $\mathbf{Z}/2$-representation V.

3) Is $Or/i : Or/T^{n-1} \longrightarrow Or/T^{n}$ for the inclusion $i : T^{n-1} \longrightarrow T^{n}$ of tori admissible?

4) Let G be a Lie group and X a proper G-CW-complex. Then X is G-homotopy equivalent to a n-dimensional G-CW-complex for $n \geq 3$ if and only if the cellular $\mathbf{Z}\Pi/(G,X)$-chain complex $C^{c}(X)$ is $\mathbf{Z}\Pi/(G,X)$-homotopy equivalent to an n-dimensional one.

5) Let G be a compact Lie group and X be a G-CW-complex. Suppose that the forgetful functor $F : \Pi/(G,X) \longrightarrow Or/G$ is an equivalence of categories and for the singular $\mathbf{Z}Or/G$-homology $H_{m}^{s}(X) = 0$ for $0 < m \leq d$ and $H_{o}^{s}(X) \cong H_{o}^{s}(G/G)$ holds. Prove that then X is G-homotopy equivalent to a G-CW-complex Y with a single G-fixed point as d-skeleton.

6) A G-homology theory H_{*} for a topological group G with values in the abelian category $\mathcal{A}$ is a collection of functors H_{n} : {pairs of G-spaces} $\longrightarrow \mathcal{A}$ satisfying

i) Homotopy invariance.

$f,g : (X,A) \longrightarrow (Y,B)$ G-homotopic $\Longrightarrow H_{n}(f) = H_{n}(g)$

ii) Exactness.

For any G-NDR-pair (X,A) there is a natural long exact sequence where i and j are the inclusions

$$\cdots \longrightarrow H_{n+1}(X,A) \xrightarrow{\partial} H_{n}(A) \xrightarrow{H_{n}(i)} H_{n}(X) \xrightarrow{H_{n}(j)} H_{n}(X,A) \longrightarrow \cdots$$

iii) Excision .

Let X be the union $A \cup B$ of two closed G-spaces such that $(A, A \cap B)$ is a G-NDR-pair. Then the inclusion i induces an isomorphism $H_{n}(A, A \cap B) \longrightarrow H_{n}(X,B)$.

iv) Dimension axiom.

$H_{n}(G/H) = 0$ for $n > 0$.

Let G be a compact Lie group. Show that the singular ROr/G-homology H^{s} defines a G-homology theory {pairs of G-spaces} $\longrightarrow$ MOD-ROr/G . Prove that the same is true for the cellular homology H^{c} : {pairs of G-CW-complexes} $\longrightarrow$ MOD-ROr/G.

7) Let H_* be any non-equivariant homology theory with values in R-MOD. Let G be a compact Lie group. Show that the composition of H_* with $X/ : Or/G \to$ {top.sp.} $G/H \longrightarrow map(G/H,X)^G/WH_o = X^H/WH_o$ yields a G-homology theory {pairs of G-spaces} $\longrightarrow$ MOD-ROr/G . Treat also the case of a cohomology theory.

8) Let G be a compact Lie group. Consider a pair of G-spaces (X,A) and a ROr/G-comodule M, i.e. a covariant functor $Or/G \longrightarrow$ {R-MOD} . Let $H_n(X;M)$ be $H_n(C^s(X) \otimes_{ROr/G} M)$. Show that we obtain a G-homology resp. G-cohomology theory with values in R-MOD .

9) Let H_* be a G-homology theory for the topological group G with values in the abelian category $\mathcal{A}$. Given a G-CW-complex, define a chain complex $C(X)$ in $\mathcal{A}$ by

$$\ldots \longrightarrow H_n(X_n,X_{n-1}) \xrightarrow{\Delta} H_{n-1}(X_{n-1},X_{n-2}) \xrightarrow{\Delta} \ldots$$

where Δ is $H_n(X_n,X_{n-1}) \xrightarrow{\partial} H_{n-1}(X_{n-1}) \xrightarrow{i} H_{n-1}(X_{n-1},X_{n-2})$.

a) Show that there is a natural isomorphism $H_*(X) \cong H_*(C(X))$

b) Prove that $C(X)$ depends only on the G-CW-complex structure and the functor $H_o : Or/G \longrightarrow \mathcal{A}$ $G/H \longrightarrow H_o(G/H)$. Here Or/G has as objects all homogeneous spaces and G-homotopy classes of G-maps $G/H \longrightarrow G/K$ as morphisms.

c) Let K_* be a second G-homology theory. If there is a natural equivalence between H_o and $K_o : Or/G \longrightarrow \mathcal{A}$ then H_n and K_n : {pairs of G-CW-compl.} $\longrightarrow \mathcal{A}$ are natural equivalent. □

.4. Comparison of geometry and algebra.

[n this section we introduce the algebraic counterparts of the geometrically defined groups $Wa^G(X)$, $Wh^G(X)$, $K^G_{-n}(X)$ and $U^G(X)$ and invariants $w^G(X)$, $\tau^G(X)$ and $^G(X)$ of chapter I. We also establish algebraic versions of the various product and restriction ormulas. We show that the algebraic and geometric approach agree. In the sequel G s a Lie group and any G-space X is required to satisfy assumption 8.13. Recall that his means for a G-CW-complex X that all isotropy groups G_x are compact.

Ve start with introducing algebraic K-groups of a G-space X . Recall that we have ssigned to X an EI-category, the discrete fundamental category $\Pi/(G,X)$ in De- inition 8.28. In section 10 we have defined for an EI-category Γ algebraic K- roups $K_n(R\Gamma)$ (Definition 10.7), $Wh(R\Gamma)$ (Definition 10.8), $U(\Gamma)$ (Definition 10.9) nd $\tilde{K}_n(R\Gamma)$ (Definition 10.28). In this context recall assumption 9.1. about the ing R that the notion of the rank of a finitely generated R-module is specified.

efinition 14.1. Let X be a G-space. Define

$$K_n^G(X;R) := K_n(R\Pi/(G,X)) \qquad n \in \mathbf{Z}$$

$$Wh^G(X;R) := Wh(R\Pi/(G,X))$$

$$U^G(X) := U(\Pi/(G,X))$$

$$\tilde{K}_n^G(X;R) := \tilde{K}_n(R\Pi/(G,X)) \qquad n \in \mathbf{Z}$$

f R is $\mathbf{Z}$ we write briefly $K_n^G(X), Wh^G(X)$ and $\tilde{K}_n^G(X)$. □

4.2. Let $f : X \longrightarrow Y$ be a G-map. We obtain by induction with $\Pi/(G,f)$ homo- orphisms $f_* : K_n^G(X) \longrightarrow K_n^G(Y)$, $f_* : Wh^G(X) \longrightarrow Wh^G(Y)$, $f_* : U^G(X) \longrightarrow U^G(Y)$, nd $f_* : \tilde{K}_n^G(X) \longrightarrow \tilde{K}_n^G(Y)$. Let $h : X \times I \longrightarrow Y$ be a G-homotopy between f and $: X \longrightarrow Y$. It induces a natural equivalence between $\Pi/(G,f)$ and $\Pi/(G,g)$ so hat $f_* = g_*$ holds by Proposition 10.6. We obtain covariant functors {G-spaces} $\rightarrow$ {abel.gr.} by K_n^G, Wh^G, U^G and $\tilde{K}_n^G$ satisfying $f \simeq_G g \Longrightarrow f_* = g_*$. □

.3. The Definitions 5.1 and 14.1 of $U^G(X)$ coincide by Remark 8.34. □

14.A. The finiteness obstruction

Now we introduce the algebraic invariants we are interested in. Consider a finitely dominated G-CW-complex Y. Let (X,r,i,h) be a finite domination. Then we obtain by 13.4 and 13.6 $\mathbf{Z}\Pi/(G,X)$-chain maps $C^c(r) : r_*C^c(X) \longrightarrow C^c(Y)$ and $r_*C^c(i) : r_*i_*C^c(Y) \longrightarrow r_*C^c(X)$, a $\mathbf{Z}\Pi/(G,X)$-chain isomorphism $C^c(\hat{h}) : C^c(Y) \longrightarrow r_*i_*C^c(Y)$ and a $\mathbf{Z}\Pi/(G,X)$-chain homotopy between $C^c(r) \circ r_*C^c(i) \circ C^c(\hat{h})$ and the identity. Hence $C^c(Y)$ is a finitely dominated $R\Pi/(G,X)$-chain complex. We have defined its finiteness obstruction $o(C^c(Y)) \in K_o(\mathbf{Z}\Pi/(G,Y))$ and its Euler characteristic $\chi(C^c(Y)) \in U(\mathbf{Z}\Pi(G,Y))$ in Definition 11.1 and 11.19.

Definition 14.4. Let X be a finitely dominated G-CW-complex. Define its algebrai finiteness obstruction resp. reduced finiteness obstruction

$$o^G(X) \in K_o^G(X) , \quad \tilde{o}^G(X) \in \tilde{K}_o^G(X)$$

and its Eulercharacteristic

$$\chi^G(X) \in U^G(X)$$

by $o(C^c(X)), \tilde{o}(C^c(X))$ and $\chi(C^c(X))$ for the cellular $\mathbf{Z}\Pi/(G,X)$-chain complex $C^c(X)$ □

Remark 14.5. This can easily be extended to a finitely dominated G-space Y. By Proposition 2.12 there is a finitely dominated G-CW-complex X and a G-homotopy equivalence $f : X \longrightarrow Y$. Now define $o^G(Y)$ by $f_*o^G(X)$. □

Theorem 14.6.

a) Obstruction property

Let G be a compact Lie group and X be a finitely dominated G-space. Then X is G-homotopy equivalent to a finite G-CW-complex if and only if $\tilde{o}^G(X) \in \tilde{K}_o^G(X)$ vanishes

b) Homotopy invariance.

Let $f : X \longrightarrow Y$ be a G-homotopy equivalence between finitely dominated G-spaces. Then we have $f_*o^G(X) = o^G(Y)$ and $f_*\chi^G(X) = \chi^G(Y)$.

c) Additivity

Consider the G-push out of finitely dominated G-spaces with i a G-cofibration

$$\begin{array}{ccc} X_o & \longrightarrow & X_1 \\ {\scriptstyle i}\downarrow & \overset{j_o}{\searrow} & \downarrow{\scriptstyle j_1} \\ X_2 & \underset{j_2}{\longrightarrow} & X \end{array}$$

Then we have

$$o^G(X) = j_{1_*} o^G(X_1) + j_{2_*} o^G(X_2) - j_{o_*} o^G(X_o)$$

and

$$\chi^G(X) = j_{1_*} \chi^G(X_1) + j_{2_*} \chi^G(X_2) - j_{o_*} \chi^G(X_o)$$

Proof. Let (Y,r,i) be a finite domination of the G-CW-complex X. By the Lemma 14.7 below we can attach finitely many cells to Y to extend r to a $(G,2)$-connected map $\bar{r} : \bar{Y} \longrightarrow X$. By restriction to the 2-skeleton we obtain a 2-dimensional finite G-CW-complex Z and a $(G,2)$-connected map $f : Z \longrightarrow X$. By Lemma 8.35 $\Pi/(G,f) : \Pi/(G,Z) \longrightarrow \Pi/(G,X)$ is an equivalence of categories. Consider the $\mathbb{Z}\Pi/(G,Z)$ chain map $C^c(f) : C^c(Z) \longrightarrow f^* C^c(X)$. We have $H_k(\mathrm{Cone}^c(f)) = 0$ for $k \leq 2$. Hence Cone($C^c(f)$) is $\mathbb{Z}\Pi/(G,Z)$-homotopy equivalent to $\Sigma^3 F$ for an appropriate projective $\mathbb{Z}\Pi/(G,Z)$-chain complex F. We get from Theorem 11.2. that $\tilde{o}(F) = -\tilde{o}(\mathrm{Cone}(C^c(f)) = -\tilde{o}(C^c(Z)) + \tilde{o}(f^* C^c(X)) = \tilde{o}(f^* C^c(X)) = f^*(\tilde{o}(C^c(X)) = 0$ so that F can be choosen as a finite free $\mathbb{Z}\Pi/(G,Z)$-chain complex. If $g : \Sigma^3 F \longrightarrow \mathrm{Cone}(C^c(f))$ is a $\mathbb{Z}\Pi/(G,Z)$-homotopy equivalence, define the following commutative diagram with exact rows by requiring that the right square is a pull-back

$$\begin{array}{ccccccccc} 0 & \longrightarrow & C^c(Z) & \longrightarrow & \mathrm{Cyl}(C^c(f)) & \longrightarrow & \mathrm{Cone}(C^c(f)) & \longrightarrow & 0 \\ & & \| & & \uparrow{\scriptstyle g'} & & \uparrow{\scriptstyle g} & & \\ 0 & \longrightarrow & C^c(Z) & \overset{i}{\longrightarrow} & D & \longrightarrow & \Sigma^3 F & \longrightarrow & 0 \end{array}$$

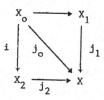

Then D is a finite free $\mathbb{Z}\Pi/(G,Z)$-chain complex with $D|_2 = C^c(Z)|_2$ and the composition $h : D \xrightarrow{g'} Cyl(C^c(f)) \xrightarrow{pr} f^*C^c(Y)$ is a $\mathbf{Z}\Pi/(G,Z)$-homotopy equivalence with $h \circ i = C^c(f)$. By Theorem 13.19. we can extend $f : Z \longrightarrow X$ to a G-homotopy equivalence $F : Y \longrightarrow X$ with a finite G-CW-complex Y as source.

b) follows from 13.4 , 13.6 and Theorem 11.2.

c) Use Lemma 2.13. to substitute the G-push out of finitely dominated G-spaces by a G-push out of finitely dominated G-CW-complexes. Then apply Theorem 11.2 and Lemma 13.7. □

Now we verify the promised lemma. Notice that only here the compactness of G is needed in the proof of Theorem 14.6.

Lemma 14.7. Let G be a compact Lie group.

a) If X is a finitely dominated G-space then $\pi_o(X^H)$ is finite and $\pi_1(X^H,x)$ is finitely presented for any $x \in X^H$ and $H \subset G$.

b) Let $f : X \longrightarrow Y$ be a G-map between G-CW-complexes of finite orbit type. Suppose that $\pi_o(X^H)$ and $\pi_o(Y^H)$ are finite, $\pi_1(X^H,x)$ is finitely generated for $x \in X^H$ and $\pi_1(Y^H,y)$ is finitely presented for $y \in Y^H$ where H runs over $\mathrm{Iso}\, X \cup \mathrm{Iso}\, Y$.

Then we can extend f to a G-map $g : Z \longrightarrow Y$ such that g is a Π-isomorphism and (Z,X) is a relatively finite pair of G-CW-complexes. □

Given $\mathcal{F} \subset \mathrm{Con}\, G$, we call a G-map $f : X \longrightarrow Y$ a $\mathcal{F}$-Π-isomorphism if for any $(H) \in \mathcal{F}$ the maps $\pi_o(f^H) : \pi_o(X^H) \longrightarrow \pi_o(Y^H)$ and $\pi_1(f^H,x) : \pi_1(X^H,x) \rightarrow \pi_1(Y^H,fx)$ for any $x \in X^H$ are bijective. If $\mathcal{F}$ is Con G we omitt $\mathcal{F}$. For the proof of Lemma 14.7. we need

Lemma 14.8. Let A be a finitely generated group and B a finitely presented group. Then the kernel of an epimorphism $f : A \longrightarrow B$ is finitely generated as normal subgroup.

Proof. Let $[X : R]$ be a finite presentation of B and Y a finite set of generators for A. Let $F(X \amalg Y)$ resp. $F(X)$ be the free group generated by the disjoint union $X \amalg Y$ resp. X. Denote by $q : F(X) \longrightarrow B$ the epimorphism induced by [X : R] . Let $p : F(X \amalg Y) \longrightarrow A$ be defined by $py = y$ for $y \in Y$ and $px = z$ for $x \in X$ where z is any element in A with $fz = qx$. If $j : F(X) \to F(X+Y)$ is the inclusion there is a retraction $r : F(X \amalg Y) \longrightarrow F(X)$ of j making the following diagram commute

$$\begin{array}{ccc} F(X \amalg Y) & \xrightarrow{\ r\ } & F(X) \\ {\scriptstyle p}\downarrow & & \downarrow{\scriptstyle q} \\ A & \xrightarrow{\ f\ } & B \end{array}$$

y Crowell-Fox [1963], p. 43+44, the finite set $M = j(R) \cup \{y \cdot (j \circ r(y))^{-1} \mid y \in Y\}$ generates kernel $q \circ r$ = kernel $f \circ p$ as a normal subgroup. Hence kernel f is generated as a normal subgroup by the finite set $N = \{p(m) \mid m \in M\}$. □

roof of Lemma 14.7

a) A retract of a finitely presented group is again finitely presented (see Wall [1965], Lemma 3.1). Hence it suffices to treat the case of a finite G-CW-complex X. Since G/K^H is a finite disjoint union of WH-orbits (see Theorem 1.33) X^H is a finite WH-CW-complex. By Lemma 7.4 X^H is homotopy equivalent to a finite CW-complex A . Obviously $\pi_o(A)$ is finite. The fundamental group $\pi_1(A,a)$ has a finite presentation by Schubert [1964], III 5.7. for any $a \in A$.

b) Since $\pi_o(Y^H)$ is finite for $(H) \in \mathcal{F} = \{(K) \mid K \in \operatorname{Iso} X \cup \operatorname{Iso} Y\}$ and $\pi_1(Y^H,y)$ is finitely generated for $y \in Y^H$ and $(H) \in \mathcal{F}$, we can make $f : X \to Y$ 1-connected. Now consider $(H) \in \mathcal{F}$, $x \in X^H$, $y = fx \in Y^H$ and the group homomorphism $\pi_1(f^H,x) : \pi_1(X^H,x) \longrightarrow \pi_1(Y^H,y)$. Its kernel is finitely generated as normal subgroup because of Lemma 14.8. Hence we can achieve by attaching finitely many cells $G/H \times D^2$ that $\pi_1(f^H,x)$ is an isomorphism. As X and Y have finite orbit type we can assume without loss of generality that f is a $\mathcal{F}$-Π-isomorphism. It remains to show that f is even a Π-isomorphism.

y attaching cells $G/H \times D^2$ for $(H) \in \mathcal{F}$ by trivial maps $G/H \times S^1 \xrightarrow{pr} G/H \longrightarrow X$ we obtain a $(\mathcal{F},2)$-connected extension $\bar{f} : \bar{X} \longrightarrow Y$. By Proposition 2.14

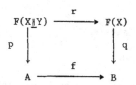

$\overline{f}$ is even (G,2)-connected. Up to homotopy $\overline{X}^K$ comes from X^K by attaching trivially two-cells D^2 and cells D^r of dimension $r \geq 3$ so that the inclusion $X \longrightarrow \overline{X}$ is a Π-isomorphism. Therefore f is a Π-isomorphism. □

Proposition 14.9. Let G be a compact Lie group and X a G-CW-complex of finite orbit type.

a) The following statements are equivalent.

i) X is finitely dominated.

ii) X is G-homotopy equivalent to a skeletal finite G-CW-complex and G-homotopy equivalent to a finite-dimensional G-CW-complex.

iii) $\pi_o(X^H)$ is finite for $H \in \text{Iso}\, X$ and $\pi_1(X^H,x)$ is finitely presented for $x \in X^H$ and $H \subset \text{Iso}\, X$ and the cellular $\mathbf{Z}\Pi/(G,X)$-chain complex $C^c(X)$ is finitely dominated.

b) X is G-homotopy equivalent to a skeletal finite G-CW-complex if and only if $\pi_o(X^H)$ is finite and $\pi_1(X^H,x)$ is finitely presented and $C^c(X)$ is $\mathbf{Z}\Pi/(G,X)$-homotopy equivalent to a finitely generated free $\mathbf{Z}\Pi/(G,X)$-chain complex.

c) X is G-homotopy equivalent to a n-dimensional G-CW-complex for $n \geq 3$ if and only if $C^c(X)$ is $\mathbf{Z}\Pi/(G,X)$-homotopy equivalent to a n-dimensional $\mathbf{Z}\Pi/(G,X)$-chain complex.

Proof.

b) The "if" statement is the non-trivial part. By Lemma 14.7 there is a finite G-CW-complex Z and a Π-isomorphism $f : Z \longrightarrow X$. Since $C^c(X)$ is up to homotopy finitely generated free, $\text{Cone}(C^c(f) : C^c(Z) \longrightarrow f^*C^c(X))$ is homotopy equivalent to a finitely generated free $\mathbf{Z}\Pi/(G,X)$-chain complex D by Lemma 11.6. Since $H_n(\text{Cone}(C^c(f)))$ is zero for $n = 0,1$ we have the exact sequence

$$0 \longrightarrow \text{kernel}(d_2 : D_2 \longrightarrow D_1) \longrightarrow D_2 \longrightarrow D_1 \longrightarrow D_o \longrightarrow 0 \quad .$$

Therefore kernel d_2 and $H_2(D) = H_2(\text{Cone}\, C^c(f))$ are finitely generated. Because of Proposition 13.10 and Theorem 13.15 $\pi_2(f)$ is a finitely generated $\mathbf{Z}\Pi/(G,X)$-module. Hence we can attach finitely many cells to Z to make f (G,2)-connected.

'e can assume without loss of generality that $f : Z \longrightarrow X$ is (G,2)-connected and , a finite two-dimensional G-CW-complex.

.s in the proof of Theorem 14.6 we construct a finitely generated free $\mathbb{Z}\Pi/(G,Z)$-hain complex C with $C|_2 = C^c(Z)$ and a homotopy equivalence $g : C \longrightarrow f^*C^c(X)$ xtending $C^c(f) : C^c(Z) \longrightarrow f^*C^c(X)$. By Theorem 13.19 we can extend $f : Z \to X$ o a G-homotopy equivalence $\bar{f} : \bar{Z} \longrightarrow X$ such that $\bar{Z}$ is skeletal finite.

) follows from Theorem 13.19.

) i) => iii) is obvious. The implication iii) => ii) follows from b) and c) and roposition 11.11. Next we prove ii) => i). Without loss of generality we can assume hat X is n-dimensional and that there is a skeletal finite G-CW-complex Y to-ether with a G-homotopy equivalence $f : Y \to X$. Let $r : Y_n \longrightarrow X$ be its estriction to the n-skeleton. Since r is (G,n)-connected $r_* : [X,Y_n] \to [X,X]$ s surjective by Proposition 2.3. Hence there is $i : X \longrightarrow Y_n$ with $r \circ i \simeq_G id$ so $_n$ dominates X. □

riterions for $R\Gamma$-chain complexes to be finitely dominated, finitely generated or n-imensional can be found in section 11. See also Wall [1966] for the case $G = \{1\}$.

y Theorem 14.2 we obtain by $o^G(X) \in K_o^G(X)$ a functorial additive invariant for he category with cofibrations and weak equivalences having finitely dominated G-paces as objects. Because of Theorem 6.9 there is a natural transformation

4.10. $$\Phi : Wa^G \oplus U^G \longrightarrow K_o^G$$

niquely determined by the property that $\Phi(X)(w^G(X), \chi^G(X)) = o^G(X)$ for any finitely ominated G-space. It induces a natural transformation

4.11. $$\tilde{\Phi} : Wa^G \longrightarrow \tilde{K}_o^G$$

heorem 14.12. Both Φ and $\tilde{\Phi}$ are natural equivalences.

roof. We must show for any finitely dominated G-CW-complex Y that (Y) : $Wa^G(Y) \oplus U^G(Y) \longrightarrow K_o^G(Y)$ is bijective. We start with surjectivity.

Let the finitely generated projective $\mathbb{Z}\Pi/(G,Y)$-module P represent $[P] \in K_o^G(Y)$. By the Eilenberg swindle we can find a based free $\mathbb{Z}\Pi/(G,Y)$-chain complex F concentrated in dimension four and five such that F is chain homotopy equivalent to $4(P)$. Consider the projection $pr : C^c(Y) \oplus F \longrightarrow C^c(Y)$. Allthough this is not a chain homotopy equivalence, the construction in Theorem 13.19. yields an extension $r : X \longrightarrow Y$ of $id : Y \longrightarrow Y$ with $r_*C^c(X) = C^c(Y) \oplus F$ since F is concentrated in two consecutive dimensions. By Proposition 14.9. X is finitely dominated and $r_*o^G(X)$ is $o^G(Y) + [P]$. Hence $\Phi(Y)$ is surjective.

Let $\eta \in Wa^G(Y) \oplus U^G(Y)$ be an element in the kernel of $\Phi(Y)$. Without loss of generality we can assume the existence of a finitely dominated G-CW-complex X and a G-map $f : X \longrightarrow Y$ such that η is $([f], f_*\chi^G(X))$. By Lemma 14.7. we can achieve that f is a Π-isomorphism so that $\Pi/(G,f)$ is an equivalence of categories by Lemma 8.35. Then $f_* : K_o^G(X) \longrightarrow K_o^G(Y)$ is a bijection and $o^G(X)$ vanishes. By Theorem 14.6. X is G-homotopy equivalent to a finite G-CW-complex so that $[f] \in Wa^G(Y)$ is zero. The homomorphism $rk : K_o^G(X) \to U^G(X)$ sends $o^G(X)$ to $\chi^G(X)$ by definition. This implies $\chi^G(X) = 0$. Hence η is zero. □

14.B. The Whitehead torsion.

Now we come to the Whitehead torsion. Consider a G-homotopy equivalence $f : X \to Y$ between finite G-CW-complexes. Then we have a $\mathbb{Z}\Pi/(G,Y)$-chain homotopy equivalence $C^c(f) : f_*C^c(X) \longrightarrow C^c(Y)$ between finite free $\mathbb{Z}\Pi/(G,Y)$-chain complexes with preferred equivalence class of bases (see Lemma 13.2 , 13.4 and 13.6) .

Definition 14.13. Define the algebraic Whitehead torsion

$$\tau^G(f) \in Wh^G(Y)$$

by the Whitehead torsion $\tau(C^c(f))$ of $C^c(f) : f_*C^c(X) \longrightarrow C^c(Y)$ (see 12.24.)

Theorem 14.14.

a) Obstruction property.

A G-homotopy equivalence between finite G-CW-complexes is simple if and only if its algebraic Whitehead torsion $\tau^G(f) \in Wh^G(X)$ vanishes.

b) Homotopy invariance.

Let f and $g : X \longrightarrow Y$ be G-homotopy equivalences between finite G-CW-complexes. Then $f \simeq_G g \Longrightarrow \tau^G(f) = \tau^G(g)$.

c) Additivity

Consider the following commutative diagram of finite G-CW-complexes such that i_1 and k_1 are inclusions of G-CW-complexes, the squares cellular G-push outs and f_o, f_1, f_2 and f are G-homotopy equivalences.

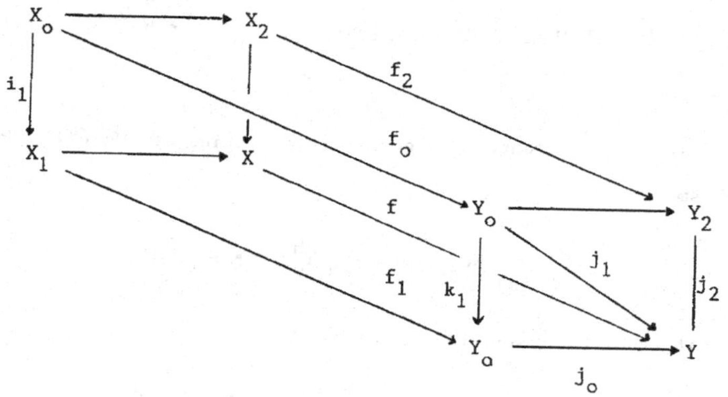

Then: $j_{1*}\tau^G(f_1) + j_{2*}\tau^G(f_2) - j_{o*}\tau^G(f_o) = \tau^G(f)$.

d) Logarithmic Property.

Let $f : X \longrightarrow Y$ and $g : Y \longrightarrow Z$ be G-homotopy equivalences between finite G-complexes. Then $\tau^G(g\circ f) = \tau^G(g) + g_*\tau^G(f)$.

Proof.

b) follows from 13.4. , 13.6 and Theorem 12.25.

c) is a consequence of Lemma 13.7 and Theorem 12.25.

d) follows from Theorem 12.25.

a) Because of b), c) and d) (Wh^G, τ^G) is a functorial additive invariant for the category $\mathcal{C}$ with cofibrations and weak equivalences specified in 6.10. Let Wh^G_{geo} be the geometric Whitehead group and τ^G_{geo} the geometric Whitehead torsion of section four. By Theorem 6.11. there is a natural transformation $\Phi : U^G \oplus Wh^G_{geo} \rightarrow Wh^G$

uniquely determined by the property that it sends $(\chi^G(Y), \tau^G_{geo}(f))$ to $\tau^G(f)$ for a G-homotopy equivalence $f : X \longrightarrow Y$ of finite G-CW-complexes. Since $\tau^G(id : X \to X)$ is zero, Φ induces a natural transformation

14.15. $$\tilde{\Phi} : Wh^G_{geo}(Y) \longrightarrow Wh^G(Y)$$

sending $\tau^G_{geo}(f)$ to $\tau^G(f)$.

Because of Theorem 4.8. we have to show $\tau^G_{geo}(f) = 0$ provided that $\tau^G(f)$ vanishes. This follows from the next Theorem. □

Theorem 14.16. The natural transformation $\tilde{\Phi} : Wh^G_{geo} \longrightarrow Wh^G$ is a natural equivalence.

Proof. We start with introducing a geometric splitting of $Wh^G(Y)$. Namely define an homomorphism

$$\psi : \bigoplus_{(H)} Wh^1_{geo}(EWH \times_{WH} Y^H) \longrightarrow Wh^G_{geo}(Y)$$

by the composition of the following maps.

$$\bigoplus_{(H)} \psi_1(H) : \bigoplus_{(H)} Wh^1_{geo}(EWH \times_{WH} Y^H) \longrightarrow \bigoplus_{(H)} Wh^{WH}_{geo}(EWH \times Y^H)$$

where $\psi_1(H)$ is the isomorphism given by the pull back construction. Its inverse is induced by dividing out the WH-action.

$$\bigoplus_{(H)} pr_* : \bigoplus_{(H)} Wh^{WH}(EWH \times Y^H) \longrightarrow \bigoplus_{(H)} Wh^{WH}(Y^H)$$

if $pr : EWH \times Y^H \longrightarrow Y^H$ is the projection.

$$\bigoplus_{(H)} \psi_2(H) : \bigoplus_{(H)} Wh^{WH}(Y^H) \longrightarrow \bigoplus_{(H)} Wh^G(G \times_{NH} Y^H)$$

where $\psi_2(H)$ is given by restriction with $NH \longrightarrow WH$ and induction with $NH \subset G$.

$$\sum_{(H)} \psi_3(H) : \bigoplus_{(H)} Wh^G(G \times_{NH} Y^H) \longrightarrow Wh^G(Y)$$

if $\psi_3(H)$ is induced by $G \times_{NH} Y^H \longrightarrow Y \quad (g,y) \longrightarrow g \cdot y$.

We get from Proposition 8.33. an identification

$$\bigoplus_{(H)} Wh^1(EWH \times_{WH} Y^H) = \text{Split } Wh^G(Y)$$

Consider the isomorphism of Theorem 10.34.

$$E : \text{Split } Wh^G(Y) \longrightarrow Wh^G(Y)$$

We obtain a commutative diagram

$$\begin{array}{ccc} Wh^G_{geo}(Y) & \xrightarrow{\tilde{\Phi}^G} & Wh^G(Y) \\ \uparrow \psi & & \cong \uparrow E \\ & & \text{Split } Wh^G(Y) \\ & & \| \\ \bigoplus_{(H)} Wh^1_{geo}(EWH \times_{WH} Y^H) & \xrightarrow{\bigoplus_{(H)} \tilde{\Phi}^1} & \bigoplus_{(H)} Wh^1(EWH \times_{WH} Y^H) \end{array}$$

The non-equivariant transformations $\tilde{\Phi}^1$ are isomorphisms by Stöcker [1970] or Cohen [1973] § 21. Hence it suffices to show that ψ is surjective since then ψ and $\tilde{\Phi}^G$ must be isomorphisms.

Consider $[X,Y] \in Wh^G(Y)$. Choose a numeration $\{(H_1),(H_2),\dots,(H_r)\}$ of $\{(H) \in \text{Con } G \mid H \in \text{Iso } X\}$ such that $(H_i) \leq (H_j)$ implies $i \geq j$. Let $X(k)$ be the G-CW-subcomplex $\bigcup_{i=1}^{k} X^{(H_i)} \subset X$ and define $Y(k) \subset Y$ analogously. As the G-homotopy equivalence $Y \longrightarrow X$ induces a G-homotopy equivalence $Y(k) \longrightarrow X(k)$, the inclusion $Y \longrightarrow X(k) \cup_{Y(k)} Y$ is a G-homotopy equivalence by Lemma 2.13. Because of Lemma 4.3 there is a G-deformation retraction $r(k) : X(k) \cup_{Y(k)} Y \longrightarrow Y$. Consider the pair $(Z(k),Y)$ with $Z(k) = (X(k) \cup_{Y(k)} Y) \cup_{r(k-1)} Y$. It defines $[Z(k),Y] \in Wh^G(Y)$. We obtain from Lemma 4.14. in $Wh^G_{geo}(Y)$

$$[X,Y] = \sum_{k=1}^{r} [Z(k),Y]$$

otice that $Z(k)$ is obtained by attaching cells of type G/H_k to Y. For $H = H_k$ the omposition $\psi_3(H) \circ \psi_2(H) : Wh^{WH}(Y^H) \longrightarrow Wh^G(Y)$ sends $[Z(k)^H,Y^H]$ to $[Z(k),Y]$. ince $(Z(k)^H,Y^H)$ is relatively WH-free and $\psi_1(H): Wh^1_{geo}(EWH \times_{WH} Y^H) \rightarrow Wh^{WH}_{geo}(EWH \times Y^H)$

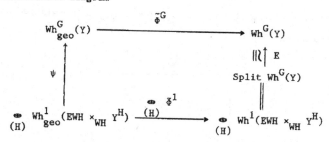

is bijective, it remains to show the following:

Assume for $[X,Y] \in Wh^G(Y)$ that (X,Y) is relatively G-free. Then $[X,Y]$ lies in the image of $pr_* : Wh^G(EG \times Y) \longrightarrow Wh^G(Y)$.

Now we construct inductively for $n = -1,0,1,2,...$ a pair of G-CW-complexes $(A(n), EG \times Y)$ such that $(A(n) \cup_{pr} Y, Y) \simeq (X_n \cup Y, Y)$ rel Y holds. The begin $n = -1$ is $A(-1) = EG \times Y$, the induction step from $(n-1)$ to n done as follows. Fix a simple G-homotopy equivalence $f : A(n-1) \cup_{pr} Y \longrightarrow X_{n-1} \cup Y$ relative Y and write $X_n \cup Y$ as the G-push out

$$\begin{array}{ccc} \coprod_{i \in I} G \times S^{n-1} & \xrightarrow{\coprod_{i \in I} q(i)} & X_{n-1} \cup Y \\ \downarrow & & \downarrow \\ \coprod_{i \in I} G \times D^n & \xrightarrow{\coprod_{i \in I} Q(i)} & X_n \cup Y \end{array}$$

We have the push out

$$\begin{array}{ccc} EG \times Y & \xrightarrow{pr} & Y \\ \downarrow & & \downarrow \\ A(n-1) & \xrightarrow{\overline{pr}} & A(n-1) \cup_{pr} Y \end{array}$$

Since pr is a weak homotopy equivalence, the same is true for $\overline{pr}$ and hence for $f \circ \overline{pr} : A(n-1) \longrightarrow X_n \cup Y$. Therefore we can find G-maps $\hat{q}(i) : G \times S^{n-1} \longrightarrow A(n-1)$ such that $f \circ \overline{pr} \circ \hat{q}(i) \simeq_G q(i)$ holds. Let $A(n)$ be the G-push out

$$\begin{array}{ccc} \coprod_{i \in I} G \times S^{n-1} & \xrightarrow{\coprod_{i \in I} \hat{q}_i} & A(n-1) \\ \downarrow & & \downarrow \\ \coprod_{i \in I} G \times D^n & \longrightarrow & A(n) \end{array}$$

Then $(A(n) \cup_{pr} Y, Y) \simeq (X_n, Y)$ rel Y follows from Lemma 4.12. For $n = \dim X$ we end up with $(A, EG \times Y)$ such that $(A \cup_{pr} Y, Y) \simeq (X,Y)$ rel Y holds.

In particular $Y \longrightarrow A \cup_{pr} Y$ is a weak homotopy equivalence. Because of Lemma 2.13 applied to

$$\begin{array}{ccc} EG \times Y & \longrightarrow & Y \\ \downarrow & & \downarrow \\ A & \longrightarrow & A \cup_{pr} Y \end{array}$$

$EG \times Y \longrightarrow A$ is a weak homotopy equivalence and by Theorem 2.4 a G-homotopy equivalence. Hence $(A, EG \times Y)$ defines an element in $Wh^G(EG \times Y)$ mapped to $[X,Y]$ under pr_*. This finishes the proof of Theorem 14.16. □

Remark 14.17. In the present proof of Theorem 14.16. we reduced the claim just to the non-equivariant case. We could also give a "global" proof using the Realization Theorem 13.19 as we did for the finiteness obstruction. □

Remark 14.18. If we want to drop the condition that G is compact in Theorem 4.6.a), one has to assume that there is a finite G-CW-complex Z and a $(G,2)$-connected G-map $Z \longrightarrow X$. Similar considerations apply to Proposition 14.9. The problem is Lemma 14.7. which is only true for compact G. □

14.C. Product and restriction formulas.

Now we come to the algebraic versions of the product formulas of section seven.

Let G and H be compact Lie groups. Let $\otimes : U^G(X) \otimes U^H(Y) \longrightarrow U^{G\times H}(X\times Y)$ and $\otimes : K_o^G(X) \otimes K_o^H(Y) \longrightarrow K_o^{G\times H}(X\times Y)$ be the composition of $\otimes$ defined in 10.23 and I_* induced by I defined in 13.32. Let $\otimes : U^G(X) \otimes Wh^H(Y) \longrightarrow Wh^{G\times H}(X\times Y)$ and $\otimes : U^H(Y) \otimes Wh^G(X) \longrightarrow Wh^{G\times H}$ be given by 10.24 and 13.32. Then we obtain from Proposition 10.46, 11.18, 12.30 and Lemma 13.34.

Theorem 14.19. Product formula.

a) Let X be a finitely dominated G-CW-complex and Y be a finitely dominated H-CW-complex. Then:

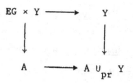

$$o^{G\times H}(X\times Y) = o^G(X) \circledast o^H(Y)$$

and

$$\chi^{G\times H}(X\times Y) = \chi^G(X) \circledast \chi^H(Y)$$

b) Let $f : X' \longrightarrow X$ be a G-homotopy equivalence between finite G-CW-complexes and $g : Y \longrightarrow Y'$ analogously over H. Then

$$\tau^{G\times H}(f\times g) = \chi^G(X) \circledast \tau^H(g) + \chi^H(Y) \circledast \tau^G(f) \qquad \square$$

Let $i : H \longrightarrow G$ be an inclusion of compact Lie groups of the same dimension. Since the functor $\Pi/(i,X) : \Pi/(H,\mathrm{res}\,X) \longrightarrow \Pi/(G,X)$ is admissible by Lemma 13.24, it induces by restriction $i^* : U^G(X) \longrightarrow U^H(\mathrm{res}\,X)$, $i^* : K_o^G(X) \longrightarrow K_o^H(\mathrm{res}\,X)$ and $i^* : Wh^G(X) \longrightarrow Wh^H(\mathrm{res}\,X)$. (see 10.18 and 10.20). We get from Lemma 13.24.

Theorem 14.20. Restriction formula for codimension zero.

Let $i : H \longrightarrow G$ be an inclusion of compact Lie groups of the same dimension.

a) If X is a finite G-CW-complex, then $i^*\chi^G(X) = \chi^H(\mathrm{res}\,X)$ holds.

b) If X is a finitely dominated G-space, then we have: $i^*o^G(X) = o^H(\mathrm{res}\,X)$.

c) If $f : X' \longrightarrow X$ is G-homotopy equivalence between finite G-CW-complexes then $i^*\tau^G(f) = \tau^H(\mathrm{res}\,f)$ is valid. $\square$

Let G be a finite group and $i : G \longrightarrow G\times G$ the diagonal map. Let $\Delta : \Pi(G,X\times Y) \longrightarrow \Pi(G,X) \times \Pi(G,Y)$ be $\Pi(G,pr_X) \times \Pi(G,pr_Y)$ if $X \times Y$ carries the diagonal action. One easily checks that the functors RES_Δ and $\mathrm{RES}_{\Pi(i,X\times Y)} \circ \mathrm{IND}_I : \text{MOD-}\mathbf{Z}\Pi(G,X) \times \Pi(G,Y) \longrightarrow \text{MOD-}\mathbf{Z}\Pi(G,X\times Y)$ agree because of $\mathrm{res}(G/K_1\times G/K_2)^H = G/K_1^H \times G/K_2^H$. Since $\Pi(i,X\times Y)$ is admissible, Δ is admissible. Let the pairings

$$\circledast : U^G(X) \circledast U^G(Y) \longrightarrow U^G(X\times Y)$$

$$\circledast : K_o^G(X) \circledast K_o^G(Y) \longrightarrow K_o^G(X\times Y)$$

and

$$\circledast : U^G(X) \circledast Wh^G(Y) \longrightarrow Wh^G(X\times Y)$$

be the compositions of the pairing of 10.23 and 10.24 with Δ^* of 10.18. The Theorems 14.19 and 14.20 imply

Theorem 14.21. Diagonal product formula for finite G .

a) If X and Y are finite G-CW-complexes we have $\chi^G(X\times Y) = \chi^G(X) \circledast \chi^G(Y)$.

b) If X and Y are finitely dominated G-spaces then $o^G(X\times Y) = o^G(X) \circledast o^G(Y)$.

c) If $f : X' \longrightarrow X$ and $g : Y' \longrightarrow Y$ are G-homotopy equivalences of finite G-complexes then: $\tau^G(f\times g) = \chi^G(X) \circledast \tau^G(g) + \chi^G(Y) \circledast \tau^G(f)$. □

Now we want to establish algebraic restriction formulas of diagonal product formulas for arbitrary inclusions of compact Lie groups $i : H \longrightarrow G$. This requires some more algebra .

Let Γ_1 and Γ_2 be EI-categories. Let ho FACC-$R\Gamma_2$ be the homotopy category of $R\Gamma_2$-chain complexes possessing a finite approximation. Homotopy category means that we take homotopy classes of chain maps as morphism. Given a covariant functor $F : \Gamma_1 \longrightarrow$ ho FACC-$R\Gamma_2$, we will define for $n = 0,1$ the transfer homomorphism of F

14.22.
$$\mathrm{trf}_F : K_n(R\Gamma_1) \longrightarrow K_n(R\Gamma_2) .$$

Obviously FBMOD-$R\Gamma_1$:= {fin.gen. based free $R\Gamma_1$-mod.} and ho FACC-$R\Gamma_2$ are R-additive categories (see section 10). Now F extends in a unique way to a functor of R-additive categories also denoted by F

14.23.
$$F : \text{FBMOD-}R\Gamma_1 \longrightarrow \text{ho FACC-}R\Gamma_2.$$

Any morphism $f : x \longrightarrow y$ in Γ_1 induces $f_* : R\,\mathrm{Hom}(?,x) \longrightarrow R\,\mathrm{Hom}(?,y)$ by composition. Each $R\Gamma_1$-homomorphism $R\,\mathrm{Hom}(?,x) \longrightarrow R\,\mathrm{Hom}(?,y)$ can be uniquely written as

$$\sum_{f \in \mathrm{Hom}(x,y)} r_f \cdot f_* .$$

An arbitrary morphism in FBMOD-$R\Gamma_1$ looks like

$$\Big(\sum_{f \in \mathrm{Hom}(x_i,x_j)} r_f \cdot f_*\Big)_{i,j} \; \bigoplus_{i=1}^{r} R\,\mathrm{Hom}(?,x_i) \longrightarrow \bigoplus_{j=1}^{s} R\,\mathrm{Hom}(?,y_j)$$

It is sent by F to

$$\left(\sum_{f \in Hom(x_i,x_j)} r_f \cdot F(f)\right)_{i,j} : \bigoplus_{i=1}^{r} F(x_i) \longrightarrow \bigoplus_{j=1}^{s} F(y_j)$$

We next define trf_F for $n = 1$. Let the automorphism $f : M \longrightarrow M$ in FBMOD-$R\Gamma_1$ represent $[f] \in K_1(R\Gamma_1)$. We have assigned to the self-equivalence $F(f)$ of the $R\Gamma_2$-chain complex $F(M)$ having a finite approximation its self-torsion $t(F(f)) \in K_1(R\Gamma_2)$ in 12.18. We define

14.24. $$trf_F([f]) = t(F(f))$$

This makes sense because of the properties of the self-torsion listed in Example 12.17.

Now we consider $n = 0$. Let the finitely generated projective $R\Gamma_1$-module P represent $[P] \in K_o(R\Gamma_1)$. Choose $p : M \to M$ with $p \circ p = p$ and image $p = P$ in FBMOD-$R\Gamma_1$. We have assigned to the homotopy projection $F(p) : F(M) \longrightarrow F(M)$ its finiteness obstruction $o(F(p)) \in K_o(R\Gamma_2)$ in Definition 11.26. Because of Theorem 11.27. we can define

14.25. $$trf_F([P]) = o(F(f)) .$$

<u>Proposition 14.26.</u> <u>Let</u> F <u>and</u> $F' : \Gamma_1 \to h_o$-FACC-$R\Gamma_2$ <u>be given. Suppose the existence of a natural transformation</u> $\Phi : F \longrightarrow F'$ <u>such that</u> $\Phi(x) : F(x) \longrightarrow F'(x)$ <u>is a weak homology equivalence for any object</u> x <u>in</u> Γ_1 . <u>Then</u>

$$trf_F = trf_{F'}$$

<u>Proof:</u> Theorem 11.27. and Example 12.17. □

Let ho FDCC-$R\Gamma_2$ be the homotopy category of finitely dominated $R\Gamma_2$-chain complexes or, equivalently of $R\Gamma_2$-chain complexes of the homotopy type of a finite projective $R\Gamma_2$-chain complex. Suppose now that F is a functor $\Gamma_1 \to$ ho FDCC-$R\Gamma_2$. Given $x \in Ob\ \Gamma_1$ and $y \in Ob\ \Gamma_2$, let $F(x,y)$ be the functor where i is the obvious inclusion of the groupoid $Aut(x)^\wedge$ associated with $Aut(x)$

$$F(x,y) : Aut(x)^\wedge \xrightarrow{i} \Gamma_1 \xrightarrow{F} \text{ho FDCC-}R\Gamma_2 \xrightarrow{S_y} \text{ho FDCC-}R[y]$$

It induces $trf_{F(x,y)} : K_n(R[x]) \longrightarrow K_n(R[y])$. Let

$$\text{Split } trf_F : \text{Split } K_n(R\Gamma_1) \longrightarrow \text{Split } K_n(R\Gamma_2)$$

e given by $(trf_{F(x,y)} \mid \bar{x} \in \text{Is } \Gamma_1 \, , \, \bar{y} \in \text{Is } \Gamma_2)$.

roposition 14.27. If E and S are the inverse isomorphisms of Theorem 10.34. e have the commutative diagram for $n = 0,1$

$$\begin{array}{ccc} K_n(R\Gamma_1) & \xrightarrow{\ trf_F\ } & K_n(R\Gamma_2) \\ \uparrow E & & \downarrow S \\ \text{Split } K_n(R\Gamma_1) & \xrightarrow{\ \text{Split } trf_F\ } & \text{Split } K_n(R\Gamma_2) \end{array}$$

□

he functor $F(x,y) : \text{Aut}(x)^{\wedge} \to \text{ho FDCC-}R[y]$ is the same as a chain homotopy repre-entation (C,U) , i.e. a finitely dominated $R[y]$-chain complex C together with homomorphism of moniods $U : \text{Aut}(x) \longrightarrow [C,C]_{R[y]}$. Namely, let C be $F(x,y)(R[x])$ nd U send $f \in \text{Aut}(x)$ to $F(x,y)$ applied to the $R[x]$-map of right $R[x]$-modules $[x] \longrightarrow R[x]$ $g \longmapsto f \circ g$. The homomorphism $trf_{F(x,y)}$ is studied in Lück [1986], 1987] .

xample 14.28. Consider a covariant functor $F : \Gamma_1 \longrightarrow \Gamma_2$. It induces a functor $_! : \Gamma_1 \to \text{ho FDCC-}R\Gamma_2$ by sending $f : x \longrightarrow y$ to $[O(F(f)_*)] : O(R\,\text{Hom}(?,Fx)) \longrightarrow$ $(\text{Hom}(?,Fy))$. Recall that $O(\ )$ denotes the corresponding chain complex concentra-ed in dimension zero. Then F_* defined by induction and $trf_{F_!}$ yield the same ap $K_n(R\Gamma_1) \longrightarrow K_n(R\Gamma_2)$.

uppose that F is admissible. Then F defines a functor $F^! : \Gamma_2 \to \text{ho FDCC-}R\Gamma_1$ y sending y to $O(F^* R\,\text{Hom}(?,y))$. Then $trf_{F!}$ and $F^* : K_n(R\Gamma_2) \longrightarrow K_n(R\Gamma_1)$ agree. □

4.29. When does $F : \Gamma_1 \to \text{ho FDCC-}R\Gamma_2$ induces also a homomorphism $trf_F : Wh(R\Gamma_1)$ $\to Wh(R\Gamma_2)$? Consider for $x \in \text{Ob } \Gamma_1$ the homomorphism $\phi(x) : \text{Aut}(x) \to Wh(R\Gamma_2)$ ending $f \in \text{Aut}(x)$ to $t(F(f))$ considered as an element in $Wh(R\Gamma_2)$. If each omomorphism $\phi(x)$ vanishes, $trf_F : K_1(R\Gamma_1) \longrightarrow K_1(R\Gamma_2)$ induces $trf_F : Wh(R\Gamma_1)$ $\to Wh(R\Gamma_2)$. We have to check for a trivial unit $u \cdot f_* : R\,\text{Hom}(?,x) \to R\,\text{Hom}(?,x)$ hat the projection of its image under trf_F into $Wh(R\Gamma_2)$ vanishes. Since $u \cdot f_*$ s $(u \cdot id) \circ f_*$ we may assume $f = id$. But $F(u \cdot id)$ is $u \cdot id : F(x) \to F(x)$ and hence

$t(F(u\cdot id))$ is zero in $Wh(R\Gamma_2)$ □

Let $i : H \longrightarrow G$ be an inclusion of compact Lie groups and X a G-space. We want to define a functor

14.30. $$F : \Pi/(G,X) \longrightarrow \text{ho FDCC-}R\Pi/(H,\text{res } X)$$

into the homotopy category of finitely dominated $R\Pi/(H,\text{res } X)$-chain complexes. Given an object $x : G/K \longrightarrow X$, define $F(y)$ by $(\text{res } x)_* C^s(\text{res } G/K)$. Consider a morphism $[\sigma,w] : x \longrightarrow y$ in $\Pi/(G,X)$ from $x : G/K \longrightarrow X$ to $y : G/L \longrightarrow X$. Choose a morphism $(\sigma,w) : x \longrightarrow y$ in $\Pi(G,X)$ representing $[\sigma,w]$. We get from res w a natural equivalence Φ_w between the induction functors $(\text{res } x)_*$ and $(\text{res } y \circ \text{res } \sigma)_*$. Let $F([\sigma,w]) : F(x) \to F(y)$ be the homotopy class of the composition, if we write G/K instead of res G/K.

14.31. $$(\text{res } x)_* C^s(G/K) \xrightarrow{\Phi_w(C^s(G/K))} (\text{res } y)_*(\text{res } \sigma)_* C^s(G/K) \xrightarrow{(\text{res } y)_*(C^s(\text{res } \sigma))} (\text{res } y)_* C^s(G/K) .$$

We leave it to the reader to check that $F([\sigma,w])$ is independent of the choice $(\sigma,w) \in [\sigma,w]$.

Example 14.32. Let X be a point. Then 14.30. reduces to a functor

$$F : Or/G \longrightarrow \text{FDCC-}ROr/H$$

sending $[\sigma] : G/K \longrightarrow G/L$ to $[C^s(\text{res } \sigma)] : C^s(\text{res } G/K) \longrightarrow C^s(\text{res } G/L)$ □

We get from 14.22. and 14.30. an homomorphism

14.33. $$trf_F : K_n^G(X) \longrightarrow K_n^H(\text{res } X)$$

for $n = 0,1$. Next we show that trf_F induces a map

14.34. $$trf_F : Wh^G(X) \longrightarrow Wh^H(\text{res } X)$$

Because of 14.29. we must show that the homomorphisms $\phi(x) : Aut(x) \longrightarrow Wh(R\Gamma_2)$ vanish. One easily checks that $\phi(x)$ of 14.29. factorizes over the geometrically defined map $\phi_{geo}(x)$ of 7.5.

$$\begin{array}{ccc}
\mathrm{Aut}_{\Pi/(G,X)}(x) & \xrightarrow{\phi(x)} & Wh^H(\mathrm{res}\ X) \\
\Big\downarrow \mathrm{pr} & & \cong \Big\uparrow \tilde{\Phi}^H(\mathrm{res}\ X) \\
\mathrm{Aut}_{\Pi_o}(G,X)(x) & \xrightarrow[\phi_{geo}(x)]{} & Wh^H_{geo}(\mathrm{res}\ X)
\end{array}$$

here pr is the epimorphism induced by the cannonical projection $\Pi/(G,X) \to \Pi_o(G,X)$
ence all the maps $\phi(x)$ vanish if all the homomorphisms $\phi_{geo}(x)$ vanish, i.e.
. is simple with respect to $H \subset G$ (Definition 7.7). But this is true by Lemma
.27.

ne easily verifies that we also get a map

4.35. $$trf_F : U^G(X) \longrightarrow U^H(\mathrm{res}\ X)$$

ince $U^G(X) = K_o^f(\mathbb{Z}\Pi/(G,X))$ and $U^H(\mathrm{res}\ X) = K_o^f(\mathbb{Z}\Pi/(H,\mathrm{res}\ X)$ by Proposition 10.42.

efinition 14.36. Let $i : H \longrightarrow G$ be an inclusion of compact Lie groups. Define
he homomorphisms

$$i^* : U^G(X) \longrightarrow U^H(\mathrm{res}\ X)$$

$$i^* : K_n^G(X) \longrightarrow K_n^H(\mathrm{res}\ X) \qquad n = 0,1$$

$$i^* : Wh^G(X) \longrightarrow Wh^H(\mathrm{res}\ X)$$

y the transfer homomorphisms trf_F associated with the functor

$$F : \Pi/(G,X) \longrightarrow \mathrm{ho}\ FDCC\text{-}\mathbb{Z}\Pi/(H,\mathrm{res}\ X) \qquad \square$$

ne easily verifies that i^* is a natural transformation between the covariant
unctors U^G and $U^H \circ \mathrm{res}$, resp. K_n^G and $K_n^H \circ \mathrm{res}$ resp. Wh^G and $Wh^H \circ \mathrm{res}$
rom {G-spaces} to {abel. gr.} . If H and G have the same dimension i^* of
efinition 14.36. agrees with i^* defined by restriction because of Example 14.28.

heorem 14.37. Restriction formula

et $i : H \longrightarrow G$ be an inclusion of compact Lie groups.

) If X is a finite G-CW-complex then $i^*\chi^G(X) = \chi^H(\mathrm{res}\ X)$.

b) *If* X *is a finitely dominated* G-*space then* $i^{*}o^{G}(X) = o^{H}(\text{res } X)$.

c) *Let* $f : X \longrightarrow Y$ *be a* G-*homotopy equivalence between finite* G-CW-*complexes*. *We have defined* $\tau^{H}_{geo}(\text{res } f) \in Wh^{H}_{geo}(\text{res } X)$ *in* 7.22. *Then its image under* $\tilde{\Phi}^{H}(\text{res } X) : Wh^{H}_{geo}(\text{res } X) \longrightarrow Wh^{H}(\text{res } X)$ is $i^{*}(\tau^{G}(f))$. □

The Theorem 14.37. is equivalent to the next result. Let Res be the natural transformation of Theorem 7.25. and $\widetilde{\text{Res}}$ the induced one. Consider the diagrams

14.38.

$$\begin{array}{ccc} U^{G}(X) \circledast Wa^{G}(X) & \xrightarrow{\text{Res } X} & U^{H}(\text{res } X) \circledast Wa^{H}(\text{res } X) \\ \big\downarrow \Phi^{G}X & & \big\downarrow \Phi^{H}(\text{res } X) \\ K_{o}^{G}(X) & \xrightarrow{i^{*}} & K_{o}^{H}(\text{res } X) \end{array}$$

14.39.

$$\begin{array}{ccc} Wh^{G}_{geo}(X) & \xrightarrow{\widetilde{\text{Res}}\ X} & Wh^{H}_{geo}(\text{res } X) \\ \big\downarrow \tilde{\Phi}^{G}X & & \big\downarrow \tilde{\Phi}^{H}(\text{res } X) \\ Wh^{G}(X) & \xrightarrow{i^{*}} & Wh^{H}(\text{res } X) \end{array}$$

Proposition 14.40. *The diagrams* 14.38. *and* 14.39. *are commutative*.

Proof. We give the proof only for 14.39. the other case is similar. The general strategy is based on the observation that one can realize any element in $Wh^{G}_{geo}(X)$ by a pair of finite G-CW-complexes (Y,X) such that Y is obtained from X by attaching cells of dimension n and $n+1$ only (Proposition 4.57.). The construction 7.10. yields a pair of finite H-CW-complexes (Y',X') together with a H-homotopy equivalence $f : X' \longrightarrow X$ satisfying $\tau^{H}(\text{res}(Y,X)) = f_{*}[Y',X']$. Then one computes that the cellular $\mathbf{Z}\Pi/(H,\text{res } X)$-chain complex $f_{*}C^{c}(Y',X')$ is based isomorphic to the mapping cone of some representative of $F(c)$ if F is the functor 14.30. and $c : C^{c}_{n+1}(Y,X) \longrightarrow C^{c}_{n}(Y,X)$ is the differential. This implies by definition of the transfer that $trf_{F} \circ \Phi^{G}(X)([Y,X]) = \Phi^{H}(\text{res } X)(\text{res}(Y,X))$ holds what is to be proved.

The main step is the computation of $f_{*}C^{c}(Y',X')$. For simplicity we assume that X^{H} is non-empty and simply connected for $H \subset G$. The general case is notationally, but not conceptually , more complicated. We leave it to the reader to derive the general proof from the one we give now. Also the proof in Lück [1986] , section 7, may serve as a pattern.

If X^H is non-empty and simply connected for all $H \subset G$, the projection $pr : X \to \{*\}$ induces an equivalence of categories $\Pi/(G,pr) : \Pi/(G,X) \to \Pi/(G,*)$ because of Lemma 8.35. so that $pr_* : Wh^G(X) \to Wh^G(*)$ and $pr_* : Wh^H(\mathrm{res}\ X) \to Wh^H(*)$ are isomorphisms. By naturality it suffices to consider $X = *$. This has the advantage that now everything takes place over the orbit category $Or/G = \Pi/(G,*)$. The proof is based on the following commutative diagram we explain now.

14.41.

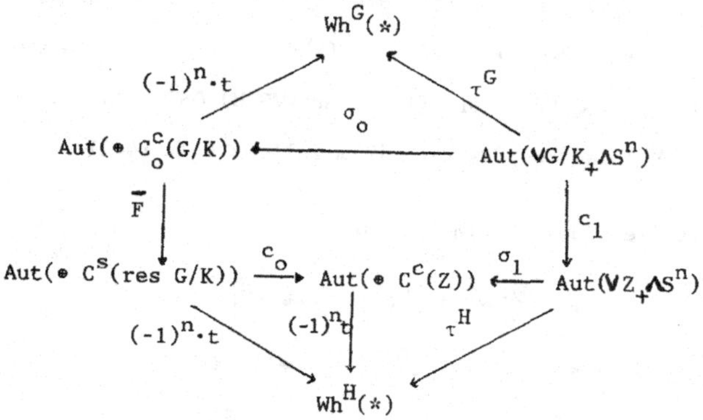

We fix $K \subset G$, an integer $n \geq 2$ and a finite index set I . All sums $\oplus$ and wedges $\vee$ run over I . We choose a finite H-CW-complex Z and a H-homotopy equivalence $z : Z \to \mathrm{res}\ G/K$. Let Z_+ resp. G/K_+ be the pointed space $Z \coprod *$ resp. $G/K \coprod *$ with $*$ as base point. Consider S^n as a pointed space. Let $Aut(\vee G/K_+ \wedge S^n)$ resp. $Aut(\vee Z_+ \wedge S^n)$ be the group of pointed G- resp. H-homotopy classes of pointed G- resp. H-self equivalences. Restriction to H and conjugation with the H-homotopy equivalence $\vee z_+ \wedge id : \vee Z_+ \wedge id \to \vee \mathrm{res}\ G/K_+ \wedge S^n$ defines c_1 . The homomorphisms τ^G and τ^H are given by the equivariant Whitehead torsion mapped into $Wh^H(*)$ by the homomorphism induced by the projection onto $*$. Let $Aut(\oplus C^s(\mathrm{res}\ G/K))$ nd $Aut(\oplus C^c(Z))$ be the group of homotopy classes of self-chain equivalences. Notice hat $C^c(G/K)$ is concentrated in dimension zero and $C_o^c(G/K) = \mathbf{Z}\ Hom(?,G/K) = \mathbf{Z}[?,G/K]^G$ is based free. Let $Aut(\oplus C_o^c(G/K))$ be the group of automorphisms of the initely generated based free $\mathbf{Z}Or/G$-module $\oplus C_o^c(G/K)$ and $\bar{F}$ be given by the functor of Example 14.32. The map σ_1 is given by applying C^c, conjugation with the suspension isomorphism $\Sigma^n C^c(Z) \cong C^c(\vee Z_+ \wedge S^n)$ and applying Σ^{-n}. Define σ_o analogously. If $\psi : \bullet\, C^c(Z) \longrightarrow \bullet\, C^s(Z)$ is the homotopy equivalence of Proposition 13.10. c_o sends the class of $\phi : \bullet\, C^s(\mathrm{res}\ G/K) \longrightarrow \bullet\, C^s(\mathrm{res}\ G/K)$ to the class of $\phi' : \bullet\, C^c(Z) \longrightarrow \bullet\, C^c(Z)$ if $\phi \circ \bullet C^s(z) \circ \psi \simeq \bullet C^s(z) \circ \psi \circ \phi'$ holds. The homomorphisms t are given by the absolute torsion. It follows from the definitions that the diagram 14.44 commutes.

By Proposition 4.57 and Theorem 10.34 any element in $Wh^G(*)$ can be written as a sum of elements of the shape $\tau^G(f)$ for some K, I and $[f] \in \mathrm{Aut}(\vee G/K_+ \wedge S^n)$. Hence it suffices to prove $\mathrm{Res}(*)(\tau^G(f)) = i^*(\tau^G(f))$. We get from the definitions that $\mathrm{Res}(*)(\tau^G(f)) = \tau^H \circ c_1(f)$ for $[f] \in \mathrm{Aut}(\vee G/H)_+ \wedge S^n)$ and $i^* \circ t(g) = t \circ \bar{F}(g)$ for $g \in \mathrm{Aut}(\bullet\, C^c_o(G/K))$ holds. Now the claim follows from 14.41 □

Let G be a compact Lie group. Define pairings

$$\bullet : K_o^G(X) \bullet K_o^G(Y) \longrightarrow K_o^G(X\times Y)$$
$$\bullet : U^G(X) \bullet U^G(Y) \longrightarrow U^G(X\times Y)$$
$$\bullet : U^G(X) \bullet Wh^G(Y) \longrightarrow Wh^G(X\times Y)$$

as the composition of the pairings appearing in Theorem 14.19. for $H = G$

$$\bullet : K_o^G(X) \bullet K_o^G(Y) \longrightarrow K_o^{G\times G}(X\times Y)$$
$$\bullet : U^G(X) \bullet U^G(Y) \longrightarrow U^{G\times G}(X\times Y)$$
$$\bullet : U^G(X) \bullet Wh^G(Y) \longrightarrow Wh^{G\times G}(X\times Y)$$

with $i^* : K_o^{G\times G}(X\times Y) \longrightarrow K_o^G(X\times Y)$, $i^* : U^{G\times G}(X\times Y) \longrightarrow U^G(X\times Y)$, and $i^* : Wh^{G\times G}(X\times Y) \longrightarrow Wh^G(X\times Y)$. We get from Theorem 14.19. and Theorem 14.37. the following generalization of Theorem 14.21.

<u>Theorem 14.42.</u> . <u>Diagonal product formula</u>

<u>Let</u> G <u>be a compact Lie group</u>.

a) <u>If</u> X <u>and</u> Y <u>are finite</u> G-CW-<u>complexes, we have</u> $\chi^G(X\times Y) = \chi^G(X) \bullet \chi^G(Y)$.

•) If X and Y are finitely dominated G-spaces then $o^G(X\times Y) = o^G(X) \otimes o^G(Y)$.

.) If $f : X' \longrightarrow X$ and $g : Y' \longrightarrow Y$ are G-homotopy equivalences of finite G-CW-omplexes then $\tau^G(f\times g) = \chi^G(X) \otimes \tau^G(g) + \chi^G(Y) \otimes \tau^G(g)$. □

emark 14.43. Similarly to Proposition 14.40. also the product and diagonal product ormulas defined geometrically in section seven and algebraically in this section gree under the natural equivalences Φ and $\tilde{\Phi}$.

e have defined in 7.34 geometrically a map $\phi(X) : Wa^G(X) \longrightarrow Wh^G_{geo}(X\times S^1)$. Since here is an obvious equivalence of categories $\Pi/(G,X) \times \hat{\mathbf{Z}} \longrightarrow \Pi/(G,X\times S^1)$, we obtain rom 10.49 an algebraic homomorphism $\tilde{B}(X) : \tilde{K}^G_o(X) \longrightarrow Wh^G(X\times S^1)$. We have defined in 10.49 such that the following holds (compare Ranicki [1986], Proposition 3.1)

roposition 14.44. The following diagram commutes

$$\begin{array}{ccc} Wa^G(X) & \xrightarrow{\phi(X)} & Wh^G_{geo}(X\times S^1) \\ \downarrow \tilde{\Phi}^G(X) & & \downarrow \tilde{\Phi}^G(X\times S^1) \\ \tilde{K}^G_o(X) & \xrightarrow{B(X)} & Wh^G(X\times S^1) \end{array}$$ □

e get from the Definitions 7.36 and 10.53.

heorem 14.45. The natural equivalence $\tilde{\Phi}^G(X\times T^{n+1}) : Wh^G_{geo}(X\times T^{n+1}) \longrightarrow Wh^G(X\times T^{n+1})$ nduces a natural equivalence

$$\tilde{\Phi}^G(X) : \tilde{K}^G_{-n}(X)_{geo} \longrightarrow \tilde{K}^G_{-n}(X) := \tilde{K}_{-n}(\mathbf{Z}\Pi/(G,X))$$ □

ow Proposition 10.54 gives the promised proof of Theorem 7.38.

4.D. Some conclusions

e mention some easy consequences of the results of this section and the Splitting heorem 10.34. for algebraic K-theory of RΓ-modules, Proposition 8.33. and Remark .34. Denote for $C \in \pi_o(X^H)$ its isotropy group under the WH-action by $WH(C)$. Then H(C) acts on C . Recall that $EWH(C) \longrightarrow BWH(C)$ is the universal principal WH(C) undle.

Theorem 14.46. We have for $n \in \mathbb{Z}$

$$K_n^G(X) \cong \bigoplus_{(H)} \bigoplus_{C \in \pi_o(X^H)/WH} K_n(\mathbb{Z}[\pi_1(EWH(C) \times_{WH(C)} C)])$$

Similar splittings hold for $\tilde{K}_o^G(X)$ and $Wh^G(X)$. □

Corollary 14.47. Let G be a compact Lie group and X be a G-space such that $\pi_o(X^H)$ and $\pi_1(X^H,x)$ are finite for all $H \subset G$ and $x \in X^H$.

a) If G is finite, $\tilde{K}_o^G(X)$ is finite.

b) $K_n^G(X)$ is zero for $n \leq -2$.

Proof. If Γ is a finite group, then $\tilde{K}_o(\mathbb{Z}\Gamma)$ is finite by Swan [1960a] and $K_n(\mathbb{Z}\Gamma)$ is $\{0\}$ for $n \leq -2$ by Carter [1980] . □

Corollary 14.48. Let G and H be compact Lie groups. Let X be a finitely dominated G-CW-complex and Y be a finitely dominated H-CW-complex. Suppose that $\pi_o(X^K)$ and $\pi_1(X^K,x)$ are finite for all $K \subset G$ and $x \in X^K$ and similar for Y. Then

$$o^{G\times H}(X\times Y) = \chi^G(X) \otimes o^H(Y) + o^G(X) \otimes \chi^H(Y) - \chi^G(X) \otimes \chi^H(Y) .$$

Proof. Theorem 11.24., Theorem 14.19. □

Theorem 14.49. Let G be a torus and Y a G-space such that Y^H is connected and non-empty for $H \subset G$. Let $i : Y^G \longrightarrow Y$ be the inclusion and res denote restriction to the trivial group.

a) If Y is finitely dominated as a G-space, we get

$$o(\text{res } Y) = i_* o(Y^G) .$$

b) If X and Y are finite G-CW-complexes and $f : X \longrightarrow Y$ is a G-homotopy equivalence, we have

$$\tau(\text{res } f) = i_* \tau(f^G) .$$

Proof. We only prove a) because b) is done similarly. We get from Theorem 14.46.

$$K_o^G(Y) = \bigoplus_{H \subset G} K_o(\mathbb{Z}\pi_1(EG/H \times_{G/H} Y^H))$$

As Y^H is connected and has a fixed point, $Y^H \longrightarrow EG/H \times_{G/H} Y^H$ induces an isomorphism on the fundamental groups and hence an isomorphism $j_H : K_o(\mathbf{Z}\pi_1(Y^H)) \longrightarrow K_o(\mathbf{Z}\pi_1(EG/H \times_{G/H} Y^H))$. Let $res_H : K_o(\mathbf{Z}\pi_1(EG/H \times_{G/H} Y^H)) \longrightarrow K_o(\mathbf{Z}\pi_1(Y))$ be the restriction of $res : K_o^G(Y) \longrightarrow K_o(res\, Y)$ (Definition 14.36.) to the summand of $H \subset G$. By Proposition 14.27. and the assumptions about Y^H the composition $res_H \circ j_H$ is given by the chain homotopy representation

$$\pi_1(X^H) \longrightarrow [\mathbf{Z}\pi_1(X) \otimes_{\mathbf{Z}} C^s(G/H), \mathbf{Z}\pi_1(X) \otimes_{\mathbf{Z}} C^s(G/H)]_{\mathbf{Z}\pi_1(X)}$$

sending $u \in \pi_1(X^H)$ to the $\mathbf{Z}\pi_1(X)$-self chain map $w \otimes c \longrightarrow i_{H*}u \cdot w \otimes c$ if $i_H : Y^H \longrightarrow Y$ is the inclusion and $C^s(G/H)$ the singular chain complex. Hence $res_H \circ i_H$ is $\chi(G/H) \cdot i_{H*} : K_o(\mathbf{Z}\pi_1(X^H)) \longrightarrow K_o(\mathbf{Z}\pi_1(X))$. As $\chi(G/H)$ is zero for $H \neq G$, we get $res_H = 0$ for $H \neq G$ and $res_G \circ j_G = i_{G*}$. Now the claim follows from Theorem 14.37. □

<u>Theorem 14.50.</u> <u>Let</u> G <u>be a compact Lie group whose component of the identity is not a product of some copies of</u> S^1 <u>and</u> SO(3) . (<u>In particular,</u> G <u>is not finite</u>). <u>Let</u> X <u>be a free finitely dominated</u> G-<u>space resp.</u> $f : X \longrightarrow Y$ <u>be a</u> G-<u>homotopy equivalence of finite free</u> G-CW-<u>complexes</u>. <u>Let</u> res <u>denote restriction to the trivial group</u>. <u>Then</u> res X <u>is homotopy equivalent to a finite</u> CW-<u>complex resp.</u> res f <u>is simple</u>.

<u>Proof</u>. Under the conditions above $res\ K_o^G(X) \longrightarrow K_o(res\, X)$ and $Wh^G(Y) \longrightarrow Wh(res\, Y)$ are zero by Lück [1987]. Now apply Theorem 14.37. □

<u>Comments 14.51.</u> With this section we have achieved one of the main goals summarized as follows. We have introduced geometrically defined groups $Wa^G(X)$, $Wh^G_{geo}(X)$, $K_n^G(X)_{geo}$ and $U^G(X)$ and identified them with algebraic objects $\tilde{K}_n^G(X)$, $Wh^G(X)$, $K_n^G(X)$. These algebraic objects are determined by algebraic K-groups of certain group rings. Moreover, we have defined geometrically and algebraically invariants like finiteness obstruction and Whitehead torsion which correspond under these identifications. They decide questions like whether a finitely dominated G-CW-complex is G-

homotopy equivalent to a finite one or whether a G-homotopy equivalence between finite G-CW-complexes is simple. Finally we have stated various formulas in the geotric and algebraic setting and proved that they coincide.

Some of the notions and results of this section exist already in the literature and are established by more or less ad hoc methods requiring less machinery. For example, one can define the splitted version of the obstruction groups and obstructions directly without mentioning the notion of a module over a category. Nevertheless it turns out that it is useful to do some efforts to develop these notions. At least our approach unifies the different treatments of these invariants one can find in the literature and applies to compact Lie groups and requires no assumptions about the connectivity of the fixed point sets. To establish the restriction and (diagonal) product formulas, however, it seems to be necessary to develop some machinery. Doing this at once on the splitted level is just too complicated whereas the language of modules over a category gives the right concept. This will be also the case in the following sections.

Here is a list of references containing material related to this section:

Andrzejewski [1986], Baglivo [1978], tom Dieck [1981], Dovermann-Rothenberg [1986], Hauschild [1978], Iizuka [1984], Illman [1974], [1985], [1986], Kwasik [1983], Lück [1983], [1987], Rothenberg [1978] .

<u>Exercises 14.52.</u>

1. Let X be a finitely dominated G-space. Show that the Definition 5.3 and Definition 14.4 of the equivariant Euler characteristic $\chi^G(X) \in U^G(X)$ agree.

2. Let X be a G-CW-complex of finite orbit type such that X^H is simply connected for $H \in \mathrm{Iso}\, X$. Prove that the following statements are equivalent:
   i) X is G-homotopy equivalent to a skeletal-finite G-CW-complex.
   ii) $H_i(X^H)$ is finitely generated for $H \in \mathrm{Iso}\, X$ and $i \geq 0$.
   iii) $H_i(X^H/WH_o)$ is finitely generated for $H \in \mathrm{Iso}\, X$ and $i \geq 0$.

3) Let X be a G-CW-complex of finite orbit type. Then $H_i(X^H)$ is finitely generated for all $H \subset G$ and $i \geq 0$ if and only if $H_i(X^H/WH_o)$ is finitely ge-

nerated for all $H \subset G$ and $i \geq 0$.

4. Let X be a G-manifold such that $\pi_o(X_H) \longrightarrow \pi_o(X^H)$ and $\pi_1(X_H,x) \longrightarrow \pi_1(X^H,x)$ are bijective for $H \subset G$ and $x \in X_H$ where X_H is $X^H \backslash X^{>H}$. Show that the forgetful map $\Phi : Wh^G_{Iso}(X) \longrightarrow Wh^G_{geo}(X)$ is a bijection (see 4.43.)

5. Let X be a finitely dominated G-space such that $\pi_1(X^H,x)$ is finite for $H \subset G$ and $x \in X^H$ and $\chi^G(X)$ vanishes. Let X' be a G'-space with the same properties. Show that $X \times X'$ is $G \times G'$-homotopy equivalent to a finite $G \times G'$-CW-complex.

6. Let X be a finitely dominated G-CW-complex such that $\pi_1(X^H,x)$ is finite for $H \in \text{Iso } X$ and $x \in X^H$. Prove that $i^* : U^G(X) \longrightarrow U^K(\text{res } X)$ sends $\chi^G(X)$ to $\chi^K(\text{res } X)$ for $K \subset G$. (Hint: $\tilde{K}_o(\mathbb{Z}\Delta) \longrightarrow \tilde{K}_o(\mathbb{Q}\Delta)$ is trivial for finite Δ) .

7. Let $H \subset G$ be a subgroup. Induction ind_H^G induces a functor $\text{ind} : \Pi/(H,X) \longrightarrow \Pi/(G,G\times_H X)$ and hence $\text{ind}_* : K_o^H(X) \longrightarrow K_o^G(G\times_H X)$ and $\text{ind}_* : Wh^H(X) \rightarrow Wh^G(G\times_H X)$ for a H-space X . Show $\text{ind}_*(o^H(X)) = o^G(G\times_H X)$ for a finitely dominated H-space X and $\text{ind}_*(\tau^H(f)) = \tau^G(G\times_H f)$ for a H-homotopy equivalence between finite H-CW-complexes $f : X' \longrightarrow X$.

8. Let X be a free connected G-CW-complex. Let $i_* : K_o^{\{1\}}(\text{res } X) \rightarrow K_o^G(X)$ be induction composed with the map induced from $G \times \text{res } X \longrightarrow X$ $g,x \longrightarrow gx$. Suppose that G is not finite. Show that $i^* \circ i_* : K_o^{\{1\}}(\text{res } X) \longrightarrow K_o^{\{1\}}(\text{res } X)$ is zero. Prove the analogous result for Wh.

9. Let X be a free connected G-CW-complex such that $\pi_1(G,1) \longrightarrow \pi_1(X,x)$ given by evaluation is trivial. Show that $i^* : K_o^G(X) \longrightarrow K_o^1(\text{res } X)$ and $i^* : Wh^G(X) \longrightarrow Wh^1(\text{res } X)$ are zero if G is connected and non-trivial.

10. Extend the definition of $trf_F : K_n(R\Gamma_1) \longrightarrow K_n(R\Gamma_2)$ for $n = 0,1$ to all integers $n \leq -1$, (see 14.22.).

11. Let G be a compact Lie group. Consider the G-push out with i_2 a G-cofibration

$$\begin{array}{ccc} X_o & \xrightarrow{i_1} & X_1 \\ {\scriptstyle i_2}\downarrow & \searrow^{j_o} & \downarrow{\scriptstyle j_1} \\ X_2 & \xrightarrow[j_2]{} & X \end{array}$$

a) If X_o, X_1, and X_2 are finitely dominated, then X is finitely dominated

b) If three of the G-spaces X_o, X_1, X_2 and X are finitely dominated and i_k is a Π-isomorphism for $k = 1,2$, i.e. $\pi_o(i_k^H)$ and $\pi_1(i_k^H, x)$ are bijective for all $H \subset G$, $x \in X_o^H$, then all four spaces are finitely dominated.

c) Statement b) becomes false if one drops the condition that i_k is a Π-isomorphism for $k = 1,2$.

12. Let G be a finite group acting topologically on the standard disk such that the equivariant finiteness obstruction is not zero. (existence of such an action is proved in Quinn [1982]). Show that the G-action cannot be smooth.

13. Extend Theorem 14.49. to $o(\operatorname{res} X) = i_* o(\operatorname{res} X^H)$ for $i : X^H \longrightarrow X$ the inclusion and $H \subset G$ arbitrary.

CHAPTER III

RΓ-MODULES AND GEOMETRY

Summary

In the first two chapters the algebraic notions were all motivated by already existing geometric models. In Chapter III the algebra of RΓ-modules is extended further and applied to geometry.

Let G be a compact Lie group. Suppose that $p : E \longrightarrow B$ is a G-fibration such that $p^{-1}(b)$ is a finitely dominated G_b-CW-complex for $b \in B$. We define *geometric transfer maps* $p^! : K_o^G(B) \longrightarrow K_o^G(E)$ and $p^! : Wh^G(B) \longrightarrow Wh^G(E)$, provided that p is simple (see 15.21. and 15.22.). If p is a G-vector bundle over a G-manifold B and $f : B_o \longrightarrow B$ a G-homotopy equivalence of G-manifolds, then p is simple and $p^!(\tau^G(f)) = \tau^G(\overline{f})$ holds where $\overline{f} : p^*E \longrightarrow E$ is given by the pull back construction and $\tau^G(f)$ and $\tau^G(\overline{f})$ are defined with respect to the simple structure 4.36. We also define *algebraic transfer maps* $p^* : K_n^G(B) \longrightarrow K_n^G(E)$ for $n = 0,1$ (see 15.11.) and show in Theorem 15.25. that $p^!$ and p^* agree. We derive from the algebraic description.

Theorem 15.28. *Let* G *be a finite group of odd order. Consider a* d-*dimensional* G-*vector bundle* $\xi \downarrow B$ *with vanishing* $w_1(\xi) \in H^1(B;\mathbf{Z}/2)$. *Let* $p : S\xi \longrightarrow B$ *be the associated* G-*sphere bundle. Suppose the existence of a* G-*representation* V *such that* res SV *and* $S\xi_b$ *are* G_b-*homotopy equivalent for* $b \in B$. *Then*

$$p_* \circ p^* = (1 - (-1)^d) \cdot \mathrm{id} \quad \square$$

In section 16. we give a second splitting of algebraic K-theory which is based on the restriction functor Res_x MOD-RΓ $\longrightarrow$ MOD-R[x] and inclusion functor I_x : MOD-R[x] $\longrightarrow$ MOD-RΓ . It is dual to the splitting of section 10. coming from the splitting functor S_x and the extension functor E_x . In geometry (Res,I) corresponds to the stratification of a G-space X given by $\{X^H \mid H \subset G\}$, whereas (S,E) corresponds to $\{X_H \mid H \subset G\}$. These two splittings are related by a kind of *K-theoretic Moebius inversion*, i.e. a pair of inverse isomorphisms ω and μ defined in

16.22. and 16.26. An EI-category Γ is finite, if the set of isomorphism classes of objects $\mathrm{Is}\,\Gamma$ and $\mathrm{Hom}(x,y)$ for all $x,y \in \mathrm{Ob}\,\Gamma$ are finite . If Γ is finite, let $m(\Gamma)$ be the smallest common multiple of all numbers $|\mathrm{Aut}(x)|$ for $x \in \mathrm{Ob}\,\Gamma$.

Theorem 16.29. The K-theoretic Moebius inversion Theorem. Let Γ be a finite EI-category and R be noetherian with $m(\Gamma) \in R^*$. Then (E,S), (Res,I) and (ω,μ) are pairs of two another inverse isomorphisms and the following diagram commutes

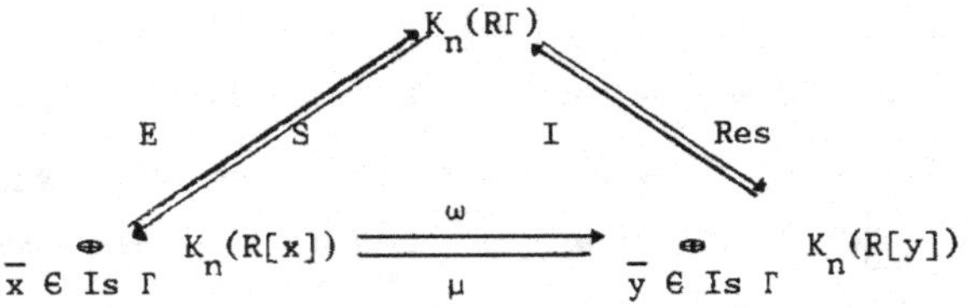

□

Its proof is based on the Filtration Theorem 16.8. for $R\Gamma$-modules which is the dual of the Cofiltration Theorem 9.39. for projective $R\Gamma$-modules.

Theorem 16.29. enables us to detect the isomorphism class of a finitely generated projective $R\Gamma$-module from the isomorphism classes of $R[x]$-modules given by $P(x)$ for $x \in \mathrm{Ob}\,\Gamma$. In particular the isomorphism type of a finitely generated free $R\Gamma$-module F is determined by $\{\mathrm{rk}_R(F(x)) \mid \bar{x} \in \mathrm{Is}\,\Gamma\}$ (see Theorem 16.36.).

Section 17. is devoted to the homological algebra of $R\Gamma$-modules. A $R\Gamma$-module M has length ℓ , if for any chain $\bar{x}_o < \bar{x}_1 < \ldots < \bar{x}_r$ of elements in $\mathrm{Is}\,\Gamma$ with $M(x_o) \neq \{0\}$ we have $r \leq \ell$. Recall that $\bar{x} < \bar{y}$ means that $\mathrm{Hom}(x,y) \neq \emptyset$ and $\bar{x} \neq \bar{y}$ is valid. We call Γ free if $\mathrm{Aut}(y)$ acts freely on $\mathrm{Hom}(x,y)$ for all $x,y \in \mathrm{Ob}\,\Gamma$.

Theorem 17.18. and 17.28. Suppose either that Γ is free or that Γ is finite with $m(\Gamma) \in R^*$. Let M and N be $R\Gamma$-modules such that M has finite length ℓ . Then there is a spectral sequence (E_r,d_r) , $r = 1,2$,... satisfying

a) (E_r,d_r) converges to $\mathrm{Ext}^n_{R\Gamma}(M,N)$.

b) The E_1-term and the first differential can be computed in terms of certain Ext-groups over $R[x]$ for $x \in \mathrm{Ob}\,\Gamma$.

c) $E^{p,q}_r \neq \{0\} \Rightarrow 0 \leq p \leq \ell$ and $0 \leq r$ □

We derive from this spectral sequence some bounds for the homological dimension of a $R\Gamma$-module. If Γ is finite and R a field with $m(\Gamma) \in R^*$, then any $R\Gamma$-module M has a homological dimension less or equal to its length (Proposition 17.31.). Let Γ be finite and free and $m(\Gamma) \in R^*$. Then M has a homological dimension $\leq p$ only if certain inequalities for the numbers $rk_R(M(x))$, $\bar{x} \in Is\,\Gamma$ are valid (Proposition 17.34.).

In section 18 we introduce equivariant Reidemeister torsion for a wide class of G-spaces for a finite group G. A round sturcture $\{\Phi\}$ for a G-space X is a collection of (stable equivalence classes of) RWH-isomorphisms $\Phi(H) : H(X^H;\mathbb{Q})_{odd} \to H(X^H;\mathbb{Q})_{ev}$ for each $H \subset G$ (Definition 18.7.). If X is a finite G-CW-complex with round structure $\{\Phi\}$, we define its equivariant Reidemeister torsion

18.10. $$\ell^G(X,\{\Phi\}) = \ell^G(X) \in Wh(\mathbb{Q}OrG) = \bigoplus_{(H)} Wh(\mathbb{Q}WH) .$$

If X is a finitely dominated G-space (not necessarily G-homotopy equivalent to a finite G-CW-complex) with round structure $\{\Phi\}$, we still can define its reduced equivariant Reidemeister torsion

18.12. $$\bar{\rho}^G(X,\{\Phi\}) = \bar{\rho}^G(X) \in K_1(\mathbb{Q}OrG)/K_1(\mathbb{Z}_{(|G|)}OrG) = \bigoplus_{(H)} K_1(\mathbb{Q}WH)/K_1(\mathbb{Z}_{(|G|)}WH) .$$

A finitely dominated G-space X admits a round structure if and only if $\chi(X^H) \in \mathbb{Z}$ is zero for all $H \subset G$ (Lemma 18.8). Notice that this class of G-spaces is closed under G-push outs and products. We do not make any assumptions about the WH-action on $H(X^H;\mathbb{Q})$ for $H \subset G$ and never divide out norm ideals as it is done in the classical case (cf. Rothenberg [1978]).

We prove sum, diagonal product and join formulas and relate these invariants to Whitehead torsion and the finiteness obstruction. Roughly speaking, the rationalized Whitehead torsion is the difference of the Reidemeister torsion (see 18.18.) and the reduced equivariant Reisemeister torsion is a refinement of the finiteness obstruction. Namely, we construct a certain boundary homomorphism $\partial : K_1(\mathbb{Q}OrG)/K_1(\mathbb{Z}_{(|G|)}OrG) \longrightarrow K_o(\mathbb{Z}OrG)$ and prove

Proposition 18.30. Let X be a finitely dominated G-CW-complex such that $\chi(X^H)=0$, WH acts trivially on $H_*(X^H;\mathbb{Q})$ and $H_*(X^H)$ contains no p-torsion for any prime number p with $(p,|G|) = 1$ for each $H \subset G$. For any $H \subset G$ fix a $\mathbb{Z}$-isomorphism $(H(X^H)/\mathrm{Tors}\, H(X^H))_{odd} \longrightarrow (H(X^H)/\mathrm{Tors}\, H(X^H))_{ev}$. Then $\bar{\rho}^G(X)$ is defined and satisfies

$$\partial(\bar{\rho}^G(X)) = -o^G(X) . \quad \square$$

We construct a map $\rho^G_{\mathbb{R}} : \mathrm{Rep}_{\mathbb{R}}(G) \longrightarrow \mathrm{Wh}(\mathbb{Q}\mathrm{Or}G)$, $[V] \longrightarrow \rho^G(S(V \otimes_{\mathbb{R}} \mathbb{C}))$ and show

Theorem 18.38. The map $\rho^G_{\mathbb{R}}$ is an injective homomorphism $\square$

Thus we reprove de Rham's result

Corollary 18.42. Two real G-representations V and W are linearly isomorphic if and only if their unit spheres are G-diffeomorphic. $\square$

We also introduce Reidemeister torsion for G-manifolds with invariant Riemannian metri (see 18.51., 18.52.) which are related to analytic torsion. Poincare-Reidemeister torsion, introduced in 18.53., measures the failure of equivariant simple Poincaré duality

In section 19. we consider the Swan homomorphism $sw : \mathbb{Z}/|G|^* \longrightarrow K_o(\mathbb{Z}G)$ sending $\bar{r}$ to $[M]$ if M is any finite abelian group of order r with trivial G-action. Let $\overline{sw} : \mathbb{Z}/|G|^* \longrightarrow K_1(\mathbb{Q}G)/K_1(\mathbb{Z}_{(|G|)}G)$ send $\bar{r}$ to the class of $\mathbb{Q} \xrightarrow{r} \mathbb{Q}$ or, equivalently, of the unit $1 + \frac{r-1}{|G|} \cdot \sum_{g \in G} g \in \mathbb{Q}G^*$. Let $\partial : K_1(\mathbb{Q}\mathrm{Or}G)/K_1(\mathbb{Z}_{(|G|)}G) \longrightarrow K_o(\mathbb{Z}G)$ be induced from the boundary map of the localization square. The following theorem was conjectured in Rothenberg [1978a] and Wall [1979] .

Theorem 19.4. The map $\overline{sw}$ is injective and a lift of sw, i.e. $\partial \circ \overline{sw} = -sw$. $\square$

Let $C(G)$ be $\prod_{(H)} \mathbb{Z}$ and $\bar{C}(G) = C(G)/|G| \cdot C(G)$. The Burnside ring $A(G)$ is a subrin of $C(G)$ and contains $|G| \cdot C(G)$. Put $\bar{A}(G) = A(G)/|G| \cdot C(G)$ and $\mathrm{Inv}(G) = \bar{C}(G)^*/\bar{A}(G)$ We define generalized Swan homomorphisms and boundary maps

19.12. $\overline{SW} : \bar{C}(G)^* \longrightarrow K_1(\mathbb{Q}\mathrm{Or}G)/K_1(\mathbb{Z}_{(|G|)}\mathrm{Or}G)$

$SW : \bar{C}(G)^* \longrightarrow K_o(\mathbb{Z}\mathrm{Or}G)$

18.28. $\partial : K_1(\mathbb{Q}\mathrm{Or}G)/K_1(\mathbb{Z}_{(|G|)}\mathrm{Or}G) \longrightarrow K_o(\mathbb{Z}\mathrm{Or}G)$

and extend Theorem 19.4. to modules over $\mathrm{Or}\,G$.

Theorem 19.13. and 19.19.

a) $\partial \circ \overline{SW} = -SW$

b) $\overline{SW}$ induces an injection $\overline{SW}_o$

$$\begin{array}{ccc} \overline{C}(G)^* & \longrightarrow & K_1(\mathbb{Q}OrG)/K_1(\mathbb{Z}_{(|G|)}OrG) \\ \downarrow \mathrm{pr} & \nearrow \overline{SW}_o & \\ Inv(G) & & \end{array}$$

□

In section 20. we apply Reidemeister torsion and the generalized Swan homomorphisms to G-homotopy representations of finite groups G. A G-homotopy representation is a finite-dimensional G-CW-complex X such that X^H is homotopy equivalent to $S^{n(H)}$ for $n(H) = \dim(X^H)$ and $H \subset G$. The dimension function of X is $Dim(X) = (n(H)+1)_{(H)} \in C(G)$. Let X and Y be two G-homotopy representations with the same dimension function. Let $[X,Y]^G$ be the set of G-homotopy classes of G-maps. After we have choosen a coherent orientation, we obtain a map $DEG : [X,Y]^G \longrightarrow C(G)$, $[f] \longrightarrow (\deg f^H)_{(H)}$.

Proposition 20.12. b. Suppose $n(L) \geq n(K)+2$ for $K,L \in Iso\, X$, $L \subset K$, $L \neq K$ or suppose G to be nilpotent. Then $DEG : [X,Y]^G \longrightarrow C(G)$ is injective with finite cokernel.

Hence we have to determine the image of DEG in order to control $[X,Y]^G$. In other words, given a collection $\{d(H) \in \mathbb{Z} \mid H \subset G\}$, we must decide whether there is a G-map $f : X \longrightarrow Y$ with $\deg f^H = d(H)$ for $H \subset G$. We say that $\{d(H) \in \mathbb{Z} \mid H \subset G\}$ satisfies the unstable conditions if

20.8. i) $d(H) = d(K)$ if $(H) = (K)$

ii) $d(H) = d(K)$ if $n(H) = n(K)$

iii) $d(H) \in \{\pm 1\}$ if $n(H) = 0$

iv) $d(H) = 1$ if $n(H) = -1$

Notice that $\{\deg f^H \mid H \subset G\}$ satisfies 20.8. We call a collection $\{d(H) \in \mathbb{Z} \mid H \subset G\}$ of integers prime to $|G|$ a weight function if and only if the following holds:

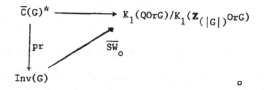

given $\{e(H) \in \mathbf{Z} \mid H \subset G\}$, there is a G-map $f : X \longrightarrow Y$ with $\deg(f^H) = e(H)$ for $H \subset G$ if and only if $\{e(H) \in \mathbf{Z} \mid H \subset G\}$ satisfies the unstable conditions 20.8. and $(d(H) \cdot e(H))_{(H)} \in C(G)$ lies in the Burnside ring $A(G)$ (Definition 20.35). Notice that $A(G) \subset C(G)$ can be described by explicit congruences. Hence $[X,Y]^G$ is determined by specifying a weight function.

<u>Theorem 20.38.</u> <u>Let</u> $\{d(H) \mid H \subset G\}$ <u>be a collection of integers prime to</u> $|G|$. <u>It is a weight function of and only if it satisfies the unstable conditions 20.8. and</u> $\overline{SW}$ <u>sends</u> $(\overline{d(H)} \in \mathbf{Z}/|G|^*)_{(H)} \in \overline{C}(G)^*$ <u>to</u> $\overline{\rho}^G(Y) - \overline{\rho}^G(X) \in K_1(\mathbb{Q}OrG)/K_1(\mathbf{Z}_{(|G|)}OrG)$ □

We get as an immediate consequence the classification of G-homotopy representations by the reduced equivariant Reidemeister torsion.

<u>Corollary 20.39.</u> <u>The following statements are equivalent for</u> G-<u>homotopy representat</u> X <u>and</u> Y <u>with</u> $Dim(X) = Dim(Y)$ <u>and a coherent orientation</u> $\Phi(X,Y)$.

a) X <u>and</u> Y <u>are oriented</u> G-<u>homotopy equivalent</u>.

b) X <u>and</u> Y <u>are stably oriented</u> G-<u>homotopy equivalent</u>.

c) $\overline{\rho}^G(X) - \overline{\rho}^G(Y) = 0$. □

We also analyse the various homotopy representation groups and examine the case of an abelian group G more closely.

15. The transfer associated with a G-fibration

We introduce the notion of a G-fibration $p : E \longrightarrow B$. Under certain assumptions about the fibres $F_b = p^{-1}(b)$ for $b \in B$, we define both algebraically and geometrically transfer maps $p^* : K_o^G(B) \longrightarrow K_o^G(E)$ and $Wh^G(B) \longrightarrow Wh^G(E)$ and show that they agree. The algebraic description is used to state some vanishing results.

Let G be a topological group and $p : E \longrightarrow B$ be a G-map. We say that p has the G-HLP (G-homotopy lifting property) for a G-space X if for any G-maps $h : X\times I \rightarrow B$ and $f : X \longrightarrow E$ satisfying $p \circ f = h \circ i_o$ for $i_o : X \longrightarrow X\times I$ $x \longrightarrow (x,o)$ there is a G-map $\bar{h} : X\times I \longrightarrow E$ such that $\bar{h} \circ i_o = f$ and $p \circ \bar{h} = h$ holds.

15.1.

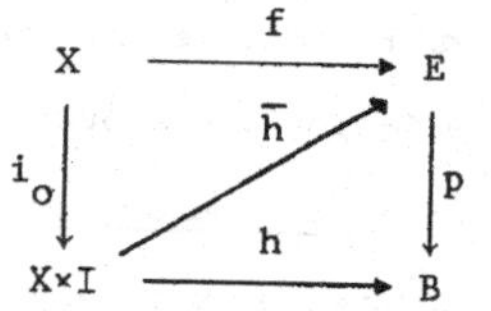

Definition 15.2. We call $p : E \longrightarrow B$ a G-fibration if p has the G-HLP for all G-CW-complexes X. □

15.3. A G-map $p : E \longrightarrow B$ is a G-fibration if and only if $p^H : E^H \longrightarrow B^H$ has the (non-equivariant) HLP for D^n, $n \geq 0$.

15.4. Let $p : E \longrightarrow B$ be a numerable locally trivial (G,α,Γ)-bundle in the sense of tom Dieck [1969] or Lashof-Rothenberg [1978] where G is a compact Lie group, Γ a topological group and $\alpha : G \longrightarrow Aut(\Gamma)$ an homomorphism. Let F be a space with left G- and left Γ-action such that $g(\gamma f) = \alpha(g)(\gamma)gf$ holds for $g \in G$, $\gamma \in \Gamma$, $f \in F$. Then we have the associated G-bundle $E \times_\Gamma F \longrightarrow B$ with fibre F. It is a G-fibration by Dold [1963] and 15.3.

Let G be a compact Lie group and M be a smooth G-manifold. Then $M_{(H)} \subset M$ is a G-submanifold with a normal G-vector bundle ν. The associated G-sphere and G-disc bundles $S\nu$ and $D\nu$ over $M_{(H)}$ are G-fibrations. □

15.5. Let (X,A) be a pair of G-CW-complexes and $p : E \longrightarrow B$ be a G-fibration. Then we can solve the relative G-homotopy lifting problem

$$\begin{array}{ccc} A\times I \cup X & \xrightarrow{\quad f \quad} & E \\ {\scriptstyle i_o}\downarrow & \overset{\bar{h}}{\nearrow} & \downarrow{\scriptstyle p} \\ X\times I & \xrightarrow[\quad h \quad]{} & B \end{array}$$ □

15.16. The pull back of a G-fibration is again one. □

A G-fibre map $(\bar{f},f) : p \longrightarrow p'$ of G-fibrations is a pair of G-maps $\bar{f} : E \longrightarrow E'$ and $f : B \longrightarrow B'$ satisfying $p' \circ \bar{f} = f \circ p$. We call $(\bar{f}_o,f_o)$ and $(\bar{f}_1,f_1) : p \longrightarrow p'$ G-<u>fibre homotopic</u> if there is a G-<u>fibre homotopy</u> $(\bar{h},h) : p\times I \longrightarrow p'$ with $(\bar{h}_i,h_i) = (\bar{f}_i,f_i)$ for $i = 0,1$. We call $(\bar{h},h)$ a <u>strong</u> G-<u>fibre homotopy</u> if h is stationary i.e. $h(b,t) = h(b,0)$ for $b \in B$. Notice that this implies $f_o = f_1$. Let $p : E \longrightarrow B$ and $p' : E' \longrightarrow B$ be G-fibrations over the same space B . We call a G-fibre map $(\bar{f},id) : p \longrightarrow p'$ a G-<u>fibre homotopy equivalence</u> if there is a G-fibre map $(\bar{g},id) : p' \longrightarrow p$ with both composites strongly G-fibre homotopic to the identity. □

15.7. Let $p : E \longrightarrow B$ be a G-fibration. Let $h : X\times I \longrightarrow B$ be a G-homotopy between f_o and $f_1 : X \longrightarrow B$. Choose a solution $\bar{h}$ of

$$\begin{array}{ccc} f_o^*E & \xrightarrow{\quad f_o^* \quad} & E \\ {\scriptstyle i_o}\downarrow & \overset{\bar{h}}{\nearrow} & \downarrow{\scriptstyle p} \\ f_o^*E\times I & \xrightarrow[\quad h \circ (p_o\times id) \quad]{} & B \end{array}$$

Denote by $g_h : f_o^*E \longrightarrow f_1^*E$ the G-fibre map given by $\bar{h}_1$, $p_o := f_o^*p$ and the pull back property of f_1^*E .

Next consider a second G-homotopy $k : X\times I \longrightarrow B$ between f_o and f_1 and define $\bar{k}$ and g_k as above. Let $M : X\times I\times I \longrightarrow B$ be a G-homotopy relative $X\times\partial I$ between h and k . Choose a solution $\bar{M}$ of

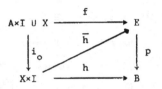

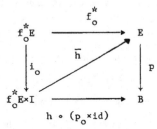

$$\begin{array}{ccc}
f_o^*E\times I\times\partial I \cup f_o^*E\times\{0\}\times I & \xrightarrow{\ \bar h\,\cup\,\bar k\,\cup(\bar f_o\circ pr)\ } & E \\
\Big\downarrow{\scriptstyle i} & \overset{\bar M}{\nearrow} & \Big\downarrow{\scriptstyle p} \\
f_o^*E\times I\times I & \xrightarrow[\ M\,\circ\,(p_o\times id\times id)\ ]{} & B
\end{array}$$

Let $g_M : f_o^*E \times I \longrightarrow f_1^*E$ be given by $\bar{M}(-,1,-)$ and $p_o \times id$. One easily checks (see Switzer [1975], p. 342)

Lemma 15.8.

a) $g_h : f_o^*E \longrightarrow f_1^*E$ is a G-fibre homotopy equivalence and $\bar{h} : f_o^*E\times I \longrightarrow E$ and $h : X\times I \longrightarrow B$ define a G-fibre homotopy between $\bar{f}_1 \circ g_h$ and $\bar{f}_o$.

b) $g_M : f_o^*E\times I \longrightarrow f_1^*E$ is a strong G-fibre homotopy between g_h and g_k. We obtain a G-fibre homotopy of G-fibre homotopies between $(\bar{f}_1 \circ g_M) * \bar{k}$ and $\bar{h}$ by $\bar{M}$, i.e. we have $\bar{M}(x,t,0) = \bar{h}(x,t)$, $\bar{M}(x,1,t) = \bar{k}(x,t)$, $\bar{M}(x,0,t) = \bar{f}_o(x)$ and $\bar{M}(x,1,t) = \bar{f}_1 \circ g_M(x,t)$ for $x \in X$, $t \in I$.

Given a G-space Z, define a category $\mathfrak{C} \downarrow Z$ as follows. An object x is a G-space X together with a G-map $x : X \longrightarrow Z$. Given two objects $x : X \longrightarrow Z$ and $y : Y \rightarrow Z$, we consider pairs (f,h) consisting of a G-map $f : X \longrightarrow Y$ and a G-homotopy $h : X\times I \longrightarrow Z$ between x and $y \circ f$. We call two such pairs (f_o,h_o) and (f_1,h_1) equivalent if there are G-maps $\bar{f} : X\times I \longrightarrow Y$ and $\bar{h} : X\times I\times I \longrightarrow Z$ satisfying $\bar{f}(u,i) = f_i(u)$, $\bar{h}(u,t,i) = h_i(u,t)$, $\bar{h}(u,0,t) = x(u)$ and $\bar{h}(u,1,t) = y \circ f(u,t)$ for $u \in X$, $t \in I$, $i \in \{0,1\}$. An equivalence class $[f,h]$ of such pairs is a morphism $x \longrightarrow y$. Composition is defined on representatives by $[g,k] \circ [f,h]$ $[g \circ f, h * (k \circ f\times id)]$ where $*$ denotes the obvious composition of G-homotopies.

From now on we suppose that assumption 8.13. is satisfied.

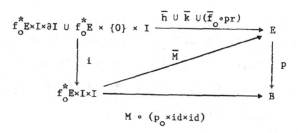

We get from Lemma 15.8. a covariant functor $\tilde{t}p_p : \Pi/(G,B) \longrightarrow \mathfrak{C} \downarrow E$. It sends an object $x : G/H \longrightarrow B$ to $\bar{x} : x^*E \longrightarrow B$ given by the pull-back. Let $[\sigma,w] : x \rightarrow y$ be a morphism in $\Pi/(G,B)$ represented by a G-map $\sigma : G/H \longrightarrow G/K$ and a G-homotopy $w : G/H \times I \longrightarrow B$ between x and $y \circ \sigma$. By Lemma 15.8 . a) we get a strong G-fibre homotopy $\bar{w} : x^*E \times I \longrightarrow E$. Define the image of $[\sigma,w] : x \longrightarrow y$ under $\tilde{tp}_p$ by the class of $(\bar{\sigma} \circ g_w, \bar{w}) : \bar{x} \longrightarrow \bar{y}$ in $\mathcal{C} \downarrow E$ where $\bar{\sigma} : \sigma^* y^* E \longrightarrow y^* E$ is given by the pull back construction. This is well defined by Lemma 15.8. b). Composition with p defines a functor $\mathcal{C} \downarrow p : \mathcal{C} \downarrow E \longrightarrow \mathcal{C} \downarrow B$. Let tp_p be $(\mathcal{C} \downarrow p) \circ \tilde{tp}_p$

<u>Definition 15.9.</u> <u>We call the covariant functor</u>

$$\tilde{tp}_p : \Pi/(G,B) \longrightarrow \mathcal{C} \downarrow E \quad \text{resp.} \quad tp_p : \Pi/(G,B) \longrightarrow \mathcal{C} \downarrow B$$

<u>the total fibre transport resp. the fibre transport of</u> p . □

Given a G-space Z , let ho $CC\text{-}R\Pi/(G,Z)$ be the homotopy category of projective $R\Pi/(G,Z)$-chain complexes. The functor

15.10. $$C^s \downarrow Z : \mathcal{C} \downarrow Z \longrightarrow \text{hoCC-}R\Pi/(G,Z)$$

sends $x : X \longrightarrow Z$ to $x_* C^s(X)$ if $C^s(X)$ is the singular $R\Pi/(G,Z)$-chain complex of X (see Definition 13.1.) . Let $[f,h] : x \longrightarrow y$ be a morphism and $(f,h) \in [f,h]$ a representative. We get a $\mathbf{Z}\Pi/(G,Y)$-chain map $C^s(f) : f_* C^s(X) \longrightarrow C^s(Y)$ and a $\mathbf{Z}\Pi/(G,Y)$-chain isomorphism $C^s(h/) : x_* C^s(X) \longrightarrow y_* f_* C^s(X)$ (see 13.4. and 13.6.). Let $C^s \downarrow Z([f,h])$ be represented by $y_* C^s(f) \circ C^s(h/) : x_* C^s(X) \longrightarrow y_* C^s(Y)$ .Notice that it is essential for the construction of $C^s \downarrow Z$ that the homotopy h is part of the structure of a morphism (f,h) . It is not sufficient to define a morphism $x \to y$ as a G-homotopy class $[f]$ of G-maps $f : X \to Y$ satisfying $y \circ f \simeq_G x$.

Suppose that for any $b \in B$ the G_b-space $F_b = p^{-1}(b)$ is finitely dominated. Then the composition $C^s \downarrow E \circ \tilde{tp}_p$ is a covariant functor $\Pi/(G,B) \longrightarrow \text{hoFDCC-}\mathbf{Z}\Pi/(G,E)$ into the homotopy category of finitely dominated projective $\mathbf{Z}\Pi/(G,E)$-chain complexes by Proposition 13.10 and 13.2 . Hence we obtain from 14.22. a <u>transfer homomorphism</u> associated with p

15.11. $$p^* := \text{trf}_{C^s \downarrow E \circ \tilde{tp}_p} : K_n^G(B) \longrightarrow K_n^G(E)$$

for $n = 0,1$. We call p <u>simple</u> if $p^* : K_1^G(B) \longrightarrow K_1^G(F)$ induces $p^* : Wh^G(B) \longrightarrow Wh^G(E)$.

Next we want to interpret p^* geometrically. Notice in the sequel the analogy with section seven. We need the following hypothesis (compare Assumption 7.3.).

<u>Assumption 15.12.</u> The G_b-space $F_b = p^{-1}(b)$ is G_b-homotopy equivalent to a finite G_b-CW-complex for any $b \in B$. □

Given $b : G/H \longrightarrow B$ in $\Pi/(G,B)$, define an homomorphism

15.13.
$$\phi(b) : \mathrm{Aut}(b) \longrightarrow Wh^G(E)$$

as follows. Choose a simple structure ξ on b^*E (see section 4). Then $\phi(b)$ sends $[\sigma,w] \in \mathrm{Aut}(b)$ to $\overline{b}_*\tau^G(\tilde{t}p_p([\sigma,w]) : (b^*E,\xi) \longrightarrow (b^*E,\xi))$. This is independent of the choice of ξ. Now one easily verifies that $\phi(b)([\sigma,w])$ is just the image of $[\sigma,w]$ under $K_1^G(B) \xrightarrow{p^*} K_1^G(E) \longrightarrow Wh^G(E)$. This implies

<u>Lemma 15.14.</u> p <u>is simple if and only if</u> $\phi(b)$ <u>is trivial for all</u> $b \in \mathrm{Ob}\Pi/(G,B)$.

Suppose that B is a finite G-CW-complex. Let I_n be the set of n-cells. In the sequel all sums run over I_n and we sometimes omitt the index n. Make the following choices (compare 7.8. and 7.9.)

15.15. For any $\overline{b} \in \mathrm{Is}\,\Pi/(G,B)$ fix a representative $b : G/K \longrightarrow B$ and a simple structure $\gamma(b)$ on b^*E. □

15.16.

i) Fix for $n \geq 0$ a G-push out

$$\begin{array}{ccc} \coprod G/H_i \times S^{n-1} & \xrightarrow{\coprod q_i^n} & B_{n-1} \\ \downarrow & & \downarrow \\ \coprod G/H_i \times D^n & \xrightarrow{\coprod Q_i^n} & B_n \end{array}$$

ii) $Q_i^n|G/H_i \times * : G/H_i \longrightarrow B$ is an object in $\Pi/(G,B)$. Let $b_i^n : G/K_i \to B$ be the representative of its isomorphism class specified in 15.15. Choose an isomorphism $[\sigma_i^n,w_i^n] : Q_i^n|G/H_i \times * \longrightarrow b_i^n$ in $\Pi/(G,B)$.

iii) Fix a representative (σ_i^n,w_i^n) for $[\sigma_i^n,w_i^n]$, that is a G-map $\sigma_i^n : G/H_i \to G/K_i$ and a G-homotopy $w_i^n : G/H_i \times I \longrightarrow B$ with $w_i^n(gH_i,x,0) = Q_i^n(gH_i,*)$ and

$w_i^n(gH_i,x,1) = b_i^n \circ \sigma_i^n(gH_i)$. Let $g_i^n : Q_i^{n*}E \longrightarrow \sigma_i^{n*}b_i^{n*}E \times D^n$ be a strong G-fibre homotopy equivalence constructed in 15.7. □

15.17. With these choices we get a unique simple structure ξ on E . Suppose ξ_{n-1} on $E_{n-1} = p^{-1}(B_{n-1})$ is already defined. We get simple structures on $b_i^{n*}E{\times}S^{n-1}$ and $b_i^{n*}E{\times}D^n$ from $\gamma(b_i^n)$. Now equip $Q_i^{n*}E$ and $q_i^{n*}E$ with the simple structures such that $(\overline{\sigma_i^n} \times \mathrm{id}) \circ g_i^n : Q_i^{n*}E \longrightarrow b_i^{n*}E \times D^n$ and $(\overline{\sigma_i^n} \times \mathrm{id}) \circ g_i^n| : q_i^{n*}E \rightarrow b_i^{n*}E{\times}S^{n-1}$ have vanishing Whitehead torsion. We get from 15.16. i) a G-push out with $\overline{i}$ a G-cofibration (compare Lemma 1.26.)

$$\begin{array}{ccc} \coprod q_i^{n*}E & \longrightarrow & E_{n-1} \\ \downarrow \overline{i} & & \downarrow \\ \coprod Q_i^{n*}E & \longrightarrow & E_n \end{array}$$

Now equip E_n with the simple structure such that this is a G-push out of G-spaces with simple structure. □

We make the following assumption (compare Assumption 7.23.)

<u>Assumption 15.18.</u> p is simple. □

Now suppose that we have made also second choices 15.15'. and 15.16'. Define an homomorphism depending only on 15.15. and 15.15'.

15.19. $$\Theta_p : U^G(B) \longrightarrow Wh^G(E)$$

as follows. Given $u \in U^G(B)$, choose an isomorphism $[\sigma,w] : b \longrightarrow b'$ between the representatives $b,b' \in u$ of 15.15. and 15.15'. Applying $\tilde{tp}_p$ gives a G-homotopy equivalence $b^*E \longrightarrow b'^*E$. Using the simple structures $\gamma(b)$ and $\gamma(b')$ it defines an element in $Wh^G(b'^*E)$. Let its image under $\overline{b}'_* : Wh^G(b'^*E) \longrightarrow Wh^G(E)$ be $\Theta_p(u)$. This is independent of the choice of $[\sigma,w]$ by Lemma 15.14. as p is simple.

<u>Lemma 15.20.</u> <u>Let</u> ξ <u>and</u> ξ' <u>be the simple structures we get from</u> 15.17. <u>according to the choices</u> 15.15., 15.16., <u>and</u> 15.15' , 15.16'. <u>Then</u>

$$\tau^G(\mathrm{id} : (E,\xi) \longrightarrow (E,\xi')) = \Theta_p(\chi^G(B)).$$

<u>Proof:</u> Lemma 15.8. guarantees that the G-homotopy class of $(\bar{G}_i^n \times id) \circ g_i^n : Q_i^{n*}E \to b_i^{n*}E \times D^n$ does not depend on the choice 15.16. iii) so that ξ is independent of the choice 15.16.iii). Now one can proceed as in the proof of Lemma 7.13. □

Let $p : E \longrightarrow B$ be a G-fibration satisfying Assumptions 15.12. and 15.18. Let B be a finite G-CW-complex. Because of Lemma 15.20. the following definition of a map

15.21. $$p^! : Wh^G(B) \longrightarrow Wh^G(E)$$

makes sense. Given $\eta \in Wh^G(B)$ choose a G-fibre homotopy equivalence $(\bar{f},f) : p_o \to p$ such that $p_o : E_o \longrightarrow B_o$ is a G-fibration over a finite G-CW-complex B_o and $\tau^G(f) = \eta$ holds. Fix choices 15.15. and 15.16. for both p_o and p. Let ξ_o and ξ be the simple structures on E_o and E defined in 15.17. By composition with $\bar{f}$ we obtain from the choice 15.15. for p_o a choice 15.15. for p. Regarding this as the first and the given choice 15.15. for p as the second, let $\Theta_{\bar{f}} : U^G(B) \to Wh^G(E)$ be the map defined in 15.19. Then we put

$$p^!(\eta) = \tau^G(\bar{f} : (E_o,\xi_o) \longrightarrow (E,\xi)) - \Theta_{\bar{f}}(\chi^G(B))$$

Under assumption 15.12. the total space E is G-homotopy equivalent to a finite G-CW-complex if B is. Hence we get an homomorphism

15.22. $$p^! : Wa^G(B) \longrightarrow Wa^G(E)$$

sending $[f : X \longrightarrow B]$ to $[\bar{f} : f^*E \longrightarrow E]$. Define

15.23. $$p^! : U^G(B) \longrightarrow U^G(E)$$

by $p^!(u) = \bar{b}_* \chi^G(b^*E)$ if $b : G/H \longrightarrow B$ is a representative of $u \in \text{Is}\,\Pi/(G,B)$. The direct sum of 15.22. and 15.23. and the natural isomorphism $U^G(B) \oplus Wa^G(B) \to K_o^G(B)$ and $U^G(E) \oplus Wa^G(E) \longrightarrow K_o^G(E)$ of 14.10. yield

15.24. $$p^! : K_o^G(B) \longrightarrow K_o^G(E)$$

<u>Theorem 15.25.</u> <u>Let</u> $p : E \longrightarrow B$ <u>be a G-fibration satisfying assumptions 15.12. and 15.18</u>. <u>Then</u> p^* <u>and</u> $p^! : K_o^G(B) \longrightarrow K_o^G(E)$ <u>and</u> $Wh^G(B) \longrightarrow Wh^G(E)$ <u>are defined and agree</u>.

Proof. We give only the sketch of a proof for Wh^G, K_o^G is done similarly. In the proof of Theorem 14.16. we have introduced a geometric splitting

$$\bigoplus_{(H)} Wh^{\{1\}}(EWH \times_{WH} B^H) \longrightarrow Wh^G(B) .$$

It corresponds under the natural isomorphisms relating geometric and algebraic Whitehead groups (Theorem 14.16.) to the algebraic splitting of Theorem 10.34. Now one easily reduces the claim to a G-fibration $p : E \longrightarrow B$ with the property that B is a connected finite CW-complex with trivial G-action and the assertion that the compositions $Wh^{\{1\}}(B) \xrightarrow{r} Wh^G(B) \xrightarrow{p^*} Wh^G(E)$ and $Wh^{\{1\}}(B) \xrightarrow{r} Wh^G(B) \xrightarrow{p^!} Wh^G(E)$ agree if r is restriction with $G \longrightarrow \{1\}$.

We leave it to the reader to carry over the proof in the case $G = \{1\}$ in Lück [1986] to this situation. Compare also Theorem 14.40. and its proof. □

Let $p : E \longrightarrow B$ be a G-fibration such that the G_b-space F_b is finitely dominated for $b \in B$. Composing the fibre transport $tp_p : \Pi/(G,B) \longrightarrow \mathcal{C} \downarrow B$ of Definition 15.9. with $C^s \downarrow B : \mathcal{C} \downarrow B \longrightarrow$ ho CC-$\mathbf{Z}\Pi/(G,B)$ of 15.10. defines a functor $\Pi/(G,B) \longrightarrow$ ho FDCC-$R\Pi/(G,B)$. Let $trf_{C^s \downarrow B \circ tp_p}$ be the transfer associated with it in 14.22. and $p_* : Wh^G(E) \longrightarrow Wh^G(B)$ be the homomorphism induced by p. We get from the definitions

<u>Proposition 15.26.</u> $p_* \circ p^* = trf_{C^s \downarrow B \circ tp_p}$.

<u>Corollary 15.27.</u> <u>Let</u> $p_o : E_i \longrightarrow B$ <u>for</u> $i = 0,1$ <u>be a</u> G-<u>fibration over</u> B <u>such that</u> $p_i^{-1}(b)$ <u>is a finitely dominated</u> G_b-<u>space for</u> $b \in B$.

a) <u>Suppose the existence of a functor</u> $\phi : \Pi/(G,E_o) \longrightarrow \Pi/(G,E_1)$ <u>such that the composition of the functor</u> ho FDCC($\mathbf{Z}\Pi/(G,E_o)$) $\longrightarrow$ ho FDCC($\mathbf{Z}\Pi/(G,E_1)$) <u>induced by induction with</u> ϕ <u>and</u> $\tilde{tp}_{p_o} : \Pi/(G,B) \longrightarrow$ ho FDCC($\mathbf{Z}\Pi/(G,E_o)$) <u>is naturally equivalent to</u> $\tilde{tp}_{p_1}$. <u>Then we have</u>

$$\phi_* \circ p_o^* = p_1^*$$

b) <u>If</u> tp_{p_o} <u>and</u> $tp_{p_1} : \Pi/(G,B) \longrightarrow \mathcal{C} \downarrow B$ <u>are naturally equivalent then</u>

$$p_{o*} \circ p_o^* = p_{1*} \circ p_1^* \qquad \square$$

Now we want to analyse the situation which is relevant for geometry. Namely, let $\xi \downarrow B$ be a G-vector bundle and $p : S\xi \longrightarrow B$ the associated G-sphere bundle. Recall from 15.4. p is a G-fibration. Consider an automorphism $[\sigma,w]$ of $b : G/H \longrightarrow B$ in $\Pi/(G,B)$. Then $\tilde{tp}_p([\sigma,w])$ is given by induction $ind_{G_b}^G$ applied to a G_b-diffeomorphism $S\xi_b \longrightarrow S\xi_b$. Hence p satisfies assumptions 15.12. and 15.18. by 4.36. We have algebraic transfer maps

$$p^* : K_o^G(B) \longrightarrow K_o^G(S\xi)$$

$$p^* : Wh^G(B) \longrightarrow Wh^G(S\xi)$$

which agree with the geometric transfer homomorphisms $p^!$ by Theorem 15.25.

<u>Theorem 15.28.</u> <u>Let</u> G <u>be a finite group of odd order</u>. <u>Consider a</u> d-<u>dimensional</u> G-<u>vector bundle</u> $\xi \downarrow B$ <u>with vanishing</u> $w_1(\xi) \in H^1(B;\mathbf{Z}/2)$. <u>Let</u> $p : S\xi \longrightarrow B$ <u>be the associated</u> G-<u>sphere bundle</u>. <u>Suppose the existence of a</u> G-<u>representation</u> V <u>such that</u> res SV <u>and</u> $S\xi_b$ <u>are</u> G_b-<u>homotopy equivalent for</u> $b \in B$. <u>Then</u>

$$p_* \circ p^* = (1 - (-1)^d)id .$$

<u>Proof.</u> The quotient V/V^G is of complex type (see Serre [1977], 13.9 b). Hence $\dim SV = \dim SV^H \bmod 2$ holds for $H \subset G$ so that $\chi^G(SV) \in A(G) \equiv U^G(*)$ is $(1 - (-1)^d)[G/G]$. Let η be the trivial G-vector bundle $q : B\times V \longrightarrow B$. Then $q_* \circ q^*$ is given by $\chi^G(SV) \in A(G)$ and the $U^G(*)$-module structure on $K_o^G(B)$ and $Wh^G(B)$ by Theorem 14.21. so that $q_* \circ q^* = (1 + (-1)^d) \cdot id$ is true. Because of Corollary 15.27. b it suffices to construct a natural equivalence $\Phi : tp_q \rightarrow tp_p$ of functors $\Pi/(G,B) \longrightarrow \mathcal{C} \downarrow B$ since then $p_* \circ p^* = q_* \circ q^*$ holds.

We can suppose without loss of generality that B is connected. Fix an object $y : G \longrightarrow B$ and a non-equivariant homotopy equivalence $\Phi_o : SV \longrightarrow S\xi_y$. Consider an arbitrary object $b : G/H \longrightarrow B$. Choose a path w from b to y in B. Let $g_w : S\xi_y \longrightarrow S\xi_b$ be the (non-equivariant) fibre homotopy equivalence defined in 15.7. for the fibration $S\xi \longrightarrow B$. By assumption there is an H-homotopy equivalence $SV \longrightarrow S\xi_b$. Since H has odd order $A(H)^* = \{\pm 1\}$ holds. (tom Dieck [1979], p. 8). Therefore two H-homotopy equivalences $SV \rightarrow S\xi_b$ are H-homotopic if and only if they are homotopic as non-equivariant maps (see tom Dieck [1987] II.4, II.8). Hence there is an H-homotopy equivalence

$\phi(b) : SV \longrightarrow S\xi_b$ uniquely determined up to H-homotopy by the property that $g_w \circ \Phi_o$ and $\phi(b)$ are homotopic. Since $w_1(S\xi)$ vanishes, the homotopy class of g_w does not depend on w. Notice for the projection $\sigma : G \longrightarrow G/H$ that the G-map $y^*S\xi \longrightarrow \sigma^*b^*S\xi$ induced from g_w together with the G-homotopy $y^*S\xi \times I \longrightarrow G \times I \xrightarrow{w} B$ represents the morphism $tp_p((\sigma,w)) : tp_p(y) \longrightarrow tp_p(b)$ in $\mathcal{C} \downarrow B$. Define a morphism $\Phi(b) : tp_q(b) \longrightarrow tp_p(b)$ in $\mathcal{C} \downarrow B$ by $G \times_H \phi(b) : b^*S\eta = G/H \times SV = G \times_H SV \to G \times_H \xi_b = b^*S\xi$ and the stationary G-homotopy $b^*S\eta \times I \longrightarrow G/H \times I \xrightarrow{pr} G/H \xrightarrow{b} B$. This defines the natural equivalence $\Phi : tp_q \longrightarrow tp_p$. □

<u>Corollary 15.29.</u> <u>Under the conditions of Theorem 15.28. the transfer</u> p^* <u>vanishes provided that</u> $\dim \xi$ <u>is even and</u> $\dim \xi_b^H \notin \{1,2\}$ <u>for</u> $H \subset G_b$, $b \in B$ <u>holds.</u>

<u>Proof.</u> For any $H \subset G$ we get a spherical fibration $p^H : S\xi^H \longrightarrow B^H$. For $b \in B^H$ we have $\dim S\xi_b^H \notin \{0,1\}$ so that $p^H|C : (p^H)^{-1}(C) \longrightarrow C$ is 2-connected or $(p^H)^{-1}(C)$ is empty for any component $C \in \pi_o(B^H)$. We get from Proposition 8.33. that $p_* : Wh^G(S\xi) \longrightarrow Wh^G(B)$ is injective. Since $p_* \circ p^*$ is zero by Theorem 15.28. the claim $p^* = 0$ follows. □

<u>Example 15.30.</u> Let G be a compact Lie group and M be a G-manifold with $Iso(M) = \{G,\{1\}\}$. Suppose that M^G and M are connected. We have introduced an homomorphism

$$\Phi : Wh^G_{Iso}(M) \longrightarrow Wh^G_\rho(M)$$

in 4.43. we want to analyse further. Recall the splitting $Wh^G_{Iso}(M) = Wh(\pi_1(M^G)) \oplus Wh(\pi_1(M_G/G))$ and $Wh^G_\rho(M) = Wh(\pi_1(M^G)) \oplus Wh(\pi_1(EG \times_G M))$. Let k be the composition $\pi_1(M_G/G) \xrightarrow{\cong} \pi_1(EG \times_G M_G) \longrightarrow \pi_1(EG \times_G M^G)$. Consider the sphere bundle $S\nu \downarrow M^G$ associated with the normal G-vector bundle ν of M^G in M. Dividing out the G-action defines a non-equivariant fibration $p : S\nu/G \longrightarrow M^G$ with typical fibre SV/G if V is the normal slice G-representation. Let $p^* : Wh(\pi_1(M^G)) \longrightarrow Wh(\pi_1(S\nu/G))$ be the associated transfer. Denote by $\partial_o : \pi_1(S\nu/G) \longrightarrow \pi_o(G)$ the boundary map in the long homotopy sequence of $G \longrightarrow S\nu \longrightarrow S\nu/G$. Let $\partial : \pi_1(S\nu/G) \to \pi_o(G) \times \pi_1(M^G)$ be $\partial_o \times p_*$ and j be the composition $\pi_o(G) \times \pi_1(M^G) = \pi_1(EG \times_G M^G) \to \pi_1(EG \times_G M)$. Then Φ is given in terms of the splitting by the matrix (see the proof of Theorem 4.51

$$\begin{pmatrix} id & 0 \\ j_* \circ \partial_* \circ p^* & k_* \end{pmatrix}$$

Firstly, we consider the case where G is not $\{1\}$ or $\mathbf{Z}/2$. If $SV \longrightarrow SV$ is a G-map, $\deg f \equiv 1 \bmod |G|$ for finite G and $\deg f = 1$ for infinite G holds. Hence any G-homotopy equivalence $SV \longrightarrow SV$ is G-homotopic to the identity so that $p : S\nu \longrightarrow B$ is untwisted in the sense of Lück [1987], 4.4. Therefore $\partial_* \circ p^*$ is given by $o(SV/G) = \chi(SV/G) \cdot [\mathbf{Z}\pi_o(G)] \in K_o(\mathbf{Z}\pi_o(G))$ and $\otimes_{\mathbf{Z}} : K_o(\mathbf{Z}\pi_o(G)) \otimes K_1(\mathbf{Z}\pi_1(M^G)) \longrightarrow K_1(\mathbf{Z}\pi_o(G) \times \pi_1(M^G))$ (see Lück [1987]). If G is finite, $\chi(SV/G)$ is zero so that $\partial_* \circ p^*$ and in particular $j_* \circ \partial_* \circ p^*$ are trivial. If G is infinite $\chi(SV/G)$ is not necessarily zero and $j_* \circ \partial_* \circ p^*$ can be non-trivial.

Secondly, let G be $\mathbf{Z}/2$. Let $w : \pi_1(M^G) \longrightarrow \{\pm 1\}$ be the first Stiefel Whitney class of $S\nu$. Denote by $Sw\,(\pi_1(M^G))$ the Grothendieck group of $\mathbf{Z}G$-modules which are finitely generated and free over $\mathbf{Z}$. It operates on $Wh(\pi_1(M^G))$ by $\otimes_{\mathbf{Z}}$ and the diagonal action. Let $\mathbf{Z}$ resp. $\mathbf{Z}^w$ be the abelian group $\mathbf{Z}$ with the trivial G-action resp. G-action determined by w. Then $p_* \circ p^* : Wh(\pi_1(M^G)) \to Wh(\pi_1(M^G))$ is multiplication with $[\mathbf{Z}] - (-1)^{\dim V}[\mathbf{Z}^w]$. If $\dim V$ is odd and w trivial, $p_* \circ p^*$ is $2\,id$ and $\partial_* \circ p^* : Wh(\pi_1(M^G)) \longrightarrow Wh(\pi_1(M^G) \times \pi_o(G))$ is just the map induced from the inclusion $\pi_1(M^G) \longrightarrow \pi_1(M^G) \times \pi_o(G)$. If $\dim V$ is even and w is trivial, $\partial_* \circ p^*$ is zero. However, for appropriate $\pi_1(M^G)$ and w it can happen that $\dim V$ is even and $p_* \circ p^*$ is not zero. This shows that $j_* \circ \partial_* \circ p^*$ does not only depend on the normal slice G-representation V. The results above follow from Lück [1986], [1987]. □

Comments 15.31. These transfer maps play an important role in the study of G-manifolds. Each NH-normal bundle $\nu(M^H,M)$ of M^H in M determines such a transfer by its sphere bundle. We have already discussed their appearence in the equivariant s- cobordism theorem when we have related isovariant and equivariant Whitehead torsion by an homomorphism $\Phi : Wh^G_{Iso}(M) \longrightarrow Wh^G_\rho(M)$.

Using the equivariant s-cobordism theorem and the map Φ above, one can define an involution $* : Wh^G_\rho(M) \longrightarrow Wh^G_\rho(M)$ by reversing equivariant h-cobordisms provided that the weak gap conditions 4.49. are satisfied. In Connolly-Lück [1988] the involution $*$ is computed in the terms of the splitting of $Wh_\rho^G(M)$. In general this is not the direct sum of algebraic involutions on the summands, certain correction terms involving transfer maps come in. However, this is the case if G is finite of odd order and TM_x is the restriction to G_x of a $NH(x)$-representation for all $x \in M$ since then the correction terms vanish because of Corollary 15.29.

Let $(f,\partial f) : (M,\partial M) \longrightarrow (N,\partial N)$ be a G-homotopy equivalence of G-manifolds. For a certain homomorphism $\Phi_f : U^G(N) \longrightarrow Wh^G(N)$ the formula

$$\tau^G(f) = -*\tau^G(f,\partial f) - * \circ \Phi_f(\chi^G(N,\partial N))$$

is proved in Connolly-Lück [1988] (see also Dovermann-Rothenberg [1988]). Under mild restrictions Φ_f is zero. Then $\tau^G(f)$, $\tau^G(f,\partial f)$ and $\tau^G(\partial f)$ all vanish if $\tau^G(f)$ or $\tau^G(f,\partial f)$ is zero. This is an important result for the proof of the equivariant π-π-theorem in the simple category (see Dovermann-Rothenberg [1988], Lück-Madsen [1988 a]). In this context we mention that there are also L-theoretic versions of such transfer maps (see Browder-Quinn [1975], Lück-Madsen [1988 b], Lück-Ranicki [1988]).

The main use of the algebraic description is the conclusion that the geometric transfer depends only on the (total) fibre transport (Corollary 15.27.). We used this fact to state a vanishing result for p^* in Theorem 15.28. Moreover, by Proposition 14.27. it reduces the computation of the geometric transfer to the study of the algebraic transfers $K_n(R) \longrightarrow K_n(S)$ for $n = 0,1$ given by a so called chain homotopy representation (C,U) of R in S in Lück [1986]. In Lück [1986] und [1987] some vanishing results and more information and references can be found. Unfortunately, it turns out that these transfer maps are very hard to compute. This can already be seen in the case of a non-equivariant fibration $S^1 \rightarrow E \xrightarrow{p} B$ with connected E and B and finite fundamental groups $\pi_1(E)$ and $\pi_1(B)$ which is extensively studied in Oliver [1985].

In Lück [1986 a] the notion of an equivariant first Stiefel-Whitney class w_ξ of a G-vector bundle ξ is defined. If w_ξ and w_η for two G-vector bundles ξ and

η over the same G-space agree then $tp_{S\xi}$ and $tp_{S\eta}$ coincide so that $p(S\xi)_* \circ p(S\xi)^* = p(S\eta)_* \circ p(S\eta)^*$ holds. Finally we mention Dovermann-Rothenberg [1988 a] where the geometric transfer of a G-fibration is analyzed, too.

Exercises 15.32.

1) Let B be a G-space with a simple structure β and a G-vector bundle $\nu \downarrow B$. Make choices 15.15. and 15.16. using in 15.15. the preferred simple structure of 4.36. for $S\nu_b$ as a G_b-manifold for $b \in B$. Show that the simple structure ξ on $S\nu$ of 15.17. is unique. Now suppose that B is a (compact) G-manifold and β is the preferred simple structure of 4.36. Then ξ agrees with the simple structure on the G-manifold $S\nu$ of 4.36.

2) Let $p : E \longrightarrow B$ be a G-fibration over a G-space B with simple structure β. Suppose that p satisfies assumptions 15.12. and 15.18. and that $\chi^G(B)$ is zero. Then we get from 15.17. a preferred simple structure ξ. How does ξ depend on β ?

3) Let $\nu \downarrow B$ be a G-vector bundle. Suppose that B^H is non-empty and connected and $w_1(S\nu^H) \in H^1(B^H;\mathbf{Z}/2)$ vanishes for all $H \subset G$. Let V be the G-representation ν_x for some $x \in B^G$. Show for $p : S\nu \longrightarrow B$ that $p_* \circ p^*$ is given by $\chi^G(SV) \in U^G(*)$ and the $U^G(*)$-module structure on $K_n^G(B)$.

4) Let G be a finite group and $\xi \downarrow B$ a complex G-vector bundle. Suppose for the complex G-representation V that V and ξ_b are isomorphic as complex G_b-representations for all $b \in B$. Prove $p_* \circ p^* = 0$ for $p : S\nu \longrightarrow B$.

5) Let $\nu \downarrow B$ be a G-vector bundle and $p : D\nu \longrightarrow B$ be the associated disc bundle. Show $p_* \circ p^* = id$.

6) Let M be a G-manifold satisfying the weak gap conditions 4.49. Show that the following definition of an involution $* : Wh_\rho^G(M) \longrightarrow Wh_\rho^G(M)$ makes sense. For $\eta \in Wh_\rho^G(M)$ choose an equivariant h-cobordism $(W;M,N)$ with $i_*^{-1}\tau^G(i) = \eta$ for $i : M \longrightarrow W$ the inclusion. Let $*(\eta)$ be $i_*^{-1}\tau^G(j)$ for $j : N \longrightarrow W$ the inclusion. Compute $*$ in terms of the splitting $Wh_\rho^G(M) = Wh(\pi_1(M^G)) \oplus Wh(\pi_1(EG\times_G M))$

if $\mathrm{Iso}\, M = \{G,\{1\}\}$ and M^G and M are connected.

7) Give an example of a pair of G-homotopy equivalences $(f,\partial f):(M,\partial M) \longrightarrow (M,\partial N)$ of (compact) G-manifolds such that $\tau^G(f)$ is zero and $\tau^G(f,\partial f)$ is not zero. Prove that for trivial G such an example does not exist.

8) Let $\xi_i \downarrow M_i$ be G-vector bundles over (compact) G-manifolds M_i for $i = 0,1$ and $(\bar{f},f) : \xi_o \longrightarrow \xi_1$ be a G-vector bundle map. Let $S\bar{f} : S\xi_o \longrightarrow S\xi_1$ be the induced G-homotopy equivalence between G-manifolds. Show $\tau^G(S\bar{f}) = p_1^*(\tau^G(f))$ for $p_1 : S\xi_1 \longrightarrow M_1$, if τ^G is taken with respect to the simple structures 4.36.

9) Let $p_1 : E_2 \longrightarrow E_1$ and $p_o : E_1 \longrightarrow E_o$ be G-fibrations satisfying assumptions 15.12. and 15.18. so that the geometric transfer maps $p_i^!$ are defined. Show that $p_o \circ p_1 : E_2 \longrightarrow E_1$ is a G-fibration satisfying assumptions 15.12. and 15.18. and that $(p_o \circ p_1)^! = p_o^! \circ p_1^!$ is true.

10) Give an example of a finite group G, and a (compact) G-manifold M of precisely two orbit types (H) and $(\{1\})$ such that M^H and M are simply connected, $\chi^H(SV) \in A(H)$ is zero for the H-normal slice and

$$\Phi : Wh^G_{Iso}(M) = Wh(\pi_1(M_H/WH)) \oplus Wh(\pi_1(M_G/G)) \longrightarrow Wh^G_\rho(M) = Wh(\pi_1(EWH \times_{WH} M^H)) \oplus Wh(\pi_1(EG \times_G M))$$

is not given by a diagonal matrix.

11) Let $p : E \longrightarrow B$ be a G-fibration over a finite G-CW-complex B. Let F be a finitely dominated G-CW-complex such that F_b and F are G_b-homotopy equivalent for all $b \in B$. Prove in $U(G)$ that $\chi^G(E) = \chi^G(B) \cdot \chi^G(F)$ is true.

12) Give an example of G-vector bundles ξ and η over the space $\mathbb{R}P^2$ with trivial G-action and a G-fibre homotopy equivalence $f : S\xi \longrightarrow S\eta$ covering the identity such that $\tau^G(f)$ is not zero.

13) Show for a G-map $f : X \longrightarrow Y$ the existence of a G-fibration $\bar{f} : \bar{X} \longrightarrow Y$ together with a G-homotopy equivalence $h : X \longrightarrow \bar{X}$ satisfying $\bar{f} \circ h \simeq_G f$.

16. A second splitting

In section 9 and 10 we have analyzed the structure of projective $R\Gamma$-modules and of $K_n(R\Gamma)$ using a splitting theorem based on the splitting functor S_x and the extension functor E_x. Now we introduce a second splitting induced from the restriction functor Res_x and the inclusion functor I_x. We relate these in a certain sense dual splittings by a kind of Moebius inversion. This enables us in appropriate situations to recognize a $R\Gamma$-module M from all its values $M(x)$, $x \in \mathrm{Ob}\,\Gamma$.

Definition 16.1. We call a category Γ finite if $\mathrm{Is}\,\Gamma$ and $\mathrm{Hom}(x,y)$ for all $x,y \in \mathrm{Ob}\,\Gamma$ are finite. Let $|x|$ be $|\mathrm{Aut}(x)|$ for $x \in \mathrm{Ob}\,\Gamma$. Denote by $m(\Gamma)$ the smallest common multiple of all numbers $|x|$, $x \in \mathrm{Is}\,\Gamma$. If $\mathrm{Aut}(y)$ operates freely on $\mathrm{Hom}(x,y)$ for $x,y \in \mathrm{Ob}\Gamma$, we call Γ free. □

Example 16.2. Let G be a compact Lie group and $\mathfrak{F} \subset \mathrm{Con}\,G$ a finite set. Let $Or/G_{\mathfrak{F}}$ be the full subcategory of the discrete orbit category Or/G (see 8.36.) consisting of all objects G/H with $(H) \in \mathfrak{F}$. Then $Or/G_{\mathfrak{F}}$ is finite. If WH is finite for all $(H) \in \mathfrak{F}$, then $Or/G_{\mathfrak{F}}$ is free by Lemma 8.26. The orbit category $Or\,G$ of a finite group is finite and free and $m(Or\,G)$ is the group order $|G|$. □

16.3. Recall the partial ordering $\leq$ on the set $\mathrm{Is}\,\Gamma$ of isomorphism classes $\bar{x}$ of objects $x \in \mathrm{Ob}\,\Gamma$ defined by $\bar{x} \leq \bar{y} \iff \mathrm{Hom}(x,y) \neq \emptyset$. We write $\bar{x} < \bar{y}$ if $\bar{x} \leq \bar{y}$ and $\bar{x} \neq \bar{y}$ holds. For a non-negative integer ℓ we define a ℓ-chain c from $\bar{x} \in \mathrm{Is}\,\Gamma$ to $\bar{y} \in \mathrm{Is}\,\Gamma$ to be a sequence $c : \bar{x} = \bar{x}_0 < \bar{x}_1 < \ldots < \bar{x}_\ell = \bar{y}$. Let $ch_\ell(\bar{x},\bar{y})$ be the set of ℓ-chains from $\bar{x}$ to $\bar{y}$. Define the length $\ell(\bar{y})$ of $\bar{y} \in \mathrm{Is}\,\Gamma$ by $\max\{\ell \in \mathbf{Z} \mid \text{there is } \bar{x} \in \mathrm{Is}\,\Gamma \text{ with } ch_\ell(\bar{x},\bar{y}) \neq \emptyset\}$. The length $\ell(\Gamma)$ of Γ is $\max\{\ell(\bar{x}) \mid \bar{x} \in \mathrm{Is}\,\Gamma\}$. Given a $R\Gamma$-module M, let its support $\mathrm{supp}\,M$ be $\{\bar{x} \in \mathrm{Is}\,\Gamma \mid M(x) \neq \{0\}\}$. Define its length $\ell(M)$ by $\max\{\ell(\bar{x}) \mid \bar{x} \in \mathrm{supp}\,M\}$ if $M \neq \{0\}$ and $\ell(\{0\}) = -1$. If $\mathrm{Is}\,\Gamma$ is not finite, it may happen that $\ell(\bar{x})$, $\ell(\Gamma)$ or $\ell(M)$ are ∞. □

e have defined the splitting functor $S_x : \text{MOD-}R\Gamma \longrightarrow \text{MOD-}R[x]$ in 9.26., the extension functor $E_x : \text{MOD-}R[x] \longrightarrow \text{MOD-}R\Gamma$ in 9.28., the restriction functor $Res_x : \text{MOD-}R\Gamma \longrightarrow \text{MOD-}R[x]$ in 9.27. and the inclusion functor $I_x : \text{MOD-}R[x] \longrightarrow \text{MOD-}R\Gamma$

in 9.29. Recall from Lemma 9.31. that (E_x, Res_x) and (S_x, I_x) are pairs of adjoint functors. Given a $R\Gamma$-module, let the $R\Gamma$-homomorphism

16.4. $$K_x M : M \longrightarrow I_x \circ S_x M$$

be the adjoint of $id : S_x M \longrightarrow S_x M$. Hence $K_x M$ evaluated at x is the projection $M(x) \longrightarrow S_x M$. For an integer ℓ we define a $R\Gamma$-homomorphism where $Is(\ell)$ is $\{\bar{x} \in Is\,\Gamma \mid \ell(\bar{x}) = \ell\}$.

16.5. $$K_\ell M = \prod_{\bar{x} \in Is(\ell)} K_x M \; : M \longrightarrow \prod_{\bar{x} \in Is(\ell)} I_x \circ S_x M$$

and a $R\Gamma$-module

16.6. $$KER_\ell M = kernel(K_\ell M)$$

If $L_\ell M : KER_\ell M \longrightarrow M$ is the inclusion, we obtain a natural exact sequence

16.7. $$0 \longrightarrow KER_\ell M \xrightarrow{L_\ell M} M \xrightarrow{K_\ell M} \prod_{\bar{x} \in Is(\ell)} I_x \circ S_x M \longrightarrow 0$$

<u>Theorem 16.8.</u> <u>Filtration theorem for $R\Gamma$-modules.</u>

<u>Let</u> Γ <u>be a EI-category and</u> M <u>be a $R\Gamma$-module of length</u> $\ell(M) \leq \ell$ <u>for some fixed integer</u> ℓ. <u>Then there is a filtration</u>

$$M = M_o \xhookrightarrow{L_o} M_1 \xhookrightarrow{L_1} M_2 \xhookrightarrow{L_2} \cdots \xhookrightarrow{L_{\ell-1}} M_\ell = M$$

<u>satisfying</u>

a) $M_{i-1} = KER_i M_i$ <u>and</u> L_i <u>is the inclusion</u>.

b) M_i <u>has length</u> $\ell(M_i) \leq i$. <u>If</u> Γ <u>is finite and</u> M <u>finitely generated</u>, M_i <u>is finitely generated</u>.

c) <u>Let</u> $L^i : M_i \longrightarrow M$ <u>be</u> $L_{\ell-1} \circ L_{\ell-2} \circ \ldots L_i$. <u>Then</u> $Res_x(L^i) = Res_x M_i \longrightarrow Res_x M$ <u>is an isomorphism for</u> $\ell(\bar{x}) \leq i$ <u>and</u> $Res_x M_i = \{0\}$ <u>for</u> $\ell(x) > i$.

d) <u>We obtain from 16.7. a natural short exact sequence</u>

$$0 \longrightarrow M_{i-1} \longrightarrow M_i \longrightarrow \prod_{\bar{x} \in Is(i)} I_x \circ Res_x M \longrightarrow 0$$

e) <u>If</u> $0 \longrightarrow M \longrightarrow N \longrightarrow P \longrightarrow 0$ <u>is an exact sequence of</u> $R\Gamma$-<u>modules of length</u> $\leq \ell$, <u>then the induced sequence</u> $0 \longrightarrow M_i \longrightarrow N_i \longrightarrow P_i \longrightarrow 0$ <u>is exact</u>.

Proof. Induction over i. Notice that $S_xM = \mathrm{Res}_xM$ holds if $\ell(M) \leq \ell(x)$ is true. We get a) from Lemma 16.10. below.

16.9. The Filtration Theorem does not give a splitting of $K_n(R\Gamma)$ in general as Res_x and I_x do not respect "projective" whereas E_x and S_x do. On the other hand Res_x and I_x are exact what is not true for S_x (see Example 9.34.). If Γ is free, E_x is exact. The Filtration Theorem gives a splitting if one considers finitely generated $R\Gamma$-modules instead of finitely generated projective ones.

We call R noetherian if any submodule of a finitely generated R-module is finitely generated.

Lemma 16.10. Let Γ be a finite EI-category.

a) A $R\Gamma$-module M is finitely generated if and only if Res_xM is finitely generated over R for all $x \in \mathrm{Ob}\,\Gamma$.

b) If R is noetherian, any submodule of a finitely generated $R\Gamma$-module is finitely generated.

Proof:

a) We use induction over $\ell = \ell(M)$. The begin $\ell(M) = -1$ is trivial, the induction step done as follows. Suppose that $M(x)$ is finitely generated over R for all $x \in \mathrm{Ob}\,\Gamma$. For $x \in \mathrm{Is}(\ell)$ choose a finitely generated free $R[x]$-module F and an epimorphism $p : F \longrightarrow \mathrm{Res}_xM$. Then E_xF is a finitely generated free $R\Gamma$-module and $E_xp : E_xF \longrightarrow E_x \circ \mathrm{Res}_xM$ is surjective. The adjoint of $\mathrm{id}: \mathrm{Res}_xM \longrightarrow \mathrm{Res}_x \circ I_x \circ \mathrm{Res}_xM$ is an epimorphism $E_x \circ \mathrm{Res}_xM \longrightarrow I_x \circ \mathrm{Res}_xM$ so that $I_x \circ \mathrm{Res}_xM$ and hence $\prod_{\bar{x} \in \mathrm{Is}(\ell)} I_x \circ \mathrm{Res}_xM$ is finitely generated. By Theorem 16.8 c) and the induction hypothesis $\mathrm{KER}_\ell M$ is finitely generated. By Theorem 16.8. d) the $R\Gamma$-module M is finitely generated. The other implication is obvious. □

Definition 16.11. If FMOD-$R\Gamma$ is the exact category of finitely generated $R\Gamma$-modules, define for $n \geq 0$ (see 10.4.):

$$Gr_n(R\Gamma) = K_n(\text{FMOD-}R\Gamma)$$

$$\text{Split } Gr_n(R\Gamma) = \bigoplus_{\bar{x} \in \text{Is }\Gamma} Gr_n(R[x]) \qquad \square$$

Let Γ be a finite EI-category. Then Res_x, I_x and KER_ℓ respect "finitely generated" by Lemma 16.10. and are exact. Define homomorphisms

16.12.
$$Res : Gr_n(R\Gamma) \longrightarrow \text{Split } Gr_n(R\Gamma)$$
$$I : \text{Split } Gr_n(R\Gamma) \longrightarrow Gr_n(R\Gamma)$$

by the sum of maps $Gr(R\Gamma) \longrightarrow Gr(R[x])$ induced from Res_x and $Gr(R[x]) \longrightarrow Gr(R\Gamma)$ induced from I_x. Analogously to Theorem 10.34. one proves

<u>Theorem 16.13.</u> Res <u>and</u> I <u>are isomorphisms</u>, <u>inverse to one another</u> $\square$

We briefly discuss the natural properties. If $F : \Gamma_1 \longrightarrow \Gamma_2$ is a functor of finite EI-categories, restriction defines an homomorphism $F^* : Gr(R\Gamma_2) \longrightarrow Gr(R\Gamma_1)$. Let $F^!_{\bar{y},\bar{x}} : Gr(R[y]) \longrightarrow Gr(R[x])$ be zero, if $\bar{y} \neq \overline{Fx}$ holds and be induced by restriction with $Aut(x) \longrightarrow Aut(y)$ $f \longrightarrow F(f)$ for $y = Fx$. The collection of the $F^!_{\bar{y},\bar{x}}$ defines

16.14.
$$F^! : \text{Split } Gr_n(R\Gamma_2) \longrightarrow \text{Split } Gr_n(R\Gamma_1)$$

Notice that $F^!$ is quite different from F^* defined in 10.32. Now Gr_n and Split Gr_n become contravariant functors on the category F-EI-CAT of finite EI-categories with values in abelian groups. One easily checks:

<u>Lemma 16.15.</u> $Res : Gr_n \longrightarrow \text{Split } Gr_n$ <u>and</u> $I : \text{Split } Gr_n \longrightarrow Gr_n$ <u>are natural equivalences</u>.

Let $Gr_o^P(R\Gamma)$ be the Grothendieck group of finitely generated $R\Gamma$-modules M for which $M(x)$ is projective over R for all $x \in Ob\,\Gamma$. Let $F_* : Gr_o^{P'}(R\Gamma) \longrightarrow Gr_o(R\Gamma)$ be induced from the forgetful functor F. If any submodule of a projective R-module is again projective, R is called <u>hereditary</u>.

<u>Lemma 16.16.</u> <u>If</u> Γ <u>is finite and</u> R <u>is hereditary and noetherian</u>, $F_* : Gr_o^P(R\Gamma) \longrightarrow Gr_o(R\Gamma$

<u>is bijective</u>.

<u>Proof.</u> We define a map $G : Gr_o(R\Gamma) \longrightarrow Gr_o^P(R\Gamma)$ as follows. If M is a finitely generated $R\Gamma$-module, choose a finite not necessarily projective resolution (C,f) of M such that $C_n(x)$ is projective over R for all $x \in Ob\ \Gamma$ and $n \geq 0$. Define $G([M])$ by $\Sigma(-1)^n[C_n]$. As R is hereditary and noetherian (C,f) exists. Namely choose a finitely generated free $R\Gamma$-module C_o together with an epimorphism $p : C_o \rightarrow M$. Let C_1 be the kernel of p, $c_1 : C_1 \longrightarrow C_o$ the inclusion and C_i be zero for $i > 1$. We leave it to the reader to verify that G is well-defined and inverse to F_*. □

Let Γ_1 and Γ_2 be finite EI-categories. If M is a finitely generated $R\Gamma_1$-module such that $M(x)$ is projective over R for $x \in Ob\ \Gamma$ then we get an exact functor $FMOD\text{-}R\Gamma_2 \longrightarrow FMOD\text{-}R\Gamma_1{\times}\Gamma_2 \quad N \longrightarrow M \otimes_R N$ (see 9.13.) inducing $Gr_n(R\Gamma_2) \rightarrow Gr_n(R\Gamma_1{\times}\Gamma_2)$. We obtain a pairing

16.17. $$Gr_o^P(R\Gamma_1) \otimes Gr_n(R\Gamma_2) \xrightarrow{\otimes_R} Gr_n(R\Gamma_1 \times \Gamma_2)$$

Using $F_* : Gr_o^P(R\Gamma_1) \xrightarrow{\cong} Gr_o(R\Gamma_1)$ we get a pairing

16.18. $$Gr_o(R\Gamma_1) \otimes Gr_n(R\Gamma_2) \xrightarrow{\otimes_R} Gr_n(R\Gamma_1 \times \Gamma_2)$$

Define

16.19. $$\text{Split } Gr_o^P(R\Gamma_1) \otimes \text{Split } Gr_n(R\Gamma_2) \xrightarrow{\otimes_R} \text{Split } Gr_n(R\Gamma_1 \times \Gamma_2)$$

$$\text{Split } Gr_o^P(R\Gamma_1) \otimes \text{Split } Gr_n(R\Gamma_2) \xrightarrow{\otimes_R} \text{Split } Gr_n(R\Gamma_1 \times \Gamma_2)$$

by the various pairings $Gr_o^P(R[x]) \otimes Gr_n(R[y]) \longrightarrow Gr_n(R[(x,y)])$ and Split F_* : Split $Gr_o^P(R\Gamma_1) \longrightarrow$ Split $Gr_o(R\Gamma_1)$. Notice that one can define 16.18. not directly using $\otimes_R$ as $M \otimes_R ?$ is not necessarily exact for an arbitrary finitely generated $R\Gamma_1$-module M.

<u>Lemma 16.20.</u> <u>Let</u> Γ <u>be finite and</u> R <u>be hereditary and noetherian.</u>

a) Res <u>and</u> I <u>are compatible with these parings</u> $\otimes_R$

b) $\otimes_R : Gr_o(R\Gamma_1) \otimes Gr_n(R\Gamma_2) \longrightarrow Gr_n(R\Gamma_1 \times \Gamma_2)$ <u>is natural with respect to</u> $F^!$.

Suppose either that Γ is finite and free or that Γ is finite and $m(\Gamma)$ (see Definition 16.1.) is a unit in R. Then $R\,\mathrm{Hom}(y,x)$ is a finitely generated projective $R[x]$-module for $x,y \in \mathrm{Ob}\,\Gamma$. Hence $E_x : \mathrm{FMOD}\text{-}R[x] \longrightarrow \mathrm{FMOD}\text{-}R\Gamma$ is exact and induces $Gr_n(R[x]) \longrightarrow Gr_n(R\Gamma)$. The sum of these maps define

16.21. $$E : \mathrm{Split}\; Gr_n(R\Gamma) \longrightarrow Gr_n(R\Gamma)$$

We want to compare E with Res and I. Let $\omega_{x,y} : Gr_n(R[x]) \longrightarrow Gr_n(R[y])$ be induced from the exact functor $\mathrm{FMOD}\text{-}R[x] \longrightarrow \mathrm{FMOD}\text{-}R[y]$ sending M to $M \otimes_{R[x]} R\,\mathrm{Hom}(y,x)$ for $x,y \in \mathrm{Ob}\,\Gamma$. We get

16.22. $$\omega : \mathrm{Split}\; Gr_n(R\Gamma) \longrightarrow \mathrm{Split}\; Gr_n(R\Gamma)$$

By definition $\mathrm{Res} \circ E$ is ω. Next we construct an inverse μ of ω.

Let $c \in ch_\ell(\bar{y},\bar{x})$ be the ℓ-chain $\bar{y} = \bar{x}_o < \bar{x}_1 < \ldots < \bar{x}_\ell = \bar{x}$. Denote by $S(c)$ the $\mathrm{Aut}(x)$-$\mathrm{Aut}(y)$-set (i.e. a set with left $\mathrm{Aut}(x)$- and right $\mathrm{Aut}(y)$-action commuting with one another)

16.23. $$S(c) = \mathrm{Hom}(x_{\ell-1},x_\ell) \times_{\mathrm{Aut}(x_{\ell-1})} \mathrm{Hom}(x_{\ell-2},x_{\ell-1}) \times_{\mathrm{Aut}(x_{\ell-2})} \cdots \times_{\mathrm{Aut}(x_1)} \mathrm{Hom}(x_o,x_1).$$

By the assumptions about Γ and R the $R[x]$-module $RS(c)$ is projective. Hence we obtain an exact functor $\mathrm{FMOD}\text{-}R[x] \longrightarrow \mathrm{FMOD}\text{-}R[y]$ $M \longrightarrow M \otimes_{R[x]} RS(c)$ inducing

16.24. $$\mu_{x,y}(c) : Gr_n(R[x]) \longrightarrow Gr_n(R[y])$$

Define $\mu_{x,y} : Gr_n(R[x]) \longrightarrow Gr_n(R[y])$ by

16.25. $$\mu_{x,y} = \sum_{\ell \geq o} (-1)^\ell \sum_{c \in ch_\ell(\bar{y},\bar{x})} \mu_{x,y}(c)$$

We obtain the Moebius inversion

16.26. $$\mu : \mathrm{Split}\; Gr_n(R\Gamma) \longrightarrow \mathrm{Split}\; Gr_n(R\Gamma)\ .$$

Theorem 16.27. Suppose either that Γ is finite and free or that Γ is finite and $m(\Gamma)$ is a unit in R. Then (ω,μ) and (Res,I) are pairs of isomorphisms, inverse to another, E is an isomorphism and the following diagram commutes

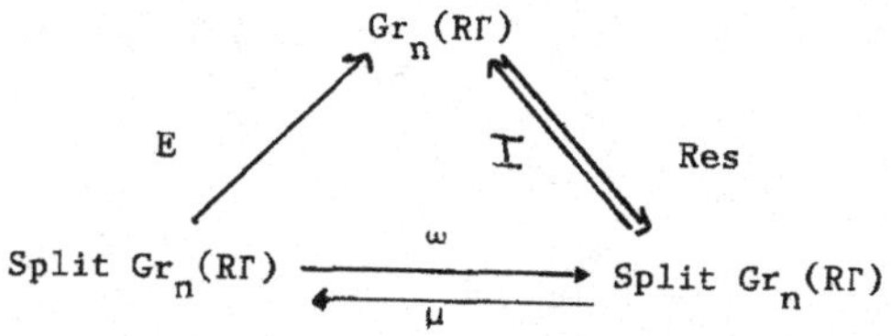

<u>Proof</u>: Since $Res \circ E = \omega$ obviously is true and $Res \circ I = I \circ Res = id$ holds by Theorem 16.13., it only remains to prove $\omega \circ \mu = \mu \circ \omega = id$. Given $\bar{x}, \bar{y} \in Is\,\Gamma$ with $\bar{x} < \bar{y}$, let $c_1(\bar{x},\bar{y})$ be the unique 1-chain from $\bar{x}$ to $\bar{y}$. If c is the ℓ-chain $\bar{x} = \bar{x}_o < \bar{x}_1 < \ldots < \bar{x}_\ell = \bar{y}$ from $\bar{x}$ to $\bar{y}$ and d the m-chain $\bar{y} = \bar{y}_o < \bar{y}_1 < \ldots < \bar{y}_m = \bar{z}$ from $\bar{y}$ to $\bar{z}$ let $c * d$ be the $(\ell+m)$-chain $\bar{x} = \bar{x}_o < \bar{x}_1 < \ldots < \bar{x}_\ell < \bar{y}_1 < \ldots < \bar{y}_m = z$ from $\bar{x}$ to $\bar{z}$. We have the following relations

i) $\omega_{x,y} = \mu_{x,y} = 0$, if $\bar{y} \leq \bar{x}$ is not true.

ii) $\omega_{x,x} = \mu_{x,x} = id$

iii) $\omega_{x,y} = \mu_{x,y}(c_1(y,x))$, if $\bar{y} < \bar{x}$ holds

iv) $\mu_{y,z}(c) \circ \mu_{x,y}(d) = \mu_{x,z}(c * d)$

v) The map $\coprod_{\substack{\bar{z} \in Is\,\Gamma \\ \bar{x} \leq \bar{z} < \bar{y}}} ch_\ell(\bar{x},\bar{z}) \longrightarrow ch_{\ell+1}(\bar{x},\bar{y})$ sending $c \in ch_\ell(\bar{x},\bar{z})$ to $c * c_1(\bar{z},\bar{y})$ is bijective. Analogously one gets a bijection

$$\coprod_{\substack{\bar{z} \in Is\,\Gamma \\ \bar{x} < \bar{z} \leq \bar{y}}} ch_\ell(\bar{z},\bar{y}) \longrightarrow ch_{\ell+1}(\bar{x},\bar{y}) \quad \text{by} \quad c \in ch_\ell(\bar{z},\bar{y}) \longmapsto c_1(\bar{x},\bar{z}) * c .$$

Because of i) and ii) it suffice to show $(\omega \circ \mu)_{\bar{x},\bar{y}} = (\mu \circ \omega)_{\bar{x},\bar{y}} = 0$ provided that $\bar{y} < \bar{x}$ is true. We verify only second equality, the first one is done similarly

$$(\mu \circ \omega)_{\bar{x},\bar{y}} = \sum_{\substack{\bar{z} \in Is\,\Gamma \\ \bar{y} \leq \bar{z} \leq \bar{x}}} \mu_{z,y} \circ \omega_{x,z} =$$

$$\sum_{\substack{\bar{z} \in Is\,\Gamma \\ \bar{y} \leq \bar{z} < \bar{x}}} \quad \sum_{\ell \geq o} (-1)^\ell \sum_{c \in ch_\ell(\bar{y},\bar{z})} \mu_{z,y}(c) \circ \mu_{x,z}(c_1(\bar{z},\bar{x})) \quad +$$

$$\sum_{\ell \geq o} (-1)^\ell \sum_{c \in ch_\ell(\bar{y},\bar{x})} \mu_{x,y}(c) \quad =$$

$$\sum_{\ell \geq o} (-1)^{\ell} \sum_{\substack{\bar{z} \in \operatorname{Is} \Gamma \\ \bar{y} \leq \bar{z} < \bar{x}}} \; \sum_{c \in ch_{\ell}(\bar{y},\bar{z})} \mu_{x,y}(c * c_1(\bar{z},\bar{x})) \quad +$$

$$\sum_{\ell \geq 1} (-1)^{\ell} \sum_{c \in ch_{\ell}(\bar{y},\bar{x})} \mu_{x,y}(c) \;=$$

$$\sum_{\ell \geq o} (-1)^{\ell} \sum_{c \in ch_{\ell+1}(\bar{y},\bar{x})} \mu_{x,y}(c) + \sum_{\ell \geq 1} (-1)^{\ell} \sum_{c \in ch_{\ell}(\bar{x},\bar{y})} \mu_{x,y}(c) = 0$$

□

In favourite situations Res and I induce also a splitting for K-theory.

Proposition 16.28. Let Γ be a finite EI-category. Suppose that $m(\Gamma)$ (see Definition 16.1.) is a unit in R and R is noetherian and hereditary. Then the forgetful functor $F : FPMOD\text{-}R\Gamma\text{-}FMOD \to R\Gamma$ induces an isomorphism

$$F_* : K_n(R\Gamma) \longrightarrow Gr_n(R\Gamma).$$

Proof. We will show in the next section that under the assumptions above any finitely generated $R\Gamma$-module has a finite projective resolution. Now apply Quillen [1973], corollary 1 in § 4. □

We derive from Theorem 10.34, Theorem 16.27. and Proposition 16.28. the main result of this section.

Theorem 16.29. The K-theoretic Moebius Inversion Theorem. Let Γ be a finite EI-category and R noetherian and hereditary with $m(\Gamma) \in R^*$. Then (E,S), (ω,μ) and (Res,I) are pairs of two another inverse isomorphisms and the following diagram commutes

$$\begin{array}{ccccc} & & K_n(R\Gamma) & & \\ & E \swarrow\nearrow S & & I \nwarrow\searrow Res & \\ \text{Split } K_n(R\Gamma) & & \underset{\mu}{\overset{\omega}{\rightleftarrows}} & & \text{Split } K_n(R\Gamma) \end{array}$$

□

From now on we suppose that we have specified the notion of a rank $rk_R M$ of a finitely generated R-module such that $rk_R M^1 - rk_R M^o + rk_R M^2 = 0$ holds for any exact sequence $0 \longrightarrow M^1 \longrightarrow M^o \longrightarrow M^2 \longrightarrow 0$ of finitely generated R-modules and $rk_R R = 1$ is true.

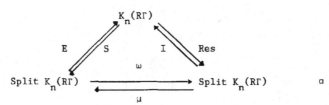

Write $C(\Gamma) = \prod_{\bar{y} \in \mathrm{Is}\,\Gamma} \mathbf{Z}$. Of course this is the same as $U(\Gamma)$ as abelian group but in some cases $U(\Gamma)$ and $C(\Gamma)$ will carry different ring structures. Define for a finitely generated $R\Gamma$-module M over a finite EI-category

16.30.
$$\widehat{\mathrm{rk}}_{R\Gamma}\, M \in C(\Gamma)$$

by $(\mathrm{rk}_R M(x))_{\bar{x} \in \mathrm{Is}\,\Gamma}$. Recall from Definition 10.38. the notion of

$$\mathrm{rk}_{R\Gamma} M \in U(\Gamma)$$

given by $(\mathrm{rk}_R(S_x M \otimes_{R[x]} R))_{\bar{x} \in \mathrm{Is}\,\Gamma}$. If $0 \longrightarrow M^1 \longrightarrow M^o \longrightarrow M^2 \longrightarrow 0$ is an exact sequence of finitely generated $R\Gamma$-modules, we have $\widehat{\mathrm{rk}}_{R\Gamma} M^1 - \widehat{\mathrm{rk}}_{R\Gamma} M^o + \widehat{\mathrm{rk}}_{R\Gamma} M^2 = 0$ but this is not true for $\mathrm{rk}_{R\Gamma}$ in general (Remark 10.39.). We get induced homomorphisms

16.31.
$$\mathrm{rk}_{R\Gamma} : K_o(R\Gamma) \longrightarrow U(\Gamma)$$
$$\widehat{\mathrm{rk}}_{R\Gamma} : Gr_o(R\Gamma) \longrightarrow C(\Gamma)$$

We want to relate $\mathrm{rk}_{R\Gamma}$ and $\widehat{\mathrm{rk}}_{R\Gamma}$. Let $\omega_{\bar{x},\bar{y}}$ be the integer $|\mathrm{Hom}(y,x)|$. Define

16.32.
$$\bar{\omega} : U(\Gamma) \longrightarrow C(\Gamma)$$

by $(n(\bar{x}))_{\bar{x} \in \mathrm{Is}\,\Gamma} \longrightarrow \left(\sum_{\bar{x} \in \mathrm{Is}\,\Gamma} \bar{\omega}_{\bar{x},\bar{y}} \cdot n(x)\right)_{\bar{y} \in \mathrm{Is}\,\Gamma}$.

__Lemma 16.33.__ __Let__ Γ __be a finite__ EI-__category__. __The map__ $\bar{\omega}$ __is injective__. __Its cokernel is finite of order__ $\prod_{\bar{x} \in \mathrm{Is}\,\Gamma} |x|$. __If__ M __is a finitely generated free__ $R\Gamma$-__module__ $\bar{\omega}(\mathrm{rk}_{R\Gamma} M) = \widehat{\mathrm{rk}}_{R\Gamma} M$ __holds__.

__Proof.__ $\bar{\omega}$ is given by a triangular matrix. The entry on the diagonal for $\bar{x} \in \mathrm{Is}\,\Gamma$ is $|x| := |\mathrm{Aut}(x)|$. The formula $\bar{\omega}(\mathrm{rk}_{R\Gamma} M) = \widehat{\mathrm{rk}}_{R\Gamma} M$ is obviously true for $M = R\,\mathrm{Hom}(?,x)$. Any finitely generated free $R\Gamma$-module is a finite direct sum of such modules $R\,\mathrm{Hom}(?,x)$. □

Given $\bar{x},\bar{y} \in \mathrm{Is}\,\Gamma$ with $\bar{y} \le \bar{x}$ and $c \in ch_\ell(\bar{y},\bar{x})$, we have defined a Aut(x)-Aut(y)-set $S(c)$ in 16.23. Let $\bar{\mu}_{\bar{x},\bar{y}}(c)$ be the integer $|\mathrm{Aut}(x)\backslash S(c)|$ and put

16.34.
$$\bar{\mu}_{\bar{x},\bar{y}} = \sum_{\ell \ge o} (-1)^\ell \sum_{c \in ch_\ell(\bar{y},\bar{x})} \bar{\mu}_{\bar{x},\bar{y}}(c) .$$

We define

16.35. $$\bar{\mu} : C(\Gamma) \longrightarrow C(\Gamma)$$

by $(n(\bar{x}))_{\bar{x} \in \mathrm{Is}\,\Gamma} \longrightarrow (\sum_{\bar{x} \in \mathrm{Is}\,\Gamma} \bar{\mu}_{\bar{x},\bar{y}} \cdot n(\bar{x}))_{\bar{y} \in \mathrm{Is}\,\Gamma}$.

Let $i : U(\Gamma) \longrightarrow C(\Gamma)$ send $(n(x))_{\bar{x} \in \mathrm{Is}\,\Gamma}$ to $(|x| \cdot n(x))_{\bar{x} \in \mathrm{Is}\,\Gamma}$. Notice that i is injective.

Theorem 16.36. Let Γ be a finite free EI-category. Then the following is true

a) $\bar{\mu} \circ \bar{\omega} = i$

b) image $\bar{\omega} = \{\eta \in C(\Gamma) \mid \bar{\mu}(\eta) \in \text{image } i\}$

c) We have for a finitely generated free $R\Gamma$-module M

$$i(\mathrm{rk}_{R\Gamma} M) = \bar{\mu}(\mathrm{rk}\hat{}_{R\Gamma} M)$$

$$\mathrm{rk}\hat{}_{R\Gamma} M = \bar{\omega}(\mathrm{rk}_{R\Gamma} M)$$

Proof. Let $j : U(\Gamma) \longrightarrow K_o(R\Gamma)$ send the base element corresponding to $\bar{x}$ to $[R\,\mathrm{Hom}(?,x)]$. Consider the diagram

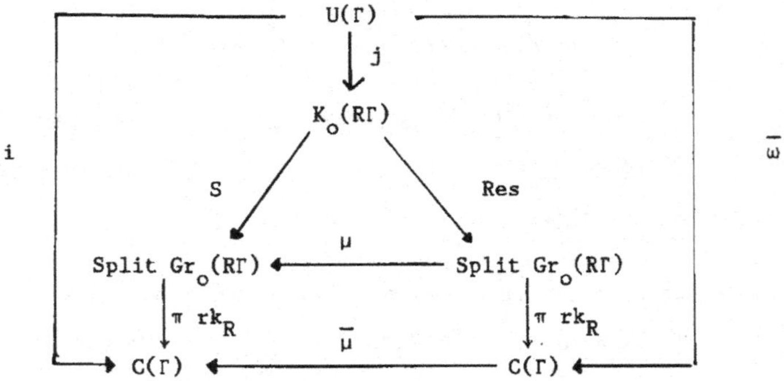

We get $\mu \circ \mathrm{Res} = S$ from Theorem 16.27. For $\bar{x},\bar{y} \in \mathrm{Ob}\,\Gamma$ with $\bar{y} \leq \bar{x}$ and $c \in ch_\ell(\bar{y},\bar{x})$ we have $\mathrm{rk}_R(P \otimes_{R[x]} R[S(c)]) = \mathrm{rk}_R P \cdot \mathrm{rk}_R R(\mathrm{Aut}(x)\backslash S(c))$ for a finitel generated $R[x]$-module P, as $\mathrm{Aut}(x)$ operates freely on $S(c)$. Hence the square commutes. One easily checks that the whole diagram commutes. This proves a). Since $\bar{\mu}$ is injective b) follows from a). Lemma 16.33. proves c) □

Remark 16.37. Theorem 16.36. gives us the possibility to determine a finitely generated free $R\Gamma$-module F by all its values $F(x)$ as i is injective and $rk_{R\Gamma}F$ determines F up to isomorphism by Lemma 10.40. Notice that image $\bar{\omega} \subset C(\Gamma)$ is described by a system of congruence. Namely, Theorem 16.36. says that $n = (n(\bar{x}))_{\bar{x} \in \mathrm{Is}\,\Gamma} \in C(\Gamma)$ lies in the image of $\bar{\omega}$ if and only if for each $\bar{y} \in \mathrm{Is}\,\Gamma$ the congruence

$$n(\bar{y}) \equiv - \sum_{\substack{\bar{x} \in \mathrm{Is}\,\Gamma \\ \bar{y} < \bar{x}}} \bar{\mu}_{\bar{x},\bar{y}} \cdot n(\bar{x}) \bmod |y|$$

holds. □

Example 16.38. Let G be a finite group and Γ be $\mathrm{Or}\, G$. We have identified $U(\mathbf{Z}\mathrm{Or}\, G)$ with the Burnside ring $A(G)$ in Example 10.47. Then $\bar{\omega} : A(G) \to \prod_{(H)} \mathbf{Z}$ is just the character map ch of 5.15. Hence we get from Remark 16.37. a set of congruences describing the image of $A(G)$ in $\prod_{(H)} \mathbf{Z}$ under ch (compare tom Dieck [1987], IV.5.10 Kratzer-Thévenaz [1984]). □

Comments 16.39. The splitting given by (Res, I) and (E,S) are dual to one another. This may be illuminated by the following list of notions and properties. The left side belongs to (Res, I), the right side to (E,S).

1)	supp M = $\{\bar{x} \in \mathrm{Is}\,\Gamma \mid \mathrm{Res}_x M \neq \{0\}\}$	Iso M = $\{\bar{x} \in \mathrm{Is}\,\Gamma \mid S_x M \neq \{0\}\}$
2)	length $\ell(M)$	cardinality of Iso M
3)	S_x and I_x are adjoint	E_x and R_x are adjoint
4)	Res_x and I_x are exact but do not respect "projective"	E_x and S_x are not exact, but respect "projective"
5)	Res_x and I_x respect "finitely generated" for finite Γ	E_x and S_x respect "finitely generated"
6)	$\mathrm{Res}_y \circ I_x = 0$ for $\bar{x} \neq \bar{y}$ and $\mathrm{Res}_x \circ I_x = \mathrm{id}$	$S_y \circ E_x = 0$ for $\bar{x} \neq \bar{y}$ and $S_x \circ E_x = \mathrm{id}$

7)	$K_xM : M \longrightarrow I_x \circ S_xM$	$J_xM : E_x \circ S_xM \longrightarrow M$
8)	$KER_xM = \ker(K_xM)$	$COK_xM = \mathrm{cok}(J_xM)$
9)	Under certain conditions we have the exact sequence $0 \to KER_xM \to M \to I_x \circ Res_xM \to 0$	Under certain conditions we have the exact sequence $0 \to E_x \circ S_xM \to M \to COK_xM \to 0$
10)	(Res,I) define inverse isomorphisms $Gr_n(R\Gamma) \to \text{Split } Gr_n(R\Gamma)$	(E,S) define inverse isomorphisms $K_n(R\Gamma) \to \text{Split} K_n(R\Gamma)$

In geometry (Res,I) resp. (E,S) corresponds to the stratification of a G-space X by $\{X^H \mid H \subset G\}$ resp. $\{X_H \mid H \subset G\}$. The Moebius inversion is related to the problem how X^H can be built from the various X_H and vice versa.

Let $(I,\leq)$ be a partially ordered set. Consider I as an EI-category $\Gamma(I)$ having as objects the elements of I and Hom(x,y) consists of precisely one element if $x \leq y$ holds and is empty otherwise. Notice that $\Gamma(I)$ is a so called Δ-category, that is a category where $|\mathrm{Hom}(x,y)| \leq 1$ holds for all objects. Any EI-category Γ determines a partially ordered set $(\mathrm{Is}\,\Gamma,\leq)$. We get a bijective correspondence between Δ-categories and partially ordered sets. If Γ is a Δ-category and R is a principal integral domain $\Pi\, rk_R$ induces an isomorphism $Gr_n(R\Gamma) \to U(\Gamma)$ so that $\omega = \overline{\omega}$ and $\mu = \overline{\mu}$ holds. Then $\overline{\mu}$ is the usual Moebius inversion of combinatoric (see Aigner [1979], IV.2).

For computations concerning Gr(RG) for finite G we refer f.e. to Curtis-Reiner [1981], Hambleton-Taylor-Williams [1988], Webb [1987]

Exercises 16.40.

1) An EI-category Γ is a groupoid if and only if $L(\Gamma)$ is zero.

2) Consider a finite group. Let n(G) be $\Sigma \log_p(|G|)$ where Σ runs over all prime numbers. Show $L(\mathrm{Or}\ G) \leq n(G)$. Find sufficient conditions on G for $L(\mathrm{Or}\ G) = n(G)$.

3) Show the equivalence of the following assertions about an EI-category Γ of bounded length $L(\Gamma)$:

i) Γ is finite

ii) An $R\Gamma$-module M is finitely generated if and only if $M(x)$ is finitely generated over R for all $x \in \mathrm{Ob}\,\Gamma$.

4) An EI-category Γ is free if and only if $E_x : \text{MOD-}\mathbf{Z}[x] \longrightarrow \text{MOD-}\mathbf{Z}\Gamma$ is an exact functor for all $x \in \mathrm{Ob}\,\Gamma$.

5) Give an example of a commutative associative ring R with unit such that $K_o(R) \longrightarrow Gr_o(R)$ is no isomorphism.

6) Let Γ be a finite EI-category and R a Dedekind domain. Show that $F_* : K_o(R\Gamma) \longrightarrow Gr_o(R\Gamma)$ is an isomorphism if and only if $m(\Gamma)$ is a unit in R.

7) Let Γ be a finite EI-category. Show that two finitely generated projective $\mathbf{Q}\Gamma$-modules P and Q are isomorphic if and only if $P(x)$ and $Q(x)$ are $\mathbf{Q}[x]$-isomorphic for $x \in \mathrm{Ob}\,\Gamma$.

8) Let C be a finite free $R\Gamma$-chain complex over the finite free EI-category Γ. Prove $i(\chi_{R\Gamma}(C)) = \sum_{n \geq o} (-1)^n \bar{\mu}(\hat{rk}_{R\Gamma} H_n(C))$. Show that this also holds for a finitely dominated $\mathbf{Z}\Gamma$-chain complex but is in general false for a finitely dominated $\mathbf{Q}\Gamma$ chain complex.

9. Let Γ be the EI-category having two objects x and y such that $\mathrm{Hom}(x,y)$ and $\mathrm{Aut}(x)$ consists of precisely one element, $\mathrm{Hom}(y,x)$ is empty and $\mathrm{Aut}(y)$ is a non-trivial finite group. Show that there is no map $\hat{\mu} : C(\Gamma) \longrightarrow C(\Gamma)$ satisfying $\hat{\mu} \circ \omega = i$ (see Theorem 16.36).

10) Suppose that the diagonal functor $\Delta : \Gamma \longrightarrow \Gamma\times\Gamma$ is admissible for the finite EI-category Γ. Equip $U(\Gamma)$ with the ring structure $U(\Gamma) \otimes U(\Gamma) \xrightarrow{\otimes_R} U(\Gamma\times\Gamma) \xrightarrow{\Delta^*} U(\Gamma)$ and $C(\Gamma) = \prod_{\mathrm{Is}\,\Gamma} \mathbf{Z}$ with the product ring structure. Show that $\bar{\omega} : U(\Gamma) \longrightarrow C(\Gamma)$ is a ring homomorphism.

11) Compute μ and ω for $\Gamma = \mathrm{Or}\,A_5$ if A_5 is the alternating group.

12) Let G be a finite abelian group and Γ be $\mathrm{Or}\,G$. Show that

$\bar{\mu}_{G/K,G/H} = \sum_{\ell \geq 0} (-1)^{\ell} |ch_{\ell}(G/H,G/K)|$ and $\bar{\omega}_{G/K,G/H} = |K/H|$ holds for $H \subset K$ and that $\bar{\mu}_{G/K,G/H}$ and $\bar{\omega}_{G/K,G/H}$ vanish otherwise. Prove that $\mu_{G/K,G/H}$ resp. $\omega_{G/K,G/H}$: $Gr_n(R[G/K]) \longrightarrow Gr_n(R[G/H])$ is $\bar{\mu}_{G/K,G/H} \cdot res(p)$ resp. $res(p)$ for p the projection $G/H \longrightarrow G/K$ if $H \subset K$ is true and is zero otherwise. Compute for $G = \mathbf{Z}/p^n$ that $\bar{\mu}_{G/K,G/H} = 1$, if $H = K$, $\bar{\mu}_{G/K,G/H} = -1$, if $H \subset K$ and $|K/H|$ is a prime number, and $\bar{\mu}_{G/K,G/H} = 0$ otherwise.

17. Homological algebra.

Since the category MOD-$R\Gamma$ of $R\Gamma$-modules is abelian, we can do homological algebra over it. We examine the homological dimension of a $R\Gamma$-module and give upper bounds in favourable cases. This is used to define generalized Swan homomorphism in section 19. if Γ is finite and R is $\mathbb{Z}$. We develop tools for computing Ext and Tor groups, f.e. a spectral sequence.

Recall from section 11 that a resolution (P,f) of a $R\Gamma$-module is a projective $R\Gamma$-chain complex P with $H_i(P) = 0$ for $i > 0$ together with an isomorphism $f : H_o(P) \longrightarrow M$. We often omitt f. If P_n is finitely generated resp. free for all $n \geq 0$, we call P finitely generated resp. free. We say, P is finite if P is finitely generated and finite-dimensional. Recall that a commutative associative ring R with unit is noetherian resp. hereditary if any submodule of a finitely generated resp. projective R-module is also finitely generated resp. projective. We have defined the notion length in 16.3. and finite resp. free EI-category in Definition 16.1.

Lemma 17.1.

a) Any $R\Gamma$-module M has a free resolution P with $\ell(P_n) \leq \ell(M)$ for $n \in \mathbb{N}$.

b) Let Γ be finite and R noetherian. Then any finitely generated $R\Gamma$-module M has a finitely generated resolution P with $\ell(P_n) \leq \ell(M)$ for $n \in \mathbb{N}$.

Proof: Let M be a $R\Gamma$-module and the Ob Γ-subset $(S,\sigma) \subset M$ be a Ob Γ-set of generators. Then the free $R\Gamma$-module F with (S,σ) as base has length $\ell(F) \leq \ell(M)$ and there is an epimorphism $F \longrightarrow M$. Now a) follows directly and b) using Lemma 16.10. □

Definition 17.2. The homological dimension hdim M of a $R\Gamma$-module $M \neq \{0\}$ is the integer d, if M has a d-dimensional but no $(d-1)$-dimensional resolution, and is ∞ if M has no finite dimensional resolution. Put $\text{hdim}\{0\} = -1$. Let $\text{HDIM}(R\Gamma)$ be $\max\{\text{hdim } M \mid M \text{ a } R\Gamma\text{-module}\}$ □

Let C be a $R\Gamma$-chain complex with differential $c_r : C_r \longrightarrow C_{r-1}$. Given a $R\Gamma$-module N, let $\text{Hom}_{R\Gamma}(C,N)$ be the $R\Gamma$-cochain complex with $\text{Hom}_{R\Gamma}(C,N)_n = \text{Hom}_{R\Gamma}(C_n,N)$ and codifferential $c^n : \mathrm{Hom}_{R\Gamma}(C,N)_n \longrightarrow \mathrm{Hom}_{R\Gamma}(C,N)_{n+1}$ given by $(-1)^n \cdot \mathrm{Hom}_{R\Gamma}(c_{n+1},N) : \mathrm{Hom}_{R\Gamma}(C_n,N) \longrightarrow \mathrm{Hom}_{R\Gamma}(C_{n+1},N)$.

Definition 17.3. Let M be a $R\Gamma$-(contra-)module and N be a $R\Gamma$-(contra-) resp. $R\Gamma$-comodule. If P is any resolution of M, define for $n \geq 0$ the R-module

$$\mathrm{Ext}^n_{R\Gamma}(M,N) = H^n(\mathrm{Hom}_{R\Gamma}(P,N))$$

$$\mathrm{Tor}^{R\Gamma}_n(M,N) = H_n(P \otimes_{R\Gamma} N) \qquad \square$$

This is independent of the choice of P by Lemma 11.7. We collect some basic properties

17.4. As $\mathrm{Hom}_{R\Gamma}(-,N)$ is left exact and $- \otimes_{R\Gamma} N$ is right exact, we have natural isomorphisms $\mathrm{Ext}^o(M,N) = \mathrm{Hom}_{R\Gamma}(M,N)$ and $\mathrm{Tor}_o(M,N) = M \otimes_{R\Gamma} N$. A map $f : M^o \longrightarrow M^1$ induces $f^* : \mathrm{Ext}^n(M^1,N) \longrightarrow \mathrm{Ext}^n(M^o,N)$ and $f_* : \mathrm{Tor}_n(M^o,N) \longrightarrow \mathrm{Tor}_n(M^1,N)$ whereas $g : N^o \longrightarrow N^1$ induces $g_* : \mathrm{Ext}^n(M,N^o) \longrightarrow \mathrm{Ext}^n(M,N^1)$ and $g_* : \mathrm{Tor}_n(M,N^o) \rightarrow \mathrm{Tor}_n(M,N^1)$

17.5 Let $0 \longrightarrow M^1 \xrightarrow{i} M^o \xrightarrow{j} M^2 \longrightarrow 0$ be an exact sequence of $R\Gamma$-modules. Because of Lemma 11.6. there is an exact sequence $0 \longrightarrow P^1 \xrightarrow{k} P^o \xrightarrow{\ell} P^1 \longrightarrow 0$ of projective $R\Gamma$-chain complexes such that P^n is a resolution of M^n for $n = 1,0,2$ and $H_o(k) = i$ and $H_o(\ell) = j$ holds. Given a $R\Gamma$-module N , also $0 \longrightarrow \mathrm{Hom}_{R\Gamma}(P^2,N) \longrightarrow \mathrm{Hom}_{R\Gamma}(P^o,N) \longrightarrow \mathrm{Hom}_{R\Gamma}(P^1,N) \longrightarrow 0$ and $0 \longrightarrow P^1 \otimes_{R\Gamma} N \longrightarrow P^o \otimes_{R\Gamma} N \longrightarrow P^1 \otimes_{R\Gamma} N \rightarrow 0$ are exact. The long homology sequences give long exact sequences

$$0 \longrightarrow \mathrm{Hom}_{R\Gamma}(M^2,N) \longrightarrow \mathrm{Hom}_{R\Gamma}(M^o,N) \longrightarrow \mathrm{Hom}_{R\Gamma}(M^1,N) \longrightarrow \mathrm{Ext}^1(M^2,N) \longrightarrow \mathrm{Ext}^1(M^o,N)$$
$$\longrightarrow \mathrm{Ext}^1(M^1,N) \longrightarrow \mathrm{Ext}^2(M^2,N) \longrightarrow \ldots$$

and

$$\cdots\rightarrow \mathrm{Tor}_2(M^2,N) \longrightarrow \mathrm{Tor}_1(M^1,N) \longrightarrow \mathrm{Tor}_1(M^o,N) \longrightarrow \mathrm{Tor}_1(M^2,N) \longrightarrow M^1 \otimes_{R\Gamma} N \longrightarrow M^o \otimes_{R\Gamma} N$$
$$\longrightarrow M^2 \otimes_{R\Gamma} N \longrightarrow 0 \qquad \square$$

17.6. Let M be a $R\Gamma$-module and n an integer. Then we have $\mathrm{hdim}_{R\Gamma} M \leq n$ if and only if $\mathrm{Ext}^{n+1}(M,N)$ vanishes for all $R\Gamma$-modules N . This follows from Proposition 11.10. $\square$

17.7. Let $0 \longrightarrow M^1 \longrightarrow M^o \longrightarrow M^2 \longrightarrow 0$ be an exact sequence of $R\Gamma$-modules. Suppose that two of them possess finite-dimensional resolutions. We conclude from 17.5. and 17.6. that then all three have finite homological dimension and we have

$$\text{hdim } M^1 \leq \max(\text{hdim } M^o, -1 + \text{hdim } M^2)$$

$$\text{hdim } M^o \leq \max(\text{hdim } M^1, \text{hdim } M^2)$$

$$\text{hdim } M^2 \leq \max(\text{hdim } M^o, 1 + \text{hdim } M^1) \quad \square$$

17.8. Let $0 \longrightarrow M \longrightarrow P_{n-1} \longrightarrow P_{n-2} \longrightarrow \dots \longrightarrow P_o \longrightarrow N \longrightarrow 0$ be an exact sequence of $R\Gamma$-modules. Suppose that P_i is projective for $i = 0,1\dots n-1$. Then M is projective if and only if $\text{hdim } N \leq n$ is true. If $\text{hdim } M > 0$ or $\text{hdim } N \geq n$ then $\text{hdim } N = n + \text{hdim } M$. This follows inductively over n from 17.7. $\square$

Next we establish a spectral sequence converging to $\text{Ext}^n_{R\Gamma}(M,N)$ whose E^1-term is given by certain Ext-groups $\text{Ext}^q_{R[y]}(M(x) \otimes_{R[x]} RS(c), N(y))$ over the group rings $R[y]$. Some preparations are needed. Let M be a $R\Gamma$-module. Define a $R\Gamma$-module

17.9. $$EM = \bigoplus_{\bar{x} \in \text{Is } \Gamma} E_x \circ \text{Res}_x M$$

The direct sum over the adjoints of $\text{id} : \text{Res}_x M \longrightarrow \text{Res}_x M$ for each $\bar{x} \in \text{Is } \Gamma$ (see Lemma 9.31.) defines an epimorphism $q : EM \longrightarrow M$. Denote

17.10. $$KM = \text{kernel}(q)$$

We obtain an exact sequence

17.11. $$0 \longrightarrow KM \xrightarrow{i} EM \xrightarrow{q} M \longrightarrow 0$$

Define inductively for $p \geq 0$

17.12. $$K^oM = M, \quad K^pM = KK^{p-1}M$$

Iterating 17.11. yields a long exact sequence

17.13. $$\dots \longrightarrow EK^pM \longrightarrow EK^{p-1}M \longrightarrow \dots EKM \longrightarrow EM \longrightarrow M \longrightarrow 0$$

Let M be a $R\Gamma$-module of finite length $\ell = \ell(M)$ and N be a $R\Gamma$-module. We define a finite filtration of R-cochain complexes

17.14. $$\{0\} \subset F^{\ell}Q \subset F^{\ell-1}Q \subset \dots \subset F^{1}Q \subset F^{o}Q = Q$$

satisfying

17.15. There are resolutions $P(p)$ of $K^{p}(M)$ and $\hat{P}(p-1)$ of $EK^{p-1}M$ such that $F^{p}Q$ is $\Sigma^{p}\mathrm{Hom}_{R\Gamma}(P(p),N)$ and $F^{p}Q/F^{p+1}Q$ is $\Sigma^{p}\mathrm{Hom}_{R\Gamma}(\hat{P}(p),N)$.

Suppose we have already constructed $\{0\} \subset F^{\ell}Q \subset \dots \subset F^{p}Q$. We get from 17.11. an exact sequence

17.16. $$0 \longrightarrow K^{p}M \longrightarrow EK^{p-1}M \longrightarrow K^{p-1}M \longrightarrow 0$$

Let $P(p)$ be the resolution of $K^{p}M$ for which $F^{p}Q$ is $\Sigma^{p}\mathrm{Hom}_{R\Gamma}(P(p),N)$. Because of Lemma 11.6. we can find an exact sequence of $R\Gamma$-chain complexes $0 \longrightarrow P(p) \xrightarrow{i} \hat{P}(p-1) \longrightarrow P' \longrightarrow 0$ such that H_o applied to it gives 17.16. and $\hat{P}(p-1)$ and P' are resolutions of $EK^{p-1}M$ and $K^{p-1}M$. If we write $P(p-1) = \mathrm{Cone}(i)$ then $P(p-1)$ is a resolution of $K^{p-1}M$ and we have an exact sequence $0 \longrightarrow \hat{P}(p-1) \longrightarrow P(p-1) \longrightarrow \Sigma P(p) \longrightarrow 0$. It induces an exact sequence

17.17. $$0 \longrightarrow \mathrm{Hom}_{R\Gamma}(\Sigma P(p),N) \longrightarrow \mathrm{Hom}_{R\Gamma}(P(p-1),N) \longrightarrow \mathrm{Hom}_{R\Gamma}(\hat{P}(p-1),N) \longrightarrow 0$$

Put $F^{p-1}Q = \Sigma^{p-1}\mathrm{Hom}_{R\Gamma}(P(p-1),N)$. This finishes the construction of 17.14. and 17.15. follows from 17.17.

To such a filtration there is assigned a spectral (cohomology) sequence (E_r,d_r) , $r = 1,2,\dots$ (see Cartan-Eilenberg [1956] XV, MacLane [1963] XI). Notice that $\ell(KM) < \ell(M)$ is true.

<u>Theorem 17.18.</u> <u>Let</u> M <u>and</u> N <u>be</u> $R\Gamma$-<u>modules. Suppose that</u> M <u>has finite length</u> ℓ . <u>Then there is a spectral sequence</u> (E_r,d_r) , $r = 1,2,\dots$ <u>satisfying</u>:

a) (E_r,d_r) <u>converges to</u> $\mathrm{Ext}^{n}_{R\Gamma}(M,N)$.

b) <u>The</u> E_1-<u>term is given for</u> $p,q \geq 0$ <u>by</u>

$$E_1^{p,q} = \mathrm{Ext}^{q}_{R\Gamma}(EK^{p}M,N)$$

<u>If</u> j <u>is the composition</u> $EK^{p+1}M \longrightarrow K^{p+1}M \longrightarrow EK^{p}M$, <u>the first differential</u> $d_1^{p,q} : E_1^{p,q} \longrightarrow E_1^{p+1,q}$ <u>is</u> $j^{*} : \mathrm{Ext}^{q}_{R\Gamma}(EK^{p}M,N) \longrightarrow \mathrm{Ext}^{q}_{R\Gamma}(EK^{p+1}M,N)$.

c) <u>We have for all</u> $r \geq 1$

$$E_r^{p,q} \neq \{0\} \Rightarrow 0 \leq p \leq \ell \quad \underline{\text{and}} \quad 0 \leq q$$

d) <u>If</u> $r > \max(p, \ell - p)$ <u>holds, we get</u>

$$E_r^{p,q} = E_\infty^{p,q} \qquad \square$$

Ve recall that a <u>spectral sequence</u> (E_r, d_r) is a collection of bigraded modules $E_r^{p,q}$ vith differentials $d_r : E_r^{p,q} \longrightarrow E_r^{p+r,q-r+1}$ such that $d_r \circ d_r$ is zero and $E_{r+1} =$ $H(E_r, d_r)$. The statement a) in Theorem 17.18. means the following: There is a fil-ration $0 \subset F^\ell \mathrm{Ext}^n_{R\Gamma}(M,N) \subset F^{\ell-1}\mathrm{Ext}^n_{R\Gamma}(M,N) \subset \ldots \; F^1\mathrm{Ext}^n_{R\Gamma}(M,N) \subset F^o\mathrm{Ext}^n_{R\Gamma}(M,N) = \mathrm{Ext}^n_{R\Gamma}(M,N)$ such that

$$F^p\mathrm{Ext}^n_{R\Gamma}(M,N)/F^{p+1}\mathrm{Ext}^n_{R\Gamma}(M,N) = E_\infty^{p,n-p} .$$

Notice that E_∞ is determined by Theorem 17.18. d). We get from Theorem 17.18. c) that the E_1-term has the following shape for $\ell = 3$

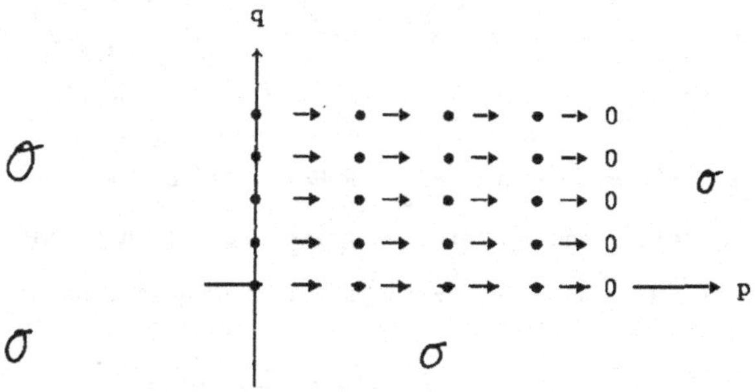

<u>xample 17.20.</u> Suppose that Γ is a EI-category having two objects x and y with $< \bar{y}$. Then we obtain from Theorem 17.18. a) the exact sequence

$$0 \longrightarrow E_\infty^{1,n-1} \longrightarrow \mathrm{Ext}^n_{R\Gamma}(M,N) \longrightarrow E_\infty^{0,n} \longrightarrow 0$$

s $E_2 = H(E_1, d_1)$ and $E_\infty = E_2$ by Theorem 17.18.c) we get from Theorem 17.18. d) he exact sequence

$$0 \longrightarrow E_\infty^{o,n} \longrightarrow \mathrm{Ext}^n_{R\Gamma}(EM,N) \xrightarrow{j^*} \mathrm{Ext}^n_{R\Gamma}(EKM,N) \longrightarrow E_\infty^{1,n} \longrightarrow 0$$

ence we obtain a long exact sequence

$$\longrightarrow \mathrm{Ext}^n_{R\Gamma}(M,N) \longrightarrow \mathrm{Ext}^n_{R\Gamma}(EM,N) \xrightarrow{j^*} \mathrm{Ext}^n_{R\Gamma}(EKM,N) \longrightarrow \mathrm{Ext}^{n+1}_{R\Gamma}(M,N) \longrightarrow$$

Notice that $EKM = KM$ holds as $\ell(KM) \leq 0$ is true. Now one easily identifies this long exact sequence with the one defined in 17.5. for $0 \longrightarrow KM \xrightarrow{j} EM \longrightarrow M \longrightarrow 0$ □

Now we want to analyze the E_1-term further.

<u>Lemma 17.21.</u> <u>Suppose either that</u> Γ <u>is free or that</u> Γ <u>is finite with</u> $m(\Gamma) \in R^*$. <u>Then the adjunction between</u> E_x <u>and</u> Res_x <u>(see Lemma 9.31.)</u> <u>defines for a</u> $R[x]$-<u>module</u> L <u>and a</u> $R\Gamma$-<u>module</u> N <u>an isomorphism</u>

$$\mathrm{Ext}^n_{R\Gamma}(E_xL,N) = \mathrm{Ext}^n_{R[x]}(L,N(x))$$

<u>Proof:</u> If P is a resolution of L, we get by E_xP a resolution of E_xL under th conditions above. We have $\mathrm{Hom}_{R\Gamma}(E_xP,N) = \mathrm{Hom}_{R[x]}(P,\mathrm{Res}_xN)$. □

Next consider the exact sequence $0 \longrightarrow KM(y) \longrightarrow EM(y) \xrightarrow{q} M(y) \longrightarrow 0$ of $R[y]$- modules obtained by evaluating 17.11. at $y \in \mathrm{Ob}\,\Gamma$. Let $q_{x,y}\colon M(x) \otimes_{R[x]} R\,\mathrm{Hom}(x,y) \longrightarrow M(y)$ send $m \otimes \phi$ to $M(\phi)(m)$.

Define $i_{x,z} : M(x) \otimes_{R[x]} R\,\mathrm{Hom}(y,x) \longrightarrow M(z) \otimes_{R[z]} R\,\mathrm{Hom}(y,z)$ by id, if $x = z$ holds, by $-q_{x,y}$, if $y = z$ holds, and by zero otherwise. Recall that $EM(y)$ is $\oplus\, M(z) \otimes_{R[z]} R\,\mathrm{Hom}(y,z)$ if $\oplus$ runs over $z \in \mathrm{Is}\,\Gamma$. The collection $i_{x,z}$ gives

17.22. $$i : \bigoplus_{\substack{\bar{x} \in \mathrm{Is}\,\Gamma \\ \bar{y} < \bar{x}}} M(x) \otimes_{R[x]} R\,\mathrm{Hom}(y,x) \longrightarrow EM(y)$$

<u>Lemma 17.23.</u> <u>The</u> $R[y]$-<u>map</u> i <u>induces a natural isomorphism of</u> $R[y]$-<u>modules</u>

$$i : \bigoplus_{\substack{\bar{x} \in \mathrm{Is}\,\Gamma \\ \bar{y} < \bar{x}}} M(x) \otimes_{R[x]} R\,\mathrm{Hom}(y,x) \longrightarrow KM(y) \qquad \blacksquare$$

We have defined the notion of a p-chain c from $\bar{y}$ to $\bar{x}$ in 16.3. and assigned to c a $\mathrm{Aut}(x)$- $\mathrm{Aut}(y)$-set $S(x)$ in 16.23. Let $ch_p(\Gamma)$ be the set of triples $(\bar{y},\bar{c},\bar{x})$ with $\bar{y},\bar{x} \in \mathrm{Is}\,\Gamma$, $c \in ch_p(\bar{y},\bar{x})$. One derives inductively over p from Lemma 17.23. (use statement v)) in the proof of Theorem 16.27.)

__Lemma 17.24.__ __We obtain for__ $z \in \mathrm{Ob}\,\Gamma$ __a natural__ R[y]-__isomorphism__

$$K^pM(z) = \bigoplus_{\substack{c \in ch_p(\Gamma) \\ \bar{z} = \bar{y}}} M(x) \otimes_{R[x]} RS(c) \qquad \square$$

Since $\mathrm{Hom}_{R\Gamma}(\bigoplus_{i \in I} L_i, N) = \prod_{i \in I} \mathrm{Hom}_{R\Gamma}(L_i, N)$ holds, we obtain from Lemma 17.21. and Lemma 17.24. isomorphisms provided that either Γ is free or that Γ is finite with $m(\Gamma) \in R^*$

17.25. $$\mathrm{Ext}^q_{R\Gamma}(EK^pM,N) \cong \prod_{ch_p(\Gamma)} \mathrm{Ext}^q_{R[x]}(M(x) \otimes_{R[x]} RS(c), N(y))$$

Fix $\eta = (\bar{v},\bar{d},\bar{u}) \in ch_p(\Gamma)$ with $d : \bar{v} = \bar{v}_o < \bar{v}_1 < \ldots < \bar{v}_p = \bar{u}$ for $p \geq 1$. For any $\xi = (\bar{y},c,\bar{x}) \in ch_{p-1}(\Gamma)$ we want to define a map

17.26. $$d_{\xi,\eta} : \mathrm{Ext}_{R[y]}(M(x) \otimes_{R[x]} RS(c), N(y)) \longrightarrow \mathrm{Ext}_{R[v]}(M(u) \otimes_{R[u]} RS(d), N(v)) .$$

Suppose $\bar{y} = \bar{v}_1, \bar{x} = \bar{u}$ and $c : \bar{v}_1 < \bar{v}_2 < \ldots < \bar{v}_p$. Let

$$\psi_1 : \mathrm{Ext}^q_{R[y]}(M(x) \otimes_{R[x]} RS(c), N(y)) \longrightarrow \mathrm{Ext}^q_{R[v]}(M(u) \otimes_{R[u]} RS(d), N(y) \otimes_{R[y]} R\,\mathrm{Hom}(v,y))$$

be given by $- \otimes_{R[y]} R\,\mathrm{Hom}(v,y)$ and the obvious identification

$$RS(c) \otimes_{R[y]} R\,\mathrm{Hom}(v,y) = RS(d) .$$

More precisely, let P be a (projective) R[y]-resolution of $M(x) \otimes_{R[x]} RS(c)$ and Q a (projective) R[v]-resolution of $M(x) \otimes_{R[x]} RS(d)$. Then $P \otimes_{R[y]} R\,\mathrm{Hom}(v,y)$ is a (not necessarily projective) resolution of $M(x) \otimes_{R[x]} RS(d)$. Let $f : Q \longrightarrow P \otimes_{R[y]} R\,\mathrm{Hom}(v,y)$ be a R[v]-chain map inducing the identity on homology. Then ψ_1 is induced from the chain map

$$\mathrm{Hom}_{R[y]}(P,N(y)) \longrightarrow \mathrm{Hom}_{R[v]}(P \otimes_{R[y]} R\,\mathrm{Hom}(v,y), N(y) \otimes_{R[y]} R\,\mathrm{Hom}(v,y))$$
$$\xrightarrow{f^*} \mathrm{Hom}_{R[v]}(Q, N(y) \otimes_{R[y]} R\,\mathrm{Hom}(v,y)) .$$

The map $q : N(y) \otimes_{R[y]} R\,\mathrm{Hom}(v,y) \longrightarrow N(v)$ sending $n \otimes f$ to $N(f)(n)$ induces

$$\psi_2 : \mathrm{Ext}^q_{R[u]}(M(u) \otimes_{R[u]} RS(d), N(y) \otimes_{R[y]} R\,\mathrm{Hom}(v,y)) \rightarrow \mathrm{Ext}^q_{R[u]}(M(u) \otimes_{R[u]} RS(d), N(v)) .$$

Let $d_{\xi,\eta}$ be $\psi_2 \circ \psi_1$.

Suppose $\bar{y} = \bar{v}$, $\bar{x} = \bar{u}$ and $c : \bar{v}_o < \bar{v}_1 < \ldots < \bar{v}_{i-1} < \bar{v}_{i+1} < \ldots < \bar{v}_p$ for $0 < i < p$. There is an obvious map

$$\psi : RS(d) \longrightarrow RS(c)$$

coming from $\mathrm{Hom}(v_{i+1},v_i) \times_{\mathrm{Aut}(v_i)} \mathrm{Hom}(v_i,v_{i-1}) \longrightarrow \mathrm{Hom}(v_{i+1},v_{i-1})$ mapping (f,g) to $g \circ f$. It induces

$$(id \otimes \psi)^* : \mathrm{Ext}_{R[x]}(M(x) \otimes_{R[x]} RS(c),N(y)) \longrightarrow \mathrm{Ext}_{R[u]}(M(u) \otimes_{R[u]} RS(d),N(v)) .$$

Let $d_{\xi,\eta}$ be $(-1)^i(id \otimes \psi)^*$.

Suppose $\bar{y} = \bar{v}$, $\bar{x} = \bar{v}_{p-1}$ and $c : \bar{v}_o < \bar{v}_1 < \ldots < \bar{v}_{p-1}$. Let $q : M(u) \otimes_{R[u]} R\,\mathrm{Hom}(x,u) \longrightarrow M(x)$ send $m \otimes f$ to $M(f)(m)$. Since $RS(d)$ is $R\,\mathrm{Hom}(x,u) \otimes_{R[x]} RS(c)$ we get by $q \otimes_{R[x]} RS(c)$ a map

$$\psi : M(u) \otimes_{R[u]} RS(d) \longrightarrow M(x) \otimes_{R[x]} RS(c)$$

Let $d_{\xi,\eta}$ be $(-1)^p \psi^*$.

For all other ξ let $d_{\xi,\eta}$ be zero. For fixed η there are only finitely many ξ with $d_{\xi,\eta} \neq 0$. Hence the collection $(d_{\xi,\eta})_{\xi,\eta}$ defines an homomorphism

17.27.
$$d : \prod_{(\bar{y},\bar{c},\bar{x}) \in ch_{p-1}(\Gamma)} \mathrm{Ext}^q_{R[x]}(M(x) \otimes_{R[x]} RS(c),N(x)) \longrightarrow \prod_{(\bar{v},d,\bar{u}) \in ch_p(\Gamma)} \mathrm{Ext}^q_{R[u]}(M(u) \otimes_{R[u]} RS(d),N(v))$$

Theorem 17.28. *Suppose either that Γ is free or that Γ is finite with $m(\Gamma) \in R^*$. Let M and N be $R\Gamma$-modules such that M has finite length $\ell = \ell(M)$. Let (E_r,d_r), $r \geq 1$ be the spectral sequence of Theorem 17.18.*

Then we obtain from 17.25. an isomorphism

$$E_1^{p,q} = \prod_{ch_p(\Gamma)} \mathrm{Ext}^q_{R[y]}(M(x) \otimes_{R[x]} RS(c),N(y))$$

Under this identification the first differential $d_1^{p,q} : E_1^{p,q} \longrightarrow E_1^{p+1,q}$ agrees with the map d of 17.27. □

Next we can define the cohomology of a category Γ with coefficients in a $\mathbb{Z}\Gamma$-module N. Let $\underline{\mathbb{Z}}$ be the constant $\mathbb{Z}\Gamma$-module with value $id : \mathbb{Z} \longrightarrow \mathbb{Z}$ on any morphism $f : x \longrightarrow y$.

Definition 17.29. If M is a $R\Gamma$-comodule, define

$$H_n(\Gamma,M) = \mathrm{Tor}_n^{\mathbb{Z}\Gamma}(\underline{\mathbb{Z}},M) = \mathrm{Tor}_n^{R\Gamma}(\underline{R},M)$$

If M is a $R\Gamma$-contramodule, define

$$H^n(\Gamma,M) = \mathrm{Ext}^n_{\mathbb{Z}\Gamma}(\underline{\mathbb{Z}},M) = \mathrm{Ext}^n_{R\Gamma}(\underline{R},M)$$

Example 17.30. Let Γ be a free EI-category having two objects x and y with $\bar{x} < \bar{y}$. Suppose that $\mathrm{Aut}(x)$ acts transitively on $\mathrm{Hom}(x,y)$. Fix $f : x \longrightarrow y$. Let $H \subset \mathrm{Aut}(x)$ be the isotropy group of the class given by f in $\mathrm{Hom}(x,y)/\mathrm{Aut}(y)$ so that $\mathrm{Hom}(x,y)/\mathrm{Aut}(y)$ is $\mathrm{Aut}(x)/H$. We get for $M = \underline{\mathbb{Z}}$ from Theorem 17.28.

$$E_1^{0,q} = H^q(\mathrm{Aut}(x),N(x)) \oplus H^q(\mathrm{Aut}(y),N(y))$$

$$E_1^{1,q} = H^q(H,N(x)) .$$

Let $res : H^q(\mathrm{Aut}(x),N(x)) \longrightarrow H^q(H,N(x))$ be restriction from $\mathrm{Aut}(x)$ to H. Define $s : H^q(\mathrm{Aut}(y),N(y)) \longrightarrow H^q(H,N(x))$ by

$$H^q(\mathrm{Aut}(y),N(y)) = \mathrm{Ext}^q_{R[y]}(R,N(y)) \longrightarrow \mathrm{Ext}^q_{R[x]}(R \otimes_{R[y]} R\,\mathrm{Hom}(x,y), N(y) \otimes_{R[y]} R\,\mathrm{Hom}(x,y))$$

$$\longrightarrow \mathrm{Ext}^q_{R[x]}(R\,\mathrm{Hom}(x,y)/\mathrm{Aut}(y),N(x)) = H^q(H,N(x)) .$$

Then we obtain a long exact sequence

$$\ldots \longrightarrow H^n(\Gamma,N) \longrightarrow H^n(\mathrm{Aut}(x),N(x)) \oplus H^n(\mathrm{Aut}(y),N(y)) \xrightarrow{-res+s} H^n(H,N(x))$$

$$\longrightarrow H^{n+1}(\Gamma,N) \longrightarrow \ldots . \qquad \square$$

Next we want to give some bounds for the homological dimension.

Proposition 17.31. Let M be a $R\Gamma$-module over the finite EI-category Γ. Suppose either that $m(\Gamma)$ is a unit in R or that M has a finite resolution. If $M(x)$

<u>is projective over</u> R <u>for all</u> $x \in \mathrm{Ob}\,\Gamma$ <u>we have</u>

$$\mathrm{hdim}\, M \leq \ell(M) .$$

<u>Proof.</u> Let $m(\Gamma)$ be a unit in R. Then for any $x \in \mathrm{Ob}\,\Gamma$ a $R[x]$-module M which is projective over R is projective over $R[x]$. Then 17.13. defines a $\ell(M)$-dimensional resolution of M

$$0 \longrightarrow EK^{\ell(M)}M \longrightarrow EK^{\ell(M)-1}M \longrightarrow \dots \longrightarrow EKM \longrightarrow EM \longrightarrow M \longrightarrow 0$$

Next we treat the case where M has a finite resolution $0 \longrightarrow P_n \longrightarrow P_{n-1} \longrightarrow \dots \longrightarrow P_o \longrightarrow M \longrightarrow 0$. We use induction over $\ell(M)$. In the induction begin $\ell(M) = 0$ we may suppose that M is a RG-module for a finite group G, that means $\Gamma = \hat{G}$ (see Example 9.5.). If N is a RG-module, let N^* be the RG-module $\mathrm{Hom}_R(N,R)$ with the G-structure given by $(fg)(n) = f(ng^{-1})$ for $f \in \mathrm{Hom}_R(N,R)$, $n \in N$, $g \in G$. There is a natural isomorphism $(N_1 \oplus N_2)^* = N_1^* \oplus N_2^*$. If the RG-module N is finitely generated and projective over R, the canonical map $N \longrightarrow N^{**}$ is a RG-isomorphism. Obviously RG^* is RG-isomorphic to RG. Hence N^* is a finitely generated projective RG-module if N is. Therefore we obtain an exact sequence $0 \longrightarrow M^* \longrightarrow P_o^* \longrightarrow P_1^* \longrightarrow \dots \longrightarrow P_n^* \longrightarrow 0$ such that P_i^* is finitely generated and projective over RG. Then M^* and hence M is a finitely generated projective RG-module. This implies $\mathrm{hdim}_{RG} M = 0$.

We come to the induction step from $\ell - 1$ to $\ell = \ell(M)$. For $x \in \mathrm{Ob}\,\Gamma$ with $\ell(x)=\ell$ we have $\mathrm{Res}_x P_n = S_x P_n$ if $\ell(P_n) \leq \ell$ holds. Hence $\mathrm{Res}_x M$ has a finite resolution and is a projective $R[x]$-module by the induction begin by Lemma 17.1. and 17.8. Let

$$q : \bigoplus_{\substack{\bar{x} \in \mathrm{Is}\,\Gamma \\ \ell(\bar{x}) = \ell}} E_x \circ \mathrm{Res}_x M \longrightarrow M$$

be the direct sum of the adjoints of $\mathrm{id} : \mathrm{Res}_x M \longrightarrow \mathrm{Res}_x M$. As $\ell(\mathrm{cok}\, q) < \ell$ holds there is a finitely generated projective $R\Gamma$-module P with $\ell(P) < \ell$ and a map $\alpha : P \longrightarrow M$ such that $q \oplus \alpha$ is surjective. If K is the kernel of $q \oplus \alpha$, we get an exact sequence of $R\Gamma$-modules such that the middle entry is finitely generated projective

$$0 \longrightarrow K \longrightarrow P \oplus \bigoplus_{\substack{\bar{x} \in \mathrm{Is}\,\Gamma \\ \ell(\bar{x}) = \ell}} E_x \circ \mathrm{Res}_x M \longrightarrow M \longrightarrow 0$$

Because of Lemma 11.6. there is a finite resolution for K. Since $\ell(K) < \ell$ the induction hypothesis and 17.8. show $\mathrm{hdim}\, M \leq \ell$ □

<u>Corollary 17.32.</u> <u>Let</u> Γ <u>be finite and</u> R <u>be a field such that</u> $m(\Gamma)$ <u>and</u> $ch(R)$ <u>are prime. Then we have</u>

$$\mathrm{HDIM}(R\Gamma) \leq \ell(\Gamma)$$

<u>Remark 17.33.</u> If $p_1,\dots,p_r$ are distinct primes with product m, we have

$$\mathrm{HDIM}(\mathbb{Q}\mathrm{Or}\mathbb{Z}/m) = \ell(\mathrm{Or}\mathbb{Z}/m) = r$$

On the other hand $\mathrm{HDIM}(\mathbb{Q}\mathrm{Or}\mathbb{Z}/p^n) = 1$ and $\ell(\mathrm{Or}\mathbb{Z}/p^n) = n$ for p a prime number. □

<u>Proposition 17.34.</u> <u>Let</u> Γ <u>be finite with</u> $m(\Gamma) \in R^*$. <u>Let</u> M <u>be a</u> $R\Gamma$<u>-module such that</u> $M(x)$ <u>is finitely generated projective over</u> R <u>for each</u> $x \in \mathrm{Ob}\,\Gamma$.

a) <u>We have</u> $\mathrm{hdim}_{R\Gamma} M \leq p$ <u>for some integer</u> p <u>if and only if</u> $K^p M$ <u>is a projective</u> $R\Gamma$<u>-module.</u>

b) <u>Suppose that</u> $\mathrm{hdim}_{R\Gamma} M \leq p$ <u>holds. Then there is for each</u> $x \in \mathrm{Is}\,\Gamma$ <u>a finitely generated projective</u> $R[x]$<u>-module</u> P_x <u>such that</u>

$$\mu \circ (\omega - \mathrm{id})^p \circ \mathrm{Res}([M]) = ([P_x])_{\bar{x} \in \mathrm{Is}\,\Gamma}$$

<u>holds if</u> μ, ω <u>and</u> Res <u>are the isomorphisms appearing in Theorem 16.27.</u>

c) <u>Suppose that</u> Γ <u>is also free. Let</u> $\bar{\mu} : C(\Gamma) \longrightarrow C(\Gamma)$ <u>be the homomorphism of 16.35. Let</u> $\hat{\omega} : C(\Gamma) \longrightarrow C(\Gamma)$ <u>send</u> $(n(\bar{x}))_{\bar{x} \in \mathrm{Is}\,\Gamma}$ <u>to</u>

$$\Big(\sum_{\bar{x} \in \mathrm{Is}\,\Gamma} n(\bar{x}) \cdot |\mathrm{Hom}(y,x)/\mathrm{Aut}(x)|\Big)_{\bar{y} \in \mathrm{Is}\,\Gamma}$$

<u>If</u> $\mathrm{hdim}_{R\Gamma} M \leq p$ <u>holds we have</u>

$$\bar{\mu} \circ (\hat{\omega} - \mathrm{id})^p (\mathrm{rk}^{\wedge}_{R\Gamma} M) \geq 0 \quad \text{for } \bar{x} \in \mathrm{Is}\,\Gamma$$

<u>if</u> $rk^{\wedge}_{R\Gamma}M$ <u>is</u> $(rk_R M(x))_{\bar{x}} \in \mathrm{Is}\,\Gamma$. <u>The maps</u> $\hat{\omega}$ and $\bar{\mu}$ <u>are inverse to one another.</u>

<u>Proof.</u> We get from 17.13. the exact sequence

$$0 \longrightarrow K^pM \longrightarrow EK^{p-1}M \longrightarrow EK^{p-2}M \longrightarrow \dots \longrightarrow EM \longrightarrow M \longrightarrow 0$$

As EK^jM is a finitely generated projective $R\Gamma$-module a) follows from 17.8. We get in $K_o(R\Gamma)$

$$[KM] = [EM] - [M] = (E \circ \text{Res-id})[M]$$

This implies inductively over p

$$S[K^pM] = \mu \circ \text{Res}[K^pM] = \mu \circ \text{Res} \circ (E \circ \text{Res-id})^p[M] = \mu \circ (\text{Res} \circ E\text{-id})^p \circ \text{Res}[M]$$

$$= \mu \circ (\omega\text{-id})^p \circ \text{Res}[M]$$

Now b) follows from a) by taking $P_x = S_x K^pM$. We obtain c) from b) using the easily verified relations $rk^{\wedge}_{R\Gamma} \circ \omega = \hat{\omega} \circ rk^{\wedge}_{R\Gamma}$ and $rk^{\wedge}_{R\Gamma} \circ \mu = \bar{\mu} \circ rk^{\wedge}_{R\Gamma}$ (see proof of Theorem 16.36.) □

<u>Example 17.35.</u> Especially the criterion in Proposition 17.34 c) is useful. If p_1 and p_2 are distinct prime numbers, let G be $\mathbb{Z}/p_1p_2$ and H_i the subgroup $\mathbb{Z}/p_i$. Then $\mathrm{hdim}_{\mathbb{Q}OrG} M \le 1$ implies for a finitely generated $\mathbb{Q}OrG$-module M

$$rk_{\mathbb{Q}}M(G/G) \le rk_{\mathbb{Q}}M(G/H_1) + rk_{\mathbb{Q}}M(G/H_2)$$

In particular $\mathrm{hdim}\, I_{G/G}Q$ must be 2 . □

The following criterion is sometimes useful. Given a functor $i : \Gamma_1 \longrightarrow \Gamma_2$ and a $R\Gamma_1$-module N, let the coinduction $i_{\#}N$ be the $R\Gamma_2$-module $\mathrm{Hom}_{R\Gamma_1}(R\,\mathrm{Hom}_{\Gamma_2}(i(?),??),N(?$

<u>Proposition 17.36.</u> <u>Suppose that</u> $i : \Gamma_1 \longrightarrow \Gamma_2$ <u>is admissible. Assume the existence of an object</u> y <u>in</u> Γ_1 <u>such that</u> Aut(y) <u>is finite</u>, $|\mathrm{Aut}(y)|$ <u>a unit in</u> R <u>and</u> Aut(y) <u>acts transitively on the non-empty set</u> Hom(x,y) <u>for all</u> $x \in \mathrm{Ob}\,\Gamma$. <u>Let</u> M <u>be a</u> $R\Gamma_2$-<u>module such that</u> M <u>is a direct summand in</u> $i_{\#}i^*M$. <u>Then</u>

$$H^n(\Gamma,M) = 0 \quad \text{for} \quad n \ge 1$$

Proof. It suffices to show $\mathrm{Ext}^n_{R\Gamma_2}(\underline{R}, i_{\#}i^*M) = 0$. As i^* respects "projective", $i^*\underline{R}$ is $\underline{R}$ and i^* and $i_{\#}$ are adjoint (see 9.21.) we have $\mathrm{Ext}^n_{R\Gamma_2}(\underline{R}, i_{\#}i^*M) = \mathrm{Ext}^n_{R\Gamma_1}(\underline{R}, i^*M)$. But $\underline{R} = E_yR$ is projective under the conditions above. □

We need for later purposes.

Lemma 17.37. Let Γ be a finite EI-category and R be noetherian. Suppose either that Γ is free or that $m(\Gamma)$ is a unit in R . Let M be a $R\Gamma$-module such that the $R[y]$-module $M(x) \otimes_{R[x]} RS(c)$ has a finite resolution for all $x,y \in \mathrm{Ob}\,\Gamma$, $p \geq 0$, $c \in ch_p(\bar{y},\bar{x})$. Then we have

a) M has a finite resolution.

b) Let $[M] \in K_0(R\Gamma)$ be the element given by M and $S : K_0(R\Gamma) \to \mathrm{Split}\, K_0(R\Gamma)$ be the isomorphism of Theorem 10.34. For $\bar{y} \in \mathrm{Is}\,\Gamma$ we get

$$S([M])_{\bar{y}} = \sum_{p \geq 0} (-1)^p \cdot \sum_{\substack{\bar{x} \in \mathrm{Ob}\,\Gamma \\ c \in ch_p(\bar{y},\bar{x})}} [M(x) \otimes_{R[x]} RS(c)]$$

c) An analogous statement holds for $[f] \in K_1(R\Gamma)$ given by an automorphism $f : M \longrightarrow M$.

Proof.

a) Consider any $R\Gamma$-module N and the spectral sequence (E_r, d_r) of Theorem 17.18. converging to $\mathrm{Ext}^n_{R\Gamma}(M,N)$. Let d be an integer with $\mathrm{hdim}_{R[y]}(M(x) \otimes_{R[x]} RS(c)) \leq d$ Then $E_1^{p,q}$ is zero by Theorem 17.28. for $q > d$. Hence $\mathrm{Ext}^n_{R\Gamma}(M,N)$ is zero for $n > d + \ell(\Gamma)$ so that M has a finite resolution by Lemma 16.10., Lemma 17.1. and 17.8.

b) If M satisfies the hypothesis for a) then also KM by Lemma 17.23. We obtain from 17.13.

$$[M] = \sum_{\ell \geq 0} (-1)^\ell [EK^\ell M]$$

Now apply Lemma 17.24. □

Comments 17.38. Homological algebra over a category is an important tool for homotopy approximations for classifying spaces BG of a compact Lie group by the homotopy push out of a system $\{BH \mid H \in \mathfrak{F}\}$ for an appropriate set $\mathfrak{F}$ of subgroups H of G. The problem whether such a system approximates BG can be reduced to the question whether $H^n(\Gamma;M)$ vanishes for appropriate subcategories $\Gamma \subset Or/G$ and $\mathbf{Z}/p\Gamma$-modules M, often given by $M(G/H) = H^*(EG \times_G G/H, \mathbf{Z}/p)$ (see Mislin [1987], Jackowski-McClure [1987]). In this context Proposition 17.36. is needed. Mackey-structures are used to construct split epimorphisms $i_{\#}i^*M \longrightarrow M$. More information and results can be found in Jackowski-McClure-Oliver [1989]. A spectral sequence converging to $\mathrm{Ext}^n_{R\Gamma}(M,N)$ is established by Jackowski (private communication) and Slominska [1980] for Γ the orbit category. Corollary 17.32. is proven in Rothenberg-Triantafillou [1984].

Exercises 17.39.

1) Let $0 \longrightarrow N^1 \longrightarrow N^o \longrightarrow N^2 \longrightarrow 0$ be an exact sequence of $R\Gamma$-modules and M be a $R\Gamma$-module. Establish a long exact sequence

$$0 \longrightarrow \mathrm{Hom}_{R\Gamma}(M,N^1) \longrightarrow \mathrm{Hom}_{R\Gamma}(M,N^o) \longrightarrow \mathrm{Hom}_{R\Gamma}(M,N^2) \longrightarrow \mathrm{Ext}^1_{R\Gamma}(M,N^1) \longrightarrow \mathrm{Ext}^1_{R\Gamma}(M,N^o)$$
$$\longrightarrow \mathrm{Ext}^1_{R\Gamma}(M,N^2) \longrightarrow \mathrm{Ext}^2_{R\Gamma}(M,N^1) \longrightarrow \cdots$$

and similar for Tor and $\otimes_{R\Gamma}$.

2) Let $F : \Gamma \longrightarrow R\text{-MOD}$ be a $R\Gamma$-module. Identify $\lim F = F \otimes_{R\Gamma} \underline{R}$ and $\mathrm{inv}\lim F = \mathrm{Hom}_{R\Gamma}(\underline{R},F)$ if $\lim$ and $\mathrm{inv}\lim$ are taken over the diagram of R-modules $F(x)$ indexed by Γ^{op}. Show that $H_i(\Gamma,-)$ and $H^i(\Gamma,-)$ are the i-th derivate functors of $\lim$ and $\mathrm{inv}\lim$ in the sense of homological algebra.

3) Let M be a $R\Gamma$-module with finite $\mathrm{supp}\, M = \{\bar{x} \in \mathrm{Is}\,\Gamma \mid M(x) \neq \{0\}\}$. Show

$$\mathrm{hdim}\, M \leq \max\{\mathrm{hdim}\, I_x \circ \mathrm{Res}_x M \mid \bar{x} \in \mathrm{supp}\, M\}$$

4) Let Γ_1 and Γ_2 be EI-categories. Consider for $i = 1,2$ a $R\Gamma_i$-module M_i such that $M_i(x)$ is projective over R for all $x \in \mathrm{Ob}\,\Gamma$. Prove

$$\mathrm{hdim}_{R\Gamma_1\times\Gamma_2}(M_1 \otimes_R M_2) = \mathrm{hdim}_{R\Gamma_1} M_1 + \mathrm{hdim}_{R\Gamma_2} M_2 .$$

5) a) Verify for two finite EI-categories Γ_1 and Γ_2

$$\mathrm{HDIM}(\mathbb{C}\Gamma_1 \times \Gamma_2) = \mathrm{HDIM}(\mathbb{C}\Gamma_1) + \mathrm{HDIM}(\mathbb{C}\Gamma_2)$$

b) Show for a cyclic group G of order $p_1^{n_1} \cdot \ldots \cdot p_r^{n_r}$ for prime numbers $p_1 \ldots p_r$:

$$\mathrm{HDIM}(\mathbb{C}\ \mathrm{Or}\ G) = r$$

7) Let Γ be a finite free EI-category and M a $\mathbb{Z}\Gamma$-module possessing a finite resolution. Show

$$\mathrm{hdim}_{\mathbb{Z}\Gamma} M \leq \ell(\Gamma) + 1$$

8) Let G be $\mathbb{Z}/p \times \mathbb{Z}/p$ for a prime number p . Show for a $\mathbb{Q}$ Or G-module M with $\mathrm{hdim}\ M \leq 1$:

$$p \cdot \mathrm{rk}_{\mathbb{Q}} M(G/G) \leq \sum_{1 \neq H \subsetneq G} \mathrm{rk}_{\mathbb{Q}} M(G/H)$$

9) The following assertions are equivalent for a EI-category Γ .

i) $H^n(\Gamma, M) = 0$ for any R-module M and $n \geq 1$.

ii) The constant module $\underline{R}$ is projective.

iii) There is an object y in Γ such that $\mathrm{Aut}(y)$ is finite, $|\mathrm{Aut}(y)|$ is a unit in R and $\mathrm{Aut}(y)$ acts transitively on the non-empty set $\mathrm{Hom}(x,y)$ for all $x \in \mathrm{Ob}\ \Gamma$.

10) Let F be a Mackey-functor for the finite group G with values in the category of $\mathbb{Z}/p$-modules. Let $\mathrm{Or}_p G$ be the full subcategory of Or G consisting of objects G/H with H a p-group. Let M be the $\mathbb{Z}/p\ \mathrm{Or}_p G$-module given by F . Show for $n \geq 1$

$$H^n(\mathrm{Or}_p G, M) = 0$$

(Hint: Apply Proposition 17.36. to $i : \mathrm{Or}\ G_p \longrightarrow \mathrm{Or}_p G$.

11) Let Γ be a finite EI-category and R a field such that $m(\Gamma)$ and ch R are prime. Then $\mathrm{Ext}^n_{R\Gamma}(M,N)$ is the n-th cohomology group of the R-cochain complex

$$0 \longrightarrow E_1^{o,o} \longrightarrow E_1^{1,o} \longrightarrow E_1^{2,o} \longrightarrow E_1^{3,o} \longrightarrow \ldots$$

Compute this R-cochain complex for $\Gamma = \mathrm{Or}\mathbf{Z}/pq$.

12) Establish a spectral sequence similar to the one of Theorem 17.18. and Theorem 17.28. converging to $\mathrm{Tor}_n^{R\Gamma}(M,N)$.

13) Compute the E_1-term of the spectral sequence of Theorem 17.18. for $\Gamma = \mathrm{Or}\,\mathbf{Z}/p^n$, p a prime number.

18. Reidemeister torsion.

We want to assign to an appropriate G-CW-complex X its equivariant Reidemeister torsion $\rho^G(X) \in Wh(QOrG)$ and its reduced equivariant Reidemeister torsion $\bar{\rho}^G(X) \in K_1(QOrG)/K_1(\mathbb{Z}_{(|G|)}G)$. We prove sum, product and join formulas and state some calculations. We relate it to other invariants we have defined above. Roughly speaking, the rationalized Whitehead torsion is the difference of the Reidemeister torsion and the reduced Reidemeister torsion is a refinement of the finiteness obstruction. We discuss various situations where Reidemeister torsion is defined and contains useful information. For example, we construct an injective homomorphism $\rho^G_{\mathbb{R}} : Rep_{\mathbb{R}}(G) \longrightarrow Wh(QOr\,G)$, thus reproving de Rham's theorem. Throughout the remainder of the book G is a finite group.

18.A. Review of modules over the orbit category Or G.

We briefly recall the algebra of RΓ-modules as far as needed for the remainder of the book. Since we restrict ourselves to finite groups and ignore fundamental groups, the notation simplifies drastically and a lot of technical difficulties, which only occur for infinite compact Lie groups, do not arise. Hence it suffices for understanding the next three sections to read the following comparatively short survey of the necessary input of the preceeding sections instead of reading themselves. Hopefully we keep the next three sections fairly self-contained in this way. Of course a reader who is somewhat familiar with the material of sections 9., 10., 14., and 16. may skip 18.A., B., and C. keeping just in mind that all our invariants, we will consider in the next sections, live over the orbit category Or G.

The orbit category Or G has as objects homogeneous spaces G/H and as morphisms G-maps. Let R be a commutative associative ring with unit. A ROr G-module M is a contravariant functor Or G ⟶ R-MOD into the category of R-modules. A morphism f : M ⟶ N is a natural transformation. The category MOD-ROr G of R Or G-modules inherits from R-MOD the structure of an abelian category. Hence notions like "direct sum", "exact sequence", "projective", "chain complex" and "homology" are defined. We often abbreviate M(G/H) by M(H) for a R Or G-module M.

There is a forgetful functor F : MOD-R Or G ⟶ Ob(Or G)-SET into the category of sets over $Ob(Or\,G)$ sending M to $(\coprod_{H \subset G} M(G/H),\beta)$ where $\beta(M(G/H))$ is G/H. We call a $R\,Or\,G$-module M together with an $Ob(Or\,G)$-set $B \subset F(M)$ free with base B if for any $R\,Or\,G$-module N and map of $Ob(Or\,G)$-sets $f : B \longrightarrow F(N)$ there is exactl one homomorphism of $R\Gamma$-modules $\overline{f} : M \longrightarrow N$ extending f. For $K \subset G$ let $R\,Hom(?,G/K)$ be the $R\,Or\,G$-module sending G/H to the free R-module generated by $Hom(G/H,G/K)$. Given an $Ob(Or\,G)$-set (B,β), let $R\,Or\,G(B)$ be the $R\,Or\,G$-module $\bigoplus_{b \in \beta} R\,Hom(?,\beta(b))$. If we identify $b \in B$ with $id : \beta(b) \longrightarrow \beta(b) \in R\,Hom(\beta(b),\beta(b))$ we derive from the Yoneda-Lemma that $R\,Or\,G(B)$ is free with base B. One easily verifies that $B \longrightarrow R\,Or\,G(B)$ is the left adjoint of the forgetful functor F (cf. 9.16.).

Two bases (B,β) and (C,γ) of the free $R\,Or\,G$-module M are called equivalent if there is a bijection of sets (not necessarily of sets over $Ob(Or\,G)$) $\psi : B \to C$ such that for any $b \in B$ there is an isomorphism $f : \beta(b) \longrightarrow \gamma \circ \psi(b)$ in $Or\,G$ and a sign $\varepsilon \in \{\pm 1\}$ satisfying $M(f)(\psi(b)) = \varepsilon \cdot b$. If we have fixed such an equivalence class $[(B,\beta)]$ of bases (B,β), we call M free with a preferred equivalence class of bases (cf. Example 12.22.).

Let X be a G-CW-complex. Its cellular $R\,Or\,G$-chain complex $C^c(X) : Or\,G \to$ {R-chain compl.} sends G/H to the cellular chain complex $C^c(X^H)$ of $X^H = map(G/H,X)^G$. It enherits from the G-CW-structure a preferred equivalence class of bases, called cellular equivalence class of bases, as follows. Fix $n \geq 0$. For any n-dimensional cell $e_i^n \in \{e_i^n \mid i \in I_n\}$ choose a characteristic map $(Q_i^n,q_i^n) : G/H_i \times (D^n,S^{n-1}) \to (X_n,X_{n-1})$ Let $b_i^n \in C_n^c(X_n^H,X_{n-1}^H) = C_n^c(X)(H)$ be the image of the generator in $H_n(D^n,S^{n-1})$ under the map induced by (Q_i^n,q_i^n) restricted to $eH_i \times (D^n,S^{n-1})$. Then $(\{b_i^n \mid i \in I_n\},\beta)$ with $\beta(b_i^n) = G/H_i$ represents the cellular equivalence class of bases. Notice that only the equivalence class is independent of the choice of the characteristic map and is an invariant of the G-CW-complex structure (cf. Definition 13.3., Example 9.18.).

A $R\,Or\,G$-module M is finitely generated if it is a quotient of a free $R\,Or\,G$-module with base (B,β) such that B is finite. Consider the subcategory {finitely generat projective $R\,Or\,G$-modules} of the abelian category MOD-$R\,Or\,G$. The standard constructions applied to it define its K-theory, denoted by $K_n(R\,Or\,G)$ for $n \in \mathbf{Z}$.

A trivial unit in $K_1(R\,Or\,G)$ is an element represented by an automorphism $R\,Hom(?,G/H) \longrightarrow R\,Hom(?,G/H)$, $g \longrightarrow \varepsilon \cdot f \circ g$ for some $H \subset G$, $f \in Aut(G/H)$, $\varepsilon \in \{\pm 1\}$. If $U \subset K_1(R\,Or\,G)$ is the subgroup of trivial units, put $Wh(R\,Or\,G) = K_1(R\,Or\,G)/U$. Let $K_o^f(R\,Or\,G)$ be K_o of {finitely generated free $R\,Or\,G$-modules}. Define $\tilde{K}_o(R\,Or\,G)$ to be the cokernel of the forgetful homomorphism $K_o^f(R\,Or\,G) \longrightarrow K_o(R\,Or\,G)$. If M and N are $R\,Or\,G$-modules, let $M \otimes_R N$ be the $R\,Or\,G$-module $G/H \longrightarrow M(G/H) \otimes_R N(G/H)$. We get parings $\otimes_R : K_o(R\,Or\,G) \otimes K_n(R\,Or\,G) \longrightarrow K_n(R\,Or\,G)$ and $\otimes_R : K_o^f(R\,Or\,G) \otimes \tilde{K}_o(R\,Or\,G) \longrightarrow \tilde{K}_o(R\,Or\,G)$ and $\otimes_R : K_o^f(R\,Or\,G) \otimes Wh(R\,Or\,G) \longrightarrow Wh(R\,Or\,G)$. In particular $K_o^f(R\,Or\,G)$ and $K_o(R\,Or\,G)$ become commutative associative rings with unit. If $i : H \longrightarrow G$ is the inclusion of a subgroup, we obtain a functor $i_* : Or\,H \longrightarrow Or\,G$, $H/K \longrightarrow ind(H/K) = G/K$. Composition with i_* defines a restriction homomorphism $i^* : K_n(R\,Or\,G) \longrightarrow K_n(R\,Or\,H)$ and similar for K_o^f, $\tilde{K}_o$, and Wh (cf. section 10.).

These K-groups can be computed by K-groups of group rings. Define for $H \subset G$ functors

$$S_H : \text{MOD-}R\,Or\,G \longrightarrow \text{MOD-}RWH$$
$$E_H : \text{MOD-}RWH \longrightarrow \text{MOD-}R\,Or\,G$$

as follows. Given a $R\,Or\,G$-module M, let $M(H)_s \subset M(H)$ be the RWH-submodule generated by all images $M(f) : M(G/K) \longrightarrow M(G/H)$ where $f : G/H \longrightarrow G/K$ runs through all non-isomorphisms in $Or\,G$ with source G/H and arbitrary target. Define $S_H M = M(H)/M(H)_s$, and $E_H N = N \otimes_{RWH} R\,Hom(?,G/H)$ for a RWH-module N. These functors induce a pair of inverse isomorphisms

$$S : K_n(R\,Or\,G) \longrightarrow \bigoplus_{(H)} K_n(RWH)$$

$$E : \bigoplus_{(H)} K_n(RWH) \longrightarrow K_n(R\,Or\,G)$$

and analogously for K_o^f, $\tilde{K}_o$ and Wh (cf. Theorem 10.34.). Let $U(G) = A(G)$ be the Burnside ring. Then $U(G) \longrightarrow K_o^f(R\,Or\,G)$, $[G/H] \longrightarrow [R\,Hom(?,G/H)]$ is a ring isomorphism. Define an homomorphism $rk_{R\,Or\,G} : K_o(R\,Or\,G) \longrightarrow U(G)$ by $[P] \longrightarrow \sum_{(H)} rk_{RWH}(S_H P) \cdot [G/H]$. We get a ring isomorphism

$$rk_{R\,Or\,G} \otimes F : K_o(R\,Or\,G) \longrightarrow U(G) \otimes \tilde{K}_o(R\,Or\,G)$$

with respect to the ring structure on the target given by $(u,v)\cdot(u',v') = (u\cdot u', u\cdot v' + u'\cdot v)$ (cf. proof of Theorem 11.24.).

There is another splitting, provided R is Q , or any field of characteristic prime to $|G|$. The functors

$$\mathrm{RES}_H : \text{MOD-}\mathbb{Q}\,\mathrm{Or}\,G \longrightarrow \text{MOD-}\mathbb{Q}WH$$
$$I_H : \text{MOD-}\mathbb{Q}WH \longrightarrow \text{MOD-}\mathbb{Q}\,\mathrm{Or}\,G$$

are given by $\mathrm{Res}_H(M) = M(H)$ and $I_H(N)(K) = N \otimes_{RWH} R\,\mathrm{Hom}(G/K,G/H)$ if $(H) = (K)$ and $I_H(N)(K) = \{0\}$, if $(H) \neq (K)$. They induce a pair of inverse isomorphisms

$$\mathrm{RES} : K_n(\mathbb{Q}\,\mathrm{Or}\,G) \longrightarrow \bigoplus_{(H)} K_n(\mathbb{Q}WH)$$
$$I : \bigoplus_{(H)} K_n(\mathbb{Q}WH) \longrightarrow K_n(\mathbb{Q}\,\mathrm{Or}\,G)$$

Tensoring with $R\,\mathrm{Hom}(G/K,G/H)$ induces an homomorphism $\omega_{H,K} : K_n(\mathbb{Q}WH) \longrightarrow K_n(\mathbb{Q}WK)$. Their direct sum defines an isomorphism

$$\omega : \bigoplus_{(H)} K_n(\mathbb{Q}WH) \longrightarrow \bigoplus_{(K)} K_n(\mathbb{Q}WK) .$$

An explicit inverse μ of ω is given in section 16. by the K-theoretic Moebius inversion. We get a commutative diagram of pairs of inverse isomorphisms (cf. Theorem 16.29.)

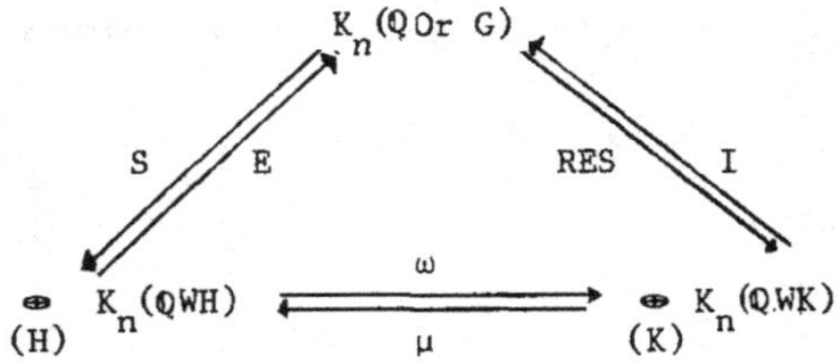

18.B. Review of invariants for R OrG-chain complexes.

Next we recall the basic invariants for R Or G-chain complexes. A R Or G-chain complex C is finite if C_n is finitely generated for $n \geq 0$ and zero for $n < 0$ and large n. We call C finitely dominated if there is a finite free R Or G-chain complex D and R Or G-chain maps $r : D \longrightarrow C$ and $i : C \longrightarrow D$ satisfying $r \circ i \simeq \mathrm{id}$. This is equivalent to the existence of a finite projective R Or G-chain complex P with

$P \simeq C$. Given a finitely dominated $R\,Or\,G$-chain complex C , define its <u>finiteness obstruction</u> $o(C) \in K_o(R\,Or\,G)$ by $\Sigma(-1)^n[P_n]$ for any finite projective $R\,Or\,G$-chain complex P chain homotopy equivalent to C . Its image under $K_o(R\,Or\,G) \to \tilde{K}_o(R\,Or\,G)$ is the <u>reduced finiteness obstruction</u> $\tilde{o}(C)$. Let the <u>Euler characteristic</u> $\chi(C) \in U(G)$ be $rk_{R\,Or\,G}(o(C))$. A finitely dominated $R\,Or\,G$-chain complex C is $R\,Or\,G$-chain equivalent to a finite free one if and only if $\tilde{o}(C)$ vanishes.

Let $U \subset K_1(R\,Or\,G)$ be a subgroup and M and N be finitely generated projective $R\,Or\,G$-modules. A U-<u>stable equivalence class of stable isomorphisms</u> $\{\phi\} : M \longrightarrow N$ is represented by an isomorphism $\phi : M \bullet X \longrightarrow N \bullet X$ for some finitely generated projective $R\,Or\,G$-module X . In the sequel π denotes permutation maps. The isomorphism $\psi : M \bullet Y \longrightarrow N \bullet Y$ represents the same U-stable equivalence class as ϕ if the torsion $t(S) \in K_1(R\,Or\,G)$ of the following square S lies in U where $t(S)$ is defined by running around in the clockwise direction.

$$\begin{array}{ccc}
M \bullet X \bullet Y & \xrightarrow{\phi \bullet id} & N \bullet X \bullet Y \\
\downarrow \pi & & \downarrow \pi \\
M \bullet Y \bullet Y & \xrightarrow{\psi \circ id} & M \bullet X \bullet Y
\end{array}$$

Consider a $R\,Or\,G$-chain equivalence of finite projective $R\,Or\,G$-chain complexes $f : C \longrightarrow D$ together with a U-stable equivalence class of stable isomorphisms $\{\phi\} : C_{odd} \bullet D_{ev} \longrightarrow D_{odd} \bullet C_{ev}$. If c is the differential and γ a chain contraction of the mapping cone $Cone(f)$ of f , define the torsion $t(f,\{\phi\}) \in K_1(R\,Or\,G)/U$ by the torsion $t(S)$ of the square

$$\begin{array}{ccc}
Cone(f)_{odd} & \xrightarrow{\{(c+\gamma)\}} & Cone(f)_{ev} \\
\pi \downarrow & & \downarrow \pi \\
D_{odd} \bullet C_{ev} & \underset{\{\phi\}}{\text{———}} & C_{odd} \bullet D_{ev}
\end{array}$$

(cf. Definition 12.4.)

We will use this definition in two special cases. Let $f : C \longrightarrow D$ be a $R\,Or\,G$-chain equivalence between finite free $R\,Or\,G$-chain complexes with preferred equivalence classes of bases. These bases give a U-stable equivalence class of stable isomorphisms

$\{\phi\} : C_{odd} \oplus D_{ev} \longrightarrow D_{odd} \oplus C_{ev}$ if U denotes the subgroup of trivial units. Define the Whitehead torsion $\tau(f) \in Wh(R\,Or\,G)$ by $t(f,\{\phi\})$. (cf. Example 12.22.).

Fix $H \subset G$. Consider a finite QWH-chain complex C. Regard $H(C)$ as a finite QWH-chain complex by the trivial differential. A round structure C is a $\{1\}$-stable equivalence classes of stable isomorphisms $\{\phi\} : C_{odd} \longrightarrow C_{ev}$. Let $f : (C,\{\phi\}) \longrightarrow (D,\{\psi\})$ be a QWH-chain equivalence of finite QWH-chain complexes with round structure. Let the $\{1\}$-stable equivalence class of stable isomorphisms $\{\Phi\}$: $C_{odd} \oplus D_{ev} \longrightarrow D_{odd} \oplus C_{ev}$ be induced by $\{\phi\}$ and $\{\psi\}$. Define the round torsion $t(f) \in K_1(QWH)$ by $t(f,\{\Phi\})$ (cf. Example 12.20.)

Let C be a finite QWH-chain complex with round structure $\{\alpha\}$. Equip $H(C)$ with the trivial differential. Suppose we are given a round structure $\{\beta\}$ on $H(C)$. Up to chain homotopy there is exactly one QWH-chain map $i : H(C) \longrightarrow C$ satisfying $H(i) = id$. Define the absolute torsion $t(C,\{\alpha\},\{\beta\}) \in K_1(QWH)$ by the round torsion $t(i)$ (cf. Example 12.21.)

18.C. Review of invariants of G-spaces.

We recall invariants we have already defined and are related to Reidemeister torsion. Consider a pair of finitely dominated G-CW-complexes. Let $C^c(X,A;R)$ be the cellular $R\,Or\,G$-chain complex. It is obtained from the cellular $R\Pi/(G,X)$-chain complex by induction with the projection $pr : X \longrightarrow \{point\}$ (cf. Definition 8.37.). It has a preferred equivalence class of bases (see Definition 13.3.) The finiteness obstruction, the reduced finiteness obstruction and the Euler characteristic of (X,A) over $R\,Or\,G$

18.1. $o^G(X,A;R) \in K_0(R\,Or\,G)$

18.2. $\tilde{o}^G(X,A;R) \in \tilde{K}_0(R\,Or\,G)$

18.3. $\chi(X,A;R) \in U(G)$

are defined by $o(C^c(X,A;R))$, $\tilde{o}(C^c(X,A;R))$ and $\chi(C^c(X,A;R))$ (cf. Definition 11.1. and Definition 11.19.). Let $(f,g) : (X,A) \longrightarrow (Y,B)$ be a G-homotopy equivalence of pairs of finite G-CW-complexes. Its (equivariant) Whitehead torsion over $R\,Or\,G$

18.4. $\tau^G(f,g;R) \in Wh(R\,Or\,G)$

is defined by $\tau(C^c(f,g))$ (see Definition 12.24.). If R is $\mathbf{Z}$, we often omitt R. For $R = \mathbf{Z}$ these invariants above are the image of the corresponding invariants of of section 14. under the homomorphisms induced by $pr : X \longrightarrow \{point\}$.

18.5. We recall the obstruction properties of these invariants. Let X be a finitely dominated G-CW-complex such that X^H is simply connected or empty for $H \subset G$. Then X is G-homotopy equivalent to a finite G-CW-complex if and only if $\tilde{o}^G(X) \in \tilde{K}_o(\mathbf{Z}Or\,G)$ vanishes (see Theorem 14.6.). Consider a G-homotopy equivalence $f : X \longrightarrow Y$ between finite G-CW-complexes such that Y^H is simply connected or empty for $H \subset G$. Then f is simple (see Definition 4.4.) if and only if $\tau^G(f) \in Wh(\mathbf{Z}Or\,G)$ is zero (see Theorem 14.4.) □

18.D. Construction of equivariant Reidemeister torsion.

We need the following lemma which is proven for group rings in Swan [1960] and hence follows from Theorem 10.34.

Lemma 18.6. *Let Γ be an EI-category and $m \in \mathbf{Z}$, $m \neq 1$. If $Aut(x)$ is finite for all $x \in Ob\,\Gamma$, then the change of ring map $\tilde{K}_o(\mathbf{Z}\Gamma) \longrightarrow \tilde{K}_o(\mathbf{Z}_{(m)}\Gamma)$ is zero* □

Suppose that G is finite and R a field of characteristic zero. In order to define equivariant Reidemeister torsion, we need the following extra structure on a finitely dominated pair (X,A).

Definition 18.7. *A round structure $\{\phi\}$ on (X,A) is a collection of stable isomorphism classes of RWH-isomorphisms (see section 12)*

$$\{\phi(H)\} : H(X^H,A^H;R)_{odd} \longrightarrow H(X^H,A^H;R)_{ev}$$

indexed by $\{H|H \subset G\}$. We require $\{c(g)\} \circ \{\phi(H)\} = \{\phi(g^{-1}Hg)\} \circ \{c(g)\}$ for $H \subset G$, $g \in G$ and $c(g)$ given by conjugation. □

We recall that $H(X^H,A^H;R)_{odd}$ is $\bigoplus_{n \geq o} H(X^H,A;R)_{2n+1}$ and similar for ev.

Lemma 18.8. *The following statements are equivalent for a pair (X,A) of finitely dominated G-CW-complexes*

i) *(X,A) has a round structure.*

ii) *$o^G(X,A;R) \in K_o(ROr/G)$ is zero*

iii) $\chi^G(X,A) \in U(G)$ <u>is zero</u>.

iv) $\chi(X^H,A^H) \in \mathbf{Z}$ <u>is zero for</u> $H \subset G$.

<u>Proof.</u> The existence of a round structure is equivalent to $\Sigma(-1)^n[H_n(X,A;R)] = 0$ in $K_o(RWH)$ for all $H \subset G$. Now i) <=> ii) follows from Proposition 11.9. and Theorem 16.29. We derive ii) <=> iii) from Lemma 18.6. and iii) <=> iv) from Theorem 16.36.

□

Let (X,A) be a pair of finite G-CW - complexes with a round structure $\{\phi\}$. We abbreviate $C = C^c(X,A)$ and write $C(H)$ for $C(G/H) = C^c(X^H,A^H)$. Let $U \subset K_1(\text{ROrG})$ be the subgroup of trivial units so that $\text{Wh(ROrG)} = K_1(\text{ROrG})/U$ (see Definition 10.8. As $\chi^G(X,A)$ is zero by Lemma 18.8., the cellular bases defines a U-round structure $\{\alpha\} : C_{odd} \longrightarrow C_{ev}$ on the **Z**OrG-chain complex C . (see Example 12.20. and Lemma 12.23) Choose a {1}-round structure $\{\beta\} : C_{odd} \longrightarrow C_{ev}$ representing $\{\alpha\}$. Then

$$\{\beta(H) \otimes_{\mathbf{Z}} R\} : C(H) \otimes_{\mathbf{Z}} R_{odd} \longrightarrow C(H) \otimes_{\mathbf{Z}} R_{ev}$$

is a {1}-round structure on the RWH-chain complex $C(H) \otimes_{\mathbf{Z}} R_{odd}$. We have defined the absolute torsion

18.9. $$t(C(H) \otimes_{\mathbf{Z}} R, \{\beta(H) \otimes_{\mathbf{Z}} R\}, \{\phi(H)\}) \in K_1(RWH)$$

in Example 12.21 if we put $\Gamma = WH^{\wedge}$ there. Their collection defines an element in Split $K_1(\text{ROrG})$ whose image under the composition of the projection $pr : K_1(\text{ROrG}) \longrightarrow \text{Wh(ROrG)}$ and the isomorphism $I : \text{Split } K_1(\text{ROrG}) \longrightarrow K_1(\text{ROrG})$ of Theorem 16.29. is denoted by $\rho^G(X,A,\{\phi\})$ or briefly $\rho^G(X,A)$. Because of Theorem 16.29. and Example 12.21. $\rho^G(X,A,\{\phi\})$ depends only on the G-CW-complex structure on (X,A) and $\{\phi\}$ but not on the choice of $\{\beta\}$ within $\{\alpha\}$.

<u>Definition 18.10.</u> <u>We call</u>

$$\rho^G(X,A) \in \text{Wh(ROrG)}$$

<u>the equivariant Reidemeister torsion</u>. □

We can still define a reduced version if (X,A) is a pair of (not necessarily finite initely dominated G-CW-complexes with a round structure $\{\Phi\}$. Let m e a multiple of $|G|$. There is a canonical ring homomorphism $\mathbf{Z}_{(m)} \longrightarrow R$. Let $K_1(\mathrm{ROrG})/K_1(\mathbf{Z}_{(m)}\mathrm{OrG})$ be the cokernel of the change of rings map $K_1(\mathbf{Z}_{(m)}\mathrm{OrG}) \longrightarrow K_1(\mathrm{ROrG})$. Because of Lemma 18.6. there is a finite free $\mathbf{Z}_{(m)}\mathrm{OrG}$-chain complex F together with a chain homotopy equivalence $f : C^c(X,A;\mathbf{Z}_{(m)}) \longrightarrow F$. For $H \subset G$ let $\{\psi(H)\} : H(F \otimes_{\mathbf{Z}(m)} R)_{odd} \longrightarrow H(F \otimes_{\mathbf{Z}(m)} R)_{ev}$ be the round structure induced from $\phi(H)$ by conjugation with $H(f \otimes_{\mathbf{Z}(m)} R)$. Because of Lemma 18.8. we can choose a round structure $\{\alpha\}$ on F. We have the absolute torsion

18.11.
$$t(F(H) \otimes_{\mathbf{Z}(m)} R, \{\alpha \otimes_{\mathbf{Z}(m)} R\}, \{\psi\}) \in K_1(\mathrm{RWH})$$

The collection of these elements defines an element in Split $K_1(\mathrm{ROrG})$ whose image under Split $K_1(\mathrm{ROrG}) \xrightarrow{pr \circ I} K_1(\mathrm{ROrG})/K_1(\mathbf{Z}_{(m)}\mathrm{OrG})$ is denoted by $\bar{\rho}^G(X,A,\{\phi\})$ or briefly $\bar{\rho}^G(X,A)$. It depends only on (X,A) and $\{\phi\}$ by Theorem 16.29. and Example .2.21.

Definition 18.12. We call

$$\bar{\rho}^G(X,A) \in K_1(\mathrm{ROrG})/K_1(\mathbf{Z}_{(m)}\mathrm{OrG})$$

the reduced equivariant Reidemeister torsion. □

18.E. Basic properties of these invariants,

We collect the basic properties of the invariants defined above.

8.13. Sum formula

Consider the G-push out of finitely dominated G-CW-complexes with i a G-cofibration

8.14.
$$\begin{array}{ccc} X_o & \longrightarrow & X_2 \\ \downarrow{\scriptstyle i} & & \downarrow \\ X_1 & \longrightarrow & X \end{array}$$

Then we have in $K_o(\mathrm{ROr\,G})$, $\tilde{K}_o(\mathrm{ROr\,G})$ resp. $U(G)$ (see Theorem 14.6.)

$$o^G(X) - o^G(X_1) - o^G(X_2) + o^G(X_o) = 0$$

$$\tilde{o}^G(X) - \tilde{o}^G(X_1) - \tilde{o}^G(X_2) + \tilde{o}^G(X_o) = 0$$

$$\chi^G(X) - \chi^G(X_1) - \chi^G(X_2) + \chi^G(X_o) = 0$$

Suppose that X_o , X_1 and X_2 carry round structures $\{\phi_o\}$, $\{\phi_1\}$ and $\{\phi_2\}$. Then we call 18.14. a G-push out of G-CW-complexes with round structure if X is equipped with round strucure $\{\phi\}$ uniquely determined by the following property for $H \subset G$: Consider the Mayer-Vietoris sequence of the WH-push out obtained from 18.14. by taking the H-fixed point set as an acyclic finitely generated (projective) RWH-chain complex $D(H)$. It enherits a round structure $\{\delta(H)\}$ from $\{\phi_o\}$, $\{\phi_1\}$, $\{\phi_2\}$ and $\{\phi\}$. We require that the round torsion $t(D(H),\{\delta(H)\}) \in K_1(RWH)$ vanish. If 18.14. is a G-push out of finitely dominated G-CW-complexes with round structure we derive from Example 12.21. in $K_1(ROrG)/K_1(\mathbf{Z}_{(m)}OrG)$

$$\bar{\rho}^G(X) - \bar{\rho}^G(X_1) - \bar{\rho}^G(X_2) + \bar{\rho}^G(X_o) = 0$$

Suppose that 18.14. is a cellular G-push out (see 4.1.) of finite G-CW-complexes with round structure. Then we have in $Wh(ROrG)$

$$\rho^G(X) - \rho^G(X_1) - \rho^G(X_2) + \rho^G(X_o) = 0$$

Similarly we get for G-homotopy equivalences f_o, f_1, f_2 and f between two cellular G-push outs of finite G-CW-complexes in $Wh(ROrG)$ (see Theorem 14.14.)

$$\tau^G(f) - \tau^G(f_1) - \tau^G(f_2) + \tau^G(f_o) = 0 \qquad \square$$

18.15. Relative formula

If (X,A) is a pair of finitely dominated G-CW-complexes we have

$$o^G(X) = o^G(A) + o^G(X,A)$$

$$\tilde{o}^G(X) = \tilde{o}^G(A) + \tilde{o}^G(X,A)$$

$$\chi^G(X) = \chi^G(A) + \chi^G(X,A)$$

If A and X carry round structures , (X,A) inherits a round structure by a construction similar to 18.13. using the long homology sequences of the pairs (X^H,A^H).

We obtain

$$\bar{\rho}^G(X) = \bar{\rho}^G(A) + \bar{\rho}^G(X,A)$$

$$\rho^G(X) = \rho^G(A) + \rho^G(X,A)$$

We have for a pair $(f,g) : (X,A) \longrightarrow (Y,B)$ of G-homotopy equivalences between finite G-CW-complexes

$$\tau^G(f) = \tau^G(g) + \tau^G(f,g) \qquad \square$$

18.16. <u>Diagonal product formula</u>

Let X and Y be finitely dominated G-CW-complexes. A round structure $\{\phi\}$ on X induces a round structure $\{\phi \times id\}$ on the G-space $X \times Y$ by

$$\{(\phi \times id)(H)\} : H(X^H \times Y^H;R)_{odd} \xleftarrow{\times} (H(X^H;R) \otimes_R H(Y^H;R))_{odd} \xrightarrow{\{\pi\}}$$

$$(H(X^H;R)_{odd} \otimes_R H(Y^H;R)_{ev}) \oplus (H(X^H;R)_{ev} \otimes_R H(Y^H;R)_{odd})$$

$$\xrightarrow{\{\phi(H) \otimes_R id\} \oplus \{\phi(H) \otimes_R id\}^{-1}} (H(X^H;R)_{ev} \otimes_R H(Y^H;R)_{ev}) \oplus (H(X^H;R)_{odd} \otimes_R H(Y^H;R)_{odd})$$

$$\xrightarrow{\{\pi\}} (H(X^H;R) \otimes_R H(Y^H;R))_{ev} \xrightarrow{\times} H(X^H \times Y^H;R)_{ev} .$$

If Y also carries a round structure $\{\psi\}$ one easily checks $\{\phi \times id\} = \{id \times \psi\}$. Hence $X \times Y$ has a canonical round structure, if $\chi^G(X) = \chi^G(Y) = 0$ holds, by Lemma 18.8. One easily checks using 18.13.

$$\rho^G(X \times Y, \{\phi\} \times id) = \rho^G(X, \{\phi\}) \otimes_R \chi^G(Y)$$

$$\bar{\rho}^G(X \times Y, \{\phi\} \times id) = \bar{\rho}^G(X, \{\phi\}) \otimes_R \chi^G(Y)$$

We derive from Theorem 14.42. and Corollary 14.48. for $f : X' \longrightarrow X$ and $g : Y' \longrightarrow Y$

$$o^G(X \times Y) = \chi^G(X) \otimes o^G(Y) + o^G(X) \otimes \chi^G(Y) - \chi^G(X) \otimes \chi^G(Y)$$

$$\tilde{o}^G(X \times Y) = \chi^G(X) \otimes \tilde{o}^G(Y) + \tilde{o}^G(X) \otimes \chi^G(Y)$$

$$\chi^G(X \times Y) = \chi^G(X) \otimes \chi^G(Y)$$

$$\tau^G(f \times g) = \chi^G(X) \otimes \tau^G(g) + \tau^G(f) \otimes \chi^G(Y) \qquad \square$$

18.17. Join formula

If X and Y carry round structures $\{\phi\}$ and $\{\psi\}$ we get from 18.13. and 18.16. a unique round structure $\{\phi*\psi\}$ on the join $X * Y$ defined in 7.41. We conclude from 18.13. and 18.16.

$$o^G(X*Y) = (1-\chi^G(X)) \bullet o^G(Y) + o^G(X) \bullet (1-\chi^G(Y)) + \chi^G(X) \bullet \chi^G(Y)$$

$$\tilde{o}^G(X*Y) = (1-\chi^G(X)) \bullet \tilde{o}^G(Y) + \tilde{o}^G(X) \bullet (1-\chi^G(Y))$$

$$\chi^G(X*Y) = \chi^G(X) + \chi^G(Y) - \chi^G(X) \bullet \chi^G(Y)$$

$$\tau^G(f*g) = (1-\chi^G(X)) \bullet \tau^G(g) + \tau^G(f) \bullet (1-\chi^G(Y))$$

$$\rho^G(X*Y) = \rho^G(X) + \rho^G(Y)$$

$$\bar{\rho}^G(X*Y) = \bar{\rho}^G(X) + \bar{\rho}^G(Y)$$

18.18. Transformation under G-homotopy equivalence.

Let $(f,g) : (X,A) \longrightarrow (Y,B)$ be a G-homotopy equivalence of pairs of finitely dominated G-CW-complexes. Then

$$o^G(X,A) = o^G(Y,B)$$

$$\tilde{o}^G(X,A) = \tilde{o}^G(Y,B)$$

$$\chi^G(X,A) = \chi^G(Y,B)$$

Suppose that (X,A) and (Y,B) carry round structures $\{\phi\}$ and $\{\psi\}$. Let $t^G(H(f,g))$ be the image of the element given by the round torsion $t(H(f^H,g^H;R),\{\phi(H)\},\{\psi(H)\}) \in K_1(RWH)$ for $H \subset G$ under $I : \text{Split}\, K_1(ROrG) \longrightarrow K_1(ROrG)$. We conclude from Example 12.21. in $K_1(ROrG)/K_1(\mathbb{Z}_{(m)}OrG)$ using Lemma 18.6.

$$t^G(H(f,g)) = \bar{\rho}^G(X,A) - \bar{\rho}^G(Y,B)$$

If (X,A) and (Y,B) are pairs of finite G-CW-complexes with round structure we get in $Wh(ROrG)$

$$\tau^G(f) - t^G(H(f,g)) = \rho^G(Y,B) - \rho^G(X,A) \qquad \square$$

In this context the following result for $SK_1(\mathbb{Z}OrG) = \text{kernel}\,(K_1(\mathbb{Z}OrG) \rightarrow K_1(\mathbb{Q}OrG))$

is interesting.

Theorem 18.19.. Let G be a finite group.

a) Let G be abelian. Suppose either that for all prime numbers p dividing $|G|$ we have $G_p \cong \mathbf{Z}/p^n$ or $G_p \cong \mathbf{Z}/p \times \mathbf{Z}/p^n$ for some n or that G is $(\mathbf{Z}/2)^k$ for some k. Then $SK_1(\mathbf{Z}\mathrm{Or}G)$ is zero.

b) If G_p is cyclic for the prime p then $SK_1(\mathbf{Z}\mathrm{Or}G)_{(p)}$ is zero.

c) $SK_1(\mathbf{Z}\mathrm{Or}G) = \mathrm{Tors}\ Wh(\mathbf{Z}\mathrm{Or}G)$.

Proof. Because of Theorem 10.34. it suffices to prove the analogous statements for integral group rings. This is done in Oliver [1988] . □

18.20. Let (X,A) be a pair of finite G-CW-complexes with two round structures $\{\phi\}$ and $\{\psi\}$. Let $[\{\psi\} \circ \{\phi\}^{-1}] \in Wh(R\mathrm{Or}G)$ be the image of the element given by the collection $\{\psi(H)\} \circ \{\phi(H)^{-1}\}$ under $\mathrm{Split}\ K_1(R\mathrm{Or}G) \xrightarrow{I} K_1(R\mathrm{Or}G) \to Wh(R\mathrm{Or}G)$. We derive from Example 12.21. in $Wh(R\mathrm{Or}G)$

$$\rho^G(X,A,\{\psi\}) - \rho^G(X,A,\{\phi\}) = [\{\psi\} \circ \{\phi\}^{-1}]$$

A similar formula holds for $\bar{\rho}^G$. □

18.F. Special G-CW-complexes.

Now we consider situations where such round structures are naturally given and which are important for applications. If A is an abelian group, let A_f be $A/\mathrm{Tors}\ A$.

Definition 18.21. We call a pair of finitely dominated G-CW-complexes (X,A) special if WH acts trivially on $H_n(X^H,A^H;\mathbf{Q})$ and $\chi(X^H,A^H) \in \mathbf{Z}$ vanishes for $H \subset G$. An orientation of X is a collection of $\{1\}$-round structures

$$\{\phi(H)\} : (H(X^H,A^H)_f)_{odd} \longrightarrow (H(X^H,A^H)_f)_{ev}$$

of $\mathbf{Z}$-modules for $H \subset G$ such that $\{c(g)\} \circ \{\phi(H)\} = \{\phi(g^{-1}Hg)\} \circ \{c(g)\}$ holds for $H \subset G$, $g \in G$ and $c(g)$ given by conjugation. □

Any special pair (X,A) can be given an orientation. As $H(X^H,A^H;\mathbf{Q}) = H(X^H,A^H)_f \otimes_{\mathbf{Z}} \mathbf{Q}$ holds, an orientation on (X,A) induces a round structure in the sense of Definition

18.7. Hence for an oriented special pair of finite resp. finitely dominated G-CW-complexes (X,A) its <u>equivariant Reidemeister torsion</u>

18.22. $$\rho^G(X,A) \in Wh(\mathbb{Q}OrG)$$

resp. <u>reduced equivariant Reidemeister torsion</u>

18.23. $$\bar{\rho}^G(X,A) \in K_1(\mathbb{Q}OrG)/K_1(\mathbb{Z}_{(m)}OrG)$$

is defined. Related to these invariants are the following ones. For any pair (X,A) of finitely dominated G-CW-complexes let

18.24. $$h\chi(X,A) \in \prod_{(H)} \mathbb{Q}^*/\mathbb{Z}^*$$

be given by $h\chi(X,A)_{(H)} = \prod_{n \geq 0} |\mathrm{Tors}\, H_n(X^H,A^H)|^{(-1)^n}$. Suppose that (X,A) is special. Since WH acts trivially on $H_n(X^H,A^H;\mathbb{Q})$, $H_n(X^H,A^H;\mathbb{Q}) \longrightarrow H_n((X^H,A^H)/WH;\mathbb{Q})$ is an isomorphism (see Bredon [1972], III.2.4.) and hence $H_n((X^H,A^H) \longrightarrow (X^H,A^H)/WH)$ is finite. Let $m\chi(X,A)_{(H)}$ be $\prod_{n \geq 0} |H_n((X^H,A^H) \longrightarrow (X^H,A^H)/WH)|^{(-1)^n}$.

<u>Definition 18.25.</u> <u>We call the element</u>

$$m\chi(X,A) = (m\chi(X,A)_{(H)})_{(H)} \in \prod_{(H)} \mathbb{Q}^*/\mathbb{Z}^*$$

<u>the multiplicative Euler characteristic of</u> (X,A) . □

Next we explain how these invariants are linked. Consider the following localization square for m a multiple of $|G|$.

18.26. $$\begin{array}{ccc} \mathbb{Z}WH & \longrightarrow & \mathbb{Z}_{1/m}WH \\ \downarrow & & \downarrow \\ \mathbb{Z}_{(m)}WH & \longrightarrow & \mathbb{Q}WH \end{array}$$

The boundary homomorphism of the exact Milnor sequence associated with 18.26. (see Milnor [1971, p. 28]) induces

18.27. $$\partial_H : K_1(\mathbb{Q}WH)/K_1(\mathbb{Z}_{(m)}WH) \longrightarrow K_o(\mathbb{Z}WH)$$

Let Split ∂ : Split $K_1(\mathbb{Q}OrG)$/Split $K_1(\mathbb{Z}_{(m)}OrG) \longrightarrow$ Split $K_o(\mathbb{Z}OrG)$ be the direct sum $\bigoplus_{(H)} \partial_H$. If S and E are the isomorphisms of Theorem 10.34. define

18.28. $$\partial : K_1(\mathbb{Q}OrG)/K_1(\mathbb{Z}_{(m)}OrG) \longrightarrow K_o(\mathbb{Z}OrG)$$

by $E \circ \text{Split } \partial \circ S$. Using Lemma 18.6. we obtain an exact sequence

18.29. $$0 \longrightarrow K_1(\mathbb{Z}_{1/m}OrG)/K_1(\mathbb{Z}OrG) \longrightarrow K_1(\mathbb{Q}OrG)/K_1(\mathbb{Z}_{(m)}OrG) \longrightarrow K_o(\mathbb{Z}OrG) \to K_o(\mathbb{Z}_{1/m}OrG)$$

<u>Proposition 18.30.</u> <u>Let</u> (X,A) <u>be an oriented special pair of finitely dominated</u> G-CW-<u>complexes</u>. <u>Suppose that</u> $H_n(X^H,A^H)$ <u>contains no</u> p-<u>torsion for any prime number</u> p <u>with</u> $(p,|G|) = 1$ <u>and</u> $n \geq 0$. <u>Then we have</u>

$$o^G(X,A) = -\partial(\overline{\rho}^G(X,A))$$

<u>Proof:</u> We can assume without loss of generality for the $\mathbb{Z}OrG$-chain complex $C = C^c(X,A)$ that C_i is finitely generated projective for odd i and finitely generated free for even i and zero for large i, as $C^c(X,A)$ is at least homotopy equivalent to such a $\mathbb{Z}OrG$-chain complex. Notice that $H_i(X^H;\mathbb{Z}_{1/m}) = H_i(X^H) \otimes_{\mathbb{Z}} \mathbb{Z}_{1/m} = H_i(X^H)_{1/m}$ is a finitely generated projective $\mathbb{Z}_{1/m}WH$-module. Going through the definition of $\overline{\rho}^G(X,A)$ one easily checks the following: There are finitely generated projective $\mathbb{Z}_{1/m}WH$-modules M(H) for $H \subset G$, a finitely generated free $\mathbb{Z}OrG$-module F and a finitely generated projective $\mathbb{Z}OrG$-module P together with $\mathbb{Z}_{1/m}WH$-automorphisms

$$\hat{\alpha}(H) : M(H) \oplus F(H)_{1/m} \longrightarrow M(H) \oplus P(H)_{1/m}$$

and a $\mathbb{Z}_{(m)}OrG$-isomorphism

$$\beta : P_{(m)} \longrightarrow F_{(m)}$$

such that $o^G(X,A) = [F] - [P]$ holds and the map

$$\text{Split } K_1(\mathbb{Q}OrG) \xrightarrow{I} K_1(\mathbb{Q}OrG) \xrightarrow{pr} K_1(\mathbb{Q}OrG)/K_1(\mathbb{Z}_{(m)}OrG)$$

sends the element in Split $K_1(\mathbb{Q}OrG)$ given by the collection of elements

$$[(id \oplus \beta(H)_{(o)}) \circ \hat{\alpha}(H)_{(o)}) : M(H)_{(o)} \oplus F(H)_{(o)} \longrightarrow M(H)_{(o)} \oplus F(H)_{(o)}] \in K_1(QWH)$$

to $\bar{\rho}^G(X)$. Namely take $M(H) = (H(C(H))_{1/m})_{odd}$, $F = C_{ev}$ and $P = C_{odd}$. Since we may add a finitely generated free $\mathbb{Z}$OrG-module to F and P we can suppose $M(H) = \{0\}$ for $H \subset G$. Moreover, we can assume by Theorem 16.29. the existence of a $\mathbb{Z}_{1/m}$OrG-isomorphism $\alpha : F_{1/m} \to P_{1/m}$ such that $\hat{\alpha}(H)$ is $\mathrm{Res}_{G/H}\alpha = \alpha(H)$. Consider the following commutative diagram

18.31.
$$\begin{array}{ccc} S_{G/H}P & \xrightarrow{S_{G/H}(\alpha^{-1}) \circ i_{1/m}} & S_{G/H}F_{1/m} \\ \downarrow S_{G/H}(\beta) \circ i_{(m)} & & \downarrow S_{G/H}(\beta_{(o)} \circ \alpha_{(o)}) \circ j_{1/m} \\ S_{G/H}F_{(m)} & \xrightarrow[j_{(m)}]{} & S_{G/H}F_{(o)} \end{array}$$

if $i_{(m)} : S_{G/H}P \to S_{G/H}P_{(m)}$, $i_{1/m} : S_{G/H}P \to S_{G/H}P_{1/m}$, $j_{1/m} : S_{G/H}F_{1/m} \to S_{G/H}F_{(o)}$ and $j_{(m)} : S_{G/H}F_{(m)} \to S_{G/H}F_{(o)}$ are the obvious inclusions. Since it is isomorphic to the obvious diagram

$$\begin{array}{ccc} S_{G/H}P & \longrightarrow & S_{G/H}P_{1/m} \\ \downarrow & & \downarrow \\ S_{G/H}P_{(m)} & \longrightarrow & S_{G/H}P_{(o)} \end{array}$$

18.31. is a pull-back of abelian groups. We get from the definition of the boundary map $\partial_H : K_1(\mathbb{Q}\mathrm{OrG})/K_1(\mathbb{Z}_{(m)}\mathrm{OrG}) \to K_o(\mathbb{Z}\mathrm{OrG})$

$$\partial_H([S_{G/H}(\beta_{(o)} \circ \alpha_{(o)})]) = [S_{G/H}P] - [S_{G/H}F]$$

Now the claim follows. □

If ε denotes the augmentation map we get an homomorphism

$$K_1(\mathbb{Q}\mathrm{OrG}) \xrightarrow{\mathrm{Res}} \mathrm{Split}\ K_1(\mathbb{Q}\mathrm{OrG}) \xrightarrow{\bigoplus_{(H)} \varepsilon_*} \bigoplus_{(H)} K_1(\mathbb{Q}) = \prod_{(H)} \mathbb{Q}^* .$$

Because of Theorem 16.29. it induces an homomorphism

18.32.
$$\alpha : K_1(\mathbb{Q}\mathrm{OrG})/K_1(\mathbb{Z}_{(m)}\mathrm{OrG}) \to \prod_{(H)} \mathbb{Q}^*/\mathbb{Z}^*_{(m)}$$

<u>Proposition 18.33.</u> <u>Let</u> (X,A) <u>be a special pair of finitely dominated</u> G-CW-

complexes. Then we have in $\prod_{(H)} \mathbb{Q}^*/\mathbb{Z}^*_{(m)}$ *if* m *is a multiple of* $|G|$

$$m\chi^G(X,A) = \alpha(\bar{\rho}^G(X,A)) \cdot h\chi^G(X,A)^{-1}$$

Proof. Choose a $\mathbb{Z}_{(m)}OrG$-homotopy equivalence $f : C^c(X,A)_{(m)} \longrightarrow P$ for a round $\mathbb{Z}_{(m)}OrG$-chain complex P. Fix $H \subset G$. Equip the $\mathbb{Q}WH$-chain complex $H(P(H)_{(o)})$ with the round structure induced by $H(f(H))_{(o)}$ and the given round structure on $H(X^H,A^H;\mathbb{Q})$. The projection $p : H(P(H)_{(o)}) \longrightarrow H(P(H)_{(o)}) \otimes_{\mathbb{Q}WH} \mathbb{Q}$ is an isomorphism as WH acts trivially on $H(X^H,A^H;\mathbb{Q})$, and we get an induced round structure on the $\mathbb{Q}$-chain complex $H(P(H)_{(o)}) \otimes_{\mathbb{Q}WH} \mathbb{Q}$. Consider the up to homotopy commutative diagram of round $\mathbb{Q}$-chain complexes where q is the projection and i up to homotopy determined by $H(i) = id$

$$\begin{array}{ccc} H(P(H)_{(o)}) & \xrightarrow{\ p\ } & H(P(H)_{(o)}) \otimes_{\mathbb{Q}WH} \mathbb{Q} \\ {\scriptstyle i}\downarrow & & \downarrow {\scriptstyle i \otimes_{\mathbb{Q}WH} \mathbb{Q}} \\ P(H)_{(o)} & \xrightarrow{\ q\ } & P(H)_{(o)} \otimes_{\mathbb{Q}WH} \mathbb{Q} \end{array}$$

We get from Example 12.20. for the round torsion in $\mathbb{Q}^*/\mathbb{Z}^*_{(m)}$

$$t(i \otimes_{\mathbb{Q}WH} \mathbb{Q}) \cdot t(p) = t(q) \cdot t(i)$$

We have by definition $t(i \otimes_{\mathbb{Q}WH} \mathbb{Q}) = \alpha_H(\bar{\rho}^G(X,A))$ and $t(p) = 0$ and derive from the next lemma $t(i) = h\chi^G(X,A)_H$ and $t(q) = m\chi^G(X,A)_H$. Now the claim follows. □

Lemma 18.34. *Let* C *be a finite free* $\mathbb{Z}$*-chain complex. Suppose that* C *and* $H(C)_f$ *have round structures* $\{\alpha\}$ *and* $\{\beta\}$. *Then we have in* $\mathbb{Q}^*/\mathbb{Z}^*$ *for the absolute torsion of Example 12.21.*

$$t(C_{(o)},\{\alpha_{(o)}\},\{\beta_{(o)}\}) = \prod_{n \geq o} |\mathrm{Tors}\, H_n(C)|^{(-1)^n}$$

Proof. We use induction over a number n such that $\mathrm{Tors}\, H_i(C) = 0$ for $i > n$. If $n = -1$ we have $t(C,\{\alpha\},\{\beta\}) \in \mathbb{Z}^*$ and the claim follows. In the induction step choose a round finitely generated $\mathbb{Z}$-chain complex D concentrated in dimension n and $n+1$ such that $H_i(D) = 0$ for $i \neq n$ and $H_n(D) = \mathrm{Tors}\, H_n(C)$. Let $f : D \longrightarrow C$ be a chain map such that $H_n(f)$ is an isomorphism onto $\mathrm{Tors}\, H_n(C)$.

Equip Cone(f) and $H(f)_f$ with round structures such that

$$t(C) = t(D) \cdot t(\mathrm{Cone}(f))$$

holds. One easily checks $t(D) = |\mathrm{Tors}\, H_n(C)|^{(-1)^n}$. Now apply the induction hypothesis to Cone(f) . □

Let X and Y be special oriented G-CW-complexes. We obtain a natural isomorphism from the Künneth formula

$$H(X^H)_f \otimes_{\mathbf{Z}} H(Y^H)_f \xleftarrow{\cong} (H(X^H) \otimes_{\mathbf{Z}} H(Y^H))_f \xrightarrow[\cong]{\times} H(X^H \times Y^H)_f$$

Hence X × Y is again special and enherits an orientation. The long Mayer-Vietoris sequences M(H) of $(X^H * Y^H; X^H \times \mathrm{Cone}\, Y^H, \mathrm{Cone}(X^H) \times Y^H)$ reduces to short split exact sequences

$$0 \longrightarrow H_{i+1}(X^H * Y^H) \longrightarrow H_i(X^H \times Y^H) \longrightarrow H_i(X^H) \oplus H_i(Y^H) \longrightarrow 0$$

for $i > 0$ and into an exact sequence of free $\mathbf{Z}$-modules

$$0 \longrightarrow H_1(X^H * Y^H) \longrightarrow H_o(X^H \times Y^H) \longrightarrow H_o(X^H) \oplus H_o(Y^H)$$
$$\longrightarrow H_o(X^H * Y^H) \longrightarrow 0$$

Thus we obtain an acyclic finite free $\mathbf{Z}$-chain complex $M(H)_f = M(H)/\mathrm{Tors}\, M(H)$. Hence $X * Y$ is again special. Equip $X * Y$ with the orientation for which the round torsion of $M(H)_f$ vanishes in $K_1(\mathbf{Z})$ for $H \subset G$ if M(H) gets the obvious round structure induced from the ones of $X^H \simeq X^H \times \mathrm{Cone}(Y^H)$, $Y^H \simeq \mathrm{Cone}(X^H) \times Y^H$, $X^H \times Y^H$ and $X^H * Y^H$.

<u>Theorem 18.35.</u> <u>Join formula</u>

<u>Let</u> X <u>and</u> Y <u>be special oriented</u> G-CW-<u>complexes. Then we have</u>

a) $\rho^G(X * Y) = \rho^G(X) + \rho^G(Y)$, <u>if</u> X <u>and</u> Y <u>are finite</u>.

b) $\bar{\rho}^G(X * Y) = \bar{\rho}^G(X) + \bar{\rho}^G(Y)$

c) $o^G(X * Y) = o^G(X) + o^G(Y)$

d) $m\chi^G(X * Y) = m\chi^G(X) \cdot m\chi^G(Y)$

e) $h\chi^G(X \star Y) = h\chi^G(X) \cdot h\chi^G(Y)$.

__Proof.__   a), b) and c) follow from 18.17.

d)   Using Lemma 18.34., Example 12.20. and Example 12.21. one proves $m\chi^G(X \times Y) = 1$ and then derives $m\chi^G(X \star Y) = m\chi^G(X) \cdot m\chi^G(Y)$ using the exact sequence

$$o \to C^c(X\times Y) \to C^c(X\times Cone(Y)) \oplus C^c(Cone(X)\times Y) \to C^c(X\star Y) \to 0$$

e)   Take $G = \{1\}$, m any prime number and use a) and Proposition 18.33. An alternative proof is based on the Künneth formula and the decomposition of the Mayer Vietoris sequence $M(H)$ defined above.   □

Now we apply this to representations. Let V be a complex G-representation. Then SV is a finite special G-CW-complex (see Example 1.8. or 4.36.) and enherits an orientation from the complex structure. Since $S(V \oplus W)$ and $SV \star SW$ are oriented G-PL-homeomorphic, we obtain an homomorphism

18.36. $\rho_{\mathbb{C}}^G : Rep_{\mathbb{C}}(G) \longrightarrow Wh(\mathbb{Q}OrG)$ $[V] \longrightarrow \rho^G(SV)$

because of Theorem 18.35. Composing it with $Rep_{\mathbb{R}}(G) \to Rep_{\mathbb{C}}(G)$ sending $[V]$ to $[V \otimes_{\mathbb{R}} \mathbb{C}]$ defines

18.37. $\rho_{\mathbb{R}}^G : Rep_{\mathbb{R}}(G) \longrightarrow Wh(\mathbb{Q}OrG)$ $[V] \longrightarrow \rho^G(S(V \otimes_{\mathbb{R}} \mathbb{C}))$

__Theorem 18.38.__   __The map__ $\rho_{\mathbb{R}}^G$ __is injective for any finite group__ G .

__Proof.__   One easily checks using Lemma 16.15. that $\rho_{\mathbb{R}}^G$ is natural with respect to restriction to subgroups. The map $\Pi \, res_C^G : Rep_{\mathbb{R}}(G) \to \Pi \, Rep_{\mathbb{R}}(C)$ is injective if Π runs over the cyclic subgroups C of G . Hence we can assume without loss of generality that G itself is cyclic.

Let $Rep_{\mathbb{R}}^f(G)$ be the subgroup of $Rep_{\mathbb{R}}(G)$ generated by all free G-representations. Given a subgroup $H \subset G$ restriction with the projection $pr : G \longrightarrow G/H$ defines $Rep_{\mathbb{R}}^f(G/H) \longrightarrow Rep_{\mathbb{R}}(G)$. We obtain an isomorphism

18.39. $\underset{H \subset G}{\oplus} Rep_{\mathbb{R}}^f (G/H) \longrightarrow Rep_{\mathbb{R}}(G)$

We have also the splitting of Theorem 10.34.

18.40. $$\bigoplus_{H \subset G} Wh(\mathbb{Q}G/H) \longrightarrow Wh(\mathbb{Q}OrG)$$

Choose a numeration $H_1, H_2, \ldots, H_k$ of the subgroups of G such that $H_i \subset H_j \Rightarrow i \geq j$ holds. Then $\rho_{\mathbb{R}}^G$ looks in terms of the splittings 18.39. and 18.40. like an upper triangular matrix. Hence $\rho_{\mathbb{R}}^G$ is injective if each entry on the diagonal is. Therefore it suffices to prove the injectivity of

18.41. $$Rep_{\mathbb{R}}^f(G) \longrightarrow Rep_{\mathbb{R}}(G) \xrightarrow{\rho_{\mathbb{R}}^G} Wh(\mathbb{Q}OrG) \xrightarrow{S_G} Wh(\mathbb{Q}G)$$

It sends the free G-representation V to $\rho^G(SV \otimes_{\mathbb{R}} \mathbb{C})$ now taken just over the group ring $\mathbb{Q}G$. Let $U \subset \mathbb{C}^*$ be the subgroup generated by $\{\pm\exp(2\pi i n/|G|) \mid n=1,2,\ldots,|G|\}$ Sending the generator of G to $\exp(2\pi i/|G|)$ defines an homomorphism $Wh(\mathbb{C}G) \longrightarrow \mathbb{C}^*/U$. The composition of 18.41. with this map is injective by the classification of Lens spaces (see Cohen [1973], Milnor [1966]) based on Franz' Lemma (see Franz [1935]). □

As a corollary we get de Rham's theorem (see de Rham [1964]).

<u>Corollary 18.42.</u> <u>Two real G-representations V and W are linearly isomorphic if and only if SV and SW are G-diffeomorphic.</u>

<u>Proof.</u> Let $f : SV \longrightarrow SW$ be a G-diffeomorphism. Then f is a simple G-homotopy equivalence by 4.36. and $\tau^G(f*f) \in Wh(\mathbb{Z}OrG)$ vanishes by 18.17. We get $\rho_{\mathbb{R}}^G(SV) = \rho_{\mathbb{R}}^G(SW)$ by 18.18. Now apply Theorem 18.38. □

18.G. Reidemeister torsion for Riemannian G-manifolds.

Next we define Reidemeister torsion in a different context. Let G be a finite group and M be a (smooth, compact) G-manifold without boundary. If C is a component of M^H for $H \subset G$, let $WH(C)$ be its isotropy group under the WH-action on $\pi_o(M^H)$. Then C is a $WH(C)$-manifold without boundary. Denote its dimension by $n(C)$. We suppose

<u>Assumption 18.43.</u> We assume for $H \subset G$ and $C \in \pi_o(M^H)$

i) C is orientable, i.e. $w_1(C) \in H^1(C;\mathbb{Z}/2)$ is zero.

ii) $n(C)$ is odd.

iii) $WH(C)$ acts trivially on $H_{n(C)}(C)$. □

We call M equivariantly oriented if we have choosen for any $C \in \pi_o(M^H)$, $H \subset G$ a fundamental class $[C] \in H_{n(C)}(C)$ such that for $H \subset G$, $C \in \pi_o(M^H)$ and $g \in G$ the map $c(g)_* : H_{n(C)}(C) \longrightarrow H_{(n(g^{-1}Cg))}(g^{-1}Cg)$ induced by conjugation sends $[C]$ to $[g^{-1}Cg]$. Suppose that M is equivariantly oriented and has an invariant Riemannian metric (see Bredon [1972] VI.2.). Then we can define

18.44. $$\rho^G(M) \in Wh(\mathbb{R}\,OrG)$$

as follows. Fix $H \subset G$ and $C \in \pi_o(M^H)$. The invariant Riemannian metric on M induces an invariant Riemannian metric on the $WH(C)$-manifold C by restriction. It induces an inner global product on the $\mathbb{R}$-vector space $\Omega^p C$ of p-forms on C . If $\Delta^p : \Omega^p C \longrightarrow \Omega^p C$ is the Laplace operator, the subspace $\ker(\Delta^p) \subset \Omega^p C$ is called the space of harmonic forms. It is a $\mathbb{R}$-vector space and inherits an inner product from $\Omega^p C$. If $H^p_{dR}(C)$ denotes the de Rham cohomology, there is a natural identification

18.45. $$H^p_{dR}(C) = \ker(\Delta^p)$$

We obtain from de Rham's theorem a natural isomorphism

18.46. $$H^p(C;\mathbb{R}) = H^p_{dR}(C)$$

Hence $H^p(C;\mathbb{R})$ enherits from $\ker \Delta^p$ an inner product. The necessary differential geometry for the construction above can be found in Gallot-Hulin-Lafontaine [1987] p. 164-167.

The Hodge-star operator induces a $\mathbb{R}WH(C)$-isomorphism $* : \ker(\Delta^{n(C)-p}) \longrightarrow \ker(\Delta^p)$. Hence we get from 18.45. and 18.46. an $\mathbb{R}G$-isomorphism (see Gilkey [1984], Lemma 1.5.3.)

18.47. $$v : H^{ev}(C;\mathbb{R}) \longrightarrow H^{odd}(C;\mathbb{R}) .$$

We obtain from the universal coefficient formula an isomorphism

18.48. $$u_p : H_p(C,\mathbb{R}) \longrightarrow Hom_{\mathbb{R}}(H^p(C);\mathbb{R}) .$$

For each C we get a $\mathbb{R}WH(C)$-isomorphism

18.49. $\Phi(C) : H_{odd}(C;\mathbb{R}) \longrightarrow H_{ev}(C;\mathbb{R})$

by $u_{ev}^{-1} \circ \mathrm{Hom}_{\mathbb{R}}(v,id) \circ u_{odd}$. Since $H_p(M^H;\mathbb{R})$ is $\oplus_C \mathbb{R}WH \times_{\mathbb{R}WH(C)} H_p(C;\mathbb{R})$, where C runs over $\pi_o(M^H)/WH$, we obtain an $\mathbb{R}WH$-isomorphism

18.50. $\Phi(H) : H_{odd}(M^H;\mathbb{R}) \longrightarrow H_{ev}(M^H;\mathbb{R})$.

The upshot is that a closed G-manifold M with invariant Riemannian metric satisfying Assumption 18.43. has a preferred round structure and determines

18.51. $\rho^G(M) \in Wh(\mathbb{R}\,\mathrm{Or}\,G)$.

Next we define a variation of this construction where we can drop Assumption 18.43. Notice that we now switch from $\mathbb{R}OrG$ to $\mathbb{R}G$-modules. Let M be a (compact, smooth) connected G-manifold with invariant Riemannian metric and possibly non-empty boundary We obtain from 18.45., 18.46. and 18.48. applied to $C = M$ an inner product on $H_p(M,\mathbb{R})$ for any p . Equip the cellular $\mathbb{R}$-chain complex $C_*(M)$ with an inner product for which the cellular $\mathbb{R}$-bases is orthonormal. This inner product is compatible with the $\mathbb{R}G$-structure on $C_*^c(M)$. Let $* : K_1(\mathbb{R}G) \longrightarrow K_1(\mathbb{R}G)$ be the involution $[f] \longrightarrow [\mathrm{Hom}_{\mathbb{R}}(f,id)]$ for $f : P \longrightarrow P$ an automorphism of a finitely generated projective $\mathbb{R}G$-module. Choose any isometric $\mathbb{R}G$-isomorphism $\phi : H(M)_{odd} \oplus C(M)_{ev} \rightarrow C^c(M)_{odd} \oplus H(M)_{ev}$ Up to $\mathbb{R}G$-chain homotopy there is exactly one $\mathbb{R}G$-chain map $i : H(M) \longrightarrow C^c(M)$ satisfying $H(i) = id$. Let $t(i,\{\phi\}) \in K_1(\mathbb{R}G)$ be the torsion introduced in Definition 12.4. We define

18.52. $\rho_{PL}^G(M) \in K_1(\mathbb{R}G)^{\mathbb{Z}/2}$

by $t(i,\{\phi\}) + *t(i,\{\phi\})$. This is independent of the choice of ϕ as for any isometric $\mathbb{R}G$-automorphism $\psi : P \longrightarrow P$ of a finitely generated projective $\mathbb{R}G$-module P with inner product compatible with the $\mathbb{R}G$-structure $\psi^* = \psi^{-1}$ and hence $[\psi]+*[\psi]=0$ holds.

Another variation measures the failure of equivariant simple Poincare duality. Let M be a G-manifold with invariant Riemannian metric. For simplicity we assume that M is orientable and connected and G acts orientation preserving. We have the

Poincaré $\mathbb{R}G$-chain equivalence $\cap[M] : C^{m-*}(M,\partial M;\mathbb{R}) \longrightarrow C_*(M;\mathbb{R})$. Let

$$\phi : C^{m-odd}(M,\partial M;\mathbb{R}) \oplus C_{ev}(M;\mathbb{R}) \longrightarrow C_{odd}(M;\mathbb{R}) \oplus C^{m-ev}(M,\partial M;\mathbb{R})$$

be any $\mathbb{R}G$-isomorphism which is isometric with respect to the inner products coming from cellular bases. Define the Poincaré torsion

18.53. $\rho^G_{PD}(M;\mathbb{R}) \in K_1(\mathbb{R}G)^{\mathbb{Z}/2}$

by $t(\cap[M],\{\phi\}) + *t(\cap[M],\{\phi\})$.

If G acts freely , $\rho^G_{PD}(M;\mathbb{R})$ is zero. If the action is not free, $\rho^G_{PD}(M;\mathbb{R})$ can be non zero. The geometric explanation for this phenomenon is that the dual cell decomposition is not compactible with the G-action.

Comments 18.54.

Reidemeister torsion is the main tool in the classification of Lens spaces (see Cohen [1973], Milnor [1966], Reidemeister [1938]) . This generalizes to de Rham's theorem that two real G-representations are linearly G-isomorphic if and only if their unit spheres are G-diffeomorphic or G-PL-homeomorphic (de Rahm [1964]). We have briefly discussed this question in the topological category already in Example 4.25.). The use of torsion invariants for transformation groups is worked out in Rothenberg [1978] where a proof of de Rham's theorem similar to the one in this section is given (see also Illman [1985]). Reidemeister torsion also plays a role in equivariant surgery theory (see e.g. Madsen [1983]). Its connection to knot theory is summarized in Turaev [1986]. We will show in section 20. that G-homotopy representations are classified up to oriented G-homotopy equivalence by the reduced equivariant Reidemeister torsion.

In Ray and Singer [1971] the analytic torsion $\rho_{an}(M)$ of a Riemannian manifold was defined using the spectrum of the Laplace operator and Zeta-functions. Some evidence is given that for a closed orientable manifold $\rho_{an}(M)$ and $\rho_{PL}(M)$ (see 18.52.) agree. This was independently proven by Cheeger [1979] and Müller [1978]. The analytic torsion plays a role in different contexts (see e.g. Fried [1986], Fried [1988], Quillen [1986], Schwarz [1978], Witten [1988]). In Lott-Rothenberg [1989] an equivariant version for analytic torsion and its relation to PL-torsion is established for closed orientable G-manifolds (see also Maumary [1987]). In the case, where M has a boundary, the conjecture of Ray-Singer fails. The general relationship between analytic and PL-torsion for arbitrary (compact) G-manifolds with boundary is worked out in Lück [1989] .

In comparison with other definitions our invariant is defined for a wide class of spaces X , we only need $\chi^G(X) = 0$ in $A(G)$. This gives some useful flexibility f.e. if one takes out tubular neighbourhoods (compare 4.36.). We make no further assumptions about the homology and never kill the homology f. e. by dividing out norm ideals. This is important for several reasons. Since we keep the information about the homology, we will be able to relate the reduced Reidemeister torsion and degrees of maps between G-homotopy representations by generalized Swan homomorphisms and classify G-homotopy representations by the reduced Reidemeister torsion. Moreover, we can switch between the two splittings (S,E) and (Res,I) and obtain invariants in $Wh(\mathbb{R}OrG)$ which can explicitely be computed by $\bigoplus_{(H)} Wh(\mathbb{R}WH)$. For example in Rothenberg [1978] the groups, where the Reidemeister torsion lives in, are only known for abelian G . We also get good properties under restriction to subgroups avoiding some technica difficulties which already occur in the free case when dividing out norm ideals (see Connolly-Geist [1982]). Further references are Ewing-Löffler-Pedersen [1985a], [1985b Lück [1988] □

<u>Exercises 18.55.</u>

1) Let G be a finite group with cyclic 2-Sylow subgroup. Let V be a real G-representation and $f : SV \longrightarrow SV$ be a G-homotopy equivalence. Prove $\tau^G(f) = 0$.

2) Compute $\rho^G_{\mathbb{R}} : Rep_{\mathbb{R}}(G) \longrightarrow Wh(\mathbb{Q}OrG)$ for $G = S_3$, the symmetric group of order 6

3) Let p be a prime number and G a cyclic group of order p^n . Denote by $G(m)$ the subgroup of order p^m for $0 \leq m \leq n$. Prove for a complex G-representation V

$$m\chi(SV)_{G(m)} = \prod_{m \leq \ell \leq n} p^{(n-\ell)(\dim_{\mathbb{C}} V^{G(\ell)} - \dim_{\mathbb{C}} V^{G(\ell+1)})}$$

) Let (X,A) be a pair of finitely dominated G-CW-complexes and R a field. Show $\chi^G(X,A) = \chi^G(X,A;R)$.

) Let G be a finite abelian group and X be a finite special G-CW-complex. Show that $C^c(X^H,X^{>H};\mathbb{Q}) \otimes_{\mathbb{Q}WH} \mathbb{Q}WH/(N)$ is an acyclic finitely generated free $\mathbb{Q}WH/(N)$-chain complex with a preferred equivalence class of bases, if (N) is the norm ideal. Let $t(X^H,X^{>H})$ be its torsion in $K_1(\mathbb{Q}WH/(N))/V$ if V is the image of the trivial units U under the canonical projection $pr : K_1(\mathbb{Q}WH) \longrightarrow K_1(\mathbb{Q}WH/(N))$. Show that for any choice of orientation of X and $H \subset G$ the map

$$Wh(\mathbb{Q}OrG) \xrightarrow{S_{G/H}} Wh(\mathbb{Q}WH) \longrightarrow K_1(\mathbb{Q}WH/(N))/V$$

sends $\rho^G(X)$ of 18.22. to $t(X^H,X^{>H})$.

) Let $i : H \longrightarrow G$ be an inclusion of finite groups. Consider an oriented special pair (X,A) of finite G-CW-complexes. Let $i^* : K_1(\mathbb{Q}OrG) \longrightarrow K_1(\mathbb{Q}OrH)$ be given by restriction. Prove that $res_H^G(X,A)$ is also an oriented special pair of finite H-CW-complexes and $i^*\rho^G(X,A) = \rho^H(res_H^G(X,A))$. Is the analogous statement true for $\bar{\rho}^G(X,A)$?

) Let (X,A) be an oriented special (non-equivariant) pair of finitely dominated CW-complexes. Show for $\rho^1(X,A) \in Q^*/\mathbf{Z}^*$ defined in 18.23. and $h\chi(X,A) \in Q^*/\mathbf{Z}^*$ defined in 18.24.

$$\rho^1(X,A) = h\chi(X,A)$$

) Let M be a connected oriented free closed G-manifold of odd dimension with an invariant Riemannian metric. Let $\rho^G(M) \in Wh(\mathbb{R}G)$ be the element given by 18.51. If $* : Wh(\mathbb{R}G) \longrightarrow Wh(\mathbb{R}G)$ is the involution induced from the involution on $\mathbb{R}G$ sending $\Sigma\lambda g \cdot g$ to $\Sigma\lambda g \cdot g^{-1}$ prove

$$* \rho^G(M) = \rho^G(M)$$

) Let M be an orientable connected closed manifold of odd dimension. Let the finite group G of odd order act smoothly on M. Prove that assumption 18.43. is satisfied.

10) Let M be a connected oriented closed (non-equivariant) Riemannian manifold. Suppose $H_*(M,\mathbb{Q}) \cong H_*(S^n,\mathbb{Q})$ for some odd n. Let $V(M) \in \mathbb{R}^*$ be its volume, $\rho(M) \in \mathbb{R}^*/\mathbb{Z}^*$ be the invariant of 18.51. and $h\chi(M) \in \mathbb{Q}^*/\mathbb{Z}^*$ the element defined in 18.24. Show in $\mathbb{R}^*/\mathbb{Z}^*$

$$\rho(M) = V(M)^{-2} \cdot h\chi(M)^{-1}$$

11) Let M be a connected (non-equivariant) manifold with disjoint submanifolds X and Y. Suppose that M, X resp. Y is a homology m-resp. k- resp. ℓ-sphere for $m = k+\ell+1$. Then the linking number $\mathrm{lk}(X,Y) \in \mathbb{Z}/\{\pm 1\}$ is defined. Show $h\chi(M\backslash X\backslash Y) = 1$ if $\mathrm{lk}(X,Y) = 0$ and $h\chi(M\backslash X\backslash Y) = \mathrm{lk}(X,Y)$ in $\mathbb{Q}^*/\mathbb{Z}^*$ otherwise.

12) Let M be a closed G-manifold with invariant Riemannian metric satisfying 18.43. Define $h : Wh(\mathbb{R}OrG) \longrightarrow K_1(\mathbb{R}G)^{\mathbb{Z}/2}$ by $[f] \longrightarrow [f(G/1)] + *[f(G/1)]$ for an automorphism $f : P \longrightarrow P$ of a finitely generated projective $\mathbb{R}OrG$-module. Show $h(\rho^G(M)) = \rho^G_{PL}(M)$

13) Let $\mathbb{Z}/2$ act on S^1 by complex conjugation. Prove $\rho^{\mathbb{Z}/2}_{PD}(S^1) \neq 0$.

14. Let M be a closed connected orientable G-manifold such that G acts orientation preserving. Show

a) If $\dim M$ is odd, then $\rho^G_{PD}(M) = 0$

b) If $\dim M$ is even, then $\rho^G_{PL}(M) = -\frac{1}{2}\rho^G_{PD}(M)$.

15. Let G be a compact Lie group and V and W orthogonal G-representations. Suppose the existence of a simple G-homotopy equivalence $f : SV \longrightarrow SW$. Show that V and W are isomorphic as G-representations.

19. Generalized Swan homomorphisms

Let $sw : \mathbf{Z}/|G|^* \longrightarrow K_o(\mathbf{Z}G)$ be the Swan homomorphism associated with the finite group G and $\partial : K_1(QG)/K_1(\mathbf{Z}_{(|G|)}G) \longrightarrow K_o(\mathbf{Z}G)$ be induced from the boundary map of the localization square. We will define an injective lift $\overline{sw} : \mathbf{Z}/|G|^* \longrightarrow K_1(QG)/K_1(\mathbf{Z}_{(|G|}G)$ of sw. We extend these notions to the orbit category and obtain an injection $\overline{SW} : Inv(G) \longrightarrow K_1(QOrG)/K_1(\mathbf{Z}_{(|G|)}OrG)$. In this section G denotes a finite group.

Let m be a multiple of the order of the finite group G. The <u>Swan homomorphism</u>

19.1.
$$sw = sw(G,m) : \mathbf{Z}/m^* \longrightarrow K_o(\mathbf{Z}G)$$

sends $\bar{r} \in \mathbf{Z}/m^*$ to the class $[M] \in K_o(\mathbf{Z}G)$ of any finite abelian group M with trivial G-action and order $|M| = r$. Recall that $[M]$ is defined by $\Sigma(-1)^n[P_n]$ for any finite resolution P of M. We have defined $\partial = \partial(G,m) : K_1(QG)/K_1(\mathbf{Z}_{(m)}G) \longrightarrow K_o(\mathbf{Z}G)$ by the map in the localization square for $\mathbf{Z}G$ and m in 18.27. Next we want to define an injective lift of sw over ∂

19.2.
$$\overline{sw} = \overline{sw}(G,m) : \mathbf{Z}/m^* \longrightarrow K_1(QG)/K_1(\mathbf{Z}_{(m)}G)$$

It sends $\bar{r} \in \mathbf{Z}/m^*$ to the element represented by $[Q \xrightarrow{r} Q] \in K_1(QG)$. By stabilizing with $QG/(N)$ for N the norm element, one checks for the unit $1 + \frac{r-1}{|G|}N \in QG^*$ in $K_1(QG)$

19.3.
$$[Q \xrightarrow{r} Q] = [1 + \frac{r-1}{|G|} N]$$

Let r be $1 + am$ for some $a \in \mathbf{Z}$. Then we get in $\mathbf{Z}_{(m)}G$

$$(1 + \frac{r-1}{|G|} \cdot N)\cdot(1 - \frac{am}{|G|\cdot(1+am)}) = 1 + (\frac{am}{|G|} - \frac{am}{|G|\cdot(1+am)} - \frac{(am)^2}{|G|\cdot(1+am)})$$

$$= 1 + \frac{am(1+am)-am-(am)^2}{|G|\cdot(1+am)} = 1$$

Hence $\overline{sw}$ is a well-defined homomorphism.

<u>Theorem 19.4.</u>

a) $-\partial \circ \overline{sw} = sw$

b) <u>If m is $|G|$, then $\overline{sw}$ is injective.</u>

Proof:

a) Consider the map between squares which both are pull-backs of abelian groups

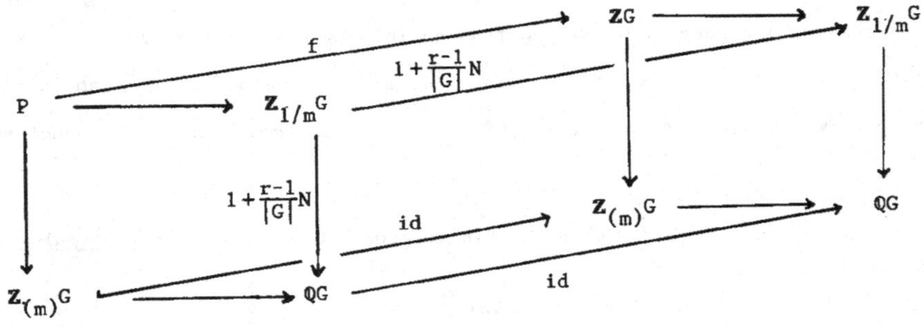

As $0 \longrightarrow \mathbf{Z}_{1/m} \xrightarrow{r} \mathbf{Z}_{1/m} \longrightarrow \mathbf{Z}/r \longrightarrow 0$ is exact, there are exact sequences of $\mathbf{Z}G$-modules

$$0 \longrightarrow \mathbf{Z}_{1/m}G \xrightarrow{1 + \frac{r-1}{|G|} N} \mathbf{Z}_{1/m}G \longrightarrow \mathbf{Z}/r \longrightarrow 0$$

and

$$0 \longrightarrow P \longrightarrow \mathbf{Z}G \longrightarrow \mathbf{Z}/r \longrightarrow 0$$

We get from the definitions

$$\partial \circ \overline{sw}(\bar{r}) = \partial([1 + \frac{r-1}{|G|} N]) = [P] - [ZG] = -sw(\bar{r})$$

b) We start with the case where G is abelian. Then the determinant induces an isomorphism $K_1(\mathbf{Q}G)/K_1(\mathbf{Z}_{(m)}G) \longrightarrow \mathbf{Q}G^*/\mathbf{Z}_{(m)}G^*$. The unit $1 + \frac{r-1}{|G|} N \in \mathbf{Q}G^*$ belongs to $\mathbf{Z}_{(m)}G$ if and only if $|G|$ divides $r-1$. Hence $\overline{sw}$ is injective for abelian G

Let $i : H \longrightarrow G$ be the inclusion of a subgroup. The canonical projection defines an homomorphism $i^* : \mathbf{Z}/|G|^* \longrightarrow \mathbf{Z}/|H|^*$. Let $i^* : K_1(\mathbf{Q}G)/K_1(\mathbf{Z}_{(|G|)}G) \longrightarrow K_1(\mathbf{Q}H)/K_1(\mathbf{Z}_{(|H|)}H)$ be induced from restriction. Then the following diagram commutes

19.5.
$$\begin{array}{ccc} \mathbf{Z}/|G|^* & \xrightarrow{\overline{sw}(G,|G|)} & K_1(\mathbf{Q}G)/K_1(\mathbf{Z}_{(|G|)}G) \\ \downarrow i^* & & \downarrow i^* \\ \mathbf{Z}/|H|^* & \xrightarrow{\overline{sw}(H,|H|)} & K_1(\mathbf{Q}H)/K_1(\mathbf{Z}_{(|H|)}H) \end{array}$$

If G_p is the p-Sylow subgroup of G, we obtain an isomorphism $\mathbf{Z}/|G|^* \to \prod_p \mathbf{Z}/|G_p|^*$. Because of 19.5. it suffices to prove injectivity of $\overline{sw}(G)$ only for p-groups. Moreover, we can assume that G is not cyclic.

Let p be an odd prime. Choose a subgroup $H \subset G$ of order p. The image of $sw : \mathbf{Z}/|G|^* \longrightarrow K_o(\mathbf{Z}G)$ has order $|G|/p$ by Taylor [1978]. Hence $i^* : \mathbf{Z}/|G|^* \longrightarrow \mathbf{Z}/|H|^*$ is injective on $\ker(sw)$. Since $\ker(\overline{sw}) \subset \ker(sw)$ holds and $\overline{sw}(H)$ is injective by the argument above, $\overline{sw}(G)$ must be injective.

Let p be 2. If G is the dihedral group $D(2^n)$, the semi-dihedral group $SD(2^m)$ or the generalized quaternion group $Q(2^n)$, we can apply Rothenberg [1978a], Theorem 1.6., Proposition 3.1. In all other cases the image of sw has order $|G|/4$ and G contains an abelian subgroup H of order 4. Because of $\{\pm 1\} = \ker(sw)$ the claim follows from an argument as above. □

19.6. Robert Oliver has pointed out to the author that he also has a proof of Theorem 19.4. We give his argument for the case that G is a non-exceptional p-group as it does not use the result of Taylor [1978]. Consider $r \in \mathbf{Z}$ and $1 + \frac{r-1}{m} N \in K_1(\hat{Q}_pG)$ such that there is $u \in K_1(\hat{\mathbf{Z}}_pG)$ mapped to $1 + \frac{r-1}{m} N$ under the change of ring map . Then $\log(u)$ (defined in Oliver [1980], p. 200) is aN for some $a \in \hat{Q}_p$. We must show $a \in \hat{\mathbf{Z}}_p$ in order to verify $\bar{r} = \bar{1}$ in $\mathbf{Z}/|G|^*$. Now p^2 divides $\text{card}\{g \in G \mid g^p = 1\}$ for a non-exceptional p-group. Let Φ and Γ be the maps introduced in Oliver [1980], p. 208. By inspecting the coefficient of 1 one recognizes that p does not divide $N - 1/p\,\Phi(N)$. We have $\Gamma(u) = a(N-1/p\,\Phi(N))$ by definition. Since $\Gamma(u)$ is integral valued by Oliver [1980], p. 208, the claim follows. □

Next we define the splitted version of generalized Swan homomorphisms for a finite free EI-category Γ. Let $c \in ch_\ell(y,x)$ be a ℓ-chain from y to x (see 16.3.). We have assigned to c a $Aut(x)$-$Aut(y)$-set $S(c)$ in 16.23. Let $\overline{S(c)}$ be the $Aut(y)$-set $Aut(x)\backslash S(c)$ representing an element in the Burnside ring of $Aut(y)$

19.7. $$[\overline{S(c)}] \in A(Aut(y)) .$$

Define

19.8. $$\mathfrak{m}(x,y) \in A(\mathrm{Aut}(y))$$

by

$$\mathfrak{m}(x,y) = \sum_{\ell \geq 0} (-1)^{\ell} \sum_{c \in ch_{\ell}(y,x)} [\overline{S(c)}]$$

We have the pairing $\otimes_R : K_n(R[y]) \otimes A(\mathrm{Aut}(y)) \longrightarrow K_n(R[y])$ induced by $? \otimes_R R(S)$ with the diagonal $\mathrm{Aut}(y)$-action for S a finite $\mathrm{Aut}(y)$-set. Hence $\mathfrak{m}(x,y)$ defines homomorphisms if m is $m(\Gamma)$ (see Definition 16.1.):

19.9. $$\mathfrak{m}_{\bar{x},\bar{y}} : K_1(\mathbb{Q}[y])/K_1(\mathbb{Z}_{(m)}[y]) \longrightarrow K_1(\mathbb{Q}[y])/K_1(\mathbb{Z}_{(m)}[y])$$
$$\mathfrak{m}_{\bar{x},\bar{y}} : K_o(\mathbb{Z}[y]) \longrightarrow K_o(\mathbb{Z}[y]) .$$

Let $(\mathrm{Split}\ \overline{SW})_{\bar{x},\bar{y}} : \mathbb{Z}/m^* \longrightarrow K_1(\mathbb{Q}[y])/K_1(\mathbb{Z}_{(m)}[y])$ be $\mathfrak{m}_{\bar{x},\bar{y}} \circ \overline{sw}(\mathrm{Aut}(y),m)$ and $(\mathrm{Split}\ SW)_{\bar{x},\bar{y}} : \mathbb{Z}/m^* \longrightarrow K_o(\mathbb{Z}[y])$ be $\mathfrak{m}_{\bar{x},\bar{y}} \circ sw(\mathrm{Aut}(y),m)$. Their collection defines for $\bar{C}(\Gamma)^* = \prod_{\mathrm{Is}\ \Gamma} \mathbb{Z}/m^*$:

19.10. $$\mathrm{Split}\ \overline{SW} : \bar{C}(\Gamma)^* \longrightarrow \mathrm{Split}\ K_1(\mathbb{Q}\Gamma)/\mathrm{Split}(K_1(\mathbb{Z}_{(m)}))$$
$$\mathrm{Split}\ SW : \bar{C}(\Gamma)^* \longrightarrow \mathrm{Split}\ K_o(\mathbb{Z}\Gamma)$$

These maps also have global descriptions. Their definition needs some preparation. Fix $y \in \mathrm{Ob}\ \Gamma$ and an integer r prime to $m = m(\Gamma)$. Equip $\mathbb{Z}/r$ and $\mathbb{Q}$ with the trivial $\mathrm{Aut}(x)$-action resp. $\mathrm{Aut}(y)$-action. For any ℓ-chain c from y to x we have identifications of $\mathrm{Aut}(y)$-modules

19.11. $$\mathbb{Z}/r \otimes_{\mathbb{Z}[x]} \mathbb{Z}(S(c)) = \mathbb{Z}/r \otimes_{\mathbb{Z}} \mathbb{Z}(\overline{S(c)})$$
$$\mathbb{Q} \otimes_{\mathbb{Q}[x]} \mathbb{Q}(S(c)) = \mathbb{Q} \otimes_{\mathbb{Q}} \mathbb{Q}(\overline{S(c)})$$

As $\mathbb{Z}/r$ equipped with the trivial $\mathrm{Aut}(y)$-action has a finite resolution over $\mathbb{Z}[y]$, the same is true for the $\mathbb{Z}[y]$-module $\mathbb{Z}/r \otimes_{\mathbb{Z}[x]} \mathbb{Z}(S(c))$ by 19.11. Let I_x be the inclusion functor of 9.29. By Lemma 17.37. the $\mathbb{Z}\mathrm{Or}G$-module $I_x\mathbb{Z}/r$ has a finite resolution and $[I_x\mathbb{Z}/r] = \mathrm{Split}\ SW(\{r(\bar{z}) \mid \bar{z} \in \mathrm{Is}\ \Gamma\})$ holds if $r(z) \equiv 1$ for $z \neq \bar{x}$ and $r(\bar{x}) \equiv \bar{r} \bmod m$. Analogously $I_x(\mathbb{Q} \xrightarrow{r} \mathbb{Q}) = \mathrm{Split}\ \overline{SW}(\{r(\bar{z}) \mid \bar{z} \in \mathrm{Is}\ \Gamma\})$ is true in $K_1(\mathbb{Q}\Gamma)/K_1(\mathbb{Z}_{(m)}\Gamma)$. We can define <u>generalized Swan homomorphisms</u>

19.12.
$$\overline{SW} : \overline{C}(\Gamma)^* \longrightarrow K_1(\mathbb{Q}OrG)/K_1(\mathbb{Z}_{(m)}OrG)$$
$$SW : \overline{C}(\Gamma)^* \longrightarrow K_o(\mathbb{Z}OrG)$$

by

$$\overline{SW}(\{r(\bar{x}) \mid \bar{x} \in Is\ \Gamma\}) = \sum_{\bar{x} \in Is\ \Gamma} [I_x(\mathbb{Q} \xrightarrow{r(x)} \mathbb{Q})]$$

and

$$SW(\{r(\bar{x}) \mid \bar{x} \in Is\ \Gamma\}) = \sum_{\bar{x} \in Is\ \Gamma} [I_x(\mathbb{Z}/r(x))] \ .$$

We get using Theorem 19.4. if ∂ is the map of 18.28.

Theorem 19.13.

a) $S \circ \overline{SW} = \text{Split}\ \overline{SW}$

b) $S \circ SW = \text{Split}\ SW$

c) $-\partial \circ \overline{SW} = SW$ □

We define commutative associative rings with unit

19.14.
$$C(G) = \prod_{(H)} \mathbb{Z}$$

19.15.
$$\overline{C}(G) = C(G)/|G| \cdot C(G) = \prod_{(H)} \mathbb{Z}/|G| \ .$$

The character map $ch : A(G) \longrightarrow C(G)$ sending $[X]$ to $(\text{card}\ X^H)_{(H)}$ is an injective ring homomorphism (see 5.15.). We use it to consider $A(G) \subset C(G)$ as a subring. By Example 16.38. we have $|G| \cdot C(G) \subset A(G) \subset C(G)$. Define

19.16.
$$\overline{A}(G) = A(G)/|G| \cdot C(G) \subset \overline{C}(G)$$

19.17.)
$$Inv(G) = \overline{C}(G)^*/\overline{A}(G)^* \ .$$

There is an obvious embedding $C(G)^* = \prod_{(H)} \mathbb{Z}^*$ into $\overline{C}(G)^*$. Hence we can define

19.18.)
$$Pic(G) = \overline{C}(G)^*/\overline{A}(G)^* \cdot C(G)^*$$

Theorem 19.19. Let G be a finite group of order m . Then $\overline{SW}$ factorizes over the canonical projection pr into an injective map $\overline{SW}_o$

$$\begin{array}{ccc} \bar{C}(G)^* & \xrightarrow{\overline{SW}} & K_1(\mathbb{Q}OrG)/K_1(Z_{(m)}OrG) \\ \text{pr}\downarrow & \nearrow_{\overline{SW}_o} & \\ Inv(G) & & \end{array}$$

□

As a detailed proof of Theorem 19.19. will be developed in section 20., we only give a sketch here. Any element in $\bar{A}(G)^*$ can be realized by $d = \{\deg(f^H) | H \subset G\}$ for a G-map $f : SV \longrightarrow SV$ and a complex G-representation V . As $\overline{SW}(d) = \bar{\rho}^G(SV) - \bar{\rho}^G(SV) = 0$ holds , $A(G)^*$ lies in the kernel of SW and the factorization follows. Given $x \in Inv(G)$ with $\overline{SW}_o(x) = 0$, choose $d = \{d(H) | H \subset G\} \in \bar{C}(G)^*$ with $pr(d) = x$. We prove inductively over the orbit types that we can choose $d(H) \equiv 1$. In the induction step we must show that d(H) can be choosen to be 1 provided that d(K) is 1 for $K \supset H$, $K \neq H$. As $SW(d) = 0$ holds the map $sw(WH, |WH|) : \mathbf{Z}/|WH|^* \longrightarrow K_1(\mathbb{Q}WH)/K_1(\mathbf{Z}_{(|WH|)}WH)$ sends d(H) to zero. By Theorem 19.4. we get $d(H) \equiv 1 \bmod |WH|$. Then there is a unit $u \in A(G)^*$ with $u(K) = 1$ for $K \supset H$, $K \neq H$ and $u(H) \equiv d(H) \bmod |G|$. Now $u^{-1} \cdot d$ has the required properties.

We consider the case of a finite abelian group G more closely. For $H,K \subset G$ define integers

19.20. $$\bar{\mu}(K,H) = \sum_{\ell \geq o} (-1)^{\ell} \cdot |ch_{\ell}(G/H,G/K)|$$

Define a map

19.21. $$\bar{\mu} : \bar{C}(G)^* = \prod_{(K)} \mathbf{Z}/m^* \longrightarrow \prod_{(H)} \mathbf{Z}/|G/H|^*$$

by

$$\{\overline{n(K)} | K \subset G\} \longrightarrow \{\sum_{(K)} \overline{n(K)}^{\,\bar{\mu}(K,H)} | H \subset G\} .$$

<u>Theorem 19.21.</u>

a) <u>The map $\bar{\mu}$ factorizes over pr into an isomorphism</u>

$$\overline{pr} : Inv(G) \longrightarrow \prod_{(H)} \mathbf{Z}/|G/H)^*$$

b) <u>If S is the isomorphism of Theorem 10.34. and q the obvious projection, the</u>

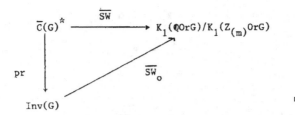

following diagram commutes

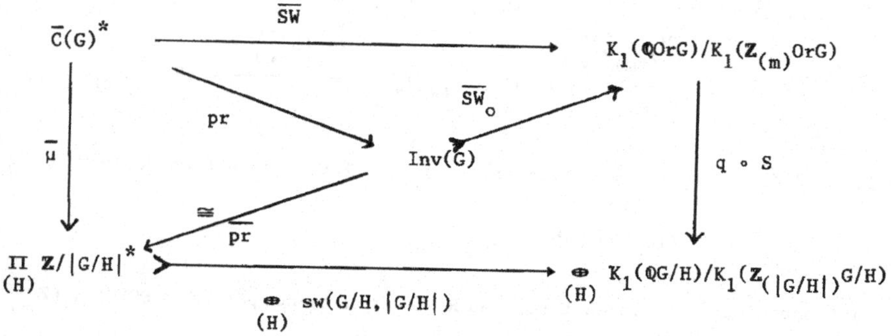

Proof: Commutatively follows from Theorem 19.13. and the identification $[\overline{S(c)}] = [\{*\}]$ for any ℓ-chain c from G/H to G/K (see 19.7.). Since $sw(G/H,|G/H|)$ is injective by Theorem 19.4. $q \circ S \circ \overline{SW}_o$ and hence $\overline{pr}$ is injective. Now the claim follows. □

19.22. Let X and Y be finitely dominated G-CW-complexes such that WH acts trivially on $H_*(X^H)_{1/m}$ for all $H \subset G$ and $m = |G|$ and $\chi(X^H) \in \mathbf{Z}$ is zero for all $H \in Iso(X)$ with $WH \neq \{1\}$ and analogously for Y. Similarly as in Lemma 18.8. one proves that X and Y are special in the sense of Definition 18.21. Fix orientations $\{\Phi_X\}$ and $\{\Phi_Y\}$, i.e. collection of $\{1\}$-stable equivalence classes of $\mathbf{Z}$-isomorphisms $\{\Phi_X(H)\} : (H(X^H)_f)_{odd} \longrightarrow (H(X^H)_f)_{ev}$ if A_f denotes $A/Tors\,A$ for a $\mathbf{Z}$-module A. Consider a G-map $f : X \longrightarrow Y$ such that $f^H : X^H \longrightarrow Y^H$ is a $\mathbf{Z}_{(m)}$-homology equivalence for all $H \subset G$. Define

19.23.
$$h\chi(X)_{1/m} \in \bar{C}(G)^* = \prod_{(H)} \mathbf{Z}/m^*$$
$$h\chi(f)_{1/m} \in \bar{C}(G)^*$$

by

$$h\chi(X)_{1/m} = \prod_{n \geq 0} |Tors(H_n(X))_{1/m}|^{(-1)^n}$$

and

$$h\chi(f)_{1/m} = h\chi(Cone(f))_{1/m} .$$

Let $\hat{\sigma}(f)_{(H)} \in \mathbf{Z}^*_{(m)}$ be the determinant of the $\{1\}$-stable isomorphism class of $\mathbf{Z}_{(m)}$-automorphisms

$$(H(X^H)_{(m)})_{odd} \xrightarrow{(H(f^H)_{(m)})_{odd}} (H(Y^H)_{(m)})_{odd} \xrightarrow{\{\Phi_Y(H)_{(m)}\}}$$
$$(H(Y^H)_{(m)})_{ev} \xrightarrow{(H(f^H)_{(m)})_{ev}^{-1}} (H(X^H)_{(m)})_{ev} \xrightarrow{\{\Phi_X(H)_{(m)}^{-1}\}} (H(X^H)_{(m)})_{odd} .$$

Let $\sigma(f)_{(H)} \in \mathbf{Z}/m^*$ be the reduction of $\hat{\sigma}(f)_{(H)}$. Their collection defines

19.24. $\qquad \sigma(f) \in \overline{C}(G)^*.$

We have introduced the equivariant reduced Reidemeister torsion $\overline{\rho}^G(X) \in K_1(\mathbf{Q}OrG)/(K_1(\mathbf{Z}_{(m)}OrG)$ in 18.22., the finiteness obstruction $o^G(X) \in K_o(\mathbf{Z}OrG)$ in 18.3. and the boundary map $\partial : K_1(\mathbf{Q}OrG)/K_1(\mathbf{Z}_{(m)}OrG) \longrightarrow K_o(\mathbf{Z}OrG)$ in 18.28.

<u>Theorem 19.25.</u> <u>We get under the conditions above</u>

a) $o^G(X) = -\partial(\overline{\rho}^G(X)) + SW(h\chi(X)_{1/m})$

b) $\overline{\rho}^G(Y) - \overline{\rho}^G(X) = \overline{SW}(\sigma(f))$

c) $o^G(Y) - o^G(X) = SW(h\chi(f)_{1/m})$

d) $h\chi(f)_{1/m} + \sigma(f) = h\chi(Y)_{1/m} - h\chi(X)_{1/m}$

<u>Proof</u>

a) is left to the reader. It is an extension of Proposition 18.30.

b) The following diagram of round $\mathbf{Q}WH$-chain complexes commutes up to homotopy if $H(i_X(H)) = H(i_Y(H)) = id$ holds

$$\begin{array}{ccc} C^c(X^H;\mathbf{Q}) & \xrightarrow{C^c(f^H,\mathbf{Q})} & C^c(Y^H;\mathbf{Q}) \\ \downarrow i_X(H) & & \downarrow i_Y(H) \\ H(X^H;\mathbf{Q}) & \xrightarrow{H(f^H,\mathbf{Q})} & H(Y^H;\mathbf{Q}) \end{array}$$

We derive from Example 12.21. for the round torsion in $K_1(\mathbf{Q}WH)$

$$t(C^c(f^H;\mathbf{Q})) - t(H(f^H;\mathbf{Q})) = t(i_Y(H)) - t(i_X(H)) .$$

By definition we have $\hat{\sigma}(f)_{(H)} = t(H(f^H;\mathbf{Q}))$ in $\mathbf{Q}^*$ for all $H \subset G$ if we consider

$(f^H;\mathbb{Q})$ as a $\mathbb{Q}$-chain equivalence. As WH acts trivially on $H(X^H;\mathbb{Q})$ and $H(Y^H;\mathbb{Q})$, we get $\overline{SW}(\sigma(f)) = pr \circ I((t(f^H;\mathbb{Q}))_{(H)})$ where $I : \bigoplus_{(H)} K_1(\mathbb{Q}WH) \longrightarrow K_1(\mathbb{Q}OrG)$ is the isomorphism of Theorem 16.29. and pr the projection. We also obtain $\overline{\rho}^G(X) = pr \circ I((t(i_Y(H))_{(H)})$ and $\overline{\rho}^G(Y) = pr \circ I((t(i_X(H))_{(H)})$ from the definitions. Because $I \circ Res = id$ by Theorem 16.29., we have $I((t(C^c(f^H;\mathbb{Q}))_{(H)}) = \tau(C^c(f;\mathbb{Q}))$ in $K_1(\mathbb{Q}OrG)$. As $C^c(f)$ is a $\mathbb{Z}_{(m)}OrG$-chain equivalence, $\tau(C^c(f;\mathbb{Q}))$ vanishes in $K_1(\mathbb{Q}OrG)/K_1(\mathbb{Z}_{(m)}OrG)$. Now the claim follows.

) follows from the exact sequence of $\mathbb{Z}OrG$-chain complexes $0 \longrightarrow \Sigma C^c(X) \longrightarrow Cone(f) \longrightarrow C^c(Y) \longrightarrow 0$, Lemma 18.34. and additivity for absolute torsion (see Example 12.21).

) The sum formula implies $o(Cone(C^c(f)) = o^G(Y) - o^G(X)$. Because of the assumption $(Cone(C^c(f))(G/H)$ is finite of order prime to $m = (G)$ and WH acts trivially, we get from Lemma 17.37. that $I_{G/H}$ applied to it has a finite resolution. We get from Proposition 11.19., Theorem 16.8. and Theorem 16.29.

$$o(Cone(C^c(f)) = \Sigma(-1)^n \cdot [H_n(Cone(C^c(f))] = SW(h\chi(f)_{1/m})$$

Notice that c) also follows directly from a), b), and d). □

<u>Comments 19.26.</u> The Swan homomorphism $sw : \mathbb{Z}/|G|^* \longrightarrow K_o(\mathbb{Z}G)$ was introduced in Swan [1960b]. Let X and Y be finitely dominated free G-CW-complexes such that X and Y are (non-equivariant) homotopy equivalent to S^n for some n. Swan showed for any G-maps $f : X \longrightarrow Y$ that its degree $\deg f$ is prime to $|G|$ and $sw(\overline{\deg f}) = o^G(Y) - o^G(X)$ holds. Notice that Theorem 19.25.c) is a generalization of this result to non-free special G-CW-complexes. Theorem 19.4.b) is equivalent to conjectures in Rothenberg [1978a], 1.4' Conjecture, and Wall [1979], p. 522.

The generalized Swan homomorphisms $\overline{SW}$ and SW are important tools for the analyses of G-homotopy representations as we will carry out in the next section. The main issue is that they relate homological information, which is usually easy to handle, with invariants in algebraic K-groups (see Theorem 19.25.). The existence of the injection $\overline{SW}_o : Inv(G) \longrightarrow K_1(\mathbb{Q}OrG)/K_1(\mathbb{Z}_{(m)}OrG)$ fits into our general pattern to translate geometric information into data with values in algebraic K-groups.

The homomorphism $Inv(G) \longrightarrow K_o(\mathbf{Z}OrG)$ induced from $SW : \bar{C}(G)^* \longrightarrow K_o(\mathbf{Z}OrG)$ is examined in tom Dieck [1985] where it is expressed by ordinary Swan homomorphisms (cf. Theorem 19.13.). Among other things multiplicative congruences for $Inv(G)$ are derived from the splitting of the inclusion of the image of $sw : \mathbf{Z}/|G|^* \longrightarrow K_o(\mathbf{Z}G)$ in ${}^oD(\mathbf{Z}G)$. One should try to do the same for $\overline{sw} : \mathbf{Z}/|G|^* \to K_1(\mathbf{Q}G)/K_1(\mathbf{Z}_{(|G|)}G)$. If $\overline{sw}$ is split injective, one would get a new proof of the result $Inv(G) = \prod_{(H)} \mathbf{Z}/|$ in tom Dieck [1985].

The Swan homomorphism detects the finiteness obstruction also in homology propagation of group actions (see Cappell-Weinberger [1987]) □

<u>Exercises 19.27.</u>

1) Show for a finite cyclic group G that $SW : \bar{C}(G)^* \longrightarrow K_o(\mathbf{Z}OrG)$ is trivial.

2) Let p and q be prime numbers with $p|q-1$. Let G be the non-abelian group of order pq. Show that $S_{G/1} \circ \overline{SW} : \bar{C}(G)^* \longrightarrow K_1(\mathbf{Q}G)/K_1(\mathbf{Z}_{(pq)}G)$ sends $(n(G/H))_{(H)}$ to the sum of $\overline{sw}(n(G)\cdot n(G/G_q)^{-1})$ and $\overline{sw}(n(G/G)\cdot n(G/G_p)^{-1}) \otimes_{\mathbf{Q}} \mathbf{Q}G/G_p$. Prove $Inv(G) = \mathbf{Z}/pq^* \times \mathbf{Z}/p^* \times \mathbf{Z}/q^*$.

3) Consider the p-group $G = \langle A,B | A^{p^2} = 1 = B^p, BAB^{-1}A^{-1} = A^p \rangle$ of order p^3. Show that the image of $S \circ \overline{SW} : \bar{C}(G)^* \longrightarrow \bigoplus_{(H)} K_1(\mathbf{Q}WH)/K_1(\mathbf{Z}_{(p)}WH)$ and the image of $\bigoplus_{(H)} \overline{sw}(WH,|WH|) : \prod_{(H)} \mathbf{Z}/|WH|^* \longrightarrow \bigoplus_{(H)} K_1(\mathbf{Q}WH)/K_1(\mathbf{Z}_{(p)}WH)$ do not agree (compare Theorem 19.21.).

4) Let V and W be complex G-representations and $f : SV \longrightarrow SW$ be a G-map. Show $\overline{SW}(\{\deg f^H \mid H \subset G\}) = \bar{\rho}^G(SV) - \bar{\rho}^G(SW)$.

5) Let $G=D(2n)$ be the dihedral group of order $2n$. Given $\bar{k} \in (\mathbf{Z}/2n)^*$, construct real G-representations V and W and a G-map $f : SV \longrightarrow SW$ such that $\deg f^H \equiv 1 \bmod 2n$ for $H \subset G$ with $H \neq \{1\}$ and $\deg f \equiv \bar{k} \bmod 2n$ holds. Conclude that $sw(D(2n)) : (\mathbf{Z}/2n)^* \longrightarrow K_o(\mathbf{Z}D(2n))$ is trivial.

6) Consider the finite free EI-category Γ having three objects x, y_o and y_1 such that $Hom(x,y_o)$, $Hom(x,y_1)$, $Aut(y_o)$, $Aut(y_1)$ consist of one element and

$\mathrm{Hom}(y_o,y_1)$ is empty. Let G be $\mathrm{Aut}(x)$ and m be $|G|$. Show that $S_{y_i} \circ \overline{SW} : \overline{C}(\Gamma)^* \longrightarrow K_1(\mathbb{Q}[y_i])/K_1(\mathbb{Z}_{(m)}[y_i])$ is trivial for $i = 0,1$ and $S_x \circ \overline{SW} : \overline{C}(\Gamma)^* \to K_1(\mathbb{Q}G)/K_1(\mathbb{Z}_{(m)}G)$ sends $(n(x),n(y_o),n(y_1))$ to $\overline{sw}(G,m)(n(x)\cdot n(y_o)^{-1}\cdot n(y_1)^{-1})$. Prove that $(1+\overline{A}(\Gamma)) \cap \overline{C}(\Gamma)^*$ is not contained in the kernel of $\overline{SW}$ (compare Theorem 19.19.).

7) Write $\mathfrak{m}(\overline{x},\overline{y}) \in A(\mathrm{Aut}(x))$, defined in 19.8., as $\sum_{j=1}^{r} [\mathrm{Aut}(y)/H_j]$. Let $i(j) : H_j \longrightarrow \mathrm{Aut}(y)$ be the inclusion and $pr_j : \mathbb{Z}/m^* \longrightarrow \mathbb{Z}/|H_j|^*$ be the projection. Show

$$(\mathrm{Split}\ \overline{SW})_{\overline{x},\overline{y}} = \sum_{j=1}^{r} i(j)_* \circ \overline{sw}(H_j,m)$$

and

$$(\mathrm{Split}\ SW)_{\overline{x},\overline{y}} = \sum_{j=1}^{r} i(j)_* \circ sw(H_j,|H_j|) \circ pr_j$$

8) Let Γ be a finite free EI-category and M a $\mathbb{Z}\Gamma$-module such that $\mathrm{Aut}(x)$ acts trivially on $M(x)$ and $M(x)$ is finite with $(\mathrm{card}\ M(x),m(\Gamma)) = 1$ for $x \in \mathrm{Ob}\ \Gamma$. Then $\mathrm{hdim}_{\mathbb{Z}\Gamma} M$ is finite and SW sends $\{\overline{\mathrm{card}\ M(x)} \mid x \in \mathrm{Is}\ \Gamma\}$ to $[M]$.

20. Homotopy representations

We apply invariants like finiteness obstruction and Reidemeister torsion to G-homotopy representations. We introduce Grothendieck groups of G-homotopy representations with certain properties. We show that G-homotopy representations are classified by their reduced Reidemeister torsion up to stable orientation preserving G-homotopy equivalence. Let G be a finite group in this section.

20.A. Review of basic facts about G-homotopy representations.

We briefly recall definitions and results about G-homotopy representations following tom Dieck [1987] as far as needed. Let S^{-1} be $\emptyset$ and $\dim \emptyset = -1$.

Definition 20.1. A G-homotopy representation X is a finite-dimensional G-CW-compl such that X^H is homotopy equivalent to $S^{n(H)}$ with $n(H) = \dim(X^H)$ for $H \subset G$. □

Lemma 20.2. A G-homotopy representation X is finitely dominated.

Proof. Because of Proposition 14.9. it suffices to show that the cellular $\mathbf{Z}\Pi(G,X)$-chain complex $C^c(X)$ is $\mathbb{Z}\Pi(G,X)$-homotopy equivalent to a finitely generated free one By Proposition 8.33. $\mathrm{Aut}(x)$ is an extension of a finite group and $\{1\}$ or $\mathbf{Z}$ for $x \in \mathrm{Ob}\,\Gamma$ so that $\mathbf{Z}[x]$ is noetherian (see Atiyah-Mac Donald [1969], ch. 7). As Is $\Pi(G,X)$ is finite, a $\mathbf{Z}\Pi(G,X)$-module M is finitely generated if and only if $M(x)$ is a finitely generated $\mathbf{Z}[x]$-module for all $x \in \mathrm{Ob}\,\Gamma$ (compare 16.8. and 16.10.). Hence any submodul of a finitely generated $\mathbf{Z}\Pi(G,X)$-module is again finitely generated. Therefore $C^c(X)$ is $\mathbf{Z}\Pi(G,X)$-homotopy equivalent to a finitely generated free $\mathbf{Z}\Pi(G,X)$-chain complex if and only if $H(\tilde{X}^H(x))$ is finitely generated over $\mathbf{Z}[x]$ for all $x \in \mathrm{Ob}\,\Gamma$. This is true because $X^H(x) \simeq S^n$ for some $n \geq -1$ holds. □

The dimension function of a G-homotopy representation X

20.3.
$$\mathrm{Dim}(X) \in C(G) = \prod_{(H)} \mathbf{Z}$$

is given by the collection $(n(H)+1)_{(H)}$ where (H) runs over the conjugacy classes of subgroups of G. We call X even if $n(H)+1$ is even for all $H \subset G$. The orientation behaviour of X at $H \subset G$ is the homomorphism

20.4. $$e(X,H) : WH \longrightarrow \{\pm 1\}$$

sending $g \in WH$ to the degree of the automorphism of $\tilde{H}^{n(H)}(X^H) \cong \mathbf{Z}$ induced from the left multiplication with g. We have defined $\chi^G(X) \in A(G)$ in 18.3. As $\chi(X^H) = 1 + (-1)^{n(H)}$ holds, one proves as in Lemma 18.8.

Lemma 20.5. The following statements are equivalent for G-homotopy representations X and Y

a) $\chi^G(X) = \chi^G(Y)$

b) $\chi(X^H) = \chi(Y^H)$ for all $H \subset G$

c) $\dim X^H \equiv \dim Y^H \bmod 2$ for all $H \subset G$

d) $e(X,H) = e(Y,H)$ for all $H \subset G$ and $\chi(X^H) = \chi(Y^H)$ for all $H \subset G$ with $|WH| = 1$

e) $e(X,H) = e(Y,H)$ for all $H \subset G$ and $\dim X^H \equiv \dim Y^H \bmod 2$ for all $H \subset G$ with $|WH| = 1$.

Next we follow Laitinen [1986] sec. 2 to get a well-defined notion of an orientation.

Lemma 20.6. Let X be a G-homotopy representation.

a) The inclusion $i : X^K \longrightarrow X^H$ is an homotopy equivalence for $H \subset K \subset G$ with $n(H) = n(K)$.

b) The set $I(H) = \{K \supset H \mid n(H) = n(K)\}$ contains a unique maximal element $\hat{H}$. □

Regard two G-homotopy representations X and Y with the same dimension function $\mathrm{Dim}(X) = \mathrm{Dim}(Y) = n+1$. A coherent orientation $\phi = \phi(X,Y)$ for them is a choice of isomorphisms

20.7. $$\phi(H) : \tilde{H}^{n(H)}(X^H) \longrightarrow \tilde{H}^{n(H)}(Y^H)$$

for $H \subset G$ satisfying

a) If $H \subset K \subset G$ and $n(H) = n(K)$ holds and $i : X^K \longrightarrow X^H$ and $j : Y^K \longrightarrow Y^H$ are the inclusions, we have $j^* \circ \phi(H) = \phi(K) \circ i^*$.

b) Let $\ell(g) : X^H \longrightarrow X^{gHg^{-1}}$ and $\ell(g) : Y^H \longrightarrow Y^{g^{-1}Hg}$ be left multiplication with $g \in G$. Then $\phi(H) \circ \ell(g)^* = \ell(g)^* \circ \phi(gHg^{-1})$ is valid.

Because of Lemma 20.5. and Lemma 20.6. such a coherent orientation exists. Let $\{d(H) \mid H \subset G\}$ be a collection of integers. They satisfy the unstable conditions 20.8. if the following holds:

20.8. i) $d(H) = d(K)$ if $(H) = (K)$

ii) $d(H) = d(K)$ if $n(H) = n(K)$

iii) $d(H) = \pm 1$ if $n(H) = 0$

iv) $d(H) = 1$ if $n(H) = -1$

Consider a G-map $f : X \longrightarrow Y$. Define the <u>degree</u> $\deg(f^H) \in \mathbf{Z}$ with respect to ϕ for $n(H) = -1$ by 1 and for $n(H) \geq 0$ by the degree of the automorphism $\phi(H) \circ (f^H)^* : \tilde{H}^{n(H)}(Y^H) \longrightarrow \tilde{H}^{n(H)}(Y^H)$ of $\tilde{H}^{n(H)}(Y^H) \cong \mathbf{Z}$. Notice that $\{\deg(f^H) | H \subset G$ satisfies the unstable conditions 20.8. Notice that 20.8. ii) implies 20.8. i).

Such a coherent orientation $\phi(X,Y)$ is a priori not given by "absolute" choices of orientations for X and Y. In general the non-triviality of the orientation behaviour causes problems. The favourite case is the one of an even G-homotopy representation X. A (coherent) <u>orientation</u> for an even G-homotopy representation X is a choice of generators

20.9. $$[X^H] \in \tilde{H}^{n(H)}(X^H)$$

for $H \subset G$ satisfying

a) $\ell(g)^*[X^{gHg^{-1}}] = [X^H]$ for $H \subset G$, $g \in G$

b) $i^*[X^H] = [X^K]$ for $H \subset K \subset G$ with $n(H) = n(K)$.

The existence of such an orientation follows from Lemma 20.5. and Lemma 20.6. as the orientation behaviour $e(X,H)$ is trivial for $H \subset G$.

If X and Y are even G-homotopy representations with $\mathrm{Dim}(X) = \mathrm{Dim}(Y)$, a choice of orientations for X and Y determines uniquely a coherent orientation $\phi = \phi(X,Y)$ by

20.10. $$\phi(H)([X^H]) = [Y^H] \quad \text{for } H \subset G .$$

The join $X * Y$ of two G-homotopy representations X and Y is again one and

$\mathrm{Dim}(X*Y) = \mathrm{Dim}(X) + \mathrm{Dim}(Y)$. Suppose that X and Y are even and oriented. Then $X * Y$ is also an even G-homotopy representation and enherits an orientation from the isomorphism

$$\text{20.11.} \qquad \tilde{H}^k(X^H) \otimes \tilde{H}^\ell(Y^H) \xrightarrow{\cup} \tilde{H}^{k+\ell}(X^H \times Y^H) \xrightarrow{\delta} \tilde{H}^{k+\ell+1}(X^H * Y^H)$$

for $H \subset G$, $k = \dim X^H$, $\ell = \dim(Y^H)$ and δ the boundary of the Mayer-Vietoris sequence of $(X^H * Y^H, X^H \times \mathrm{Cone}(Y^H), \mathrm{Cone}(X^H) \times Y^H)$. Notice that $(X * Y)^H = X^H * Y^H$ holds. If $\phi(X,Y)$ is a coherent orientation for X and Y and $\phi(X',Y')$ for X' and Y' , then we get an induced coherent orientation $\phi(X * X', Y * Y')$ for $X * Y$ by 20.11. In particular $\phi(X,Y)$ induces $\phi(X*Z, Y*Z)$ for any G-homotopy representation Z . If X and Y are not necessarily even G-homotopy representations, there seems to be no notion of an "absolute" orientation of X and Y such that, firstly, orientations on X and Y induce a coherent orientation $\phi(X,Y)$ and, secondly, $X * Y$ enherits an "absolute" orientation in a canonical way.

Let X and Y be two G-homotopy representations with the same dimension functions and a coherent orientation $\phi(X,Y)$. Let m be a multiple of $|G|$. A map $f : X \to Y$ has *invertible degrees* if $\deg(f^H)$ and m are prime for all $H \subset G$. We will frequently use the following facts proved in tom Dieck [1987], II.4 + 10 and Laitinen [1986].

Proposition 20.12. *Fix* $H \subset G$.

a) *If* $f : X \to Y$ *is a* G-*map*, $H = \hat{H}$ *and* $n(H) \geq 1$, *then for any integer* k *there is a* G-*map* $g : X \to Y$ *such that* $\deg(g^H) = \deg(f^H) + k \cdot |WH|$ *and* $g|X^{>H} = f|X^{>H}$ *holds*. ($\hat{H}$ *was defined in Lemma 20.6.*).

b) *Suppose that* G *is nilpotent or that* $n(L) \geq n(K)+2$ *holds for all* $K, L \in \mathrm{Iso}(X)$, $L \subset K$, $L \neq K$. *Then* G-*maps* f *and* $g : X \to Y$ *are* G-*homotopic if and only if* $\deg(f^K) = \deg(g^K)$ *holds for all* $K \subset G$.

c) *Let* $f : X \to Y$ *be a* G-*map with* $(\deg(f^K), m) = 1$ *for all* $K \supset H$, $K \neq H$. *Then there is a* G-*map* $g : X \to Y$ *with invertible degrees and* $f|X^{>H} = g|X^{>H}$.

d) *Let* $f : X \to Y$ *be a* G-*map with invertible degrees and* $g : Y \to X$ *be a* G-*map with* $\deg(f^K) \cdot \deg(g^K) \equiv 1 \bmod m$ *for* $K \supset H$, $K \neq H$. *Then there is a* G-*map* $h : Y \to X$ *with* $\deg(f^K) \cdot \deg(h^K) \equiv 1 \bmod m$ *for all* $K \subset G$ *and* $g|X^{>H} = h|X^{>H}$.

20.B. Homotopy representation groups.

Next we define homotopy representation groups of the finite group G in analogy to the representation ring. Let c be a subcategory of the category of G-homotopy representations resp. of oriented even G-homotopy representations and e an equivalence relation on the objects on c. Suppose that c is closed under the join $*$ and that $X * Z$ and $Y * Z$ are equivalent under e if X and Y are. We call X and $Y \in c$ stably equivalent under e if for some $Z \in c$ $X * Z$ and $Y * Z$ are equivalent under e. Let $V_e^c(G)^+$ be the set of stable equivalence classes of objects in c under e. It enherits the structure of a commutative semi-group from the join operation $*$.

Definition 20.13. The homotopy representation group $V_e^c(G)$ of G with respect to c and e is the Grothendieck group associated with $V_e^c(G)^+$. □

We collect the basic examples. We write G-hr for G-homotopy representation and G-he for G-homotopy equivalence. A G-map $f : X \longrightarrow Y$ between G-homotopy representations is a G-homotopy equivalence if and only if $\deg(f^H) \in \{\pm 1\}$ for $H \subset G$ holds (see Theorem 2.4.) If $\deg(f^H) = 1$ is true for all $H \subset G$, we call f an oriented G-homotopy equivalence.

c	e	$V_e^c(G)$
all G-hr.	G-he.	$V(G)$
finite G-hr.	G-he.	$V^f(G)$
finite G-hr.	simple G-he.	$V_s^f(G)$
even G-hr.	G-he.	$V^{ev}(G)$
oriented even G-hr.	oriented G-he.	$V_{or}^{ev}(G)$
finite even G-hr.	G-he.	$V^{f,ev}(G)$
finite even G-hr.	simple G-he.	$V_s^{f,ev}(G)$
oriented finite even G-hr.	oriented G-he.	$V_{or}^{f,ev}(G)$
oriented finite even G-hr.	oriented simple G-he.	$V_{or,s}^{f,ev}(G)$

In the sequel f denotes all the 'forgetful' homomorphisms like $f : V^{ev}(G) \longrightarrow V(G)$ The dimension function Dim satisfies $\operatorname{Dim}(X * Y) = \operatorname{Dim}(X) + \operatorname{Dim}(Y)$. Hence we get homomorphisms

20.14. $$\mathrm{Dim} : V_e^c(G) \longrightarrow C(G) = \prod_{(H)} \mathbf{Z}$$

Let $V_e^c(G,\mathrm{Dim})$ be its kernel so that we have the exact sequence

20.15. $$0 \longrightarrow V_e^c(G,\mathrm{Dim}) \longrightarrow V_e^c(G) \xrightarrow{\mathrm{Dim}} C(G)$$

The image of Dim is studied in Bauer [1988], tom Dieck-Petrie [1982] , tom Dieck [1986], tom Dieck [1987], III.5.

Next we relate the various homotopy representation groups. Notice that $1-\chi^G(X) \in A(G)$ is a unit for any G-homotopy representation X . Let $\hat{o}^G(X) \in \tilde{K}_o(\mathbf{Z}\mathrm{Or}G)$ be $(1-\chi^G(X))^{-1} \cdot \tilde{o}^G(X)$ if $\tilde{o}^G(X)$ is the reduced finiteness obstruction of 18.2. We have $\hat{o}^G(X * Y) = \hat{o}^G(X) + \hat{o}^G(Y)$ by 18.17. We obtain an exact sequence (see also tom Dieck [1985], sec. 3)

20.16. $$0 \longrightarrow V^f(G) \xrightarrow{f} V(G) \xrightarrow{\hat{o}^G} \tilde{K}_o(\mathbf{Z}\mathrm{Or}G)$$

$$0 \longrightarrow V^f(G,\mathrm{Dim}) \xrightarrow{f} V(G,\mathrm{Dim}) \xrightarrow{\hat{o}^G} \tilde{K}_o(\mathbf{Z}\mathrm{Or}G)$$

Let $[X]-[Y]$ be in the kernel of $\hat{o}^G$. Then $[X]-[Y] = [X * \Sigma^3 X] - [Y * \Sigma^3 Y]$ and $\hat{o}^G(X * \Sigma^3 X) = \hat{o}^G(Y * \Sigma^3 X) = 0$. As $X * \Sigma^3 X$ has simply connected fixed point sets and $\tilde{o}^G(X * \Sigma^3 X)$ vanishes, $X * \Sigma^3 X$ is up to G-homotopy equivalence finite (see 18.5.). The same argument applies to $Y * \Sigma^3 Y$ .Hence $[X]-[Y] \in$ image f . This shows exactness at $V(G)$. Now suppose that $[X]-[Y] \in V^f(G)$ lies in the kernel of f . Then $X * Z$ and $Y * Z$ are G-homotopy equivalent for some G-homotopy representation Z . Since we can choose Z to be finite by the next lemma, we get $[X]-[Y] = o$.

Lemma 20.17. *Let X be a G-homotopy representation. Then there is a G-homotopy representation Y and a complex G-representation V such that $X * Y$ and SV are G-homotopy equivalent.*

Proof. Since $\tilde{K}_o(\mathbf{Z}WH)$ is finite by Swan [1960a] , the group $\tilde{K}_o(\mathbf{Z}\mathrm{Or}G)$ is finite by Theorem 10.34. Hence $\hat{o}^G(\underset{n}{*} X) = n\,\hat{o}^G(X)$ is zero for appropriate n so that we can assume without loss of generality by 18.5. that X is finite. Choose an embedding of X in SV for some complex G-representation V . If U is an open equivariant regular neighborhood of X, let Y be the compact G-manifold $SV\backslash U$. It is a G-homotopy representation and $X * Y$ is G-homotopy equivalent to SV by equivariant Spanier-Whitehead duality (see Lewis-May-Steinberger [1986] , Wirthmüller [1974] . □

Since $1-\chi^G(X)$ is a unit in $A(G)^*$, $1-o^G(X)$ is a unit in $K_o(\mathbb{Z}OrG)$ under the multiplication given by $\otimes_{\mathbb{Z}}$ by Theorem 11.24. We get exact sequences from 18.17. (see also tom Dieck [1985], sec. 3)

20.18.
$$0 \longrightarrow V^{ev}(G) \longrightarrow V(G) \xrightarrow{1-\chi^G} A(G)^*$$
$$0 \longrightarrow V^{ev,f}(G) \longrightarrow V(G) \xrightarrow{1-o^G} K_o(\mathbb{Z}OrG)^*$$

Next we establish exact sequences

20.19.
$$A(G)^* \xrightarrow{\omega} Wh(\mathbb{Z}OrG) \xrightarrow{i} V_s^f(G) \xrightarrow{f} V^f(G) \longrightarrow 0$$
$$A(G)^* \xrightarrow{\omega} Wh(\mathbb{Z}OrG) \xrightarrow{i} V_{s,or}^{f,ev}(G) \xrightarrow{f} V_{or}^{f,ev}(G) \longrightarrow 0$$

The map ω was defined in 7.39. The map i sends $\eta \in Wh(\mathbb{Z}OrG)$ to any element $[X] - [Y] \in V_s^f(G)$ resp. $V_{s,or}^{f,ev}(G)$ such that there is an (oriented) G-homotopy equivalence $f : X \longrightarrow Y$ with $\hat{\tau}^G(f) := (1-\chi^G(Y))^{-1} \cdot \tau^G(f) = \eta$. If $[X'] - [Y']$ is another element with $\hat{\tau}^G(f') = \eta$ for some (oriented) G-homotopy equivalence $f' : X' \longrightarrow Y'$ we have $\hat{\tau}^G(f' * f^{-1}) = 0$ and hence $[X * Y'] = [Y * X']$ what implies $[X] - [Y] = [X'] - [Y']$. Obviously $i(\eta-\xi) = i(\eta) - i(\xi)$ so that i is a well defined homomorphism. One easily checks that 20.19. is exact.

If V is a real G-representation, SV has a simple structure by 4.36. Hence we get an homomorphism

$$S : Rep_{\mathbb{R}}(G) \longrightarrow V_s^f(G) \qquad [V] \longmapsto [SV]$$

<u>Theorem 20.20.</u> <u>The map</u> S <u>is injective</u>

<u>Proof:</u> Let $j : V_s^f(G) \longrightarrow V_{s,or}^{f,ev}(G)$ send [X] to $[X * X]$ and $\rho^G : V_{s,or}^{f,ev}(G) \longrightarrow Wh(\mathbb{Q}OrG)$ send [X] to $\rho^G(X)$. Notice that $X * X$ has a canonical orientation. Then the composition $\rho^G \circ j \circ S$ is $Rep_{\mathbb{R}}(G) \longrightarrow Wh(\mathbb{Q}OrG)$ $[V] \longrightarrow [SV \otimes_{\mathbb{R}} \mathbb{C}]$ and hence injective by Theorem 18.38.

Next we introduce an exact sequence

20.21.
$$\{1\} \longrightarrow A(G)^* \xrightarrow{ch} C(G)^* \xrightarrow{j} V_{or}^{ev}(G) \xrightarrow{f} V^{ev}(G) \longrightarrow \{0\}$$
$$\{1\} \longrightarrow A(G)^* \xrightarrow{ch} C(G)^* \xrightarrow{j} V_{or}^{ev}(G,Dim) \xrightarrow{f} V^{ev}(G,Dim) \longrightarrow \{0\}$$

The map j sends $\eta \in C(G)^*$ to any element $[X] - [Y]$ such that there is a G-homotopy equivalence $f : X \longrightarrow Y$ with $\deg f^H = \eta(H)$ for $H \subset G$. This is well-defined and gives exact sequences because of the following two results.

Lemma 20.22. There is a real G-representation V with $\mathrm{Iso}(SV) = \{H \mid H \subset G\}$

Proof: Bredon [1972] p. 24. □

Lemma 20.23. Let X be a G-homotopy representation with $\dim X^G \geq 1$. Suppose either that G is nilpotent, or $H,K \in \mathrm{Iso}(X)$, $H \subset K$, $H \neq K \Rightarrow \dim(X^H) \geq 2+\dim(X^K)$ holds. If $f : X \longrightarrow X$ is a G-self-map, let $\mathrm{Deg}(F) \in C(G)$ be given by $\{\deg f^H \mid H \subset G\}$ taken with respect to the coherent orientation $\phi(X,X)$ given by the identity. We get a map

$$\mathrm{Deg} : [X,X]^G \longrightarrow C(G)$$

Equip $[X,X]^G$ and $C(G)$ with the monoid structure induced by composition resp. multiplication.

Then Deg is an injection of commutative monoids . It has the same image as $ch : A(G,\mathrm{Iso}(X)) \longrightarrow C(G)$ (see section 5)

Proof: follows from tom Dieck [1987], II.4, II.8. See also Laitinen [1986] and Lück [1986 a]. □

We have defined $\overline{C}(G) = C(G)/|G|\cdot C(G)$, $\overline{A}(G) = A(G)/|G|\cdot C(G)$, $\mathrm{Inv}(G) = \overline{C}(G)^*/\overline{A}(G)^*$ and $\mathrm{Pic}(G) = \overline{C}(G)^*/\overline{A}(G)^*C(G)^*$ in section 19. Next we introduce isomorphisms

20.24.
$$D : V_{or}^{ev}(G,Dim) \longrightarrow \mathrm{Inv}(G)$$
$$\overline{D} : V(G,Dim) \longrightarrow \mathrm{Pic}(G)$$

Consider $[X] - [Y]$ in $V_{or}^{ev}(G,Dim)$ resp. $V(G,Dim)$. Choose a coherent orientation

$\phi(X,Y)$ compatible with the given orientations on X and Y resp. choose any coherent orientation $\phi(X,Y)$. Define $D([X]-[Y])$ by the element determined by $\{\deg f^H \mid H \subset G\}$ for any G-map $f : X \longrightarrow Y$ with invertible degrees. The proof that D and $\bar{D}$ are well-defined isomorphisms can be found in tom Dieck-Petrie [1982], 6.5.

Theorem 20.25. *The following diagram commutes and has exact rows. The maps* D *and* $\bar{D}$ *are isomorphisms*

$$\begin{array}{ccccccccc}
\{1\} & \longrightarrow & A(G)^* & \longrightarrow & C(G)^* & \longrightarrow & V^{ev}_{or}(G,Dim) & \longrightarrow & V(G,Dim) & \longrightarrow & 0 \\
 & & \Vert & & \Vert & & \downarrow D & & \downarrow \bar{D} & & \\
\{1\} & \longrightarrow & A(G)^* & \longrightarrow & C(G)^* & \longrightarrow & Inv(G) & \longrightarrow & Pic(G) & \longrightarrow & 0
\end{array}$$

□

Next we compute these homotopy representation groups at least rationally. Let

20.26. $$r(G),q(G),c(G) \in \mathbb{N}$$

be the number of irreducible real resp. rational resp. complex G-representations. Then $r(G)$ is the number of equivalence classes of elements $g \in G$ under the equivalence relation $g_1 \sim g_2 \iff g_1$ and g_2 or g_1 and g_2^{-1} are conjugated, whereas $q(G)$ is the number of conjugacy classes of cyclic subgroups and $c(g)$ the number of conjugacy classes of elements in G (see Serre [1977], p. 19, p. 96, p. 103 and p. 106

Theorem 20.27.

a) $V^c_e(G,Dim)$ *is* $Tors(V^c_e(G))$ *and in particular finite, if* $e \neq s$ *and* $e \neq (s,or)$.

b) *The rank of the abelian groups* $V^c(G)$ *and* $V^c_{or}(G)$ *is cardinality of* $B = \{(H) \mid H/[H,H] \text{ is cyclic}\}$. *If* G *is nilpotent* $card\, B = q(G)$ *holds.*

c) *The abelian groups* $V^c_s(G)$ *and* $V^c_{s,or}(G)$ *have rank* $card\, B + \sum_{(H)} r(WH)-q(WH)$.

d) *The rank of* $Wh(\mathbb{Z}OrG)$ *is* $\sum_{(H)} r(WH)-q(WH)$.

Proof.

d) By Theorem 10.34. it suffices to show $rk\, Wh(\mathbb{Z}G) = r(G) - q(G)$. This is done in Bass [1964].

a) follows from 20.15., 20.16., 20.18., 20.21., and Theorem 20.25.

b) The groups $V^c(G)$ and $V^c_{or}(G)$ have all the same rank because of 20.16 and 20.18. and 20.21. Now the claim follows from tom Dieck [1987], III.5.2.

c) follows from a), d) and 20.19. □

We mention the following result of tom Dieck [1985]

<u>Theorem 20.28.</u> $\mathrm{Inv}(G) = \prod_{(H)} \mathbb{Z}/|WH|^*$ □

20.C. <u>Classification of G-homotopy representations by reduced Reidemeister torsion</u>.

Next we want to apply the Reidemeister torsion invariants of section 18. to the classification of G-homotopy representations.

Let X and Y be two not-necessarily even G-homotopy representations with Dim(X) = Dim(Y) and a coherent orientation $\phi(X,Y)$. Since we do not require them to be even they do not define elements in $V^{ev}_{or}(G,\mathrm{Dim})$. However, we can define a relative element $v(X,Y,\phi) \in V^{ev}_{or}(G,\mathrm{Dim})$ as follows. Let Z be any G-homotopy representation with $\chi^G(X) = \chi^G(Y) = \chi^G(Z)$. Then $X * Z$ and $Y * Z$ are even. Choose orientations on $X * Z$ and $Y * Z$ such that the induced coherent orientation is just $\phi(X*Z,Y*Z)$. Define $v(X,Y,\phi(X,Y))$ by $[X*Z]-[Y*Z]$. This is independent of the choice of Z and the orientations on $X * Z$ and $Y * Z$. Now suppose that X and Y are even oriented, so that $[X]-[Y] \in V^{ev}_{or}(G,\mathrm{Dim})$ is defined. If $\phi(X,Y)$ is the induced coherent orientation we have $[X]-[Y] = v(X,Y,\phi(X,Y))$. Hence we will often write for two G-homotopy representations X and Y with Dim(X) = Dim(Y) and a coherent orientation $\phi(X,Y)$

20.29. $$v(X,Y,\phi(X,Y)) = [X]-[Y] \in V^{ev}_{or}(G,\mathrm{Dim}) .$$

If X and Y are moreover finite, we get

20.30. $$[X]-[Y] \in V^{f,ev}_{s,or}(G,\mathrm{Dim}) .$$

In the sequel let m be the order of the finite group G. Let the homomorphisms

20.31. $$\bar{\rho}^G : V^{ev}_{or}(G) \longrightarrow K_1(\mathbb{Q}\mathrm{Or}G)/K_1(\mathbb{Z}_{(m)}\mathrm{Or}G)$$
$$\bar{\rho}^G : V^{ev}_{or}(G,\mathrm{Dim}) \longrightarrow K_1(\mathbb{Q}\mathrm{Or}G)/K_1(\mathbb{Z}_{(m)}\mathrm{Or}G)$$

$$\rho^G : V^{f,ev}_{or,s}(G) \longrightarrow Wh(\mathbb{Q}OrG)$$

$$\rho^G : V^{f,ev}_{or,s}(G,Dim) \longrightarrow Wh(\mathbb{Q}OrG)$$

send the class [X] to $\bar{\rho}^G(X)$ resp. $\rho^G(X)$. We will write using 20.30.

20.32. $\bar{\rho}^G(X) - \bar{\rho}^G(Y) := \bar{\rho}^G([X]-[Y]) \in K_1(\mathbb{Q}OrG)/K_1(\mathbb{Z}_{(m)}OrG)$

$\rho^G(X) - \rho^G(Y) := \rho^G([X]-[Y]) \in Wh(\mathbb{Q}OrG)$, if X and Y are finite.

<u>Proposition 20.33.</u> <u>The following diagram commutes if</u> SW <u>is the generalized Swan homomorphism of 19.12.</u>

$$\begin{array}{ccc} \bar{C}(G)^* & \xrightarrow{\overline{SW}} & K_1(\mathbb{Q}OrG)/K_1(\mathbb{Z}_{(m)}OrG) \\ \downarrow{\scriptstyle pr} & & \uparrow{\scriptstyle \bar{\rho}^G} \\ Inv(G) & \xrightarrow[\cong]{D^{-1}} & V^{ev}_{or}(G,Dim) \end{array}$$

<u>Proof.</u> Theorem 19.25. □

Consider a collection $\{d(H) \mid H \subset G\}$ of integers with $(d(H),m) = 1$ for $H \subset G$.

<u>Proposition 20.34.</u> <u>There is a</u> G-<u>map</u> $f : X \longrightarrow Y$ <u>with</u> $\deg(f^H) = d(H)$ <u>for</u> $H \subset G$ <u>if and only if</u> $\{d(H) \mid H \subset G\}$ <u>satisfies the unstable conditions 20.8. and</u> $\overline{SW}$ <u>send</u> $\{\overline{d(H)} \mid H \subset G\} \in \bar{C}(G)^*$ <u>to</u> $\bar{\rho}^G(X) - \bar{\rho}^G(Y)$.

<u>Proof.</u> The "only if"-statement is a consequence of Proposition 20.33. In the proof of the "if"-statement we construct inductively over the orbit types the desired G-map $f : X \longrightarrow Y$. In the induction step we are given a G-map $f : X \longrightarrow Y$ with invertibl degrees and $H \subset G$ such that $\deg(f^K) = d(K)$ for $K \supset H$, $K \neq H$ holds. We must change f relative $X^{>H}$ such that $\deg(f^H)$ is $d(H)$. If $H = \hat{H}$ is valid there is a $K \supset H$, $K \neq H$ with $\dim(X^H) = \dim(X^K)$. Then we are done as $\deg(f^H) = \deg(f^K) = d(K) = d(H)$ holds by the unstable conditions 20.8. and the induction hypothesis. Hence we can assume $H \neq \hat{H}$. Because of Proposition 20.12. it suffices to prove $\deg(f^H) \equiv d(H) \bmod |WH|$. We get from the induction hypothesis and Theorem 19.13. that $\overline{sw} : \mathbb{Z}/|WH|^* \longrightarrow K_1(\mathbb{Q}WH)(K_1(\mathbb{Z}_{(|WH|)}WH)$ sends the classes of $\deg(f^H)$ and $d(H)$ to the same element. Now apply Theorem 19.4. □

efinition 20.35. We call a collection $\{d(H) \mid H \subset G\}$ of integers with $(d(H),m)=1$ or each $H \subset G$ a weight function for X and Y, if the following holds.

et $\{e(H) \mid H \subset G\}$ be a collection of integers. There is a G-map $f : X \longrightarrow Y$ with $\deg(f^H) = e(H)$ for $H \subset G$ if and only if $\{e(H) \mid H \subset G\}$ satisfies the unstable conditions 20.8. and $\{e(H)\cdot d(H) | H \subset G\}$ lies in the image of $ch : A(G) \longrightarrow C(G)$ □

0.36. Any collection of integers $\{d(H)|H \subset G\}$ satisfying 20.8.i) determines an lement in $C(G)$, denoted in the same way. Recall that $\{d(H)|H \subset G\} \in C(G)$ lies n the image of the character map $ch : A(G) \longrightarrow C(G)$ if and only if certain congruences are satisfied (see Example 16.38.) □

e get from Proposition 20.12 that a weight function determines $[X,Y]^G$ (cf. Lemma 0.23.):

heorem 20.37. Let X and Y be G-homotopy representations with the same diension function $n = \mathrm{Dim}(X) = \mathrm{Dim}(Y)$ and a coherent orientation. Consider the map

$$\mathrm{Deg} : [X,Y]^G \longrightarrow C(G), \quad [f] \longrightarrow \{\deg(f^H)|H \subset G\}.$$

) If G is nilpotent or if $n(L) \geq n(K)+2$ holds for all $K,L \in \mathrm{Iso}(X)$, $L \subset K$, $L \neq K$, then DEG is injective.

) Let $\{d(H)|H \subset G\}$ be a weight function. Then $\mathrm{image}(\mathrm{DEG}) \subset C(G)$ is the subset of elements $\{e(H)|H \subset G\}$ for which $\{e(H) \cdot d(H)|H \subset G\}$ lies in the image of $ch : A(G) \longrightarrow C(G)$ and the unstable conditions 20.8. hold. □

neorem 20.38. Let $\{d(H) \mid H \subset G\}$ be a collection of integers prime to $|G|$. It is a weight unction if and only if it satisfies the unstable conditions 20.8. and $\overline{SW}$ sends $\overline{d(H)} \mid H \subset G\} \in \overline{C}(G)^*$ to $\overline{\rho}^G(Y) - \overline{\rho}^G(X)$.

roof. Because of Proposition 20.12. and Lemma 20.23. the collection $\{d(H)|H \subset G\}$ s a weight function if and only if there is a G-map $g : Y \longrightarrow X$ with $\deg(g^H) = d(H)$. ow apply Proposition 20.36. □
weight function exists by Proposition 20.33. (cf. Laitinen [1986]).

Corollary 20.39. The following statements are equivalent for G-homotopy representations X and Y with $\mathrm{Dim}(X) = \mathrm{Dim}(Y)$ and a coherent orientation $\phi(X,Y)$.

a) X and Y are oriented G-homotopy equivalent.

b) X and Y are stably oriented G-homotopy equivalent.

c) $\bar{\rho}^G(X) - \bar{\rho}^G(Y) = 0$.

Corollary 20.40. The map $\bar{\rho}^G : V^{ev}_{or}(G,\mathrm{Dim}) \longrightarrow K_1(\mathbb{Q}OrG)/K_1(\mathbb{Z}_{(m)}OrG)$ is injective.

Notice that Proposition 20.33. together with Corollary 20.40. give the complete proof of Theorem 19.19. Corollary 20.39. is shown for $X = SV$ and abelian G in Rothenberg [1978].

Let $\kappa(G)$ be the cokernel of the composition $C(G)^* \longrightarrow \bar{C}(G)^* \xrightarrow{\overline{SW}} K_1(\mathbb{Q}OrG)/K_1(\mathbb{Z}_{(m)}OrG)$. Because of 20.21. the homomorphisms of 20.31. induce

20.41.
$$\bar{\rho}^G_\kappa : V^{ev}(G) \longrightarrow \kappa(G)$$
$$\bar{\rho}^G_\kappa : V^{ev}(G,\mathrm{Dim}) \longrightarrow \kappa(G)$$

Theorem 20.42. Let X and Y be G-homotopy representations with $\mathrm{Dim}(X) = \mathrm{Dim}(Y)$. Then X and Y are stably G-homotopy equivalent if and only if $\bar{\rho}^G_\kappa(X) - \bar{\rho}^G_\kappa(Y) := \bar{\rho}^G_\kappa([X]-[Y])$ vanishes. The map $\bar{\rho}^G_\kappa : V^{ev}(G,\mathrm{Dim}) \longrightarrow \kappa(G)$ is injective.

Proof. Because of Lemma 20.22. we can assume $\mathrm{Iso}(X) = \{H \mid H \subset G\}$ so that $H = \hat{H}$ holds for any $H \subset G$. Choose $\{d(H) \mid H \subset G\} \in C(G)^*$ such that $\overline{SW}(\{\overline{d(H)} \mid H \subset G\}) = \bar{\rho}^G(X) - \bar{\rho}^G(Y)$ holds. Then there is a G-map $f : X \longrightarrow Y$ with $\deg(f^H) = d(H)$ for $H \subset G$ by Proposition 20.36. □

In general "stably G-homotopy equivalent" does not imply "G-homotopy equivalent" for G-homotopy representations. A counter example is given in Laitinen [1986] where also the next result is proved.

Proposition 20.43. Let G be nilpotent and the 2-Sylow subgroup G_2 be abelian. Then stably G-homotopy equivalent G-homotopy representations are G-homotopy equivalent. □

Recall that we have assumed $\mathrm{Dim}(X) = \mathrm{Dim}(Y)$ until now. We drop this condition and only require $\chi^G(X) = \chi^G(Y)$. Then $\rho_\kappa^G(X) - \rho_\kappa^G(Y)$ can still be defined by $\rho_\kappa^G([X]-[Y])$ defining $[X]-[Y]$ similarly to 20.29. One checks directly that the composition of $\overline{SW} : \overline{C}(G)^* \longrightarrow K_1(\mathbb{Q}OrG)/K_1(\mathbb{Z}_{(m)}OrG)$ (see 19.12.) and $\alpha : K_1(\mathbb{Q}OrG)/K_1(\mathbb{Z}_{(m)}OrG) \longrightarrow \prod_{(H)} \mathbb{Q}^*/\mathbb{Z}_{(m)}^*$ of 18.32. is zero. Hence α induces

20.44.
$$\alpha_\kappa : \kappa(G) \longrightarrow \prod_{(H)} \mathbb{Q}^*/\mathbb{Z}_{(m)}^*$$

We have defined $m\chi(X)$ in Definition 18.25.

<u>Proposition 20.45.</u> <u>Let</u> X <u>and</u> Y <u>be</u> G-<u>homotopy representations with</u> $\chi^G(X) = \chi^G(Y)$.

a) $\alpha_\kappa(\overline{\rho}_\kappa^G(X)-\overline{\rho}_\kappa^G(Y)) = m\chi(X) \cdot m\chi(Y)^{-1}$

b) $\mathrm{Dim}(X) = \mathrm{Dim}(Y) \Longrightarrow m\chi(X) = m\chi(Y)$.

c) <u>Suppose that</u> G <u>is nilpotent and</u> $\dim(X^G) = \dim(Y^G)$. <u>Then we have</u> $\mathrm{Dim}(X) = \mathrm{Dim}(Y) \iff m\chi^G(X) = m\chi^G(Y)$.

<u>Proof.</u> a) follows from Proposition 18.33. and b) from Proposition 20.33. In the proof of c) one has to show in the induction step for $H \subset G$, $H \neq G$ that $\dim(X^H) = \dim(Y^H)$ is valid if $\dim(X^K) = \dim(Y^K)$ for $K \subset G$ with $H \subset K$, $H \neq K$ and $m\chi(X)_H = m\chi(Y)_H$ holds. As G is nilpotent, WH is not trivial for $H \neq G$ (see Huppert [1967], p. 260). Hence we can assume $H = \{1\}$ and $G \neq \{1\}$. Suppose that $n = \dim(Y) - \dim(X)$ is positive. Then there is a G-map $f : X \longrightarrow Y$ such that $(\deg(f^H),m) = 1$ for $H \subset G$, $H \neq \{1\}$ holds (compare Proposition 20.12.). Hence $\mathrm{Cone}(C^c(f))_{(m)} \longrightarrow \mathrm{Cone}(C^c(f,f^{>1}))_{(m)}$ is a $\mathbb{Z}_{(m)}G$-chain homotopy equivalence. Since $C^c(X,X^{>1})$ and $C^c(Y,Y^{>1})$ are finitely dominated and free $\mathbb{Z}G$-chain complexes, the same is true for $\mathrm{Cone}(C^c(f,f^{>1}))_{(m)}$. Hence there is a n-dimensional periodic (projective) $\mathbb{Z}_{(m)}G$-resolution P of the trivial $\mathbb{Z}_{(m)}G$-module $\mathbb{Z}_{(m)}$

$$\{0\} \longrightarrow \mathbb{Z}_{(m)} \longrightarrow P_n \longrightarrow \dots \longrightarrow P_o \longrightarrow \mathbb{Z}_{(m)} \longrightarrow \{0\} .$$

Let $m\chi(P)$ be $\prod_{n \geq o} | H_n(P \longrightarrow P \otimes_{\mathbb{Q}G} \mathbb{Q})|^{(-1)^n} \in \mathbb{Q}^*/\mathbb{Z}_{(m)}^*$. We get from the long homology sequences of $C^c(f)_{(o)}$ and $C^c(f \otimes_{\mathbb{Z}G} \mathbb{Z})_{(o)}$ that $m\chi(P) = m\chi(X)_1 \cdot (m\chi(Y)_1)^{-1} = 1$ holds. This contradicts the next lemma. □

Lemma 20.46. Let $G \neq \{1\}$ be nilpotent and P be a periodic projective $\mathbb{Z}_{(m)}G$-resolution of $\mathbb{Z}_{(m)}$. If n is the dimension of P then n is odd and we have

$$m\chi(P) = |G|^{-(n+1)/2}$$

Proof. The nilpotent group G is the product of its p-Sylow subgroups G_p which must be cyclic or generalized quaternionic (see Wall [1979]). The projection $\mathbb{Q}^*/\mathbb{Z}^*_{(m)} \longrightarrow \mathbb{Q}^*/\mathbb{Z}^*_{(p)}$ sends $m\chi^G(P)$ to $m\chi^{G_p}(\mathrm{res}_{G_p} P)$. Hence we can assume that G is a p-group, cyclic or generalized quaternionic. One easily checks

$$m\chi(P) = |G|^{-1} \cdot \prod_{k=1}^{n-1} |H_k(G;\mathbb{Z})|^{(-1)^k}$$

and the claim follows. □

Corollary 20.47. Let X and Y be G-homotopy representations with $\chi^G(X) = \chi^G(Y)$ and $\dim(X^G) = \dim(Y^G)$. Assume that G is nilpotent and G_2 is abelian. Then X and Y are G-homotopy equivalent if and only if $\bar{\rho}^G_\kappa(X) - \bar{\rho}^G_\kappa(Y) = 0$ holds.

Proof. Theorem 20.42., Proposition 20.43, and Proposition 20.45. □

Now we deal with the classification up to simple G-homotopy equivalence and the equivariant Reidemeister torsion. We have defined $SK_1(\mathbb{Z}\mathrm{Or}G)$ as the kernel of $K_1(\mathbb{Z}\mathrm{Or}G) \longrightarrow K_1(\mathbb{Q}\mathrm{Or}G)$ in section 18.

Theorem 20.47.

a) There is an exact sequence

$$A(G)^* \xrightarrow{\omega} SK_1(\mathbb{Z}\mathrm{Or}G) \xrightarrow{i} V^{f,ev}_{s,or}(G,\mathrm{Dim}) \xrightarrow{\rho^G} \mathrm{Wh}(\mathbb{Q}\mathrm{Or}G)$$

b) Let G be cyclic. Consider oriented even finite G-homotopy representations X and Y such that $\dim(X^G) = \dim(Y^G)$ is valid. Then X and Y are oriented simple G-homotopy equivalent if and only if $\rho^G(X) = \rho^G(Y)$ holds.

Proof.

a) This follows from 18.18., Theorem 18.19., 20.19., and Corollary 20.39.

b) If $\rho^G(X) = \rho^G(Y)$ holds there is an oriented G-homotopy equivalence $f: X \rightarrow Y$ by 20.39. and 20.47. Its Whitehead torsion $\tau^G(f)$ must be zero by 18.18. and Theorem 18.1

□

20.D. <u>The special case of a finite abelian group</u>.

As an illustration we examine a special case more closely. Namely, let G be a finite abelian group of order m for the remainder of the section. We have defined integers

$$\bar{\mu}(K,H) = \sum_{\ell \geq 0} (-1)^{\ell} \; |\, ch_{\ell}(G/H, G/K) \,|$$

in 19.20. for $K,H \subset G$. Put $\hat{\omega}(H,K) = 1$ for $H = K \subset G$ and $\hat{\omega}(H,K) = 0$ for $H,K \subset G$, $H \neq K$.

Define maps

20.48.
$$\bar{\mu} : C(G) = \prod_{(K)} \mathbb{Z} \longrightarrow C(G) = \prod_{(H)} \mathbb{Z}$$
$$\hat{\omega} : C(G) = \prod_{(H)} \mathbb{Z} \longrightarrow C(G) = \prod_{(K)} \mathbb{Z}$$
$$\bar{\mu} : \bar{C}(G)^* = \prod_{(K)} \mathbb{Z}/m^* \longrightarrow \prod_{(H)} \mathbb{Z}/|G/H|^*$$

by requiring that the component of $\bar{\mu}$ resp. $\hat{\omega}$ for K and H resp. H and K is given by multiplication with $\mu(K,H)$ resp. $\omega(H,K)$ if all groups are written additively. Notice that this agrees with 16.34. and 17.34.c. We get from Theorem 19.21. a commutative diagram where $\overline{pr}$ and D are isomorphisms and $\oplus\, \overline{sw}(G/H), \bar{\rho}^G$ and $q \circ S \circ \bar{\rho}^G$ injections

20.49.

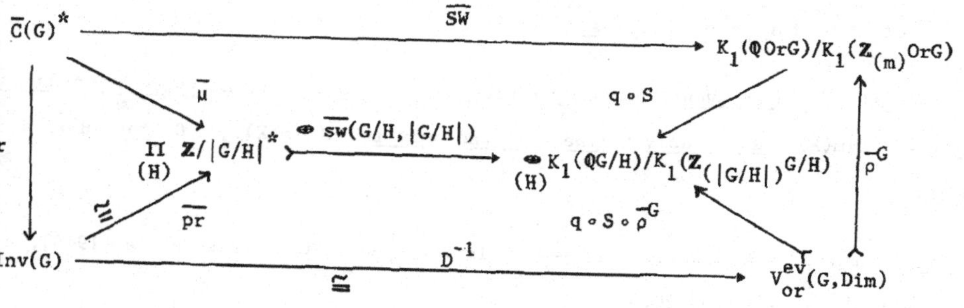

ext we want to measure the difference between G-homotopy representations and unit pheres of G-representations. Let C be $\{H \subset G \mid G/H \text{ is cyclic}\}$ and $D =$ $H \subset G \mid G/H$ is not cyclic$\}$. Consider a G-representation V over $F = \mathbb{R}, \mathbb{C}$. We can rite V as a direct sum $\bigoplus_{H \in C} V(H)$ where $V(H)$ is a free G/H-representation regarded as G-representation by restriction with $G \longrightarrow G/H$. Put $n(H) = \dim_{\mathbb{R}} (V(H))$ for $H \in C$ and $n(H) = 0$ for $H \in D$ and let $n \in C(G)$ be $(n(H))_{(H)}$. By Theorem 17.34. $\bar{\mu}$ and $\hat{\omega}$ are inverse isomorphisms. One easily checks

20.50. $$\bar{\mu}(\mathrm{Dim}(SV)) = n \quad \text{and} \quad \hat{\omega}(n) = \mathrm{Dim}(SV)$$

Consider $n \in C(G)$. We call n <u>even</u> if $n(H)$ is even for $H \subset G$. If there are real resp. complex G-representations V and W with $n = \mathrm{Dim}(SV) - \mathrm{Dim}(SW)$, we call n <u>stably</u> $\mathbb{R}$-resp. $\mathbb{C}$-<u>linear</u>. We say that n is <u>unstably</u> $\mathbb{R}$- resp. $\mathbb{C}$-<u>linear</u> if n is $\mathrm{Dim}(SV)$.

<u>Theorem 20.51.</u> <u>Let</u> G <u>be finite abelian</u>.

a) <u>The following statements are equivalent</u>

i) n <u>is stably</u> $\mathbb{R}$-<u>linear</u>.

ii) n <u>lies in the image of</u> $\mathrm{Dim} : V(G) \longrightarrow C(G)$.

iii) $\bar{\mu}(n)_H = 0$ <u>for</u> $H \in D$ <u>and</u> $\bar{\mu}(n)_H$ <u>is even for</u> $H \in C$, $|G/H| > 2$.

b) n <u>is stably</u> $\mathbb{C}$-<u>linear if and only if</u> n <u>is even and</u> $\bar{\mu}(n)_H = 0$ <u>for</u> $H \in D$.

c) n <u>is unstably</u> $\mathbb{R}$-<u>linear if and only if</u> $\bar{\mu}(n)_H = 0$ <u>for</u> $H \in D$, $\bar{\mu}(n)_H \geq 0$ <u>for</u> $H \in C$ <u>and</u> $\bar{\mu}(n)_H \equiv 0 \bmod 2$ <u>for</u> $H \in C$ <u>with</u> $|G/H| > 2$ <u>holds.</u>

d) n <u>is unstably</u> $\mathbb{C}$-<u>linear if and only if</u> n <u>is even and</u> $\bar{\mu}(n)_H = 0$ <u>for</u> $H \in D$ <u>and</u> $\bar{\mu}(n)_H \geq 0$ <u>for</u> $H \in C$ <u>holds.</u>

e) <u>Let</u> X <u>be an even</u> G-<u>homotopy representation. Then</u> Dim X <u>is stably</u> $\mathbb{C}$-<u>linear</u>. <u>Moreover</u> Dim(X) <u>is unstably</u> $\mathbb{C}$-<u>linear if and only if</u> $\bar{\mu}(\mathrm{Dim}(X))_H \geq 0$ <u>for</u> $H \in C$ <u>holds.</u>

<u>Proof.</u> In a) the implication ii) => i is verified in tom Dieck-Petrie [1978]. Now the rest follows from 20.50. □

Let X be an oriented even G-homotopy representation. By Theorem 20.51.e there are complex G-representations V and W with $\mathrm{Dim}(X*SV) = \mathrm{Dim}(SW)$. For $H \subset G$ let $\bar{\rho}^G(X)_H$ be the image of $\bar{\rho}^G(X)$ under the map $K_1(\mathbb{Q}\mathrm{Or}G)/K_1(\mathbb{Z}_{(m)}\mathrm{Or}G) \longrightarrow K_1(\mathbb{Q}G/H)/K_1(\mathbb{Z}_{(|G/H|)}G/H)$ induced by $S_{G/H}$. One easily checks as in 20.50. that

$\bar{\rho}^G(SV)_H = \bar{\rho}^G(SW)_H = 0$ holds for $H \in D$. By 20.49. there is for $H \in D$ a unique element $\lambda^G(X)_H \in \mathbf{Z}/|G/H|^*$ such that $\overline{sw}(G/H)$ sends $\lambda^G(X)_H$ to $\rho^G(X)_H = \rho^G(X*SV)_H - \rho^G(SW)_H$. Define

20.52.
$$\lambda^G(X) \in \prod_{H \in D} \mathbf{Z}/|G/H|^*$$

by $\{\lambda^G(X)_H \mid H \in D\}$. Then $\lambda^G(X)$ depends only on the oriented G-homotopy type and satisfies $\lambda^G(X*Y) = \lambda^G(X) + \lambda^G(Y)$ by Theorem 18.35. We get from Corollary 20.39. and Theorem 20.51.

Theorem 20.53. *Let G be a finite abelian group and X an even oriented G-homotopy representation.*

a) *There are complex G-representations V and W such that $X * SV$ and SW are oriented G-homotopy equivalent if and only if $\lambda^G(X)$ vanishes.*

b) *There is a complex G-representation V such that X and SV are oriented G-homotopy equivalent if and only if $\bar{\mu}(\mathrm{Dim}\ X)_H \geq 0$ for $H \in C$ holds and $\lambda^G(X)$ vanishes.* □

Let X be a G-homotopy representation. Choose a real G-representation V such that $X * SV$ is even. Choose any orientation on $X * SV$ and define

20.54.
$$\lambda^G_{\kappa}(X) \in \prod_{H \in D} \mathbf{Z}/|G/H|^*/\{\pm 1\}$$

by the reduction of $\lambda^G(X*SV)$. This does not depend on the choice of SV and the orientation on $X * SV$.

Theorem 20.54. *Let G be finite abelian.*

a) *There are real G-representations V and W such that $X * SV$ and SW are G-homotopy equivalent if and only if $\lambda^G_{\kappa}(X)$ vanishes.*

b) *There is a real G-representation V such that X and SV are G-homotopy equivalent if and only if $\lambda^G_{\kappa}(X)$ vanishes, $\bar{\mu}(\mathrm{Dim}(X))_H \geq 0$ for $H \in C$ and $\bar{\mu}(\mathrm{Dim}(X))_H$ is even for $H \in C$ with $|G/H| > 2$.*

c) *The reduced finiteness obstruction $\tilde{o}^G(X)$ is the image of $\lambda^G_{\kappa}(X)$ under the composition*

$$\prod_{H \in D} \mathbf{Z}/|G/H|^{*}/\{\pm 1\} \xrightarrow{\oplus_{H \in D} sw(G/H)} \bigoplus_{H \in D} \tilde{K}_0(\mathbf{Z}G/H) \xrightarrow{\oplus_{H \in D} E_{G/H}} \tilde{K}_0(\mathbf{Z}OrG$$

Comments 20.55.

The theory of homotopy representations was founded in tom Dieck-Petrie [1982]. The problem which dimension functions can occur as dimension functions of homotopy representations, smooth homotopy representations or unit spheres of linear representations is treated in Bauer [1988], tom Dieck [1982], tom Dieck-Petrie [1982], tom Dieck [1986], tom Dieck [1987] III.5, Madsen-Raußen [1985] . The relations between homotopy representations and the Picard group of the Burnside ring are examined in tom Dieck-Petrie [1978], tom Dieck [1984], tom Dieck [1985], tom Dieck [1986a]. The degrees of maps between G-homotopy representations are studied in tom Dieck [1986b], tom Dieck [1987], II.4 and 5, Laitinen [1986], Lück [1986a], Tornehave [1982]. Furthe references in the context of homotopy representations are tom Dieck-Löffler [1985], Lück [1988], Rothenberg [1978a].

Exercises 20.56.

1) Let X and Y be G-homotopy representations with $\dim(X^H) = \dim(Y^H)$ for $H \subset G$, $H \neq 1$. Consider a G-map $f : X \longrightarrow Y$ such that $H_*(f^H, \mathbf{Z}_{(|G|)})$ is an isomorphism for $H \neq 1$. Prove

a) If $\dim(X) = \dim(Y)$ holds and d is $|\deg(f)|$ then the trivial $\mathbf{Z}G$-module $\mathbf{Z}/d$ has a finite resolution. Conclude that $\deg(f)$ is prime to $|G|$.

b) If $n = \dim(X) - \dim(Y)$ is positive and $|G| > 2$, then n is even and G has periodic cohomology.

2) Show using Exercise 1 for the dimension function n of a G-homotopy representation.

i) If $H \triangleleft K \subset G$, $K/H \cong \mathbf{Z}/p$ and p is odd, then $n(H) - n(K)$ is even.

ii) If $H \triangleleft K \subset G$, $K/H \cong \mathbf{Z}/p \times \mathbf{Z}/p$, H_i/H for $i = 0,1,\ldots,p$ are the subgroups of order p in K/H, then $n(H) - n(K) = \sum_{i=o}^{p} (n(H_i) - n(K))$.

iii) Given $H \triangleleft L \triangleleft K \subset G$ with $L/H \cong \mathbf{Z}/2$, $n(H)-n(L)$ is even if $K/H \cong \mathbf{Z}/4$ and i divisible by 4 if K/H is a generalized quaternion group of order 2^k for $k \geq 3$.

) Let X and Y be G-homotopy representations with the same dimension function. Show using only Proposition 20.12.a) and Exercise 1 that there is a G-map $f : X \to Y$ with invertible degrees.

) Prove that $S : \mathrm{Rep}_{\mathbf{R}}(G) \longrightarrow V_s^f(G)$ sending $[V]$ to $[SV]$ has a finite cokernel if and only if G is $\mathbf{Z}/p$ for a prime number p .

) Let $f : X \longrightarrow X$ be a G-self map of a G-homotopy representation. Let $\lambda(f^H, f^{>H}) \in \mathbf{Z}$ be the Lefschetz number $\Sigma(-1)^n \, \mathrm{trace}_{\mathbf{Z}}(C_n^c(f^H, f^{>H}))$. Show that it is divisible by $|WH|$. Then one can define

$$\lambda^G(f) = \sum_{(H)} \frac{1}{|WH|} \lambda(f^H, f^{>H}) \cdot [G/H] \in A(G)$$

Prove that $ch : A(G) \longrightarrow C(G)$ maps $(1-\chi^G(X))(\lambda^G(f)-1)$ to $(\deg(f^H))_{(H)}$.

) Let G be a solvable group and X and Y be oriented even G-homotopy representations. Show that X and Y are oriented G-homotopy equivalent if and only if $\bar{\rho}^G(X) = \bar{\rho}^G(Y)$ and $\dim(X^G) = \dim(Y^G)$ holds .

) Let G be the non-abelian group of order pq for odd prime numbers p and q with $p|q-1$. Let $f : X \longrightarrow Y$ be a G-map between free even oriented G-homotopy representations with $\dim(X) = \dim(Y)$ such that $\deg f \equiv -1 \bmod p$ and $\deg f \equiv 1 \bmod q$ holds (The existence of f follows from Swan [1960b]). Show that X and Y are not G-homotopy equivalent but are stably G-homotopy equivalent. (see Laitinen [1986],4.11.)

) Let G be $H \times K$ for $H = K = \mathbf{Z}/p$ for p an odd prime. Let X be an even oriented G-homotopy representation with $\mathrm{Iso}\, X = \{1,H,K\}$. Then $\dim(X) = \dim(X^H) + \dim(Y^H) + 1$ holds so that the linking number ℓ of X^H and X^K in X is defined. Show for $\lambda^G(X) \in \mathbf{Z}/p^{2*}$ defined in 20.52. $\lambda^G(X) \equiv \ell \bmod p^2$.

) Let G be $\mathbf{Z}/m$ and $a_1, a_2, \ldots, a_r$ and $b_1, \ldots, b_r$ be integers prime to m . Let V be the G-representation $G \times \mathbf{C}^r \longrightarrow \mathbf{C}^r$ sending $t^n, (z_j \,|\, j=1,\ldots,r)$ to $(\exp(2\pi i a_j n/m) \cdot z_j \,|\, j=1,\ldots,r)$. Define W using the b_j-s analogously. Prove

) SV and SW are oriented G-homotopy equivalent if and only if $\prod_{i=1}^{r} b_i \equiv \prod_{i=1}^{r} a_i \bmod m$ holds.

b) SV and SW are G-diffeomorphic if and only if $b_i \equiv \pm a_{\sigma(i)} \bmod m$ for some permutation $\sigma \in S_r$.

10) Let G be nilpotent and X and Y be oriented even G-homotopy representations Suppose for any prime p that the restrictions of X and Y to the p-Sylow subgroup G_p are oriented G_p-homotopy equivalent. Are X and Y oriented G-homotopy equivalent

11) Are two oriented even G-homotopy representations oriented G-homotopy equivalent if their restrictions to any cyclic subgroup C are oriented C-homotopy equivalent?

12) Show that G is isomorphic to $\mathbb{Z}/p^n$ for some prime number p if and only if any G-homotopy representation is G-homotopy equivalent to SV for some real G-representation V .

13) For which groups G are there for each G-homotopy representation X real G-representations V and W such that $X * SV$ and SW are G-homotopy equivalent?

14) Let G be abelian of order m . Define

$$\bar{\mu} : \prod_{(K)} \mathbb{Q}^*/\mathbb{Z}^*_{(m)} \longrightarrow \prod_{(H)} \mathbb{Q}^*/\mathbb{Z}^*_{(m)}$$

by

$$(r(K))_{(K)} \longrightarrow \Big(\prod_{(K)} r(K)^{\bar{\mu}(K,H)}\Big)_{(H)}$$

for $\bar{\mu}(K,H)$ defined in 19.20. Show for an even G-homotopy representation X that Dim(X) is unstably $\mathbb{C}$-linear if and only if $\bar{\mu}(m\chi(X))_H = |G/H|^{-n(H)}$ with $n(H) \in \mathbb{Z}$, $n(H) \geq 0$ holds for all $H \subset G$ with cyclic G/H .

ibliography

igner, M. [1979]: "Combinatorial Theory", Grundlehren der math. Wissenschaften 234, pringer

lmkvist, G. [1978]: "K-theory of endomorphisms", J. of Algebra 55, 308-340

nderson, D.R. [1982] : "Torsion invariants and actions of finite groups", ichigan Math. J. 29, 27-42

nderson, D. and Munkholm, H. J. [1988]: "Foundations of boundedly controlled al- ebraic and geometric topology", lect. notes in math. 1323, Springer

ndrzejewski, P. [1986]: "The equivariant Wall finiteness obstruction and White- ead torison", in "Transformation Groups", Proceedings, Poznań 1985, lect. not. in ath. vol. 1217, Springer, 11-25

raki, S. [1986]: "Equivariant Whitehead groups and G-expansion categories", Advanced tudies in Pure Mathematics 9, 1-25

raki, S. and Kawakubo, K. [1988]: "Equivariant s-cobordism theorems", J. Math. oc. Jap. 40, 349-367

rmstrong, M.A. [1982]: "Lifting homotopies through fixed points", Proc. of the oyal Soc. of Edinburgh 93 A, 123-128

tiyah, M.F. [1967]: "K-Theory", Benjamin, New York-Amsterdam

tiyah, M.F. and MacDonald, I.G. [1969]: "Introduction to Commutative Algebra", ddison-Wesley

aglivo, J. [1978]: "An equivariant Wall obstruction theory", Trans. AMS 256, 05-324

arden, D. [1963]: "The structure of manifolds", Phd. thesis, Cambridge University

ass, H. [1964]: "K-theory and stable algebra", I.H.E.S. Publications math. 22, -60

ass, H. [1968]: "Algebraic K-theory", Benjamin

ass, H., Heller, A. and Swan, R.G. [1964]: "The Whitehead group of a polynomial xtension", Publ. Math. IHES 22

auer, S. [1988]: "Dimension functions of homotopy representations for compact Lie roups", Math. Ann. 280, 247-265

ourbaki, N. [1961]: "Topologie genérale I-III Paris, Hermann 3 éd.

ousfield, A.K. and Kahn, D.M. [1972]: "Homotopy limits, completions and locali- ations", lect. not. in math. 304, Springer

redon, G.E. [1967]: "Equivariant Cohomology Theories", lect. not. in math. 34, ringer

redon, G.E. [1972]: "Introduction to transformation groups", Academic Press

röcker, T. and Janich, K. [1973]: "Einführung in die Differentialtopologie", eidelberger Taschenbücher, Band 143

rowder, W. and Hsiang, W.C. [1978]: "Some problems on homotopy theory manifolds d transformation groups", Proc. Symp. Pure Math. 32 II, AMS, 251-267

Browder, W. and Quinn, F. [1975]: "A surgery theory for G-manifolds and stratified sets" in "Manifolds", Tokyo 1973, 27-36

Brown, R. [1968]: "Elements of modern topology", Mc Graw-Hill

Brown, K.S. [1974]: "Euler characteristics of discrete groups and G-spaces", Inv. Math. 27, 229-264

Brown, K.S. [1975]: "Euler characteristics of groups", Inv. math. 29, 1-5

Brown, K.S. [1982]: "Cohomology of groups", grad. texts in math. 87, Springer

Cappell, S.E. and Shaneson, J.L. [1981]: "Non-linear similarity", Ann. of Math. 113, 315-355

Cappell, S.E. and Shaneson, J.L. [1985]: "On 4-dimensinal s-cobordisms", J. Diff. Ges. 22, 97-115

Cappell, S.E., Shaneson, J.L., Steinberger, M., Weinberger, S., West, J. E. [1988] "The classification of non-linear similarities of $Z/2^t$.", research announcement

Cappell, S.E. and Weinberger, S. [1987]: "Homology propagation of group actions." Comm. Pure and Appl. Math. 40, 723-744

Cartan, H. and Eilenberg, S. [1956]: "Homological algebra", Princeton University Press, Princeton

Carter, D. [1980]: "Lower K-theory of finite groups", Comm. Alg. 8, 1927-1937

Chapman, T. [1973]: "Hilbert cube manifolds and the invariance of the Whitehead torsion", Bull. Amer. Math. Soc. 79, 52-56

Chapman, T. [1983]: "Controlled simple homotopy theory and applications", lect. notes in math 1009, Springer

Cheeger, J. [1979]: "Analytic torsion and the heat equation", Annals of Math. 109, 259-322

Cohen, M.M. [1973]: "A course in simple homotopy theory", graduate texts in math. 10, Springer

Connolly, F. and Geist, R. [1982]: "On extending free group actions on spheres and a conjecture of Iwasawa", Transactions of the AMS 274, 631-640

Connolly, F. and Kozniewski, T. [1986]: "Finiteness properties of classifying spaces of proper Γ-actions", Journal of Pure and Applied Algebra 41, 17-36

Connolly, F. and Kozniewski, T. [1988]: "Rigidity and crystallographic groups I", preprint

Connolly, F. and Lück, W. [1988]: "The involution on the equivariant Whitehead group", Math. Gott., Heft 21, to appear in K-Theory

Connolly, F. and Prassidis, S. [1987]: "Groups which act freely on $\mathbb{R}^m \times S^{n-1}$", preprint, Notre Dame, to appear in Topology

Crowell, R.H. and Fox, R.H. [1963]: "Introduction to knot theory", Ginn and Company

Curtis, C.W. and Reiner, I. [1981]: "Methods of representation theory, Vol 1", Wiley, New York

urtis, C.W. and Reiner, I. [1987]: "Methods of representation theory, Vol II", iley, New York

om Dieck, T. [1969]: "Faserbündel mit Gruppenoperation", Arch. Math. Vol 20, 36-143

om Dieck, T. [1974]: "On the homotopy type of classifying spaces", Manusc. math. 1, 41-46

om Dieck, T. [1975]: "The Burnside ring of a compact Lie group I", Math. Annalen 15, 235-250

om Dieck, T. [1979]: "Transformation groups and representation theory", lect. not. n math. 766, Springer

om Dieck, T. [1981]: "Über projektive Moduln und Endlichkeitshindernisse bei ransformationsgruppen", Manuscripta math. 34, 135-155

om Dieck, T. [1982]: "Homotopiedarstellungen endlicher Gruppen: Dimensionsfunk- ionen", Invent. math. 67, 231-252

om Dieck, T. [1984]: "Die Picard-Gruppe des Burnside-Ringes", algebraic topo- ogy conference, proceed., Arhus 1982, lect. not. in math. 1051, 573-586

om Dieck, T. [1985]: "The Picard group of the Burnside ring", J. für die reine u. ngew. Math. 361, 174-200

om Dieck, T. [1986]: "Dimension functions of homotopy representations", Bull. e La Soc. Math. de Belg. 38

om Dieck, T. [1986 a]: "Die Picard-Gruppe des Burnside-Ringes einer kompakten ieschen Gruppe I", Math. Gott. Heft 46

om Dieck, T. [1986 b]: "Kongruenzen zwischen Abbildungsgraden im äquivarianten atz von Hopf", Math. Gott., Heft 45

om Dieck, T.: [1987]: "Transformation groups", Studies in math. 8, de Gruyter

om Dieck, T. and Kamps, K.H. and Puppe, D. [1970]: "Homotopietheorie", lect. ot. in math. 157, Springer

om Dieck, T. and Löffler, P. [1985]: "Verschlingungen von Fixpunktmengen in Dar- ellungsformenI", Alg. top. conf. Göttingen 1984, Proc., lect. notes in math. 72, 167-187

m Dieck, T. and Petrie, T. [1978]: "Geometric modules over the Burnside ring", ventiones math. 47, 273-287

m Dieck, T. and Petrie, T. [1982]: "Homotopy representations of finite groups", bl. math. I.H.E.S. 56, 129-169

ld, A. [1963]: "Partitions of unity in the theory of fibrations", Ann. of Math. , 223-255

ld, A. [1972]: "lectures on algebrais topology", Grundlehren der mathematischen ssenschaften 200, Springer

naldson, S. [1987]: "Irrationality and the h-cobordism conjecture", J. of Diff. o. 26, 141-168

vermann, K.H. and Petrie, T. [1982]: "G-surgery II", Memoirs of the AMS, vol 37, . 260

Dovermann, K.H. and Rothenberg, M. [1986]: "An algebraic approach to the generalized Whitehead group", in: "Transformation Groups", Proceedings, Poznań 1985, lect. not. in math. vol. 1217, Springer, 92-114

Dovermann, K.H. and Rothenberg, M [1988]: "An equivariant surgery sequence and equivariant diffeomorphism and homeomorphism classification", Memoirs of AMS, vo. 70, no 389

Dovermann, K.H. and Rothenberg, M. [1988 a]: "The equivariant Whitehead torsion of a G-fibre homotopy equivalence", preprint, to appear in the Prodeedings of the Osaka Conference 1987, Springer lecture notes

Dress, A. [1969]: "A characterization of solvable groups", Math. Z. 110, 213-217

Dress, A. [1973]: "Contributions to the theory of induced representations", Alg. K-theory, Proc. Conf., Seattle 1972, lect. notes in math. 342, 182-240

Dress, A. [1975]: "Induction and structure theorems for orthogonal representations of finite groups", Annals of Math. 102, 291-326

Elmendorf, A.D. [1983]: "Systems of fixed point sets", Transactions of the AMS, vol. 277, 275-284

Ewing, J. and Löffler, P. and Pedersen, E.K. [1985 a]: "A local approach to the finiteness obstruction", Math. Gott. Heft 40, to appear in Oxford Journal of Mathematics

Ewing, J. and Löffler, P. and Pedersen, E.K. [1985 b]: "A rational torsion invariant", Math. Gott. Heft 43, to appear in Proc. of the AMS

Ferry, S.C. [1981 a]: "A simple-homotopy approach to the finiteness obstruction", Proc., Dubrovnik Conf. on Shape Theory 1981, lect. not. in math. 870, Springer, 73-81

Ferry, S.C. [1981 b]: "Finitely dominated compacta need not have finite type", Proc., Dubrovnik Conf. on Shape Theory 1981, lect. not. in math. 870, Springer, 1-5

Ferry, S.C. and Pedersen, E.K. [1989]: "Epsilon Surgery I", preprint

Franz, W. [1935]: "Über die Torsion einer Überdeckung", J. für reine und angew. Math. 173, 245-254

Freedman, M. [1982]: "The topology of 4-manifolds", J. Diff. Geo. 17, 357-453

Freedman, M. [1983]: "The disk theorem for four-dimensional manifolds", Proc. Int. Cong. of Math. Warsaw 647-663

Freyd, P. [1966]: "Splitting homotopy idempotents", Proc. La Jolla Conf. on Cat. Alg. 1965, 173-176, Springer

Fried, D. [1986]: "Analytic torsion and closed geodesics on hyperbolic manifolds", Inv. math. 84, 523-540

Fried, D. [1988]: "Torsion and closed geodesics on complex hyperbolic manifolds", Invent. math. 91, 31-51

Gallot, S. and Hulin, D. and Lafontaine, J. [1987]: "Differential geometry", Springer

Gersten, S. [1966]: "A product formula for Wall's obstruction", Amer. J. of Math. 88, 337-346

ersten, S. [1967]: "The torsion of a self-equivalence", Topology 6, 411-414

ersten, S. [1973]: "Higher K-theory of rings", in "Higher K-theories I", Proc. eattle 72, lect. notes in math. 342, 3-43

iffen, C.H. [1966]: "The generalized Smith conjecture", Amer. J. Math. 88, 187-198

ilkey, P. [1984]: "Invariance theory, the heat equation, and the Atiyah-Singer ndex theorem", Publish or Perish

reenberg, N.J. [1967]: "Lectures on Algebraic Topology", Benjamin

ambleton, I. and Madsen, I. [1986]: "Actions of finite groups on R^{n+k} with fixed set R^k", Can. J. Math. 38, 781-860

ambleton, I., Taylor, L., and Willians, B. [1988]: "On G_n(RG) for G a finite nil-otent group", to appear in J. of Algebra

astings, H. and Heller, A. [1981]: "Splitting homotopy indempotents", Proc. Du-rovnik Conf. on Shape theory 1981, lect. not. in math. 870, 23-36

astings, H. and Heller, A. [1982]: "Splitting homotopy idempotents of finite-di-ensional CW-complexes", Proc. of AMS 85, 619-622

auschild, H. [1978]: "Äquivariante Whitehead-Torsion", Manuscripta Math. 26, 63-82

siang, W.C. and Pardon, W. [1982]: "When are topologically equivalent orthogonal epresentations linearly equivalent?", Invent. math. 275-317

uppert, B. [1967]:"Endliche Gruppen I", Grundlehren der math. Wissenschaften, and 134, Springer

usemöller, D. [1966]: "Fibre bundles", Mc Graw-Hill, New York

llman, S. [1974]: "Whitehead torsion and group actions", Ann. Acad. Sci. Fenn. er. AI 588, 1-44

llman, S. [1975]: "Equivariant singular homology and cohomology", Mem. Amer. Math. oc. 156, 1-74

llman, S. [1978]: "Smooth equivariant triangulations of G-manifolds for G a finite roup", Math. Ann. 233, 199-220

llman, S. [1983]: "The equivariant triangulation theorem for actions of compact ie groups", Math. Ann. 262, 487-501

llman, S. [1985]: "Equivariant Whitehead torsion and actions of compact Lie groups", "Group actions on Manifolds", Contemp. Math. AMS 36, 91-106

lman, S. [1986]: "A product formula for equivariant Whitehead torsion and geometric plications" in "Transformation Groups", Poznań 1985, lect. notes in math. 1217, 3-142

zuka, K. [1984]: "Finiteness conditions for G-CW-complexes", Japan, J. Math. vol. 10 . 1, 55-69

ckowski, S. and McClure, J.E. [1987]: "Homotopy approximations for classifying spaces compact Lie groups", Math. Gott., to appear in the Proc. of the conf. on alg. top, cata 1986, lect. not. in math., Springer

Jackowski, S., McClure, J.E. and Oliver, R. [1989]: "Self maps of classifying spaces", Århus preprint

James, I.M. and Segal, G.B. [1978]: "On equivariant homotopy type", Topology 17, 267-272

Jaworowski, J.W. [1976]: "Extensions of G-maps and Euclidian G-retracts", Math. Z. 146, 143-148

Kawakubo, K. [1986]: "Stable equivalence of G-manifolds", Advanced Studies in Pure Math. 9, 27-40

Kawakubo, K. [1988]: "An s-cobordism theorem for semi-free S^1-manifolds", A fête of topology, Pop. Dedic. Itiro Tamura, 565-583

Kirby, R.C. and Siebenmann, L.C. [1977]: "Foundational essays on topological manifolds, smoothings and triangulations", Ann. of Math. Studies no. 88, Princeton Univ. Press, Princeton

Kratzer, Ch. and Thévenaz, J. [1984]: "Fonction de Möbius d'un groupe fini et anneau de Burnside", Comm. Math. Helv. 59, 425-438

Kwasik, S. [1983]: "On equivariant finiteness", Comp. Math. 48, 363-372

Kwun, K.W. and Szczarba, R.H. [1965]: "Product and sum theorems for Whitehead torsion", Ann. of Math. 82, 183-190

Laitinen, E. [1986]: "Unstable homotopy theory of homotopy representations", in Transformation Groups", Poznań 1985, lect. not. in math. 1217, 210-248, Springer

Laitinen, E. and Lück, W. [1987]: "Equivariant Lefschetz classes", Math. Gott. Heft 46, to appear in Osaka Journal

Lamotke, K. [1968]: "Semisimpliziale algebraische Topologie", Grundlehren der math. Wissenschaften in Einzeldarstellungen, Band 147, Springer

Lashof, R.K. and Rothenberg, M. [1978]: "G-smoothing theory", Proceedings of Symposia in Pure Math. vol. 32, part 1, 211-266

Lewis, L.G., May, J.P. and Steinberger, M. [1986]: "Equivariant Stable Homotopy Theory", lect. not. in math. vol. 1213, Springer

Lott, J. and Rothenberg, M. [1989]: "Analytic torsion for group actions", preprint

Lück, W. [1983]: Seminarbericht "Transformationsgruppen und algebraische K-Theorie", Göttingen

Lück, W. [1986]: "The transfer maps induced in the algebraic K_o- and K_1-groups by a fibration I", Math. Scand. 59, 93-121

Lück, W. [1986 a]: "The equivariant degree", Math. Gott., Heft 69 and in the Proceedings of the topology conference Göttingen 1987, lect. notes in math. 1361, 123-166, Springer (1988)

Lück, W. [1987]: "The transfer maps induced in the algebraic K_o- and K_1-groups by a fibration II", J. of Pure and Appl. Algebra 45, 143-169

Lück, W. [1987 a]: "Equivariant Eilenberg MacLane spaces K(G, μ, 1) for possibly nonconnected or empty fixed point sets", manusc. math. 58, 67-75

Lück, W. [1987 b]: "The geometric finiteness obstruction", Proc. of the LMS 54, 367-384 ück, W. [1988]: "Equivariant Reidemeister torsion and homotopy representations", ath. Gott., Heft 15

ück, W. [1989]: "Analytic and topological torsion for manifolds with boundary nd symmetries." Math. Gott.

ück, W. and Madsen, I. [1988 a]: "Equivariant L-theory I", Arhus, preprint, to ppear in Math. Z.

ück, W. and Madsen, I. [1988 b]: "Equivariant L-theory II", Arhus preprint, o appear in Math. Z.

ück, W. and Ranicki, A. [1986]: "Chain homotopy projections", Math. Gott. Heft 73, nd in J. of algebra 120, 361-391 (1989)

ück, W. and Ranicki, A. [1988]: "Surgery Transfer", Math. Gott., and in Proc. of he topology conf. in Göttingen 1987, lect. not. in math. 1361, 167-246, Springer

undell, A.T. and Weingram, S. [1969]: "The topology of CW-complexes", Van Nostrand einhold Company

acLane, S. [1963]: "Homology". Die Grundlehren der Mathematischen Wissenschaften in inzeldarstellungen, Band 114, Springer

acLane, S. [1971]: "Categories for the working mathematician", grad. texts in math. , Springer

adsen, I. [1983]: "Reidemeister torsion, surgery invariants and spherical space orms", Proc. of the LMS 46, 193-240

adsen, I. and Raußen, M. [1985]: "Smooth and locally linear G-homotopy represen- ations", Alg. top. conf. Göttingen 1984, Proc. lect. not. in math. 1172, 130-156

adsen, I. and Rothenberg, M. [1985 a]: "On the classification of G-spheres I: quivariant transversality", preprint, Arhus and in Acta math. 160, 65-104 (1988)

adsen, I and Rothenberg, M. [1985 b]: "On the classification of G-spheres II: PL- utomorphism groups", preprint, Arhus

adsen, I. and Rothenberg, M. [1985 c]: "On the classification of G-spheres III: OP automorphism groups", preprint, Arhus

adsen, I., Thomas, C.B. and Wall, C.T.C. [1976]: "The topological spherical space orm problem II", Topology 15, 375-382

ather, M. [1965]: "Counting homotopy types of manifolds", Topology 4, 93-94

atumoto, T. [1971]: "On G-CW-complexes and a theorem of J.H.C. Whitehead", J. Fac. ci. Univ. Tokyo, Sect. IA Math. 18, 363-374

atumoto, T. [1984]: "A complement to the theory of G-CW-complexes", Japan. J. Math. , 353-374

atumoto, I. and Shiota, M. [1987]: "Unique triangulation of the orbit space of a fferentiable transformation group and its applications", Advanced Studies in Pure ath. 9, Kinokuniya, Tokyo, 41-55

umary, S. [1987]: "The analytic and de Rham torsion", preprint, Lausanne

y, P. [1967]: "Simplicial objects in algebraic topology", D. van Nostrand Company, c. Princeton, New Jersey

Mazur, B. [1963]: "Differential topology from the point of view of simple homotopy theory, Publ. IHES 15, 5-93

Michael, E. [1956]: "Continuous selections I", Ann. Math. 63, 361-382

Milnor, J. [1957]: "The geometric realization of a semi-simplicial complex", Annals of Math. 65, 357-362

Milnor, J. [1959]: "Spaces having the homotopy type of a CW-complex", Transactions of the AMS 90, 272-280

Milnor, J. [1961]: "Two complexes which are homeomorphic but combinatorial distinct" Ann. of Math. 74, 575-590

Milnor, J. [1962]: "On axiomatic homology theory", Pacific J. Math. 12, 337-341

Milnor, J. [1966]: "Whitehead torsion", Bull. AMS 72, 358-426

Milnor, J. [1971]: "Introduction to algebraic K-theory", Princeton University Press

Mislin, G. [1987]: "The homotopy classification of self-maps of infinite quaternionic projective space", Quaterly Journal of Mathematics Oxford 38, 245-257

Mitchell, B. [1972]: "Rings with several objects", Adv. in Math. 8, 1-161

Montgomery, D. and Yang, C.T. [1957]: "The existence of a slice", Ann. of Math. 65, 108-116

Montgomery, D. and Zippin, L. [1955]: "Topological Transformation Groups", Wiley (Interscience), New York

Mostow, G.D. [1957]: "Equivariant embeddings in Euclidean spaces", Ann. of Math. 65, 432-446

Müller, W. [1978]: "Analytic torsion and R-torsion of Riemannian manifolds", Adv. in Math. 28, 233-305

Murayama, M. [1983]: "On G-ANR-s and their G-homotopy types", Osaka J. Math. 20, 479-512

Okonek, C. [1983]: "Bemerkungen zur K-Theorie äquivarianter Endomorphismen", Arch. Math. 40, 132-138

Oliver, R. [1975]: "Fixed point sets of group actions on finite acyclic complexes", Commentarii Mathematici Helv. 50, 155-177

Oliver, R. [1976]: "Smooth compact Lie groups actions on disks", Math. Zeitschrift 149, 79-96

Oliver, R. [1977]: "G-actions on disks and permutation representations II", Math. Zeitschrift 157, 237-263

Oliver, R. [1978]: "G-actions on disks and permutation representations", J. of Algebra 50, 44-62

Oliver, R. [1980]: "SK_1 for finite group rings II", Math. Scand. 47, 195-231

Oliver, R. [1985]: "The Whitehead transfer homomorphism for oriented S^1-bundles", Math. Scand. 57, 51-104

Oliver, R. [1988]: "Whitehead groups of finite groups", Cambridge Univ. Press.

liver, R. and Petrie, T. [1982]: "G-CW-surgery and $K_o(\mathbb{Z}G)$", Math. Zeitschrift 179, 1-42

alais, R.S. [1961]: "On the existence of slices for actions of non-compact Lie roups", Ann. of Math. 73, 295-323

edersen, E.K. [1984]: "On the k_{-i}-functors", Journal of algebra 90, 461-475

edersen, E.K. [1986]: "On the bounded h-cobordism theorem" in "Transformation roups", Poznań 1985, lect. notes in math. 1217, 306-320, Springer

edersen, E.K. and Weibel, C.A. [1985]: "Non-connective deloopings of algebraic -theory", in "Algebraic and Geometric Topology", Proceedings 1983, Rutgers Uni- ersity, New Brunswick, lect. notes in math. 1126, Springer

etrie, T. [1982 a]: "One fixed point actions on spheres I", Advances in Math. ol. 46, 3-14

etrie, T. [1982 b]: "One fixed point actions on spheres II", Advances in Math. ol 46, 15-70

etrie, T. and Randall, J. [1984]: "Transformation groups on manifolds", Dekker eries in Pure and Applied Math. 82, Dekker, New York

uillen, D. [1973]: "Higher algebraic K-theory I", in "Higher K-theories I". roceedings, Seattle 1972, lect. not. in math. 341, 85-147

uillen, D.G. [1983]: "Determinants of Cauchy-Riemannian operators on a Riemann urface", Functional Anal. Appl. 19 (1), 31-34

uinn, F. [1979]: "Ends of maps I", Ann. of Math. 110, 275-331

uinn, F. [1982]: "Ends of maps II", Inv. Math. 68, 353-424

uinn, F. [1988]: "Homotopically stratified sets", Journal of the AMS vol 1, o 2, 441-499

anicki, A. [1985]: "The algebraic theory of finiteness obstruction", Math. Scand. 7, 105-126

anicki, A. [1985 a]: "The algebraic theory of torsion", in "Algebraic and Geom. opology", Proc. Conf. Rutgers Univ., New Brunswick 1983, lect. not. in math. 1126, pringer, 199-237

anicki, A. [1986]: "The algebraic and geometric splittings of the K- and L-groups f polynomial extension", Proc. Symp. on Transformation groups, Poznań 1985, lect. otes in math. 1217, 321-364

anicki, A. [1987]: "The algebraic theory of torsion II: Products", K-Theory 1, 5-170

anicki, A. [1987 a]: "The algebraic theory of torsion III: Lower K-theory", pre- int

anicki, A. and Weiss, M. [1987]: "Chain complexes and assembly", Math. Gott. ft 28

y, D. and Singer, I. [1971]: "R-torsion and the Laplacian on Riemannian mani- lds", Adv. in Math. 7, 145-210

idemeister, K. [1938]: "Homotopieringe und Linsenräume", Hamburger Abhandlungen 1

de Rham, G. [1964]: "Reidemeister's torsion invariant and rotations of S^n", Differential Analysis (published for the Tata Institute of Fundamental Research, Bombay). Oxford Univ. Press, London 1964, pp 27-36, MR 32 # 8355

Rothenberg, M. [1978]: "Torsion invariants and finite transformation groups", in "Algebraic and geometric topology", part 1, AMS, 267-311

Rothenberg, M. [1978 a]: "Homotopy type of G-spheres", Algebraic topology conference Arhus 1978, Proc. lect. not. in math. 763, 573-590, Springer

Rothenberg, M. and Triantafillou, G. [1984]: "An algebraic model for G-simple homotopy types". Math. Ann. 269, 301-331

Rothenberg, M and Weinberger, S. [1987]: "Group actions and equivariant Lipschitz analysis", Bulletin of the AMS 17, 109-111

Rourke, C.P. and Sanderson, B.J. [1972]: "Introduction of Piecewise Linear Topology", Ergebnisse der Mathematik und Ihrer Grenzgebiete, Band 69, Springer

Rubinsztein, R.L. [1973]: "On the equivariant homotopy of spheres", Dissertation, preprint 58, Polish Academy of Sciences, Warscawa

Schubert, H. [1964]: "Topologie", 4. Auflage, Teubner

Schubert, H. [1970 a]: "Kategorien I", Heidelberger Taschenbücher, Band 65, Springer

Schubert, H. [1970 b]: "Kategorien II", Heidelberger Taschenbücher, Band 66, Springer

Schwarz, A. [1978]: "The partition function of degenerate quadratic functional and Ray-Singer invariants", Lett. Math. Phys. 2, 247

Segal, G.H. [1971]: "Equivariant stable homotopy theory", Actes Congr. int. Math. 2, 59-63

Serre, J.P. [1977]: "Linear representations of finite groups", graduate texts in mathematics 42, Springer

Siebenmann, L. [1965]: "The obstruction of finding a boundary for an open manifold of dimension greater than five", Ph. d. thesis, Princeton Univ.

Silvester, J.R. [1981]: "Introduction to algebraic K-theory", Chapman and Hall,

Slomińska, J. [1980]: "Equivariant Bredon cohomology of classifying spaces of families of subgroups", Bull. Ac. Sc. Pol. Sc. Math. XXVIII no 9-10, 503-508

Smith, J.R. [1986]: "Topological Realizations of Chain Complexes I- The General Theory" Topology and its applications 22, 301-313

Smith, J.R. [1987]: "Topological Realization of Chain Complexes II. The rational case", preprint

Stallings, J. [1968]: "Notes on Polyhedral Topology", Tata Institute

Steenrod, N.E. [1967]: "A convenient category of topological spaces", Mich. Math. J. 14, 133-152

Steinberger, M. [1988]: "The equivariant topological s-cobordism theorem", Inv. math. 91, 61-104

Steinberger, M and West, J. [1984]: "Controlled finiteness is the obstruction to equivariant handle decomposition", preprint

teinberger, M. and West, J. [1985]: "Equivariant h-cobordism and finiteness obtruction", Bull. AMS 12, 217-220

töcker, R. [1970]: "Whiteheadgruppe topologischer Räume", Inventiones Math. 9, 71-278

tröm, A. [1966]: "Note on cofibrations", Math. Scand. 19, 11-14

ummers, D.W. [1975]: "Smooth Z/p-actions on spheres which leaves knots pointwise ixed", Trans. AMS 205, 193-203

vensson, J.A. [1985]: "Lower equivariant K-theory", Arhus preprint, and in Math. cand 60, 179-201 (1987)

wan, R.G. [1960 a]: "Induced representations and projective modules", Ann. of Math. 1, 552-578

wan, R.G. [1960 b]: "Periodic resolutions for finite groups", Ann. of Math. 72, 67-291

wan, R.G. [1968]: "Algebraic K-theory", lect. not. in math. 76, Springer

witzer, R.M. [1975]: "Algebraic topology - homotopy and homology", Grundlehren der athematischen Wissenschaften in Einzeldarstellungen, Band 212, Springer

aylor, M.J. [1978]: "The locally free class group of prime power order", Journal of lgebra 50, 463-487

homason, R.W: [1982]: "First quadrant spectral sequences in algebraic K-theory via omotopy colimits", Communications in Algebra 10 (15), 1589-1668

ornehave, J. [1982]: "Equivariant maps of spheres with conjugate orthogonal acions", Alg. top. Conf. London Ont. 1981, Canad, Math. Soc. Conf. Proc. vol 2 part 2, 75-301

riantafillou, G.V. [1982]: "Equivariant minimal models", Transactions of the AMS 74, 509-532

riantafillou, G. [1983]: "Rationalization of Hopf G-spaces" Math. Z. 182, 485-500

uraev, V.G: [1986]: "Reidemeister torsion in knot theory", Russ. Math. Surveys 1:1, 119-182

aldhausen, F. [1985]: "Algebraic K-theory of spaces", in "Algebraic and Geometric opology", Proc. Conf. Rutgers Univ., New Brunswick 1983, lect. not. in math. 1126, pringer, 318-419

all, C.T.C. [1965]: "Finiteness conditions for CW-complexes", Annals of Math. 81, 5-69

all, C.T.C. [1966]: "Finiteness conditions for CW-complexes II", Proc. of the oyal Soc. A 295, 129-139

all, C.T.C: [1979]: "Periodic projective resolutions", Proc. of the London Math. oc. 39, 509-553

allace, A.H. [1970]: "Algebraic Topology, Homology and Cohomology", Benjamin, New ork

aner, S. [1980a] : "Equivariant homotopy theory and Milnor's theorem", Trans. AMS 58, 351-368

aner, S. [1980b]: "Equivariant fibrations and transfer", Trans. of AMS 258, 369-384

Webb, L.D. [1987]: "G-theory of group rings for groups of square-free order", K-theory 1, 417-422

Weiss, M. and Williams, B. [1987]: "Automorphisms of manifolds and algebraic K-theory II", Math. Gott. Heft 48, to appear in J. of Pure and Appl. Algebra

Whitehead, J.H.C. [1939]: "Simplicial spaces, nucleii and m-groups", Proc. of LMS 45, 243-327

Whitehead, J.H.C. [1941]: "On incidence matrices, nucleii and homotopy type", Ann. of math. 42, 1197-1239

Whitehead, J.H.C. [1949]: "Combinatorial homotopy I", Bull. Amer. Math. Soc. 55, 213-245

Whitehead, J.H.C. [1952]: "Simple homotopy types", Amer. J. Math. 72, 1-57

Whitehead, G.W. [1978]: "Elements of homotopy theory", grad. texts in math. 61, Springer

Willson, S.J. [1975]: "Equivariant homology theories on G-CW-complexes", Trans. of AMS 212, 155-171

Wirthmüller, K. [1974]: "Equivariant homology and duality", Manusc. math. 11, 373-390

Witten, E. [1988]: "Quantum field theory and the Jones polynomial", preprint

Index

Symbols

Roman letters

$X^{>(H)} = \{x \in X \mid (G_x) \supset (H),\ (G_x) \neq (H)\}$		11
$X_{(H)} = \{x \in X \mid (G_x) = (H)\}$		11
$X^H(x)$		146
$X^{>H}(x)$		146
$\tilde{X}^H(x)$		146
$\tilde{X}^{>H}(x)$		146
$X^H/(x) = X^H(x)/WH(x)_o$		150
$X^{>H}/(x)$		150
$\tilde{X}^H/(x) = \tilde{X}^H(x)/Aut(x)_o$		150
$\tilde{X}^{>H}/(x)$		150
$\tilde{X}: \Pi(G,X) \to \{top.\ sp.\}$	universal covering functor	146
$\tilde{X}/ : \Pi/(G,X) \to \{top.\ sp.\}$	discrete universal covering functor	149
X/G	orbit space	
$X \nearrow Y$	elementary expansion	62
$X \searrow Y$	elementary collaps	63
$X \wedge\!\!\!\cdot\ Y$	formal deformation	63
$\lvert x\rvert = \lvert Aut(x)\rvert$		325
$\bar{x}$	isomorphism class of objects represented by x	170
$\bar{x} \leqq \bar{y} \iff Hom(x,y) \neq \emptyset$		170

<u>Greek letters</u>

∂	boundary homomorphism of a localization square	368,369
Δ	Laplace operator	375
$\chi^G(X,A)$	Euler characteristic	100,278,360
$\chi(C)$	Euler characteristic	227
$\{\psi\} \# \{\phi\}$		244
μ	Moebius inversion	330
$\pi_o(x)$	set of components	
$\pi_1(X,x)$	fundamental group	
$\pi_n(X,x)$	n-th homotopy group	

LECTURE NOTES IN MATHEMATICS

Edited by A. Dold and B. Eckmann

Some general remarks on the publication of monographs and seminars

n what follows all references to monographs, are applicable also to ultiauthorship volumes such as seminar notes.

1. Lecture Notes aim to report new developments - quickly, informally, and at a high level. Monograph manuscripts should be reasonably self-contained and rounded off. Thus they may, and often will, present not only results of the author but also related work by other people. Furthermore, the manuscripts should provide sufficient motivation, examples and applications. This clearly distinguishes Lecture Notes manuscripts from journal articles which normally are very concise. Articles intended for a journal but too long to be accepted by most journals, usually do not have this "lecture notes" character. For similar reasons it is unusual for Ph.D. theses to be accepted for the Lecture Notes series.

   Experience has shown that English language manuscripts achieve a much wider distribution.

2. Manuscripts or plans for Lecture Notes volumes should be submitted either to one of the series editors or to Springer-Verlag, Heidelberg. These proposals are then refereed. A final decision concerning publication can only be made on the basis of the complete manuscripts, but a preliminary decision can usually be based on partial information: a fairly detailed outline describing the planned contents of each chapter, and an indication of the estimated length, a bibliography, and one or two sample chapters - or a first draft of the manuscript. The editors will try to make the preliminary decision as definite as they can on the basis of the available information.

3. Lecture Notes are printed by photo-offset from typed copy delivered in camera-ready form by the authors. Springer-Verlag provides technical instructions for the preparation of manuscripts, and will also, on request, supply special staionery on which the prescribed typing area is outlined. Careful preparation of the manuscripts will help keep production time short and ensure satisfactory appearance of the finished book. Running titles are not required; if however they are considered necessary, they should be uniform in appearance. We generally advise authors not to start having their final manuscripts specially tpyed beforehand. For professionally typed manuscripts, prepared on the special stationery according to our instructions, Springer-Verlag will, if necessary, contribute towards the typing costs at a fixed rate.

   The actual production of a Lecture Notes volume takes 6-8 weeks.

.../...

§4. Final manuscripts should contain at least 100 pages of mathematical text and should include

- a table of contents
- an informative introduction, perhaps with some historical remarks. It should be accessible to a reader not particularly familiar with the topic treated.
- a subject index; this is almost always genuinely helpful for the reader.

§5. Authors receive a total of 50 free copies of their volume, but no royalties. They are entitled to purchase further copies of their book for their personal use at a discount of 33.3 %, other Springer mathematics books at a discount of 20 % directly from Springer-Verlag.

Commitment to publish is made by letter of intent rather than by signing a formal contract. Springer-Verlag secures the copyright for each volume.

LECTURE NOTES

ESSENTIALS FOR THE PREPARATION OF CAMERA-READY MANUSCRIPTS

Springer
Springer-Verlag
Berlin Heidelberg New York
London Paris Tokyo Hong Kong

he preparation of manuscripts which are to be reproduced by photo-ffset require special care. Manuscripts which are submitted in tech-ically unsuitable form will be returned to the author for retyping. here is normally no possibility of carrying out further corrections fter a manuscript is given to production. Hence it is crucial that he following instructions be adhered to closely. If in doubt, please end us 1 - 2 sample pages for examination.

eneral. The characters must be uniformly black both within a single haracter and down the page. Original manuscripts are required: pho-ocopies are acceptable only if they are sharp and without smudges.

n request, Springer-Verlag will supply special paper with the text rea outlined. The standard TEXT AREA (OUTPUT SIZE if you are using a 4 point font) is 18 x 26.5 cm (7.5 x 11 inches). This will be scale-educed to 75% in the printing process. If you are using computer ypesetting, please see also the following page.

ake sure the TEXT AREA IS COMPLETELY FILLED. Set the margins so that hey precisely match the outline and type right from the top to the ottom line. (Note that the page number will lie outside this area). ines of text should not end more than three spaces inside or outside he right margin (see example on page 4).

ype on one side of the paper only.

pacing and Headings (Monographs). Use ONE-AND-A-HALF line spacing in he text. Please leave sufficient space for the title to stand out learly and do NOT use a new page for the beginning of subdivisons of hapters. Leave THREE LINES blank above and TWO below headings of uch subdivisions.

pacing and Headings (Proceedings). Use ONE-AND-A-HALF line spacing n the text. Do not use a new page for the beginning of subdivisons f a single paper. Leave THREE LINES blank above and TWO below hea-ings of such subdivisions. Make sure headings of equal importance re in the same form.

he first page of each contribution should be prepared in the same ay. The title should stand out clearly. We therefore recommend that he editor prepare a sample page and pass it on to the authors ogether with these instructions. Please take the following as an xample. Begin heading 2 cm below upper edge of text area.

MATHEMATICAL STRUCTURE IN QUANTUM FIELD THEORY

John E. Robert
Mathematisches Institut, Universität Heidelberg
Im Neuenheimer Feld 288, D-6900 Heidelberg

lease leave THREE LINES blank below heading and address of the uthor, then continue with the actual text on the same page.

ootnotes. These should preferable be avoided. If necessary, type hem in SINGLE LINE SPACING to finish exactly on the outline, and se-arate them from the preceding main text by a line.

Symbols. Anything which cannot be typed may be entered by hand i BLACK AND ONLY BLACK ink. (A fine-tipped rapidograph is suitable fc this purpose; a good black ball-point will do, but a pencil wil not). Do not draw straight lines by hand without a ruler (not even i fractions).

Literature References. These should be placed at the end of each pa per or chapter, or at the end of the work, as desired. Type them wit single line spacing and start each reference on a new line. Follc "Zentralblatt für Mathematik"/"Mathematical Reviews" for abbreviate titles of mathematical journals and "Bibliographic Guide for Edito and Authors (BGEA)" for chemical, biological, and physics journal Please ensure that all references are COMPLETE and ACCURATE.

IMPORTANT

Pagination. For typescript, number pages in the upper right-hand co ner in LIGHT BLUE OR GREEN PENCIL ONLY. The printers will insert t final page numbers. For computer type, you may insert page numbe (1 cm above outer edge of text area).

It is safer to number pages AFTER the text has been typed and corre ted. Page 1 (Arabic) should be THE FIRST PAGE OF THE ACTUAL TEXT. T Roman pagination (table of contents, preface, abstract, acknowledg ments, brief introductions, etc.) will be done by Springer-Verlag.

If including running heads, these should be aligned with the insi edge of the text area while the page number is aligned with the ou side edge noting that right-hand pages are odd-numbered. Runni heads and page numbers appear on the same line. Normally, the runni head on the left-hand page is the chapter heading and that on t right-hand page is the section heading. Running heads should not b included in proceedings contributions unless this is being done co sistently by all authors.

Corrections. When corrections have to be made, cut the new text fit and paste it over the old. White correction fluid may also b used.

Never make corrections or insertions in the text by hand.

If the typescript has to be marked for any reason, e.g. for provisi nal page numbers or to mark corrections for the typist, this can done VERY FAINTLY with BLUE or GREEN PENCIL but NO OTHER COLOR: the colors do not appear after reproduction.

COMPUTER-TYPESETTING. Further, to the above instructions, please no with respect to your printout that
- the characters should be sharp and sufficiently black;
- it is not strictly necessary to use Springer's special typi paper. Any white paper of reasonable quality is acceptable.

If you are using a significantly different font size, you shou modify the output size correspondingly, keeping length to bread ratio 1 : 0.68, so that scaling down to 10 point font size, yields text area of 13.5 x 20 cm (5 3/8 x 8 in), e.g.

Differential equations.: use output size 13.5 x 20 cm.

Differential equations.: use output size 16 x 23.5 cm.

Differential equations.: use output size 18 x 26.5 cm.

Interline spacing: 5.5 mm base-to-base for 14 point characters (sta dard format of 18 x 26.5 cm).
If in any doubt, please send us 1 - 2 sample pages for examinatio We will be glad to give advice.

ɔl. 1232: P.C. Schuur, Asymptotic Analysis of Soliton Problems. VIII, 80 pages. 1986.

ɔl. 1233: Stability Problems for Stochastic Models. Proceedings, 985. Edited by V.V. Kalashnikov, B. Penkov and V.M. Zolotarev. VI, 3 pages. 1986.

ɔl. 1234: Combinatoire énumérative. Proceedings, 1985. Edité par Labelle et P. Leroux. XIV, 387 pages. 1986.

ɔl. 1235: Séminaire de Théorie du Potentiel, Paris, No. 8. Directeurs: Brelot, G. Choquet et J. Deny. Rédacteurs: F. Hirsch et G. okobodzki. III, 209 pages. 1987.

ɔl. 1236: Stochastic Partial Differential Equations and Applications. oceedings, 1985. Edited by G. Da Prato and L. Tubaro. V, 257 ges. 1987.

ɔl. 1237: Rational Approximation and its Applications in Mathematics d Physics. Proceedings, 1985. Edited by J. Gilewicz, M. Pindor and Siemaszko. XII, 350 pages. 1987.

l. 1238: M. Holz, K.-P. Podewski and K. Steffens, Injective Choice nctions. VI, 183 pages. 1987.

l. 1239: P. Vojta, Diophantine Approximations and Value Distribu- n Theory. X, 132 pages. 1987.

l. 1240: Number Theory, New York 1984–85. Seminar. Edited by V. Chudnovsky, G.V. Chudnovsky, H. Cohn and M.B. Nathanson. 324 pages. 1987.

l. 1241: L. Gårding, Singularities in Linear Wave Propagation. III, 5 pages. 1987.

l. 1242: Functional Analysis II, with Contributions by J. Hoffmann- gensen et al. Edited by S. Kurepa, H. Kraljević and D. Butković. VII, 2 pages. 1987.

l. 1243: Non Commutative Harmonic Analysis and Lie Groups. oceedings, 1985. Edited by J. Carmona, P. Delorme and M. Vergne. 309 pages. 1987.

l. 1244: W. Müller, Manifolds with Cusps of Rank One. XI, 158 ges. 1987.

l. 1245: S. Rallis, L-Functions and the Oscillator Representation. I, 239 pages. 1987.

l. 1246: Hodge Theory. Proceedings, 1985. Edited by E. Cattani, F. illén, A. Kaplan and F. Puerta. VII, 175 pages. 1987.

l. 1247: Séminaire de Probabilités XXI. Proceedings. Edité par J. éma, P.A. Meyer et M. Yor. IV, 579 pages. 1987.

l. 1248: Nonlinear Semigroups, Partial Differential Equations and ractors. Proceedings, 1985. Edited by T.L. Gill and W.W. Zachary. 185 pages. 1987.

l. 1249: I. van den Berg, Nonstandard Asymptotic Analysis. IX, 187 ges. 1987.

. 1250: Stochastic Processes – Mathematics and Physics II. ceedings 1985. Edited by S. Albeverio, Ph. Blanchard and L. Streit. 359 pages. 1987.

. 1251: Differential Geometric Methods in Mathematical Physics. ceedings, 1985. Edited by P.L. García and A. Pérez-Rendón. VII, 0 pages. 1987.

l. 1252: T. Kaise, Représentations de Weil et GL_2 Algèbres de ision et GL_n. VII, 203 pages. 1987.

. 1253: J. Fischer, An Approach to the Selberg Trace Formula via Selberg Zeta-Function. III, 184 pages. 1987.

. 1254: S. Gelbart, I. Piatetski-Shapiro, S. Rallis. Explicit Construc- s of Automorphic L-Functions. VI, 152 pages. 1987.

. 1255: Differential Geometry and Differential Equations. Proceed- s, 1985. Edited by C. Gu, M. Berger and R.L. Bryant. XII, 243 es. 1987.

1256: Pseudo-Differential Operators. Proceedings, 1986. Edited H.O. Cordes, B. Gramsch and H. Widom. X, 479 pages. 1987.

. 1257: X. Wang, On the C*-Algebras of Foliations in the Plane. V, 5 pages. 1987.

1258: J. Weidmann, Spectral Theory of Ordinary Differential erators. VI, 303 pages. 1987.

Vol. 1259: F. Cano Torres, Desingularization Strategies for Three-Dimensional Vector Fields. IX, 189 pages. 1987.

Vol. 1260: N.H. Pavel, Nonlinear Evolution Operators and Semi-groups. VI, 285 pages. 1987.

Vol. 1261: H. Abels, Finite Presentability of S-Arithmetic Groups. Compact Presentability of Solvable Groups. VI, 178 pages. 1987.

Vol. 1262: E. Hlawka (Hrsg.), Zahlentheoretische Analysis II. Seminar, 1984–86. V, 158 Seiten. 1987.

Vol. 1263: V.L. Hansen (Ed.), Differential Geometry. Proceedings, 1985. XI, 288 pages. 1987.

Vol. 1264: Wu Wen-tsün, Rational Homotopy Type. VIII, 219 pages. 1987.

Vol. 1265: W. Van Assche, Asymptotics for Orthogonal Polynomials. VI, 201 pages. 1987.

Vol. 1266: F. Ghione, C. Peskine, E. Sernesi (Eds.), Space Curves. Proceedings, 1985. VI, 272 pages. 1987.

Vol. 1267: J. Lindenstrauss, V.D. Milman (Eds.), Geometrical Aspects of Functional Analysis. Seminar. VII, 212 pages. 1987.

Vol. 1268: S.G. Krantz (Ed.), Complex Analysis. Seminar, 1986. VII, 195 pages. 1987.

Vol. 1269: M. Shiota, Nash Manifolds. VI, 223 pages. 1987.

Vol. 1270: C. Carasso, P.-A. Raviart, D. Serre (Eds.), Nonlinear Hyperbolic Problems. Proceedings, 1986. XV, 341 pages. 1987.

Vol. 1271: A.M. Cohen, W.H. Hesselink, W.L.J. van der Kallen, J.R. Strooker (Eds.), Algebraic Groups Utrecht 1986. Proceedings. XII, 284 pages. 1987.

Vol. 1272: M.S. Livšic, L.L. Waksman, Commuting Nonselfadjoint Operators in Hilbert Space. III, 115 pages. 1987.

Vol. 1273: G.-M. Greuel, G. Trautmann (Eds.), Singularities, Representation of Algebras, and Vector Bundles. Proceedings, 1985. XIV, 383 pages. 1987.

Vol. 1274: N.C. Phillips, Equivariant K-Theory and Freeness of Group Actions on C*-Algebras. VIII, 371 pages. 1987.

Vol. 1275: C.A. Berenstein (Ed.), Complex Analysis I. Proceedings, 1985–86. XV, 331 pages. 1987.

Vol. 1276: C.A. Berenstein (Ed.), Complex Analysis II. Proceedings, 1985–86. IX, 320 pages. 1987.

Vol. 1277: C.A. Berenstein (Ed.), Complex Analysis III. Proceedings, 1985–86. X, 350 pages. 1987.

Vol. 1278: S.S. Koh (Ed.), Invariant Theory. Proceedings, 1985. V, 102 pages. 1987.

Vol. 1279: D. Ieşan, Saint-Venant's Problem. VIII, 162 Seiten. 1987.

Vol. 1280: E. Neher, Jordan Triple Systems by the Grid Approach. XII, 193 pages. 1987.

Vol. 1281: O.H. Kegel, F. Menegazzo, G. Zacher (Eds.), Group Theory. Proceedings, 1986. VII, 179 pages. 1987.

Vol. 1282: D.E. Handelman, Positive Polynomials, Convex Integral Polytopes, and a Random Walk Problem. XI, 136 pages. 1987.

Vol. 1283: S. Mardešić, J. Segal (Eds.), Geometric Topology and Shape Theory. Proceedings, 1986. V, 261 pages. 1987.

Vol. 1284: B.H. Matzat, Konstruktive Galoistheorie. X, 286 pages. 1987.

Vol. 1285: I.W. Knowles, Y. Saitō (Eds.), Differential Equations and Mathematical Physics. Proceedings, 1986. XVI, 499 pages. 1987.

Vol. 1286: H.R. Miller, D.C. Ravenel (Eds.), Algebraic Topology. Proceedings, 1986. VII, 341 pages. 1987.

Vol. 1287: E.B. Saff (Ed.), Approximation Theory, Tampa. Proceedings, 1985–1986. V, 228 pages. 1987.

Vol. 1288: Yu. L. Rodin, Generalized Analytic Functions on Riemann Surfaces. V, 128 pages, 1987.

Vol. 1289: Yu. I. Manin (Ed.), K-Theory, Arithmetic and Geometry. Seminar, 1984–1986. V, 399 pages. 1987.

Vol. 1290: G. Wüstholz (Ed.), Diophantine Approximation and Transcendence Theory. Seminar, 1985. V, 243 pages. 1987.

Vol. 1291: C. Mœglin, M.-F. Vignéras, J.-L. Waldspurger, Correspondances de Howe sur un Corps p-adique. VII, 163 pages. 1987

Vol. 1292: J.T. Baldwin (Ed.), Classification Theory. Proceedings, 1985. VI, 500 pages. 1987.

Vol. 1293: W. Ebeling, The Monodromy Groups of Isolated Singularities of Complete Intersections. XIV, 153 pages. 1987.

Vol. 1294: M. Queffélec, Substitution Dynamical Systems – Spectral Analysis. XIII, 240 pages. 1987.

Vol. 1295: P. Lelong, P. Dolbeault, H. Skoda (Réd.), Séminaire d'Analyse P. Lelong – P. Dolbeault – H. Skoda. Seminar, 1985/1986. VII, 283 pages. 1987.

Vol. 1296: M.-P. Malliavin (Ed.), Séminaire d'Algèbre Paul Dubreil et Marie-Paule Malliavin. Proceedings, 1986. IV, 324 pages. 1987.

Vol. 1297: Zhu Y.-l., Guo B.-y. (Eds.), Numerical Methods for Partial Differential Equations. Proceedings. XI, 244 pages. 1987.

Vol. 1298: J. Aguadé, R. Kane (Eds.), Algebraic Topology, Barcelona 1986. Proceedings. X, 255 pages. 1987.

Vol. 1299: S. Watanabe, Yu.V. Prokhorov (Eds.), Probability Theory and Mathematical Statistics. Proceedings, 1986. VIII, 589 pages. 1988.

Vol. 1300: G.B. Seligman, Constructions of Lie Algebras and their Modules. VI, 190 pages. 1988.

Vol. 1301: N. Schappacher, Periods of Hecke Characters. XV, 160 pages. 1988.

Vol. 1302: M. Cwikel, J. Peetre, Y. Sagher, H. Wallin (Eds.), Function Spaces and Applications. Proceedings, 1986. VI, 445 pages. 1988.

Vol. 1303: L. Accardi, W. von Waldenfels (Eds.), Quantum Probability and Applications III. Proceedings, 1987. VI, 373 pages. 1988.

Vol. 1304: F.Q. Gouvêa, Arithmetic of p-adic Modular Forms. VIII, 121 pages. 1988.

Vol. 1305: D.S. Lubinsky, E.B. Saff, Strong Asymptotics for Extremal Polynomials Associated with Weights on ℝ. VII, 153 pages. 1988.

Vol. 1306: S.S. Chern (Ed.), Partial Differential Equations. Proceedings, 1986. VI, 294 pages. 1988.

Vol. 1307: T. Murai, A Real Variable Method for the Cauchy Transform, and Analytic Capacity. VIII, 133 pages. 1988.

Vol. 1308: P. Imkeller, Two-Parameter Martingales and Their Quadratic Variation. IV, 177 pages. 1988.

Vol. 1309: B. Fiedler, Global Bifurcation of Periodic Solutions with Symmetry. VIII, 144 pages. 1988.

Vol. 1310: O.A. Laudal, G. Pfister, Local Moduli and Singularities. V, 117 pages. 1988.

Vol. 1311: A. Holme, R. Speiser (Eds.), Algebraic Geometry, Sundance 1986. Proceedings. VI, 320 pages. 1988.

Vol. 1312: N.A. Shirokov, Analytic Functions Smooth up to the Boundary. III, 213 pages. 1988.

Vol. 1313: F. Colonius, Optimal Periodic Control. VI, 177 pages. 1988.

Vol. 1314: A. Futaki, Kähler-Einstein Metrics and Integral Invariants. IV, 140 pages. 1988.

Vol. 1315: R.A. McCoy, I. Ntantu, Topological Properties of Spaces of Continuous Functions. IV, 124 pages. 1988.

Vol. 1316: H. Korezlioglu, A.S. Ustunel (Eds.), Stochastic Analysis and Related Topics. Proceedings, 1986. V, 371 pages. 1988.

Vol. 1317: J. Lindenstrauss, V.D. Milman (Eds.), Geometric Aspects of Functional Analysis. Seminar, 1986–87. VII, 289 pages. 1988.

Vol. 1318: Y. Felix (Ed.), Algebraic Topology – Rational Homotopy. Proceedings, 1986. VIII, 245 pages. 1988

Vol. 1319: M. Vuorinen, Conformal Geometry and Quasiregular Mappings. XIX, 209 pages. 1988.

Vol. 1320: H. Jürgensen, G. Lallement, H.J. Weinert (Eds.), Semi groups, Theory and Applications. Proceedings, 1986. X, 416 pages 1988.

Vol. 1321: J. Azéma, P.A. Meyer, M. Yor (Eds.), Séminaire c Probabilités XXII. Proceedings. IV, 600 pages. 1988.

Vol. 1322: M. Métivier, S. Watanabe (Eds.), Stochastic Analysis Proceedings, 1987. VII, 197 pages. 1988.

Vol. 1323: D.R. Anderson, H.J. Munkholm, Boundedly Controlle Topology. XII, 309 pages. 1988.

Vol. 1324: F. Cardoso, D.G. de Figueiredo, R. Iório, O. Lopes (Eds Partial Differential Equations. Proceedings, 1986. VIII, 433 page 1988.

Vol. 1325: A. Truman, I.M. Davies (Eds.), Stochastic Mechanics ar Stochastic Processes. Proceedings, 1986. V, 220 pages. 1988.

Vol. 1326: P.S. Landweber (Ed.), Elliptic Curves and Modular Forms Algebraic Topology. Proceedings, 1986. V, 224 pages. 1988.

Vol. 1327: W. Bruns, U. Vetter, Determinantal Rings. VII,236 page 1988.

Vol. 1328: J.L. Bueso, P. Jara, B. Torrecillas (Eds.), Ring Theo Proceedings, 1986. IX, 331 pages. 1988.

Vol. 1329: M. Alfaro, J.S. Dehesa, F.J. Marcellan, J.L. Rubio Francia, J. Vinuesa (Eds.): Orthogonal Polynomials and their Applic tions. Proceedings, 1986. XV, 334 pages. 1988.

Vol. 1330: A. Ambrosetti, F. Gori, R. Lucchetti (Eds.), Mathemati Economics. Montecatini Terme 1986. Seminar. VII, 137 pages. 198

Vol. 1331: R. Bamón, R. Labarca, J. Palis Jr. (Eds.), Dynamic Systems, Valparaiso 1986. Proceedings. VI, 250 pages. 1988.

Vol. 1332: E. Odell, H. Rosenthal (Eds.), Functional Analysis. P ceedings, 1986–87. V, 202 pages. 1988.

Vol. 1333: A.S. Kechris, D.A. Martin, J.R. Steel (Eds.), Cabal Semi 81–85. Proceedings, 1981–85. V, 224 pages. 1988.

Vol. 1334: Yu.G. Borisovich, Yu. E. Gliklikh (Eds.), Global Analy – Studies and Applications III. V, 331 pages. 1988.

Vol. 1335: F. Guillén, V. Navarro Aznar, P. Pascual-Gainza, F. Puer Hyperrésolutions cubiques et descente cohomologique. XII, 1 pages. 1988.

Vol. 1336: B. Helffer, Semi-Classical Analysis for the Schröding Operator and Applications. V, 107 pages. 1988.

Vol. 1337: E. Sernesi (Ed.), Theory of Moduli. Seminar, 1985. VIII, 2 pages. 1988.

Vol. 1338: A.B. Mingarelli, S.G. Halvorsen, Non-Oscillation Doma of Differential Equations with Two Parameters. XI, 109 pages. 1988.

Vol. 1339: T. Sunada (Ed.), Geometry and Analysis of Manifol Procedings, 1987. IX, 277 pages. 1988.

Vol. 1340: S. Hildebrandt, D.S. Kinderlehrer, M. Miranda (Ed Calculus of Variations and Partial Differential Equations. Proceedin 1986. IX, 301 pages. 1988.

Vol. 1341: M. Dauge, Elliptic Boundary Value Problems on Cor Domains. VIII, 259 pages. 1988.

Vol. 1342: J.C. Alexander (Ed.), Dynamical Systems. Proceedin 1986–87. VIII, 726 pages. 1988.

Vol. 1343: H. Ulrich, Fixed Point Theory of Parametrized Equivar Maps. VII, 147 pages. 1988.

Vol. 1344: J. Král, J. Lukeš, J. Netuka, J. Veselý (Eds.), Poter Theory – Surveys and Problems. Proceedings, 1987. VIII, 271 pag 1988.

Vol. 1345: X. Gomez-Mont, J. Seade, A. Verjovski (Eds.), Holomor Dynamics. Proceedings, 1986. VII, 321 pages. 1988.

Vol. 1346: O. Ya. Viro (Ed.), Topology and Geometry – Ro Seminar. XI, 581 pages. 1988.

Vol. 1347: C. Preston, Iterates of Piecewise Monotone Mappings an Interval. V, 166 pages. 1988.

Vol. 1348: F. Borceux (Ed.), Categorical Algebra and its Applicati Proceedings, 1987. VIII, 375 pages. 1988.

Vol. 1349: E. Novak, Deterministic and Stochastic Error Bound Numerical Analysis. V, 113 pages. 1988.